FISH LOCOMOTION
AN ECO-ETHOLOGICAL PERSPECTIVE

FISH LOCOMOTION

AN ECO-ETHOLOGICAL PERSPECTIVE

Editors

Paolo Domenici
CNR-IAMC
Torregrande (Oristano)
Italy

B.G. Kapoor
Formerly Professor of Zoology
Jodhpur University
India

CRC Press
Taylor & Francis Group
Boca Raton London New York

CRC Press is an imprint of the
Taylor & Francis Group, an **informa** business

A SCIENCE PUBLISHERS BOOK

First published 2010 by Science Publishers Inc.

Published 2019 by CRC Press
Taylor & Francis Group
6000 Broken Sound Parkway NW, Suite 300
Boca Raton, FL 33487-2742

© 2010,Copyright Reserved
CRC Press is an imprint of Taylor & Francis Group, an Informa business

First issued in paperback 2019

No claim to original U.S. Government works

ISBN 13: 978-0-367-45241-4 (pbk)
ISBN 13: 978-1-57808-448-7 (hbk)

Visit the Taylor & Francis Web site at
http://www.taylorandfrancis.com

and the CRC Press Web site at
http://www.crcpress.com

Library of Congress Cataloging-in-Publication Data

Fish locomotion : an eco-ethological perspective / editors
Paolo Domenici, B.G. Kapoor.
        p. cm.
 Includes bibliographical references and index.
 ISBN 978-1-57808-448-7 (hardcover)
1. Fishes--Locomotion. I. Domenici, P. (Paolo) II. Kapoor,
B. G.
 QL639.4.F57 2009
 597.157--dc22
                                        2009000070

Cover Illustration:

Common thresher shark (*Alopias vulpinus*) affixed with a pop-up satellite
tag off the Southern California Coast.
Photo by Scott Aalbers.

Preface

Fish accomplish most of their basic behaviors by swimming. Swimming is fundamental in a vast majority of fish species for feeding, avoiding predation, finding food, mating, migrating and finding optimal physical environments. Fish exhibit a wide variety of swimming patterns and behaviors. Some fish, such as tuna, never stop swimming. Others, such as sit-and-wait predators, swim only for a relatively small proportion of time. However, even in these species, swimming performance is critical for their fitness, since it determines feeding success. Within their performance boundaries, fish choose how to swim (e.g., at what speed), and these choices have profound consequences for their energetics, feeding and vulnerability, and therefore for their overall fitness.

Fish swimming also underlies many ecological processes. At its most basic, ecology is the study of the abundance and distribution of organisms—with movement, energetics, reproduction, predator-prey relationships and migration, all contributing. Locomotion plays a major role in all of these activities. Locomotor performance plays a role in defining the boundaries of the realized niche, in terms of defining, for example, habitats with current speeds that fish can endure, given their swimming capabilities. Locomotion is also fundamental in determining the smaller realized niche, i.e., the food items that can be caught by using certain locomotor patterns or performance during feeding events. In addition, the physical properties of the environment, in terms of temperature, oxygen level and current speed, are also important determinants of fish distribution, largely through their effect on fish swimming performance.

While swimming is fundamental for the fish's ecology and behavior, historically fish swimming has been studied mainly from biomechanical and physiological perspectives (Webb, 1975; Blake, 1983; Videler, 1993). More recent work has highlighted the importance of integrating these perspectives within a holistic approach that includes ecology and behavior. The aim of this book is to look at fish swimming from behavioral and ecological perspectives. Authors from various backgrounds, all interested in fish swimming from different points of view, were invited to contribute to this book with the goal of integrating various approaches while stressing the ecological and behavioral relevance of specific aspects of fish locomotion.

The first two Chapters of the book focus on how fish deal with flow. This issue is relevant for the ecology of fish, since their ability to stabilize body posture and swimming trajectories perturbed by turbulent flows affects fish species distributions and densities, and hence fish assemblages in various habitats (Chapter 1, Webb *et al.*). Webb *et al.* introduce the subject by focusing first on the physics of fish-flow interactions, followed by a review of methods used to measure turbulence. They discuss fish swimming performance in unsteady flow, in terms of energetics and behavior, and put these issues within the context of fish assemblages in typical flow environments such as rivers, shorelines and coral reefs. Blake and Chan (Chapter 2) review the biomechanics of rheotactic behavior (defined as the orientation to flow and associated behavior), discussing morphological, biomechanical, energetic and behavioral aspects. Rheotaxis is a common phenomenon found in benthic fish that live in currents, and it allows fish to avoid displacement in a flow, thus remaining in their habitat for feeding, finding mates and protection. Adaptations for rheotaxis include morphological structures (e.g., suckers) and behavioral strategies (e.g., burying). The Chapter discusses the ecomorphology of rheotaxis while reviewing the basic physics of displacement by slipping and lift-off, and gives a simple dimensional model of the scaling of these speeds with size and in relation to current speeds. The metabolic cost of station holding is discussed in various case studies. Specific examples of interaction with a substrates are also given regarding fast starting in flatfish and the attachment behavior of remoras.

A fish's choice to swim in a certain mode and at a certain speeds, has fundamental energetic and survival consequences (Chapters 3 and 4). While flow in natural systems implies a number of adaptations in fishes, in recent years a further issue has become quite relevant, that of habitat modification in streams. This has often causes habitat fragmentation, which is a major factor contributing to reductions in biodiversity and species abundance worldwide (Chapter 3, Castro Santos and Haro). These authors discuss the ability of fish to pass barriers during both up- and downstream migrations. Fishway performance is quantified as the rate of passage past barriers, which is affected by both swimming performance and behavioral factors. Castro-Santos and Haro describe motivating and orienting cues and discuss how these can affect passage performance. The authors conclude that the challenge of fish passage is both physiological and behavioral, and neither component is sufficient on its own to produce consistent and acceptable fishway performance. In addition, fish have developed various behavioral adaptations for minimizing the cost of swimming, whether this implies holding station in a flow or swimming in still water (Chapter 4, Fish). Chapter 4 discusses a number of ways fish can minimize energy expenditure while swimming, including locomotor strategies, such as optimizing swimming speed (i.e., for minimal cost of transport), burst and coast swimming (minimizing drag during the coast phase), schooling (swimming in the wake of neighbors), hitch hiking (as in remoras), drafting (taking advantage of the wakes created by object upstream). These strategies are discussed within a biomechanical, behavioural and ecological context.

In addition to dealing with the physical environment, swimming is used by fish during interactions with other animals, including conspecifics, predators and prey (Chapters 5–8). A common behaviour used by most fish species to avoid being preyed upon is the escape response. In Chapter 5, Domenici reviews escape responses from the standpoint of kinematics, ecomorphology and behavior. Escape success is not only determined by swimming performance but also by behavioral variables such as readiness and directions of escape. Intra and interspecific variability in escape response is also discussed. The links between escape performance, body design and habitat types is discussed within an ecological context, focusing on of the significance of escape responses during predator-prey interactions. For most fish species, locomotion plays integral roles in the two fundamental phases of energy acquisition: searching for food and feeding (Chapter 6, Rice and Hale). These authors discuss the locomotor demands for different feeding modes. In addition to the characteristics of the predator and prey, Hale and Rice discuss the effect of the physical environment of feeding, since feeding patterns differ between structurally complex and open habitats. Various behavioral patterns of feeding are described, and the functional and ecological significance of the different locomotor patterns employed for feeding and foraging in fishes are outlined and discussed. Langerhans and Reznick (Chapter 7) put the issues raised in the preceding Chapters on predator-prey interactions and water flow within an evolutionary context. They discuss some of the major ecological factors that might have shaped the evolution of locomotor performance in fishes. Using an integrative approach, they show how morphology and swimming performance are shaped by other factors that affect fitness, such as escape from predators or reproduction. By combining analyses among species with analyses among populations within species, they show how morphology and performance can evolve as local adaptations to risk of predation, but also how the evolution of performance can also interact with or be shaped by the evolution of life histories. Swimming is an integral part of courtship of males of various fish species. Sexual selection acts on male swimming performance to modify size and shape of the body and fins. In Chapter 8 , Kodric-Brown reviews the importance of sexual selection in the elaboration of fins and courtship behavior in male fishes. The Chapter discusses how sexual selection, acting within the framework of ecological, morphological and physiological constraints, can affect swimming performance. In addition, scaling issues, such as interactions between body size, sexual size dimorphism, secondary sexual traits, courtship behavior, and metabolism also affect swimming performance.

Environmental factors such as temperature, oxygen and salinity can have a profound effect on swimming performance and , as a consequence, on all those behaviors that involve swimming (Chapters 9 and 10). Wilson *et al.* (Chapter 9) review the effect of these factors focusing on unsteady swimming manoeuvres, and discuss the potential consequences of these effects for predator-prey and mating encounters. They show that a number of environmental factors, such as hypoxia and pollutants which are present with increasing trends in coastal areas, can have profound effects on fish swimming behavior. They conclude that an integrated approach is necessary to predict impacts of both natural and anthropogenic global change on the behaviour of fish. McKenzie and Claireaux (Chapter 10) discuss the effect of environmental factors on the physiology of aerobic exercise, i.e., endurance swimming such as that used for migration and searching for food. Their Chapter shows that temperature, salinity, oxygen and pollutants can all have a significant impact on swimming performance, with important potential consequences for fish in their natural environment. Similarly, the behavioral responses to hypoxia may have major ecological implications: some species reduce their swimming activity until conditions improve, and other species may migrate away towards less hypoxic areas.

A dual approach using both field and laboratory observations of swimming speed is fundamental for our understanding of how fish swimming performance can have ecological consequences (Chapters 11 and 12). This approach is used in Chapter 11 by Fisher and Leis, in which they discuss the importance of swimming performance in the ecology of larval fishes. Evaluation of the swimming performance of various species of larval fishes reveals considerable variance. This has major consequences for their energetics and feeding ecology, their overall dispersal and connectivity patterns, and the degree to which recruitment is an active processes involving specific habitat choice and timing. Although larvae are not the passive particles they were once considered, neither are all larval fish equivalent in their behaviour and ecology, even if they may occupy a similar habitat. Fisher and Leis conclude that intraspecifc differences in larval swimming performance need to be carefully considered when developing ecological and bio-physical models involving larvae. Fulton discusses the role of swimming performance in the ecology of reef fishes in Chapter 12. It appears that one group, the labriform-swimming fishes, have emerged as the dominant occupants of coral and rocky reefs around the world. Using solely their pectoral fins to produce thrust, around 60% of reef fish taxa utilize labriform locomotion during their daily activities. Chapter 12 discusses the ecomorphological characteristics of this group using laboratory and field-based measures of swimming performance and patterns of habitat-use across micro-, meso- and macro-ecological scales. Fulton concludes that labriform swimming combines high speed performance with efficiency and flexibility to provide a highly versatile mode of swimming suited to the challenges of a reef-associated lifestyle.

The next two Chapters (Chapter 13 and Chapter 14) deal with telemetry observations of fish movements in the field. Sims (Chapter 13) discusses recent developments in the swimming behavior and energetics of sharks. The movements of sharks, their relative behavior and distribution patterns have remained largely unknown for many species. Chapter 13 discusses a new approach for analysing and interpreting the movement patterns of free-ranging sharks. Sims describes the typical generalised movement patterns of free-ranging sharks recorded using electronic tags, and how this new technology has revolutionised shark behavioural ecology. The Chapter identifies how movement types can be linked to habitat types and how foraging models can be used to test habitat selection processes in sharks. Furthermore, the Chapter discusses a new approach of analysing shark movement data that uses methods from statistical physics (Levy walk) to evaluate behavioural performance in relation to environment. Chapter 14 (Bernal *et al.*) discusses the field observations of the movement of large pelagic fishes by putting them within an ecophysiological context. This Chapter describes the species-specific swimming and movement patterns of tunas, billfishes, and large pelagic sharks derived from extensive data sets obtained using acoustic telemetry and electronic data-archiving tags. The results are then interpreted based on the current understanding of the physiological abilities of each species, while integrating these concepts with other important ecological factors such as prey movements and availability. Bernal *et al.* conclude that although groups of large pelagic predators may occupy similar geographic areas (exploited by the same fisheries), they really occupy largely separate ecosystems. This can have major implications for effective fishery management policies, resource conservation, and for the population assessments upon which they must ultimately be based. These species-specific differences also provide a wealth of opportunities for developing a fundamental understanding of the ecophysiology of these fishes.

For commercial marine fish species, swimming performance is not only important for their foraging, migration, predator-prey interactions, but also in predicting their vulnerability to fishing gears. The final Chapter (Chapter 15, He) discusses the swimming capacity of marine fishes and its role in capture by fishing gears. Swimming performance is relevant for capture processes in both active (such as trawls and purse seines) and passive (such as gillnets, longlines and traps) fishing gears. Swimming ability varies among species and sizes of the fish and can be influenced by environmental factors such as water temperature. As a result, swimming ability of fish may affect size and species selectivity of the gear and affect seasonal and spatial differences in availability and catchability, with important implications in commercial fisheries and stock assessment surveys.

This book aims at filling a gap in the literature, by adding behavioral and ecological viewpoints to the more traditional biomechanics, ecomorphology and physiological perspectives used in studies of fish swimming. The book is therefore largely integrative by its own nature, and it includes considerations related to fisheries, conservation and evolution. This book is aimed at students and researchers interested in fish swimming from any organismal background, be it biomechanics, ecomorphology, physiology, behavior or ecology.

We wish to thank Christel LeFrançois for editorial help, and our reviewers who contributed their time and efforts and helped improve this volume: Mark Westneat, John Steffensen, Shaun Killen, Ted Castro-Santos, Frank Fish, Chris Fulton, Brian Langerhans, Paul Webb, Guy Claireaux, David McKenzie, Melina Hale, Aaron Rice, Astrid Kodric-Brown, David Reznick, Richard Brill, Diego Bernal, Michael Musyl, Chugey Sepulveda, Christel Lefrancois, Rebecca Fisher, Robbie Wilson.

REFERENCES

Webb, P.W. 1975. Hydrodynamics and energetics of fish propulsion. Bulletin of the Fisheries Research Board of Canada, pp. 1–159.
Blake, R.W. 1983. Fish Locomotion. Cambridge University Press, Cambridge, UK.
Videler, J.J. 1993. Fish Swimming. Chapman and Hall, London, UK.

Paolo Domenici
B.G. Kapoor

Contents

List of Contributors

Bernal, Diego
Department of Biology, Univesity of Massachussetts, Dartmouth, 285 Old Westport Road North Dartmouth, MA 02747–2300, USA.

Blake, R.W.
Department of Zoology, University of British Columbia, Vancover, British Columbia, V6T1Z4, Canada.

Brill, Richard
Cooperative Marine Education and Research Programme, Northeast Fisheries Science Centre, National Marine Fisheries Service, mailing address: Vivginia Institute of Marine Science, P.O. Box 1346, Gloucester Point, VA 23062, USA.

Castro-Santos, Theodore
S.O. Conte Anadromous Fish Research Centre, USGS—Leetown Science Centre, P.O. Box 796, One Migratory Way, Turner Falls, MS 01376, USA.

Chan, K.H.S.
Department of Zoology, University of British Columbia, Vancover, British Columbia, V6T1Z4, Canada.

Claireaux, Guy
ORPHY, Université Européenne de Bretagne—Campus de Brest, UFR Sciences et Technologies, 6, avenue Le Gorgeu, 29285 Brest, France.

Cotel, Aline
University of Michigan, Department of Civil and Environmental Engineering, Ann Arbor, MI 48109–2125, USA.

Domenici, Paolo
CNR-IAMC, Località Sa Mardini 09072 Torregrande (Or) Italy.

Fischer, Rebecca
Australian Institute of Marine Science, University of Western Australia, 35 Stirling Highway, Crawley, 6009, Australia.

Fish, Frank E.
Department of Biology, West Chester University, West Chester, PA19383, USA.

Fulton, Christopher J.
School of Botany and Zoology The Australian National University, Canberra, ACT 0200, Australia.

Hale, Melina E.
Department of Organismal Biology, University of Chicago, Chicago, IL 60637, USA.

Haro, Alex
S.O. Conte Anadromous Fish Research Center, USGS—Leetown Science Centre, P.O. Box 796, One Migratory Way, Turner Falls, MA 01376, USA.

He, Pingguo
University of New Hampshire, Institute for the Study of Earth, Oceans and Space, Durham, NH 03824, USA.

Johnston, Ian A.
School of Biology, Scottish Oceans Institute, University of St. Andrews, St. Andrews, Fife KY16 8LB, Scotland, UK.

Kodric-Brown, Astrid
Department of Biology, University of New Mexico, Albuqurque, NM 87131, USA.

Langerhans, R. Brian
Museum of Comparative Zoology and Department of Organismic and Evolutionary Biology, Harvard University, Cambridge, MA 02138, USA.

Present Address: Biological Station and Department of Zoology, University of Oklahoma, Norman, OK 73019, USA.

Lefrançois, Christel
LIENSs-(UMR 6250, CNRS-University of La Rochelle)—2, rue Olympe de Gouges, La Rochelle, 17000, France.

Leis, Jeffrey M.
Ichthyology, Australian Museum, 6 College St, Sydney, NSW, 2010, Australia.

McKenzie, D.J.
Université Montpellier 2, Institut des Sciences de l'Evolution, UMR 5554 CNRS-UM2, Station Méditerranéenne de l'Environnement Littoral, 1 quai de la Daurade, 34200 Sète, France.

Meadows, Lorelle A.
University of Michigan, College of Engineering, Ann Arbor, MI 40109–2102, USA.

Musyl, Michael
Joint Institute of Marine and Atmospheric Research, University of Hawaii, 1125B Ala Moana Blvd, Nonolulu, HI 96815, USA.

Reznick, David N.
Department of Biology, University of California, Riverside, CA 92521, USA.

Rice, Aaron N.
Department of Neurobiology and Behavior , Cornell University, Ithaca, NY 14853, USA.

Sepulveda, Chugey
Pfleger Institute of Environmental Research, 315N Clementine, Oceanside, CA 92054, USA.

Sims, David W.
Marine Biological Association of the UK, The Laboratory, Citadel Hill, Plymouth PL12PB, UK. and School of Biological Sciences, University of Plymouth, Drake Circus, Plymouth PL48AA, UK.

Webb, Paul W.
University of Michigan, School of Natural Resources and Environment, Ann Arbor, MI 48109–1041, USA.

Wilson, R.S.
School of Biological Science, The University of Queensland, St Lucia, QLD 4072 Australia.

1

Waves and Eddies: Effects on Fish Behavior and Habitat Distribution

Paul W. Webb,[1,*] Aline Cotel[2] and Lorelle A. Meadows[3]

INTRODUCTION

The natural habitats of fishes are characterized by water movements driven by gravity, wind, and other animals, including human activities such as shipping. The velocities of these water movements typically fluctuate, and the resultant unsteadiness is exacerbated when the flow interacts with protruding objects, such as corals, boulders, and woody debris, as well as with surfaces, such as the bottom and banks. The importance of these ubiquitous unsteady water movements is reflected in increasing annual numbers of papers considering their impacts on performance and behavior of fishes swimming in "turbulent flows" (Tritico, 2009) or "altered flows" (Liao, 2007). The ability of fishes to stabilize body postures and their swimming trajectories when these are perturbed by turbulent flows affects

Authors' addresses: [1]Paul W. Webb University of Michigan, School of Natural Resources and Environment, Ann Arbor, MI 48109-1041. E-mail: pwebb@umich.edu

[2]Aline Cotel, University of Michigan, Department of Civil and Environmental Engineering, Ann Arbor, MI 48109-2125. E-mail: acotel@umich.edu

[3]Lorelle A. Meadows, University of Michigan, College of Engineering, Ann Arbor, MI 48109-2102. E-mail: lmeadows@umich.edu

*_Corresponding author_. E-mail: pwebb@umich.edu

species distributions and densities, and hence fish assemblages in various habitats (Pavlov *et al.*, 2000; Fulton *et al.*, 2001, 2005; Cotel *et al.*, 2004; Depczynski and Bellwood, 2005; Fulton and Bellwood, 2005). Understanding impacts of turbulence on fishes is also important as human practices modify water movements, and as turbulence-generating structures become increasingly common, such as propeller wash, boat-created waves, shoreline hardening to control erosion, fish deterrents, and fish passageways (see Chapter 3 by Castro-Santos and Haro, this book; Wolter and Arlinghaus, 2003; Castro-Santos *et al.*, 2008).

Unsteady water movements have many effects on fishes but the mechanical basis for understanding the nature of responses is poorly known (Liao, 2007). At high levels, turbulence can result in mechanical injuries that injure and kill fishes. Here, we concentrate on a framework to explore interactions between fishes and turbulent flows relevant to sub-lethal effects. We argue that the distribution and strength of structures (orbits and eddies) in turbulent flow that would encompass a fish-like body are essential for the evaluation and prediction of locations chosen by fishes and their paths through turbulent flows. Thus, we first consider how fish-flow interactions may be approached as a physical phenomenon. Second, we discuss methods that have been used to quantify levels of turbulence. These discussions set the stage to revisit studies on swimming performance, behavior, habitat choices, and hence fish assemblages.

Turbulent Flow Structure and Frames of Reference

An important generalization in evaluating fish interactions with turbulent flows is recognizing that there are recurring probabilistic structures within unsteady water movements. We define these as "*orbits*" for periodic trajectories of water particles driven by non-breaking waves, and use *vortices* or *eddies* for structures in turbulent flowing water and breaking waves (Table 1.1, Figs. 1.1 and 1.2). It has been postulated that positive, negative and neutral impacts of unsteady water movements on fish performance and behavior reflect scale effects whereby only orbits and eddies of certain sizes present stability challenges and cause disturbances (Pavlov *et al.*, 2000; Odeh *et al.*, 2002; Nikora *et al.*, 2003; Webb, 2006a; Liao, 2007). However, specific data are lacking, and much of the evidence is correlative or anecdotal.

Orbits and eddies vary widely in their distribution and in their sizes in natural habitats. Fish occupy much smaller domains than the entire water body, and hence will experience some subset of available structures. It is, of course, to be expected that fish choose habitat locations, or choose trajectories when swimming through the general flow, that minimize potential negative effects or maximize potential benefits. Thus consideration of fish interactions with unsteady water movements involves two linked systems with very different

Table 1.1   Unsteady water-movements in the incident flow of a habitat derived from wave-based and from flow-based drivers. Each induces different characteristic water movement patterns. Real habitat flows are a combination of water movements from these two sources. Based on Kolomogorov (1941), Panton (1984), Vogel (1994), and Denny (1988).

Wave-induced water movements	Flow-induced water movements
Groups of waves occurring at density discontinuities (mainly the air/water interface) travel across the surface formed by the discontinuity.	Water flows within a lake or stream bathymetry.
In unbounded situations (lacking physical structures), a particle of water follows a circular path (here called orbits).	
In bounded situations, the circular path becomes elliptical parallel to the bottom.	At high current speeds and/or large systems (high Reynolds numbers, high Froude numbers), the flow changes from laminar to turbulent. The turbulent flow is made up of eddies of many sizes. Surfaces and physical objects are sources of eddies due to the viscous nature of the fluid.
A water particle returns very close to its starting position as each wave passes.	Water particles can follow various circular trajectories in eddies with a superimposed net downstream displacement, and hence do not return to their starting positions.
Wavelength decreases as waves move into shallow water, eventually breaking creating turbulent water flows, i.e. flows containing vortices or eddies.	Turbulent eddies engulf surrounding water over time (as they age) and increase in size to eventually span the physical limits determined by the bathymetry of a system.
	Eddies calve smaller eddies as they age creating a range of eddy sizes and an energy cascade. The smallest eddies are defined as the Kolmogorov eddy size in flows typically encountered in fish habitats. These small eddies dissipate the energy of the turbulent flow as heat through viscous processes.
Common drivers: wind, density gradients of thermoclines and haloclines, boats.	Common drivers for vorticity (leading to the creation of eddies/vortices) are gravity (flow-water systems), wind and shear, viscosity (associated with flow effects and variations with temperature and density), baroclinic effects, breaking waves, and vortices shed by organisms during locomotion and feeding.

The Reynolds number represents the ratio of inertial to viscous forces in flow. At small values, disturbances tend to be damped, and the flow is laminar. At high values. inertial effects tend to amplify disturbances and flow tends to be turbulent. The Reynolds number also provides the ratio of eddy sizes for a particular flow situation.

Detailed discussion of Froude number is beyond the scope of this review (see Denny, 1988). The Froude number is a ratio of inertial to gravitational forces, providing information of free-surface dynamics; cf. the Reynolds number, which is representative of the flow within the interior of the water column. As described in the text, fishes tend to be found in the lower part of the water column, when the Froude number will be less important than Reynolds number in determining the relevant flow.

properties. The first is water movements in the environment. These are independent of the fish, and are determined by the factors creating waves and currents, and interactions of waves and currents with other habitat structures. The result is *incident water movements* that characterize aquatic habitats. Second, a fish (or other organism) responds as an *embedded body* within the *incident water movements*, experiencing a sub-set of these incident water movements.

The presence of an embedded body will modify the details of the incident water movements in the vicinity of that body. In many situations, the presence of a fish body will not have a large impact on those incident water movements, although vorticity shed during swimming may have a large effect when it meets downstream propulsors within the length of the body of a fish, or those of nearby fishes. Given the need for a conceptual framework towards understanding fish-turbulence interactions, especially given apparently contradictory results, and the likely small impacts where flow overtakes a fish, we chose to simplify the problem of fish-turbulence interactions, considering incident water movements independently of the fish presence.

Flow in Fish Habitat—The Incident Flow

Water movements in fish habitat are complex. Two major factors underlie this complexity, with different types of contribution from waves and from flow (Table 1.1; Figs. 1.1 and 1.2). The relative importance of the wave induced orbits within the water, and flow-induced eddies varies among habitats, although real-world flows in fish habitats are some combination of both.

Non-Breaking Wave-Induced Water Movements

In the realm of interest to this discussion, surface (and less often subsurface) gravity waves are created on the water surface by the action of wind as well as human-induced disturbances, such as boats. In unbounded water, a traveling wave is a progressive wave form emanating from a source. Waves are described in terms of: (a) wavelength, λ, the distance between crests or between troughs, (b) period, τ, the time taken for a recurring displacement of the wave to pass a location in an environmental frame of reference, and (c) amplitude, A the vertical displacement above and below mean water level. Height, H, is also used in describing waves, where H = 2A. The wave form travels with speed, or celerity, λ/τ. Shorter wavelengths travel more slowly than longer waves. The energy in waves also is gradually dissipated over time, this occurring more rapidly for shorter waves than for longer waves. Hence longer waves propagate farther than shorter waves (Denny, 1988).

Wave trains induce the periodic, essentially closed motions in water particles in the water column that constitute the orbits. A water particle in an

Direction of wave

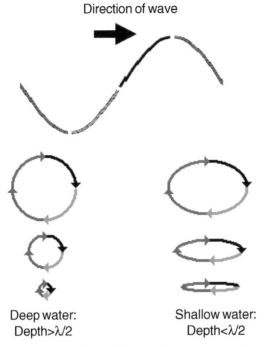

Deep water:
Depth>λ/2

Shallow water:
Depth<λ/2

Fig. 1.1 Diagrammatic representations of orbital trajectories, or orbits (circles and ellipses), followed by water particles as a non-breaking surface waves travels from left to right in deep and in shallow water. The vertical scale is exaggerated for clarity. (Based on Denny, 1988).

Direction of Flow

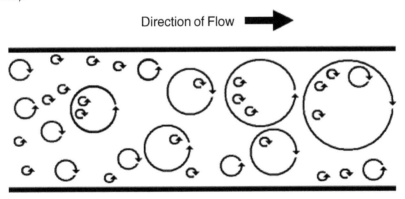

Fig. 1.2 Diagrammatic representations of eddies in eddy dominated turbulent flow in a channel flow with net flow from left to right. Eddies are illustrated by circles with arrows showing the direction of rotation. The delineations of eddies as discrete entities can be derived from streamlines, for example in Particle Image Velocimetry, with boundaries delineating regions of finite vorticity differing from that of the general flow. The figure shows that eddies vary in size, can calve new eddies, and as eddies age (i.e., as they travel downstream) eddies entrain additional water and grow.

unbounded aquatic system affected by a simple linear wave traces out a circular trajectory as a full wavelength passes by. The circular orbit attenuates exponentially with depth from the water surface, until it essentially vanishes at a depth of $\lambda/2$ (Fig. 1.1). In a real system, the circular orbits are not quite closed and there is some net translocation of the water particle in the direction of wave travel.

When a traveling wave approaches a boundary, the nature of the wave changes and hence so does the flow pattern of water affected by the wave. As the bottom shoals, the frictional influence of this boundary begins to affect the shape of the orbital trajectory when water depth $<\lambda/2$. The trajectories of water particles become elongated, and hence elliptical, with the displacement of a water particle decreasing in the vertical plane. Close to the bottom in such shallow water, the motion of water particles essentially becomes a surge, and at the bottom itself, the no-slip condition applies (Fig. 1.1).

In addition, as waves encounter shallow water, interaction with the bottom causes the wave speed to decrease, while the period remains the same. As a result, the wavelength decreases and the wave profile becomes steeper. As this process continues the wave eventually becomes unstable, resulting in a breaking wave.

A shoaling bottom also influences the direction that a wave travels. A wave typically has a long crest, this being well-known by any of us sea-gazing along a shoreline seeing serried ranks of parallel waves coming onshore. As a long-crested wave travels in varying water depth, the portion of the wave in shallower water travels more slowly. This variation in wave speed along the crest results in the wave crest bending towards alignment with the bottom contours. The process, called wave refraction, has a significant influence on the distribution of wave height and thus wave energy along a stretch of coastline. Wave height increases over topographic "highs" as waves converge, and wave height decreases over topographic depressions as waves diverge. The resultant distribution of wave energy along an irregular shoreline contributes to the alteration of the bathymetry, or bottom topography, through its effect on the erosion and deposition of sediments. The resulting sediment distribution, in turn, affects substrate for plants, sorts rocky and woody materials, both in turn affecting fish habitat.

Another factor affecting wave patterns at a shoreline scale is wave diffraction, which occurs when the progression of a wave is interrupted by a structure or body. In the absence of diffraction, the water immediately behind a structure would be calm. Instead, as a wave passes by an obstruction in the water, the wave propagates into the "shadow zone" behind the structure. Diffraction effects play an important role in the lee of a shore protection structure, such as a detached seawall, and within coastal harbors. These effects also influence sediment transport and habitat conditions for fishes.

Finally, at a shoreline-level, obstruction to wave propagation becomes important when such structures reflect a wave. The level to which the wave is reflected is dependent upon the configuration and composition of the reflecting boundary. A sloping sandy beach has a small capacity to reflect incident waves, while a vertical steel sheet-pile wall is an almost perfect reflector. A reflected wave can interact with the incoming waves and can set up a standing wave pattern such as those commonly seen within coastal harbors (Denny, 1988).

These various features of shorelines that affect wave forms and fish assemblages generally receive insufficient attention in habitat descriptions (Murphy and Willis, 1996).

Throughout the coastal community, there is a growing awareness of the deleterious impact of boat-generated wakes on the shorelines of lakes, estuaries and rivers. In such restricted waterways, these waves can greatly exceed the natural wind-induced surface waves, producing storm-sized waves at un-naturally high rates of occurrence (Cwikiel, 1996; Asplund *et al.*, 1997). With mean boat size and ownership rates increasing, this new factor introduces the potential for a higher overall wave climate and energy regime at the shoreline. These high forced-wave conditions have been shown to produce a variety of wave patterns, including solitary waves (Wu, 1987), turbulent bores (Gourlay, 2001), as well as driving riverbank erosion (Washington State Ferries, 2001) and producing erosion in environmentally sensitive wetlands (Brown and Root, 1992; Good *et al.*, 1995). Thus, during periods of high boat traffic, such as warm summer days, shorelines along restricted yet accessible waterways experience boat-induced "storms." Complicating matters for the underwater fauna, these "storms" are not accompanied by the environmental signatures typical of natural storms and are thus unpredictable.

The foregoing discussion treats waves as if they were simple sinusoidal shapes. Real-world waves are created and affected by a variety of processes (not all of which are understood), so that actual wave forms are more complex. Thus, real wave fields are composed of intermingled waves of different shapes and sizes. As a result, turbulence created by the so-called random sea (Denny, 1988) will be much more variable as a function of depth, and the variation in the size and rotation rate of orbits will resemble that of eddies seen in turbulent flowing water.

Flow-Induced Turbulent Flow: Eddy-Dominated Flow

Eddy-dominated flow is undoubtedly more common in fish habitat than non-breaking wave-induced orbits. Eddies are areas of the flow where the trajectories of water particles, or streamlines, curve, leading to circular motions, such as a whirlpool. Vorticity is defined as the curl of the velocity vector, ω, i.e., a form of angular velocity. Therefore, eddies are directly linked to vorticity and in fact are best defined as regions of finite vorticity. Modern measurement techniques such as Particle Image Velocimetry provide direct measurements of vorticity, which can be used to define the physical limits and hence sizes of eddies (e.g., Drucker and Lauder, 1999; Adrian *et al.*, 2000).

Vorticity is generated by many physical processes (see appendix A), some of which are especially important in typical fish habitats: (1) viscous dissipation due to the presence of boundary layers along the slopes and bottom of a river or lake environment, (2) baroclinic torque in marine environments where gravity acting on temperature and salinity gradients can create significant flows, and (3) local stretching of vorticity in turbulent flows due to differences in mean velocity, which stretch a vortex line and increases vorticity just like stretching a material increases its length (Panton, 1984). While wakes created behind objects in the flow or from shear layers due to velocity gradients are widely recognized as common sources of eddies, baroclinic torque and local stretching further increase the amount of vorticity in the flow, and may prove important in some situations.

Eddies created by the first mechanism, from flow interactions with habitat boundaries and objects in the water, will be the major source of eddies affecting fish swimming. The overall bathymetry of stream beds, ponds, lakes, estuaries, oceans etc. variously constrict flow, create expansion areas, and cause waves to break, all affecting eddy formation and their subsequent growth as eddies travel downstream. Objects protruding into currents are another important source of eddies commonly encountered by fishes in their habitats. Such protuberances include substratum ripples, corals and macrophytes, rocky materials, woody debris, sunken ships and many structures used in stream and lake improvement projects.

Just as the random sea is comprised of waves with many different heights and periods, real-world eddy-dominated flow is also comprised of a wide range of eddies (Figs. 1.2 and 1.3). The largest eddy size in fully developed flow, δ, is determined by the physical constraints of the system, such as the gyre filling the N. Pacific Ocean delineated by the American and Asian continents, stream width and depth, and the pipe diameter in engineering applications. Over time, and further downstream from the source, the eddy composition of a flow develops finer- and finer-scale turbulence, until the smallest eddy size, λ_o, reaches the Kolmogorov eddy size (Kolmogorov, 1941).

Eddies affecting fishes are in the "inertial sub-range", and cover a wide range of sizes, from the large eddies on the order of kilometers to Kolmogorov size eddies. Within the inertial range, energy is passed from one eddy to the next of smaller size in an inviscid fashion (i.e., no energy is lost due to viscous effects) until the Kolmogorov size is reached. At this lower limit of the eddy size range, eddies are affected by viscosity, so that Kolmogorov eddies are eventually damped by viscosity and their energy is dissipated as heat.

Eddy composition of fully developed flow can be described as a frequency distribution of eddy sizes, with eddy size decreasing logarithmically from many small-sized eddies to few large-size eddies (Fig. 1.3). The ratio of the largest to smallest eddy sizes is a function of Reynolds Number (Fig. 1.3), and $\lambda_o/\delta = Re^{-0.75}$ (Kolmogorov, 1941).

Data on the distribution of eddy sizes in natural fish habitats are lacking. As noted above, eddies arise from numerous sources in many locations. As such, the actual eddy distribution will vary locally depending not only on Reynolds number, but also upstream conditions, and details of local physical structures and bathymetry. Nevertheless, irrespective of such local effects, fishes in natural habitats will encounter eddies of sizes within the limits of λ_o and δ.

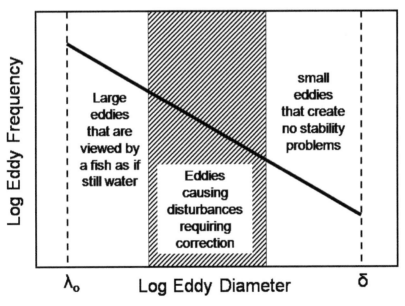

Fig. 1.3 In fully developed turbulent flow, eddy sizes vary over a wide range. The smallest size is the Kolmogorov eddy size, λ_o, where turbulent energy is dissipated as heat. The largest eddies span the physical boundaries of a system, with diameter δ. The ratio of the eddy size range in fully developed flow is a function of Reynolds number, Re, so that $\lambda_o/\delta = Re^{-0.75}$. There are many more small eddies than large eddies, and the relationship between the log of eddy-size frequency is negatively related to eddy size, such as diameter. Not all eddies will have effects on fishes. Large eddies and very small eddies may lack substantial displacement effects on fish posture and trajectories. Intermediate sized eddies may require stabilizing corrections or may overwhelm correction abilities of a fish. (Based on Kolmogorov,1941; Bell and Terhune, 1970; Pavlov *et al.*, 2000; Odeh *et al.*, 2002; Webb, 2002; Nikora *et al.*, 2003; Galbraith *et al.*, 2004.)

It is worth noting that the inertial range idea is a theoretical construct that has worked very well in engineering applications by providing a framework to compare eddy characteristics and to quantify turbulent flows. We also find it useful to in our attempt to classify turbulent situations to which fish react. However, by no means is this topic a closed subject.

Embedded Body—Fish-View of Incident Water Movements

The orbits of wave-induced water movements and eddies in currents are perturbations with the capability of destabilizing postures and trajectories of fishes embedded in the incident water movements. The magnitude of posture or trajectory displacements due to flow perturbations is expected to depend on the spatial and temporal scales of unsteady elements relative to the capabilities of a fish to damp or make corrections (Pavlov *et al.*, 2000; Odeh *et al.*, 2002; Webb, 2002; Nikora *et al.*, 2003). Damping and correction of displacements depend on: (1) the size of the unsteady water-movement elements relative to the dimensions of the fish, (2) the force and energy of these unsteady water-movement elements relative to fish inertia, momentum or energy and (3) the periodicity and predictability of unsteady water movement elements relative to response latencies of fish control systems (Webb, 2002).

The fish-view of the incident flow, however, is also affected by the motion of the fish embedded body as it passes through the water. When a fish is at rest relative to the ground, it experiences the water movements as seen by any observer in an environmental frame of reference. However, when a fish is in motion, the nature of incident flow it encounters can be affected by its velocity. We define *perceived water movements* as those affected by the fish's velocity. Thus the incident water-movement elements are perceived as compressed by a fish traveling upstream, making orbit and eddy dimensions and periods appear smaller. Water movement elements that might cause a displacement for a fish may become *apparently* small enough to be ignored. Conversely, orbits and eddies that may have been large enough to ignore may become important sources of displacements requiring stabilization of postures and/ or locomotor trajectories. An analogy familiar to readers will be the change in sound due to Doppler shifts as a listener moves towards or away from a sound source.

The impact of eddies in the incident water movements on an embedded fish will depend on where a fish is relative to all the turbulent flow elements. Hence the impacts of basin-wide or channel-wide incident water movements will vary as a fish changes location. Determining effects of incident water movements for the fish-based frame of reference therefore requires consideration of velocities, forces and moments through the limited, local domain occupied by a fish within of the coarse-scale flow (Standen *et al.*, 2004).

Quantifying Flow-Fish Interactions—A Proposal

A useful starting point for evaluating turbulent-flow impacts on a fish is circulation, which is measured by integrating (i.e. summing up) the vorticity, ω, over a surface area, A, or by integrating the velocity, V, around a certain region of the flow, l (which here can be represented by an embedded fish). Thus circulation, Γ, is defined as:

$$\Gamma = \oint \omega \, .dA \ = \ \oint \overline{V}.dl \tag{1}$$

Normalizing Γ with the product of fish velocity, V_{fish} and fish size, L_{fish}, gives a non-dimensional measure of the impact of circulation on a fish. When this ratio is small, displacements of the fish will be small, and may be small enough to neglect. Intermediate values will require active control behaviors, while large values will overwhelm these control systems and fish posture or trajectories will become highly irregular. In addition, other significant parameters are functions of circulation depending on the type and duration of interactions between fish and eddies, thrust, etc. These are primarily: (1) linear or angular impulse (Saffman, 1992) and (2) linear or angular momentum.

It was noted above that length scales of the incident water movement elements—eddy sizes—and the embedded fish body are essential for understanding fish-turbulence interactions. It was also noted that apparent size as modified by the velocity of the embedded body through the flow affects these apparent element dimensions. The normalization of equation 1 with both fish size and velocity introduces another aspect concerning the importance of the velocity of the embedded body for stability. As pointed out elsewhere, a faster object is able to withstand larger perturbations before stabilizing corrections are necessary (Weihs, 1989, 1993, 2002; Webb, 1993, 1997b, 2000, 2002). In general, the increased momentum and kinetic energy increase the damping of disturbances, while the flow over control surfaces increases both damping and corrective forces.

Thus linear (and angular impulse) or momentum are directly related to eddy circulation (Saffman, 1992). Velocity and pressure can be back-calculated from these quantities and the flow surrounding the embedded body estimated. Pressure can be integrated along the body, for example, and lift and drag forces calculated (i.e., horizontal and vertical forces), and assumptions about lift or drag coefficients are not required. This step in analyzing fish-turbulence interactions has yet to be made, but has been used in other situations (e.g., Schultz and Webb, 2002).

At this time, the analysis of the characteristics of eddies and orbits relative to those of a fish are rare, and essentially limited to size. Size alone is insufficient to defines the attributes of eddies that challenge posture and trajectory control. Hence the above discussion is a suggestion to begin considering fish-turbulence interactions in terms of the momentum and forces that determine the strength of perturbations, which working against the damping and corrective properties of a fish determine the magnitude and direction of displacements.

We are cognizant that analytical steps to go from vorticity or circulation to perturbing forces are not simple. It might seem attractive to use velocity profiles for vorticity with drag equations, analogous to attempts to use such semi-empirical, quasi-static approaches to model undulatory propulsion (Gray, 1968). However, there were few alternatives when those early studies were performed. The uncertainties concerning appropriate drag coefficients, and accommodation of unsteady effects make such as approach even less suitable today.

Methods of Measurement for Incident Flow

Habitat Classification

Historically, the ability to measure rapid variations in flow to quantify turbulence, especially in the field, has been limited (see methods in Denny, 1988; Vogel, 1994). Consequently, the simplest, and historically the most common approach to reporting the magnitude of turbulent flows and their effects on fishes has been to compare distribution patterns and behavior of fishes across habitats that differ in some measure of exposure, current speed, and habitat structural complexity. Wave-exposed reefs, riffles, and torrential streams are more energetic habitats, reasonably considered to contain more turbulent flow conditions (e.g., Hora, 1935; Hinch and Rand, 1998; MacLaughlin and Noakes, 1998; Nikora and Goring, 2000; Pavlov, 2000; Bellwood and Wainwright, 2001; Fulton et al., 2001, 2005; Nikora et al., 2003; Cotel and Webb, 2004; Roy et al., 2004; Depczynski and Bellwood, 2005; Fulton and Bellwood, 2005; Cotel et al., 2006; Smith and Brannon, 2006; Smith et al., 2006). For example, a river system is divided into various zones by speed, and inferred turbulence ranges from highly turbulent, high current headwaters, through immediate zones of lower flow, to low-flow meanders, presumably with overall low-turbulence (Allan and Castillo, 2007).

Similarly, reasonable expectations relating to current and physical habitat structure are used to classify the turbulent nature of flow in reef habitats (Friedlander and Parrish, 1998; Bellwood and Wainwright, 2001; Fulton et al., 2001; Friedlander et al., 2003; Depczynski and Bellwood, 2005; Fulton and

Bellwood, 2005; Fulton *et al.*, 2005). Thus Depczynski and Bellwood (2005) classified coral-reef habitats based on differences in exposure, which served as a surrogate for measured turbulence. Waves break over the shallow outer reef flat, and hence this zone or habitat experiences the highest flows and turbulence (Bellwood and Wainwright, 2001; Fulton *et al.*, 2001; Depczynski and Bellwood, 2005; Fulton and Bellwood, 2005). Slightly more seaward to this area of breaking waves is the crest, usually somewhat deeper, and although waves may be approaching instability, they have not usually broken. Water movements are likely to be lower than on the flat, and while reef structure would induce turbulence, the additional turbulence from wave breaking is absent. The backreef is most distant from the impact of the sea, and experiences least flow. The slope to seaward, a rapidly shoaling region of the reef, also experiences less water motion. Analogous habitat classification is also used to identify risk areas for fishes passing through turbines (Odeh *et al.*, 2002) and fishways (Castro-Santos and Haro, 2006).

Wave Gauges

Direct measurements of wave heights are provided by various wave gauges. Mechanical devices can measure extremes of waves and crests. Much more effective are impedance wave gauges that continuously record wave heights. However, these devices often pierce the surface, creating their own fine-scale flow characteristics. Pressure sensors may be employed as bottom mounted wave measurement devices. With these devices, changes in hydrostatic pressure are translated into a time series of water surface elevation measurements. In addition, recent developments in altimetry provide methods for non-surface-piercing measurement of water surface elevation from the air. In each of these cases, a time-series of water surface elevations is obtained which may be analyzed to determine the gross statistical characteristics of the water surface, as well as the detailed spectral distribution of wave energies (see below and Denny, 1988).

Dissolution of Balls

Habitat-scale estimations of net water movement and anticipated turbulence can be crudely quantified using dissolution of gypsum balls. Weight-loss from balls is calibrated by exposing balls of known starting weight and of similar dimensions to known currents (Jokiel and Morrissey, 1993). Differences in the rate of loss in different habitats are consistent with the habitat classification of expected levels of turbulence (Fulton and Bellwood, 2002, 2005; Fulton *et al.*, 2005).

Flow Meters

Various mechanical and electromagnetic flowmeters have been used to determine mean flow speeds (Denny, 1988; McMahon *et al.*, 1996). Among these, acoustic Doppler velocimetry, ADV, is growing in use because velocities can be measured with high resolution (Nikora and Goring, 1998). In ADV, three sound sources converge on a small volume of water that is interrogated at a high rate, usually up to 50 Hz. Sound is reflected from naturally-occurring or seeded, suspended fine particulate matter, and the Doppler shift in the reflected sound is used to calculate instantaneous velocities u_i, v_i and w_i along the *x*, *y*, and *z* axes.

The resultant velocity-time traces show the variation in velocity attributed to turbulence orbit and eddy structures passing the sample point (Fig. 1.4). The same pattern is seen in plotting wave height as a function of time for the open ocean (Denny, 1988). Indeed, both variation in wave height and variation in flow velocity are analyzed in similar ways to obtain statistical descriptions of the unsteadiness. The explanations of these measures and their derivations are especially well described for waves by Denny (1998).

The resultant instantaneous velocity, u_{ires}, can be calculated from u_i, v_i and w_i:

$$u_{ires} = \sqrt{u_i^2 + v_i^2 + w_i^2} \qquad (2)$$

Average velocities, $\overline{u}, \overline{v}, \overline{w}$ and u_{ires} are calculated from a time series of measurements over a long period, commonly two or more minutes.

Various statistical descriptions relating to flow are calculated from the instantaneous u_i, v_i and w_i velocities. Engineers interested in turbulence most commonly use the root mean square (rms) velocity, u_{rms}, as a measure of the flow variation, and hence of turbulence. This is calculated from the mean square deviation in velocity from the mean, numerically the statistical variance. The result is numerically the same as the standard deviation, σ. A non-dimensional measure of turbulence is derived from σ and $\overline{u}$ as turbulence intensity, TI (Sanford, 1997; Pavlov *et al.*, 2000; Odeh *et al.*, 2002), where:

$$\text{TI} = \sigma / \overline{u} \qquad (3)$$

There is some disagreement as to the value and use of TI (Smith and Brannon, 2005; Cotel *et al.*, 2006; Smith *et al.*, 2006) because $\overline{u}$ may affect fishes independently of σ. In physical terms, stabilizing posture and trajectory by fishes depends on both $\overline{u}$ and σ. However, as noted above, as u_{ires} increases, increasing fish momentum promotes stability (Webb, 2002, 2005), over a larger range of σ. Thus TI takes into account this speed-dependent aspect of stability. Nevertheless, responses of fishes to differences in TI should be made for similar values of $\bar{u}$ (Cotel *et al.*, 2006).

Statistical analysis of time-series data has also been used to estimate a length scale. This is obtained using autocorrelation analysis to determine the "integral length", which can be considered an estimate of average eddy size. Thus, Pavlov *et al.* (2000) estimated the average eddy size for his flume experiments as 0.66L, where L was the total length of the fish being studied. We suggest that conceptual and measurement problems are sufficient to make estimations of integral length of little utility in relating turbulent flow structure to embedded organisms. For example: (1) an autocorrelation analysis provides an average length scale for any flow, while a range of specific dimensions of local orbits and eddies is expected to be biologically relevant. (2) Interpretation of time-series data by autocorrelation to obtain the length scale over a large spatial domain includes assumptions such as Taylor's frozen turbulence, i.e., the flow has not changed significantly between each datum sample, which is often likely to be overly optimistic in natural flows occupied by fishes. (3) In terms of methodology, autocorrelation analyses must be applied over a domain that is large relative to the integral scale to

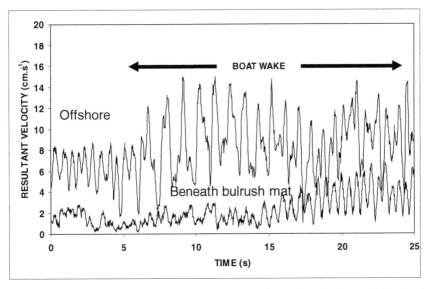

Fig. 1.4 Variation in flow velocities induced by waves from the wake from a 6.7 m boat traveling at 18 knots at a distance of 30 m parallel to the shoreline. The upper trace is from an ADV set at a height of 0.93 m from the bottom in water 1.5 m deep. The lower trace is from an ADV set midway in a column of water 0.35 m deep, within a bulrush patch (*Schoenoplectus arcutus*) with a mat of stems on the surface, showing the damping of wave induced flow. Note the decrease in mean current velocity and reduced velocity amplitude. The lag in the appearance of the boat wake effect under the bulrushes starting at about 15 s reflects the time for the wave generating the flows to travel from the offshore to the onshore sensor.

be estimated, which is usually not practical, or practiced (see O'Neill *et al.*, 2004).

Another parameter commonly used to quantify turbulent flows is the Turbulent Kinetic Energy (TKE) which measures the increase in kinetic energy due to turbulent fluctuations in the flow. Then:

$$TKE = 0.5 \ (\sigma_u^2 + \sigma_v^2 + \sigma_w^2) \tag{4}$$

The physical processes responsible for the presence of TKE in a given flow are the same as those mentioned earlier in the description of the sources of vorticity, i.e., viscous effects such as friction along bottom and sides, wakes behind protruding objects, etc. TKE provides a statistical way to evaluate the contribution of these turbulent fluctuations at a single point. Smith *et al.* (2006) found that TKE was a good predictor of juvenile rainbow trout densities in flumes.

The energy associated with any velocity occurring at a given frequency is proportional to the square of the magnitude of the velocity. Plotting a frequency distribution for the energy associated with velocities of various magnitudes gives a power spectrum (Fig. 1.5), the area under which sums to the overall mean-square velocity. The power spectrum thus indicates the contribution of velocities occurring with various periods to the overall energy in a fluid, a method to assessing the importance of various eddies from their velocity signatures. In the flow, the rate of energy dissipation in the inertial subrange for an energy cascade from large eddies to smaller and smaller

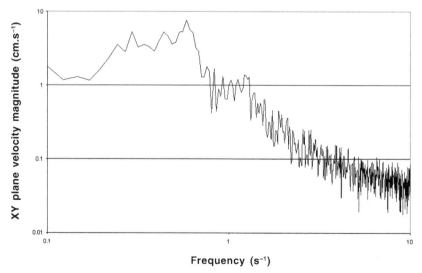

Fig. 1.5 Power spectrum for the planar XY velocity of a flow composed of waves and background turbulence in Lake Huron, MI. As the frequency increases, which corresponds to smaller and smaller length scales, the energy decreases, demonstrating that most of the energy is contained within larger scale flow features, as postulated by Kolmogorov (1941).

eddies is a constant (Kolmogorov, 1941). How the energy in the flow relates to fish kinematics is a topic of active research, and to our knowledge there are no simple relationships between the two. The energy in the flow is obviously coupled to the energy expanded by the fish but the specific functional relationship is unknown.

Ultrasound

The presence of turbulence in flow affects the delay for ultrasound traveling through the water. Sound emitters and receivers in a square formation are used and average vorticity and circulation can be determined within a sample space, which can be as small as 2 cm². This method has not been used in studying fishes or fish habitat to our knowledge, but transducers are fairly robust and could be deployed in the field and laboratory (Johari and Durgin, 1998).

Particle Image Velocimetry (PIV)

Methods of dissolution of plaster balls, ADV and similar methods have had, and will continue to have an important place in the study of the effects of unsteady incident water movements on fishes. However, these methods cannot *explain* the mechanisms of how turbulence affects fishes, nor why resulting interactions of fishes with unsteadiness in flows may be positive, negative or neutral. Tracking dyes or particles has been used to mitigate these difficulties. A modern derivative of older dye methods is Laser Induced Fluorescence (LIF) (Breidenthal, 1981) A dye is induced to fluoresce as is passes through a narrow laser beam, thereby giving an instantaneous 2-D image of the flow and any structure to the flow. Eddy periodicities also can be calculated from the average flow rate, or measured from multiple images at known time periods. A strength of LIF is its ability to rapidly assay flow before using more intensive methods, such as Particle Image Velocimetry, PIV.

PIV records the trajectories of neutrally buoyant microspheres or naturally occurring suspended particles as they pass through a laser sheet. If the video feed or the laser is pulsed at known rates, successive positions of particles can be determined, their velocity magnitude and direction determined, and hence the flow region can be mapped in terms of streamlines, velocity fields or vorticity.

From these data, the size of eddies can be determined and then related to fish size (Liao *et al.*, 2003b; Tritico, 2009). Such information would have provided the basis to compare sometimes conflicting results from various studies that have used widely different methods to induce unsteadiness into flow. In laboratory experiments, turbulence is commonly induced by in-current screens, bricks and similar objects (Pavlov *et al.*, 2000; Roy *et al.*, 2004; Smith and Brannon, 2005). A variant was used by Nikora *et al.* (2003), in which turbulence was induced by corrugated plastic walls and blocking some openings in an upstream grid. PIV also is needed to improve evaluation of designs for hydraulic structures such a fish ladders and barriers.

Systematic studies that compare the structure of incident water movements with fish behavior are still lacking but are essential to determine the size limits of eddies that create problems for fishes. Tritico (2009) measured swimming performance of creek chub, total length 15.5 cm, in the turbulent wakes created downstream of vertical and horizontal cylinder arrays. In these arrays, gaps between cylinders were equal to cylinder diameters, and observations were made for cylinder diameters from 0.4 to 8.9 cm. PIV was used to measure eddy diameters. The cylinder arrays essentially added successively larger eddies to the flow. There was no significant effect on swimming speed when eddies with diameters of 2.6 and 5.4 cm were added to the flow. Adding eddies of 8.4 cm resulted in ~10–20% reductions in swimming performance, as well as frequent failures to stabilize posture or hold position in the flow.

In practice, it is difficult to obtain simultaneous images of the turbulence incident flow and fish responses (but see Liao *et al.*, 2003b), especially when fish maneuver through a large three-dimensional space as occurs when turbulence overcomes the control capability of fish (Tritico, 2009). Current PIV methods do lend themselves to prediction of where fish would be expected in a turbulent flow based on criteria such as energy minimization (Standen *et al.*, 2004) or displacement forces and torques.

Computational Fluid Dynamics

The practical problems of explicitly determining incident-flow/embedded-body forces suggest that modeling may be a desirable complement, or alternative, to direct observations. In particular, Computational Fluid Dynamics (CFD) has the capability to describe turbulent flow components in complex flows, and to evaluate the consequences for objects embedded at various places in the flow. Paik and Sotiropoulos (2005) analyzed the flow behind a large rectangular block in a large aspect-ratio channel, and found slow moving large-scale structures interacting with the shear layer behind the block. This type of flow dynamics is important in designing river restoration projects and understanding fish behavior behind such structures. In general, CFD can model the composition of turbulent flows with and without fish-like embedded bodies, and can rapidly estimate forces and torques that such bodies would experience. This provides opportunities for rapid evaluation of fish-flow interactions.

Swimming Performance of Fishes in Unsteady Flows

Performance and Locomotor Gaits

Swimming performance is quantified in a variety of ways. In the past two decades, concepts of gaits (Alexander, 1989) have been applied to fishes, and ideally, performance levels should be compared for the same swimming gaits (Webb, 1994a, b, 1995; Drucker, 1996). Gaits range from station holding to avoid swimming by interacting with the bottom, through hovering in still or slow water, to transient, high-acceleration fast-starts. Between these extremes, aerobic muscle is used at cruising speeds sustainable indefinitely. The aerobic muscle powers paired fins at low speeds, transitioning to the body and tail swimming at higher speeds. Cruising speeds are succeeded by anaerobic-muscle powered high-speed sprints driven by body and tail motions, considered to last <15s. As fish transition into cruising swimming at low speeds and from cruising to sprinting at higher speeds, unsteady twitch-and-coast and burst-and-coast gaits, respectively, promote endurance (Webb, 1994a, b, 1995, 1997; Drucker; 1996; Gordon *et al.*, 1996, 2001).

A large body of experimental studies have probed performance limits at the cruising-sprinting transition zone as critical swimming speeds, u_{crit}. These speeds are usually assayed as those at which fish fatigue using an increasing velocity test (Blazka *et al.*, 1960; Brett, 1964; Beamish, 1978, but see Nikora *et al.*, 2003). Experiments are performed in a flume or water tunnel. It has become apparent that the cross-sectional size and length of the chamber in which these tests are made also affect performance limits (Peake and Farrell, 2004; Tudorache *et al.*, 2008), so that attention needs to be paid to methodology in comparing results from these tests. In addition, the nature of turbulence in real-world situations may diminish the application of results to natural habitat (Jenkins, 1969; Puckett and Dill, 1985; MacLaughlin and Noakes, 1998; Enders *et al.*, 2003).

Swimming Performance in Wave-Dominated Flow

Fish hover in still water, and attempt to do so in orbits induced by surface waves. When the period and amplitude of these orbits are large, with diameter>>fish length, L, fish surge with the flow and no control movements are apparent (P.W. Webb and A.J. Cotel unpublished observations). Then fish ignore the flow component, behaving as if in still water. It seems likely that very small orbits, with diameter<<L, also would be ignored.

Orbits with diameters of the order of fish total length, however, cause displacements of fishes that can exceed their stability control capabilities (Cotel and Webb, 2002). Yellow perch, *Perca flavescens*, spottail shiner, *Notropis hudsonius,* and bluegill sunfish, *Lepomis macrochirus,* tethered in a flow field dominated by wave-induced orbits, experienced large displacements to posture such that bluegill were unable to correct for postural displacements, and could not their control position in the water column (Fig. 1.6). Sometimes spottail shiners failed to control surge. Perch stability was intermediate between that of bluegill and spottail shiners. The differences in ability to control stability by perch and bluegill, both spiny-rayed fishes compared to spottail shiners, a soft-rayed fish, were compatible with laboratory observations on stability. Acanthopterygian fishes so far have been found to have longer response latencies (Webb, 2004), lower abilities to entrain in the turbulent wake of cylinders (Webb, 1998), and lower thresholds to control posture following increases in rolling torque (Eidietis *et al.,* 2002).

Fishes swimming voluntarily through the same incident flow field were observed (Fig. 1.6). These fish swam near the bottom where the orbits were reduced in size (Fig. 1.1). Compared to fishes also tethered near the bottom, free swimming fish never lost control of posture, but did surge relative to the bottom, reflecting the horizontal component of the orbits (Cotel, A.J. and Webb, P.W. unpublished observations). The absence of displacement problems for the free-swimming fishes is consistent with the improved stability capabilities expected with motion of embedded bodies through the incident turbulent water movements described above.

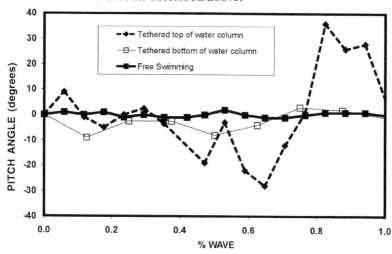

Fig. 1.6 Pitch angles of spottail shiners tethered in the top third and bottom third of 38 cm-deep water, and free-swimming through the same incident flow in the lower third of the water column. Displacements are reduced when tethered near the bottom where orbits are attenuated and by translocation. Modified from Cotel and Webb (2002).

Swimming Performance in Eddy-Dominated Flow

Very large eddies as basin-wide gyres can span oceans and lakes. It seems improbable that these affect swimming performance, as fish probably treat them in the same way as steadily flowing or still water, similar to large diameter orbits. Fish appear to use these large gyres as conveyor belts for migration (Brett, 1995; Höök *et al.*, 2003, 2004).

Very small eddies also are thought to have no effects on swimming performance. It might be considered that even small eddies in the incident flow would induce turbulence in the boundary layer and hence increase energy losses, presumably reducing performance. However, boundary layer energy losses are thought to be a small part of total energy losses (Webb, 1975; Schultz and Webb, 2002), and the boundary layer may be turbulent over much of the fish body in any case (Webb, 1975; Blake, 1983). Such small-scale eddies create smaller induced velocity or resultant forces than larger-scale eddies with larger circulation or vorticity (Biot-Savart or Kutta-Joukowski's laws; Panton, 1984). A key development in low-volume water tunnels (Blazka *et al.*, 1960; Brett, 1964) was the induction of micro-turbulence to create rectilinear flow profiles in test sections, with eddy sizes considered too small for detection (Bell and Terhune, 1970).

Pavlov and his collaborators have amassed an impressive range of results on performance of fishes in eddy-dominated turbulent flows (Pavlov *et al.*, 2000). As current speed increases through very low values, fish first orient to the flow (rheotaxis), which reduces drag and promotes station holding (Arnold and Weihs, 1978; Webb, 1989). Pavlov *et al.* (2000) found that threshold current speeds at which roach, *Rutilus rutilus*, first oriented to the flow increased when turbulence intensities increased.

As current speed increased above speeds stimulating rheotaxis, fish holding station on the bottom eventually slipped, this being the limit for the gait transition from such station holding to free swimming (Arnold and Weihs, 1978). Pavlov *et al.* (2000) found gudgeon, *Gobio gobio*, slipped at lower speeds as TI increased.

Maximum cruising speed also decreased as TI increased for perch, *Perca fluviatilis*, and roach. Similarly, prolonged swimming speeds were reduced with increasing TI for perch, roach and gudgeon (Pavlov *et al.*, 2000). Of particular interest in these experiments was the finding that performance of fishes from still-water habitats was reduced more with TI than fishes from populations from streams and rivers. Maximum burst speeds also were reduced with increasing TI for roach and again, the effects of TI were larger on fish from quieter-water habitats (Pavlov *et al.*, 2000). Finally, Reynolds stresses >30 N.m^{-2} for 10 minutes in turbulent flows that visibly buffeted fishes, reduced startle responses of hybrid bass (striped bass, *Morone saxatilis* × white bass, *M. chrysops*), and Atlantic salmon parr, *Salmo salar* (Odeh *et al.*, 2002).

While most studies to date have shown reduced swimming performance in all gaits with increasing TI, there are some notable exceptions of neutral effects. For example, Ogilvy and DuBois (1981) found elevated turbulence had no effect on cruising speeds of bluefish, *Pomatomus saltatrix*, and Nikora *et al.* (2003) found similar results for sprints by inanga, *Galaxias meculatus*. Odeh *et al.* (2002) found startle responses of juvenile rainbow trout, *Onchorhynchus mykiss*, were unaffected by turbulence levels that did affect this behavior in hybrid striped bass and Atlantic salmon parr (Odeh *et al.*, 2002). The authors of these studies have suggested that differences relate to eddy size relative to fish size, such that eddies experienced by fishes were too small to create displacements large enough to require conscious control. Alternatively, differences might result from offsetting effects of different eddy sizes in the incident water movements, some facilitating swimming, while others interfere with swimming. The discrepancies between these studies and those showing negative effects of turbulence underscore the importance of improved methods to visualize flow, as discussed above, and the need for more experiments on fish performance in known incident flow fields (Tritico, 2009).

Finally, swimming performance may be *improved* by appropriate eddy structures (Liao *et al.*, 2003a, b). General capabilities for turbulent flows to enhance performance have been the topic of much speculation, supported by a variety of observations on behavioral changes in unsteady flow (see below). However, at this time, specific data from controlled experiments are lacking to adequately evaluate the ability of fish to improve speed and endurance, or to reduce transit times by exploiting flow variations.

Energetics

An important component of behavior of fishes in turbulent flows is the deployment of control surfaces. There is increased use of various median and paired fins by fishes faced with perturbing flows (MacLaughlin and Noakes, 1998; Webb, 2004). Tail-beat frequencies of salmonids swimming in the field are often irregular compared to swimming in a flume, and the difference has been attributed to turbulence (Jenkins, 1969; Puckett and Dill, 1985; MacLaughlin and Noakes, 1998). McLaughlin and Noakes also described deployment of the pectoral fins by young-of-the-year brook trout, *Salvelinus fontinalis*, swimming in turbulent flows. Paired fin use declined with increasing swimming speed, consistent with the expectation that stability control is enhanced at higher swimming speeds.

These additional fin motions are expected to increase energy costs of swimming compared to those in steady flows. The only direct measure of increased costs for fish swimming in turbulence were obtained by Enders *et al.* (2003) for Atlantic salmon, *Salmo salar*, swimming in a water tunnel with imposed wave-like pulses. These experiments are unusual in that the flow apparently was intended to create surge perturbations. The shear forces developed in the system, however, would undoubtedly have calved eddies, probably creating a flow dominated by a limited size range of eddies. A power spectrum based on ADV measurements of the incident flow reveals some dominant flow structures, but PIV would be needed to determine their nature. The approach deserves further study as a possible method to impose more controlled flow perturbations on fishes than most those created with current methods.

Costs of swimming of fish in the surge-generated incident water movements increased by 30% as TI increased from 0.28 to 0.36 when swimming at an average speed of ~18 cm.s^{-1}, and 1.6-fold as TI increased from 0.21 to 0.30 when swimming at ~23 cm.s^{-1} (Enders *et al.*, 2003). Metabolic rates for the salmon swimming in turbulent flows was 1.9 to 4.2-fold larger than expected from data obtained by others using water tunnels with typical microturbulent flow conditions.

Stability control involves inertial corrections, and hence stability is mechanically similar to maneuvers. Energy costs associated with maneuvers, therefore, illustrate the potential for increased energy costs involved in controlling stability. Maneuvers can increase energy expenditure compared to traveling at the same average translocation speed by over 10-fold (Blake, 1979; Weatherley *et al.*, 1982; Puckett and Dill, 1984, 1985; Webb, 1991; Boisclair and Tang, 1993; Krohn and Boisclair, 1994; Boisclair, 2001; Enders *et al.*, 2003).

In contrast to these increases in swimming costs in more turbulent flows, there is a potential for fish to extract energy from turbulent flow elements and thereby *decrease* energy consumption. In flume experiments, Standen *et al.* (2004) found swimming speeds of sockeye salmon, *Oncorhynchus nerka*, migrating through a turbulent river were 1.4 to 76 times higher than those expected based on tail-beat frequencies of fishes swimming in rectilinear, steady incident flow. Liao *et al.* (2003a) and Liao (2004) found muscle activity was reduced in rainbow trout using the Kármán gait (see below).

Behavior of Fishes in Turbulent flows

Choosing Where to Swim

Performance measures are made at transitions from one gait to another. Within gaits, fish behavior also may be affected by turbulence. Most observations have been for gaits supporting lower, sustainable speeds and show turbulence levels affect locations chosen by fishes in the laboratory and the field. The compilation by Pavlov *et al.* (2000) showed that choices vary among species, with those from slower-current habitats such as carp, preferring lower TI, and more riverine species, such as gudgeon, grayling and chub choosing higher turbulence. Within species effects were also found, with populations from streams and rivers preferring higher TI than those from still-water lakes and ponds, reflecting different histories (Coutant, 1998).

There is also an interaction between current speed and TI, such that fishes tend to avoid high TI conditions at both low and at high $\overline{u_{ires}}$, but choose higher TI at intermediate current speeds (Pavlov *et al.*, 2000). This dome-shaped response pattern for TI chosen by fishes, seen as a function of average speed in the incident flow, also was found for the ability of fishes to flow-refuge in the turbulent wake behind vertical and horizontal cylinders (Webb, 1998; Tritico, 2009), and in feeding efficacy (McKenzie and Kiorobe, 1995; Galbraith *et al.*, 2004). Such dome-shaped relationships probably reflect trades-off in control capabilities and energy costs at various speeds. The low momentum and kinetic energy of embedded bodies at low speeds reduce a fish's ability to control stability, such that energy costs of swimming in turbulent regions may be higher than for swimming in less turbulent areas. Under these circumstances, fish would be expected to avoid high TI (Webb, 1998, 2002, 2006). At higher $\overline{u_{ires}}$, the energy costs for translocation will be high, perhaps leaving little surplus for other activities, such as controlling stability, especially if energy costs are as large as many studies suggest (Blake, 1979; Weatherley *et al.*, 1982; Puckett and Dill, 1984, 1985; Webb, 1991; Boisclair and Tang, 1993; Krohn and Boisclair, 1994; Boisclair, 2001; Enders *et al.*, 2003).

Unfed fish also select higher levels of TI than fed fish (Pavlov *et al.*, 2000). Pavlov *et al.* (2000) attributed this to the potential for higher levels of turbulence to increase the probability of bringing food to fish through such mechanisms as increased shear stress displacing prey organisms (Boisclair, 2001). As with TI choices, contagion rates of food also increase at intermediate levels, with lower levels supporting lower feeding, probably through reduced delivery rates, and higher levels interfering with feeding because of the cost of swimming in such environments (McKenzie and Kiorobe, 1995; Pavlov *et al.*, 2000; Galbraith *et al.*, 2004).

Similar results have been found where fishes have a wider range of choices of positions in complex flows. Smith *et al.* (2005) recorded locations chosen by juvenile rainbow trout in a flume with several hydraulic options created by various combinations of bricks. The juvenile trout chose locations with below-average current speeds, lower TI and also smaller integral lengths as determined from ADV velocity time-series data. When current speeds were high, trout chose regions with below-average current speeds, in locations where TI tended to be higher. In addition, the density of rainbow trout supported in flumes with various bricks also varied with turbulence levels (Smith *et al.*, 2005). In this situation, TKE proved the best predictor of density, with highest numbers at intermediate TKE. In the field, low current speeds may provide less usable habitat than higher flows, while high flows displace fish and also increase energy costs of holding position (Boisclair 2001; Enders *et al.*, 2003; Fulton and Bellwood, 2005; Fulton *et al.*, 2005; Smith and Brannon, 2005; Smith *et al.*, 2006). Observations in a natural trout stream showed that brown trout, *Salmo trutta* occupy habitat with intermediate levels of turbulence compared to those available (Cotel *et al.*, 2006). These situations also have dome-shaped relationships between turbulence levels chosen *versus* current speed, consistent with choices made by individual fishes in flumes described above.

Fish also appear able to choose routes through complex flows, reducing current speeds experienced, or reducing upstream passage time (Hinch and Rand, 1998; McLaughlin and Noakes, 1998). Fish often swim close to banks (Hinch and Rand, 1998), or near the bottom (MacLaughlin and Noakes, 1998; Standen *et al.*, 2004) where current speeds are lower. Standen *et al.* (2004) found that sockeye salmon, *Oncorhynchus mykiss*, migrating up the Seton River in British Columbia chose routes through unsteady flow fields with below average current speeds, but when flow rates were high, fish swam at high speeds which would instead minimize transit time. Salmonids also use reduced flow associated with in-stream structures (Fausch, 1984).

Neither the composition of the turbulent flows over the whole domain of a passageway nor eddy characteristics along the paths chosen by a fish are known. Therefore, eddy properties in incident water movements cannot be related to the size of the embedded body. Consequently, although structures in the incident turbulent flow are implicated as causing disturbances that affect behavior, definitive evidence in lacking. A notable exception is studies on a novel use of eddies whereby fish "surf" the Kármán Vortex Street shed by D-cylinders (Liao *et al.*, 2003a, b; Liao, 2004). Liao and colleagues not only recorded the behavior of the fish, but also obtained unique data on muscle use and energetics, while using PIV to visualize eddies.

The observations by Liao and colleagues are also exemplary of an important additional aspect of behavioral interactions with turbulent flow elements whereby fish extract energy from the incident flow. There has long been discussion on the likely ability of migrants to extract energy to decrease migration times or energy costs (Enders *et al.*, 2003; Standen *et al.*, 2004). Fish in schools also may be able to extract energy from thrust-eddies shed by adjacent fishes (Weihs, 1973). Strict geometric position relationships are necessary among school members to take maximum advantage of flow components in eddies that have a velocity component in the direction of motion of the benefiting fishes. In addition, not all fish benefit from thrust-eddies at the same time, such that competition would seem likely for preferred conditions. Experimental evidence for appropriate spacing of school members is equivocal (Partridge and Pitcher, 1980), but does suggest that some fishes of some species benefit from energy extraction from eddies at some times (Herskin and Steffensen, 1998; Svendsen *et al.*, 2003). In addition, energy may be extracted from eddies shed by the body and fins of a fish by downstream propulsor surfaces (Lighthill,1969; Triantafylou *et al.*, 1993; Weihs, 1993; Nauen and Lauder, 2000, 2001; Alben *et al.*, 2006; Beal *et al.*, 2006; Standen *et al.*, 2006).

Avoiding Swimming

The ultimate response to turbulence is avoidance. There are no explicit studies on such behavior, but avoidance has been described in habitats where flow is undoubtedly turbulent. For example, many reef fishes shelter in reef structure to avoid strong peak ebb or flow currents (Potts, 1970; Popper and Fishelson, 1973; Fulton *et al.*, 2001). Johansen *et al.* (2007) describe the use of shelters by labrids and pomacentrids to avoid high-current, wave-swept locations of coral reefs, but from which they make forays to feed. Fishes in streams use refuges at night in high-flow, high-turbulence runs and riffles (Webb, 2006b). All these refuging behaviors not only avoid turbulent flows, but also are postulated to make locomotor energy savings.

Fish Assemblages in Turbulent Flows

River Continuum

Current is the defining feature of running waters, with the river continuum being divided into various zones by speed, and inferred turbulence (Lagler *et al.*, 1977; Bond, 1996; Allan and Castillo, 2007). Different species of fishes dominate in various zones, with community membership consistent with the small amount of experimental data on differences in response latencies and stabilizing capabilities among groups of fishes as noted above. Near the origin of a river system, with slopes >1/20, current speeds are high. In the Himalayas, there is a recognized torrential fauna of a few species capable of attaching or refuging in streams with the highest flows (Hora, 1935). Less extreme high-current high-turbulence habitats are characterized by salmonids and cottids. At intermediate gradients, flowing waters are characterized by riffle-pool systems, populated by grayling, dace, darters and suckers, and at the lower-slopes of such systems, by shiners and other minnows. As the current in a river system becomes sluggish, fish diversity increases and suckers, bullheads and catfishes, various minnows and shiners, pike and centrarchids become abundant. Levels of turbulence have not been described for fish habitat in the middle and lower reaches of rivers, but the scale of these systems, and low current speeds compared with higher reaches suggest that variability in flow due to turbulence probably decreases along the length of the river continuum.

Shorelines

Still-water lakes and ponds have no unifying principles such as provided by the River Continuum Concept. Much open water is essentially unbounded, and unsteadiness in the incident flow is primarily associated with winds, with turbulence arising from wind-driven wave orbits to wave driven currents creating eddies (Denny, 1988). We suggest an exposure-derived flow-based classification for lacustrine and oceanic shoreline habitat may be possible, combining regional and local data on frequencies of wind direction and velocity with fetch and shoreline bathymetry. Development of such a model could draw on the experience and principles of the River Continuum Concept. The sorting of substratum resulting from shoreline currents, the formation of erosion and deposition regions, and creation of habitat for macrophytes all have direct relevance to fish assemblages. In addition, shorelines are major sites of anthropogenic disturbance, making physical models relevant to biotic questions extremely desirable.

Interactions between wave patterns and vegetation and fish communities were examined over a range of bathymetries, using a 6.7 m boat traveling at 10 and 18 knots at distances of 60 and 90 m parallel to the shoreline as a deterministic source of waves (A.J. Cotel, L. Meadows, and P.W. Webb, unpublished observations). Surface wave patterns were measured using wave gauges along transects perpendicular to the shoreline for water depths <0.5m to the drop-off where water depth rapidly increased. Currents in the lower third of the water column where fish were most abundant were measured with ADV simultaneously with wave measurements. Macrophyte species and the percentage of the bottom covered by macrophytes were measured along the same transects. Fishes were assayed in an area of ±50 m along the transects using direct snorkeling observations and two gangs each with 5 minnow traps. Shorelines varied from natural depositional regions, erosional regions, and aquatic-beds, to armored rocky shorelines and sheet metal sea-walls. For similar driving forces, wave heights were greater for steeper shorelines and human-modified shorelines. Natural shorelines had a greater capacity to absorb wave energies than human-developed and reinforced shorelines. Vegetated sites were more effective in absorbing wave energies than rocky or sandy shorelines.

Fishes are often classified into functional groups on the basis of functional-morphological studies (Aleyev, 1977; Blake, 1983; Videler, 1993; Webb and

Gerstner, 2000; Fulton *et al.*, 2001, 2005; Fulton and Bellwood, 2005; Webb, 1997, 1998, 2002, 2004, 2006). Here fishes were classified as benthic, pelagic or slow-water groups (Fig. 1.7). Pelagic fishes would be most impacted by wave energies in open water. Benthic fishes would be impacted by these and currents created as waves travel into shallow water and break. Significant relationships between the abundance of fishes in these functional groups were found with bottom bathymetry and substratum composition (Fig. 1.8). The importance of bathymetry and substratum for benthic fishes is well known, as refuges from flow are important for such fishes (Depczynski and Bellwood, 2005). Pelagic fishes are more affected by the orbital flows in non-breaking waves and also tend to avoid breaking waves.

However, vegetation patterns were the most important correlates with the abundance of slow-water fishes and pelagic fishes where waves would break. Low-slope shorelines, where the slope dissipated much wave energy, supported bulrush marsh, *Schoenoplectus arcutus*, reducing turbulence for pelagic fishes. When water depths exceeded λ/2 various pondweeds, *Potamogeton sp.* dominated, which with the reduced size of orbits at these depths correlated with high abundance of slow-water forms.

In other studies, we found that the passage of storms affected local assemblage composition (P.W. Webb, unpublished observations), as have also

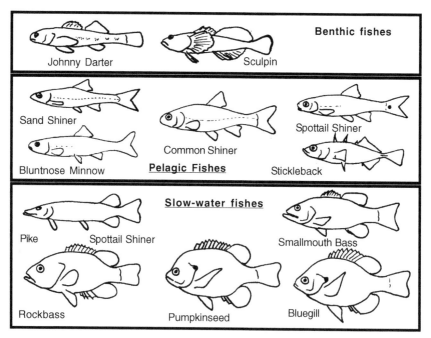

Fig. 1.7 Functional groups for fishes that comprise assemblages of mid-latitude freshwater shoreline fishes.

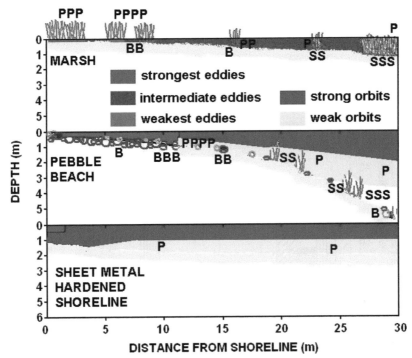

Fig. 1.8   A schematic representation of fish assemblages along three different wave-exposed shorelines. Three guilds of fishes are represented (Fig. 1.7); B = benthic fishes, P = pelagic fishes, and S = slow water fishes. One letter shows the fishes in a guild was a rare occurance, and four letters show that fishes were abundant. Prinipal turbulence features are indicated, as determined by waved gauges and PIV. Red tints representing areas where turbulence is largely derived from breaking waves, and blue tints indicate the strength of wave-driven orbits. Emergent macrophytes in marshes attenuate waves and reduce turbulence and shoreline assemblages include substantial numbers of fishes from the three guilds. Gradually sloping shorelines also dissipate wave energies, and in the breaking waves erode small substrate particles leaving primarily rocky materials. These provide habitat for benthic fishes. Pelagic fishes are common near shore where waves can suspend food items, but are found in the lower portion of the water column where orbit sizes are reduced by both depth and bottom effects (Fig. 1.1). Slow-water fishes are found in deeper water where wave effects become negligible and among macrophytes where orbits are attenuated. Sheet-pile reflects waves which interact with incoming waves. Wave effects result in bottom scour with little habitat attractive to fishes. Pelagic fishes were occasionally seen (P.W. Webb, A.J. Cotel and L. Meadows, unpublished observations).

been described for other ecosystem components (Gallucci and Netto, 2004). On a sandy beach, 7 m wide, leading to a sharp weedy drop-off, various cyprinids were most prevalent inshore along the beach during both calm conditions, and during storm winds averaging 15 km.h^{-1}, driving 40 cm-high waves onshore. Spiny-rayed fishes, primarily members of the slow-water functional group illustrated in Fig. 1.7, were found foraging along the beach habitat under calm conditions. However, few remained during storms, but rather these fishes became more abundant in deeper, weedy water, where effects of waves and currents were rapidly attenuated (Fig. 1.9).

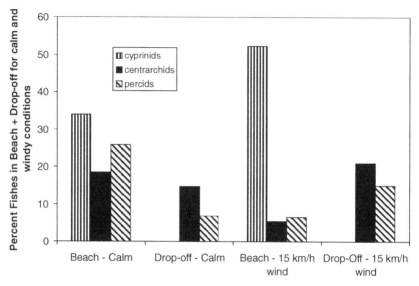

Fig. 1.9 Distribution of fishes on a sandy beach and adjacent weedy drop-off during calm and on-shore moderately windy conditions. Data shown are percentages of all fish sampled for each of the two conditions, calm and windy.

Coral Reefs

Recently, several studies have examined how exposure to turbulent flows affects coral reef fish assemblages. These studies used the habitat-classification approach to identify various coral reef zones with different exposure and imputed turbulence levels, with aggregate flows quantified through dissolution of gypsum balls. On Lizard Island, Great Barrier Reef, the small, largely benthic, cryptic component of the overall assemblage was strongly affected by substrate, with greater richness and abundance in sand/rubble habitats (Depczynski and Bellwood, 2005). Superimposed on the substrate effect was that of exposure, with higher exposure associated with more energetic and undoubtedly turbulent flows. Species richness, primarily of gobies and blennies, and the abundance of fish of each species decreased with increasing exposure above the reef base. Shallow, wave-exposed reef zones had fewest species and lowest abundance. On the sheltered side of the reef, analogous topography had no effect on assemblage composition and fish abundance. In addition, individuals on the exposed side tended to have a higher mass, a mechanisms that improves station holding for current-exposed benthic fishes (Arnold and Weihs, 1978).

Exposure also affected assemblages of pelagic Great Barrier Reef fishes. Observations were made on the Acanthuridae, Chaetodontidae, Labridae, Pomacanthidae, Pomacentridae, Serranidae, and Siganidae, families that includes >50% of reef species, and >70% of individuals (Fulton et al., 2005; Fulton and Bellwood, 2005). Fish were observed in exposed and sheltered zones, and their swimming patterns and capabilities determined. The variation in swimming gaits could be captured by assigning each species to one of three functional groups; pectoral-fin swimmers, pectoral-caudal-fin swimmers and body-caudal-fin swimmers. Pectoral-fin swimmers were most common across reef zones, but became dominant as exposure and net-current speed increased. Caudal-fin swimmers were more commonly demersal, and found more commonly in lower-flow, less energetic flows, with the pectoral-caudal-fin swimmers tending to be intermediate.

The pectoral-fin swimmers that were more common in higher turbulence habitats tend to have gibbose bodies, while body shape is more fusiform among the caudal-fin swimmers. Fishes from temperate freshwaters with deep bodies tend to be found in lower turbulence situations, while the fusiform swimmers are more common in more turbulent situations (Figs. 1.7 and 1.8). Thus the observations on coral reefs may appear contradictory. However, comparisons among freshwater species have been made for a much larger phylogenetic span, with the fusiform fishes being salmonids or cyprinids and the deep-bodied forms being spiny rayed fish such as those studied on reefs. The cursory evidence so far suggests that self-correction is more highly developed in the soft-rayed fishes, while the spiny-rayed fishes are more dependent on powered control.

The key difference among the spiny-rayed coral reef fishes proved to be swimming power in sustainable gaits. Thus, higher power and higher efficiency were achieved by more lift-based pectoral fin propulsion, which was most common in more turbulent conditions. The caudal-fin swimmers tended to be the slower-speed swimmers among the reef fishes. Pectoral-fin propulsion does not appear to be as highly developed among freshwater spiny-rayed fishes, and the power and stability advantage appears to pass to the fusiform versus gibbose forms.

If the above speculation is correct, then a common denominator suggests itself: swimming speed and power are key factors in swimming in turbulent flows. As noted above, swimming speed affects the perceived flow, which could become more beneficial in shifting more eddies into apparent sizes with reduced capabilities for perturbing fishes. Similarly, increased momentum appears to confer greater stability. Depczynski and Bellwood (2005) found that fish with greater mass, and hence greater momentum and kinetic energy at a given speed, were more prevalent in more turbulent reef zones.

CONCLUSIONS

This chapter focuses on the hydrodynamic environment experienced by fishes—and other biota—specifically on turbulence, a ubiquitous feature of aquatic habitats that has largely been neglected. It is driven by two considerations: first a lack of consensus in the literature as to the nature of effects of turbulence, negative, positive or neutral, and second a belief that turbulence has an important role in the ecology and management of salt and freshwaters. We propose a framework to link turbulent flow and biota that is a physically rigorous, but also parsimonious, testable approach to place disparate studies on the same basis. We also see it as providing a basis towards the improvement, restoration, mitigation, creation of fishways and similar practical challenges of management for aquatic systems.

The approach is consistent with the limited observations and experiments performed to date. Nonetheless, given the paucity of these sources, the content of this chapter is nonetheless speculative and requires independent validation, and undoubtedly adaptation and modification. Therefore, we suggest that the immediate need for future studies is experiments to determine fish, and other organismic responses to different types and levels of turbulence such as Tritco (2009), for various periods and magnitudes of orbits and eddies in various flows for organisms of different sizes. Such a multivariate experimental approach is no small undertaking, but we suggest that there is much fruit for many, many dissertations and theses to make a difference for a considerable time!

Another area of future study involves placing turbulence in a wider context of ecological factors that determine the distribution and numbers of animals such as fishes. The hydrodynamic flow signatures are among the physical features of a habitat. With response capabilities of organisms, such physical features determine the boundaries of a multidimensional space where an animal could live, or the "fundamental" niche. The "realized" niche, where an animal actually lives includes other factors, notably biotic factors such as competition, predation, mutualism, disease etc. The few studies to date suggest that turbulence is an important contributor to both the fundamental and realized niche. As such, legions of questions suggest themselves. For example: How might incident flow turbulence alter abilities of fish to school? What would be an energy-minimizing optimal path through an eddy field, and can fish find and use it? How does turbulence affect the behavior and success of fish predator-prey interactions? Does turbulence affect the distribution of predators and do scale effects of animal-eddy-orbit create a new category of predator refuges? Does disease and parasites affects response capabilities and choices of turbulent habitats? There is much space to further explore how turbulence affects food distribution and foraging (e.g., McKenzie and Kiorobe, 1995; Galbraith *et al.*, 2004). Functional morphologists have not tested their conclusions from laboratory often enough in the harsher laboratory of the real world. Consequently emerging technologies provide rich opportunities for field studies.

ACKNOWLEDGEMENTS

Research reported in this chapter was supported by grants from Sea Grant R/GLF-53. NSF Career 0447427 and technical support from the faculty and staff of the University of Michigan Marine Hydrodynamics Laboratories. Conversations with Dr. H. Tritico during the execution of his dissertation research were useful in clarifying principles and especially difficulties in describing turbulent flow.

REFERENCES

Adrian, R.J., K.T. Christensen and Z.C. Liu. 2000. Analysis and interpretation of instantaneous turbulent velocity fields. *Experiments in Fluids* 29: 275–290.

Alben, S., P.G. Madden and G.V. Lauder. 2006. The mechanics of active fin-shape control in ray-finned fishes. *Journal of the Royal Society of London Interface*: DOI 10.1098/rsif.2006.0181.

Alexander, R. McN. 1989. Optimization and gaits in the locomotion of vertebrates. *Physiological Reviews* 69: 1199–1227.

Aleyev, Y.G. 1977. *Nekton*. W. Junk, The Haque.

Allan, J.D. and M.M. Castillo. 2007. *Stream Ecology: Structure and Function of Running Waters*. 2nd Edition. Springer, Dordrecht, The Netherlands.

Arnold, G.P. and D. Weihs. 1978. The hydrodynamics of rheotaxis in the plaice (*Pleuronectes platessa*). *Journal of Experimental Biology* 75: 147–169.

Asplund, T.R. and C.M. Cook. 1997. Effects of motor boats on submerged aquatic macrophytes. *Lake Research Management* 13: 1–12.

Beal, D.N., F.S. Hover, M.S. Triantafyllou, J. Liao and G.V. Lauder. 2006. Passive propulsion in vortex wakes. *Journal of Fluid Mechanics* 549: 385–402.

Beamish, F.W.H. 1978. Swimming capacity. In: *Fish Physiology*, W.S. Hoar and D.J. Randall (Eds.). Academic Press, New York. Vol. 7: Locomotion, pp. 101–187.

Bell, W.H. and L.D.B. Terhune. 1970. Water tunnel design for fisheries research. *Fisheries Research Board of Canada Technical Report* 195: 1–69.

Bellwood, D.R. and P.C. Wainwright. 2001. Locomotion in labrid fishes: implications for habitat use and cross-shelf biogeography on the Great Barrier Reef. *Coral Reefs* 20: 139–150.

Blake, R.W. 1979. The energetics of hovering in the mandarin fish (*Synchropus picturatus*). *Journal of Experimental Biology* 82: 25–33.

Blake, R.W. 1983. *Fish Locomotion*. Cambridge University Press, Cambridge.

Blazka, P., M. Volf and M. Cepala. 1960. A new type of respirometer for the determination of the metabolism of fish in an active state. *Physiologia Bohemoslovenica* 9: 553–558

Boisclair, D. and M. Tang. 1993. Empirical analysis of the swimming pattern on the net energetic cost of swimming in fishes. *Journal of Fish Biology*, 42: 169–183.

Bond, C.E. 1996. *Biology of Fishes*, W.B. Saunders Company, New York.

Boisclair, D. 2001. Fish habitat models: From conceptual framework to functional tools. *Canadian Journal of Fisheries and Aquatic Sciences* 58: 1–9.

Breidenthal, R.E. 1981. Structure of turbulent mixing layers and wakes using a chemical reaction. *Journal of Fluid Mechanics* 109: 1–24.

Brett, J.R. 1964. The respiratory metabolism and swimming performance of young sockeye salmon. *Journal of the Fisheries Research Board of Canada* 21: 1183–1226.

Brett, J.R. 1995. Energetics. In: *Physiological-Ecology of Pacific Salmon*, J. R. Brett and C. Clark (Eds.). Gov. Canada, Dept. Fish. Oceans, Ottawa, Ontario. pp. 3–6.

Brown and Root, Inc. 1992. *Conceptual Engineering Report for Freshwater Bayou Canal Stabilization, Vermilion Parish, Louisiana*, Report prepared for the Coastal Restoration Division, Louisiana Department of Natural Resources.

Castro-Santos, T. and A. Haro. 2006. Biomechanics and fisheries conservation. In: *Fish Physiology*, R.E. Shadwick and G.V. Lauder (Eds.). Academic Press, New York, Vol. 23: Fish Biomechanics, pp. 469–523

Castro-Santos, T., A.J. Cotel and P.W. Webb. 2008. Fishway evaluations for better bioengineering—an integrative approach. In: Challenges for Diadromous Fishes in a Dynamic Global Environment. A.J. Haro, R.A. Rulifson, C.M. Moffitt, M.J. Dadswell, R.A. Cunjak, R.J. Klauda, K. Smith, T. Avery, J.E. Cooper, and K.L. Beal (Eds.). 2nd International Symposium on Diadromous Fishes, Halifax, N.S. 18–21 June, 2008. AFS Symposium (In Press).

Cotel, A.J. and P.W. Webb. 2004. Why won't fish wobble? *Proceedings of the 17th ASCE Engineering Mechanics Conference*, June 13–16, 2004, University of Delaware.

Cotel, A.J., P.W. Webb and H. Tritico. 2006. Do trout choose habitats with reduced turbulence? *Transactions of the American Fisheries Society* 135: 610–619.

Coutant, C.C. 1998. Turbulent attraction flows for juvenile salmonids passage at dams. *Oak Ridge National Laboratory, ORNL/TM-13608, Oak Ridge Tennessee.*

Cwikiel, W. 1996. *Living with Michigan's wetlands: A landowner's guide.* Tip of the Mitt Watershed Council, Conway, Michigan.

Denny, M. 1988. *Biology and the Mechanics of the Wave-Swept Environment.* Princeton University Press, Princeton.

Depczynski, M. and D.R. Bellwood. 2005 Wave energy and spatial variability in community structure of small cryptic coral reef fishes. *Marine Biology Progress Series* 303: 283–293.

Drucker, E.G. 1996. The use of gait transition speed in comparative studies of fish locomotion. *American Zoologist* 36: 555–566.

Drucker, E.G. and G.V. Lauder. 1999. Locomotor forces on a swimming fish: three-dimensional vortex wake dynamics quantified using digital particle image velocimetry. *Journal of Experimental Biology* 202: 2393–2412.

Eidietis, L., T.L. Forrester and P.W. Webb. 2002. Relative abilities to correct rolling disturbances of three morphologically different fish. *Canadian Journal of Zoology* 80: 2156–2163.

Enders, E.C., D. Boisclair and A.G. Roy. 2003. The effect of turbulence on the cost of swimming for juvenile Atlantic salmon (*Salmo salar*). *Canadian Journal of Fisheries and Aquatic Sciences* 60: 1149–1160.

Farlinger, S. and F.W.H. Beamish. 1977. Effects of time and velocity increments in the critical swimming speeds of largemouth bass (*Micropterus salmoides*). *Transactions of the American Fisheries Society* 106: 436–439.

Fausch, K.D. 1984. Profitable stream positions for salmonids: Relating specific growth rate to net energy gain. *Canadian Journal of Zoology* 62: 441–451.

Friedlander, A.M. and J.D. Parrish. 1998. Habitat characteristics affecting fish assemblages on a Hawaiian reef. *Journal of Experimental Marine Biology and Ecology* 224: 1–30.

Friedlander, A.M., E.K. Brown, P.L. Jokiel, W.R. Smith and K.S. Rodgers. 2003. Effects of habitat, wave exposure, and marine protected status on coral reef fish assemblages in the Hawaiian archipelago. *Coral Reefs* 22: 291–305.

Fulton, C.J. and D.R. Bellwood. 2002. Ontogenetic habitat use in labrid fishes: an ecomorphological perspective. *Marine Biology Progress Series* 236: 255–262.

Fulton, C.J. and D.R. Bellwood. 2005. Wave-induced water motion and the functional implications for coral reef fish assemblages. *Limnology and Oceanography* 50: 255–264.

Fulton, C.J., D.R. Bellwood and P.C. Wainwright. 2001. The relationship between swimming ability and habitat use in wrasses (Labridae). *Marine Biology* 139: 25–33.

Fulton, C.J., D.R. Bellwood and P.C. Wainwright. 2005. Wave energy and swimming performance shape coral reef fish assemblages. *Proceedings of the Royal Society of London*, B 272: 827–832.

Galbraith, P.S., H.I. Browman, R.G. Racca, A.B. Skiftesvik and J. Saint-Pierre. 2004. Effect of turbulence on the energetics of foraging in Atlnatic cod *Gadus morhua* larvae. *Marine Ecology Progress Series* 201: 241–257.

Gallucci, F. and S. A. Netto. 2004. Effects of the passage of cold fronts over a coastal site: an ecosystem approach. *Marine Biology Progress Series* 281: 79–92.

Good, B., D. Buchtel, J. Meffert, K. Rhinehart Radfort and R. Wilson. 1995. *Louisiana's Major Coastal Navigation Channels*, Unpublished Report, Coastal Restoration Division, Louisiana Department of Natural Resources.

Gordon, M.S., J.R. Hove, P.W. Webb and D. Weihs. 2001. Boxfishes as unusually well controlled autonomous underwater vehicles. Physiological and Biochemical Zoology 73: 663–671.

Gordon, M. S., I. Plaut and D. Kim. 1996. How puffers (TeleoteiL Tetraodontidae) swim. *Journal of Fish Biology* 49: 319–328.

Gourlay, T.P. 2001. The supercritical bore produced by a high-speed ship in a channel. *Journal of Fluid Mechanics*, 434: 399–409.

Gray, J. 1968. *Animal Locomotion*. Weidenfeld and Nicolson, London, England.

Herskin, J. and J.F. Steffensen. 1998. Energy savings in sea bass swimming in a school: measurements of tail beat frequency and oxygen consumption at different swimming speeds. *Journal of Fish Biology* 53: 366–376.

Höök, T.O., E.S. Rutherford, S.J. Brines, D.J. Schwab and M.J. McCormick. 2004. Relationships between surface water temperature and steelhead distributions in Lake Michigan. *North American Journal of Fisheries Management* 24: 211–221.

Höök, T.O., E.S. Rutherford, S.J. Brines, D.M. Mason, D.J. Schwab, M.J. McCormick, G.W. Fleischer, and T.J. DeSorcie. 2003. Spatially explicit measures of production of young alewives in Lake Michigan: Linkage between essential fish habitat and recruitment. *Estuaries* 26: 21–29.

Hora, S.L. 1935. Ancient Hindu concepts of correlation between form and locomotion of fishes. *Journal of the Asiatic Society of Bengal Science* 1: 1–7.

Huler, S. 2004. Defining the wind: the Beaufort scale, and how a nineteenth-century admiral turned science into poetry. Crown Publishers, New York.

Johansen, J.L., C.J. Fulton and D.R. Bellwood. 2007. Avoiding the flow: refuges expand the swimming potential of coral reef fishes. *Coral Reefs* doi: 10.1007/s00338-007-0217-y

Johari, H. and W.W. Durgin. 1998. Direct measurement of circulation using ultrasound. *Experiments in Fluids* 25: 445–454.

Jokiel, P.L and J.I. Morrissey. 1993. Water motion in coral reefs: evaluation of the clod-card technique. *Marine Ecology Progress Series* 93: 175–181.

Kolmogorov, A.N. 1941. Local structure of turbulence in an incompressible viscous fluid at very high Reynolds numbers, *Dolk. Akad. Nauk* SSSR 30, 299, reprinted in Usp. Fix. Nauk 93, 476–481 (1967), transl. in *Soviet Physics Uspecki* 10: 734–736 (1968).

Krohn, M.M. and D. Boisclair. 1994. Use of a stereo-video system to estimate the energy expenditure of free-swimming fish. *Canadian Journal of Fisheries and Aquatic Sciences* 51: 1119–1127.

Lagler, K.F., J.E. Bardach, R.R. Miller and D.R.M. Pasino. 1977. *Ichthyology.* John Wiley and Sons, New York.

Liao, J.C. 2004. Neuromuscular control of trout swimming in a vortex street: implications for energy economy during the Kármán gait. *Journal of Experimental Biology* 207: 3495–3506.

Liao, J.C. 2008. A review of fish swimming mechanics and behavior in perturbed flows. *Philosophical Transactions of the Royal Society,* B. (In Press).

Liao, J.C., D.N. Beal, G.V. Lauder and M.S. Trianyafyllou. 2003a. Fish exploiting vortices decrease muscle activity. *Science* 302: 1566–1569.

Liao, J., D.N. Beal, G.V. Lauder and M.S. Trianyafyllou. 2003b. The Kármán gait: Novel body kinematics of rainbow trout swimming in a vortex street. *Journal of Experimental Biology* 206: 1059–1073.

Lighthill M.J. 1969. Hydromechanics of aquatic animal propulsion. *Annual Review of Fluid Mechanics* 1: 413–45

MacKenzie, B. and T. Kiorobe. 1995. Encounter rates and swimming behavior of pause-travel and cruise larval fish predators in calm and turbulent laboratory environments. *Limnology and Oceanography* 40: 1278–1289.

McLaughlin, R.L. and D.L. Noakes. 1998. Going against the flow: An examination of the propulsive movements made by young brook trout in streams. *Canadian Journal of Fisheries and Aquatic Sciences* 55: 853–860.

McMahon, T.E., A.V. Zale and D. J. Orth. 1996. Aquatic habitat measurements. In: *Fisheries Techniques,* B. Murphy and D. W. Willis (Eds.), *American Fisheries Society,* pp. 83–120.

Murphy, B. and D.W. Willis. 1996. *Fisheries Techniques.* American Fisheries Society, Bethesda, MD.

Nauen, J.C. and G.V. Lauder. 2001. Locomotion in scombrid fishes: visualization of flow around the caudal peduncle and finlets of the chub mackerel *Scomber japonicas. Journal of Experimental Biology* 204: 2251–2263.

Nauen, J.C and G.V. Lauder. 2000. Locomotion in scombrid fishes: morphology and kinematics of the finlets of the chub mackerel *Scomber japonicus. Journal of Experimental Biology* 203: 2247–2259.

Nikora, V.I. and D.G. Goring. 1998. ADV turbulence measurements: Can we improve their interpretation? *Journal of Hydraulic Engineering* 124: 630–634.

Nikora, V.I. and D.G. Goring. 2000. Flow turbulence over fixed and weakly mobile gravel beds. *Journal of Hydraulic Engineering* 126: 79–690.

Nikora, V.I., J. Aberle, B.J.F. Biggs, I.G. Jowett and J.R.E. Sykes. 2003. Effects of size, time-to-fatigue and turbulence on swimming performance: a case study of *Galaxias maculatus. Journal of Fish Biology* 63: 1365–1382.

Odeh, M., J.F. Noreika, A. Haro, A. Maynard, T. Castro-Santos and G.F. Cada. 2002. Evaluation of the effects of turbulence on the behavior of migratory fish. *Final Report 2002, Report to Bonneville Power Administartion, Conract No., 00000022, Project No. 200005700,* pp. 1–55.

Ogilvy, C.S. and A.B. DuBois. 1981. The Hydrodynamics of swimming bluefish (*Pomatomus saltatrix*) in different intensities of turbulence: variation with changes in buoyancy. *Journal of Experimental Biology* 92: 67–85.

O'Neill, P.L., D. Nicolaides, D. Honnery and J. Soria. 2004. Autocorrelation functions and the determination of integral scale with reference to experimental and numerical data. *15th Australasian Fluid Mechanics Conference.*

Paik, J. and F. Sotiroupolos. 2005. Coherent structure dynamics upstream of a long rectangular block at the side of a large aspect ratio channel. *Physics of Fluids* 17: 115104.

Panton, R.L. 1984. Incompressible Flow. Wiley Interscience.

Partridge, B.L. and T.J. Pitcher. 1980. Evidence against a hydrodynamic function for fish schools. *Nature* (London) 279: 418–419.

Pavlov, D.S., A.I. Lupandin and M.A. Skorobogatov. 2000. The effects of flow turbulence on the behavior and distribution of fish. *Journal of Ichthyology* 40, (Supplement 2). S232–S261.

Peake, S.J. and A.P. Farrell. 2004. Fatigue is a behavioural response in respirometer-confined smallmouth bass. *Journal of Fish Biology* 68: 1742–1755.

Popper, D. and L. Fishelson. 1973. Ecology and behavior of *Anthias squamipinnis* (peters, 1855) (Anthiidae, Teleostei) in the coral habitat of eilat (Red Sea). *Journal of Experimental Zoology* 184: 409–423.

Popper, D. and L. Fishelson. 1973. Ecology and behavior of *Anthias squamipinnis* (peters, 1855) (Anthiidae, Teleostei) in the coral habitat of eilat (Red Sea). *Journal of Experimental Zoology* 184: 409–423.

Potts, J.A. 1970. The schooling ethology of *Lutianus monostigma* (Pisces) in the shallow reef environment of Aldabra. *Journal of Zoology* 161: 223–235.

Puckett, K.J. and L.M. Dill. 1984. Cost of sustained and burst swimming of juvenile coho salmon (*Oncorhynchus kisutch*). *Canadian Journal of Fisheries and Aquatic Sciences* 41: 1546–1551.

Puckett, K.J. and L.M. Dill. 1985. The energetics of feeding territoriality in juvenile coho salmon (*Oncorhynchus kisutch*). *Behaviour* 92: 97–111.

Roy, A.G., T. Buffin-Bélanger, H. Lamarre and A. Kirkbride. 2004. Size, shape and dynamics of large-scale trubulent flow structures in a gravel-bed river. *Journal of Fluid Mechanics* 500: 1027.

Saffman, P.G. 1992. *Vortex dynamics*. Cambridge University Press, Cambridge, UK.

Sanford, L.P. 1997. Turbulent mixing in experimental ecosystem studies. *Marine Biology Progress Series* 161: 265–293.

Schultz, W.W and P.W. Webb. 2002. Power requirements of swimming: Do new methods resolve old questions? *Integrative and Comparative Biology* 42: 1018–1025.

Smith, D.L. and E.L. Brannon. 2005. Response of juvenile trout to turbulence produced by prismatoidal shapes. *Transactions of the American Fisheries Society* 134: 741–753.

Smith, D.L., E.L. Brannon, B. Shafii and M. Odeh. 2006. Use of the average and fluctuating velocity components for estimation of volitional rainbow trout density. *Transactions of the American Fisheries Society* 135: 431–441.

Svendsen, J.C., J. Skov, M.J. Bildsoe and F. Steffensen. 2003. Intra-school positional preference and reduced tail beat frequency in trailing positions in schooling roach under experimental conditions. *Journal of Fish Biology* 62: 834–846.

Standen, E.M. and G.V. Lauder. 2007. Hydrodynamic function of dorsal and anal fins in brook trout (*Salvelinus fontinalis*). *Journal of Experimental Biology* 210: 325–339.

Standen, E.M., S.G. Hinch and P.S. Rand. 2004. Influence of river speed on path selection by migrating adult sockeye salmon (*Oncorhynhus mykiss*). *Canadian Journal of Fisheries and Aquatic Sciences* 61: 905–912.

Triantafyllou, G.S., M.S. Triantafyllou and M.A. Gosenbaugh. 1993. Optimal thrust development in oscillating foils with application to fish propulsion. *Journal of Fluids and Structures* 7: 205–224.

Tritico, H. 2009. *Understanding the importance of Turbulent Regimes for Fish Habitat: Beyond Turbulence Intensity*. Ph.D. Dissertaion. University of Michigan, Ann Arbor, MI.

Tudorache, C., P. Viaenen, R. Blust and G. de Boeck. 2008. Longer flumes increase critical swimming speeds by increasing burst and glide swimming duration in carp (*Cyprinus carpio, L*). *Journal of Fish Biology*, (In Press).

Videler, J.J. 1993. *Fish Swimming*. Chapman and Hall, New York.

Vogel, S. 1994. *Life in Moving Fluids*. Princeton University Press, Princeton.

Washington State Ferries Corporation. 2001. Rich Passage Wave Action Study. Monitoring Program Summary Report.

Weatherley, A.H., S.C. Rogers, D.G. Pinock and J.R. Patch. 1982. Oxygen consumption of active rainbow trout, *Salmo gairdneri* Richardson, derived from electromyograms obtained by radiotelemetry. *Journal of Fish Biology* 20: 479–489.

Webb, P.W. 1989. Station holding by three species of benthic fishes. *Journal of Experimental Biology* 145: 303–320.

Webb, P.W. 1991. Composition and mechanics of routine swimming of rainbow trout, *Oncorhynchus mykiss*. *Canadian Journal of Fisheries and Aquatic Sciences* 48: 583–590.

Webb, P.W. 1993. Is tilting at low swimming speeds unique to negatively buoyant fish? Observations on steelhead trout, *Oncorhynchus mykiss*, and bluegill, *Lepomis macrochirus*. *Journal of Fish Biology* 43: 687–694.

Webb, P.W. 1994a. Exercise performance of fish. In: *Advances in Veterinary Science and Comparative Medicine*, J.H. Jones (Ed.), Academic Press, Orlando, pp. 1–49.

Webb, P.W. 1994b. The biology of fish swimming. In: *Mechanics and Physiology of Animal Swimming*, L. Maddock, Q. Bone, and J.M.V. Rayner (Eds.), Cambridge University Press, Cambridge, pp. 45–62.

Webb, P.W. 1995. Locomotion. In: *Physiological-Ecology of Pacific Salmon*, C. Groot, L. Margolis and W. C. Clark (Eds.). UBC Press, Vancouver, pp. 70–99.

Webb, P.W. 1997. Designs for Stability and Maneuverability in Aquatic Vertebrates: What can we learn? *Proceedings of the Tenth International Symposium on Unmanned Untethered Submersible Technology*, pp. 86–108, Autonomous Undersea Systems Institute, Lee, NH.

Webb, P.W. 1997a. Swimming. In: *The Physiology of Fishes*, 2nd Edition, D.H. Evans (Ed.). CRC Press, Boca Raton, pp. 3–24.

Webb, P.W. 1998. Entrainment by river chub, *Nocomis micropogon*, and smallmouth bass, *Micropterus dolomieu*, on cylinders. *Journal of Experimental Biology* 201: 2403–2412.

Webb, P.W. 2000. Maneuverability versus stability? Do fish perform well in both? *Proc. 1st International Symposium on Aqua Bio-Mechanisms/ International Seminar on Aqua Bio-Mechanisms (ISABMEC 2000)*, August, Tokai University Pacific Center, Honolulu, Hawaii.

Webb, P.W. 2002. Control of posture, depth, and swimming trajectories of fishes. *Integrative and Comparative Biology* 42: 94–101.

Webb, P.W. 2004. Response latencies to postural disturbances in three species of teleostean fishes. *Journal of Experimental Biology* 207: 955–961.

Webb, P.W. 2006a. Stability and maneuverability. In: *Fish Physiology*, R.E. Shadwick and G.V. Lauder (Eds.). Elsevier Press, San Diego, pp. 281–332.

Webb, P.W. 2006b. Use of fine-scale current refuges by fishes in a temperate warm-water stream. *Canadian Journal of Zoology* 84: 1071–1078.

Webb, P.W. and C.L. Gerstner. 2000. Swimming behaviour: predictions from biomechanical principles. In: *Biomechanics in Animal Behaviour*, P. Domenici and R.W. Blake (Eds.), Bios Scientific Publishers Ltd., Oxford, pp. 59–77

Weihs, D. 1973. Hydromechanics of fish schooling. *Nature* (London) 241: 290–291.

Weihs, D. 1989. Design features and mechanics of axial locomotion in fish. *American Zoologist* 29: 151–160.

Weihs, D. 1993. Stability of aquatic animal locomotion. *Contemporary Mathematics* 141: 443–461.

Weihs, D. 2002. Stability *versus* maneuverability in aquatic animals. *Integrative and Comparative Biology* 42: 127–134.

Wolter, C. and R. Arlinghaus. 2003. Navigation impacts on freshwater fish assemblages: the ecological relevance of swimming performance. *Reviews in Fish Biology and Fisheries* 13: 63–89.

Wu, T.Y.T. 1987 .Generation of Upstream Advancing Solitons by Moving Disturbances. *Journal of Fluid Mechanics* 184: 75–99.

APPENDIX A:

VORTICITY EQUATION

The vorticity equation provides the source terms for generating vorticity, and describes all the complexity of turbulence in the most general situations.

$$\frac{\overrightarrow{d\omega}}{Dt} = v\nabla^2\omega + \frac{\nabla\rho \times \nabla p}{\rho^2} + \vec{\omega}.\nabla\vec{U} + \nabla \times \vec{f} + \vec{\omega}(div\vec{U}) \tag{1}$$

where v = kinematic viscosity, U =a velocity vector, p = pressure, ρ = density, and f represents any non-conservative force. $divU$ is the divergence of the velocity vector. ∇ represents the gradient of a given vector, that is, a measure of the variation of U in space, D/Dt is the total derivative, representing both the time derivative and the advection term.

$v\nabla^2\omega$ represents the diffusion of vorticity due to viscous effects, once vorticity is generated, it will diffuse throughout the flow just like cream diffusing in a cup of coffee. If one does not stir it, the only difference will be the rate at which this diffusion process occurs. For cream, the rate of diffusion is the mass diffusivity coefficient, whereas for vorticity, the rate of diffusion is characterized by the kinematic viscosity of the fluid in question.

$\dfrac{\nabla\rho \times \nabla p}{\rho^2}$ is the generation of vorticity from baroclinic torques. For example, in stratified flows, there are situations such as a tilted thermocline in the ocean where the gradients of pressure and density are not aligned therefore creating a torque which in turn generates vorticity.

$\vec{\omega}.\nabla\vec{U}$ is the stretching or tilting term due to velocity gradients.

$\nabla \times \vec{f}$ is the torque due to non-conservative body forces such as friction or non elastic stresses.

$\vec{\omega}(div\vec{U})$ represents the dilatation term or stretching of vorticity due to flow compressibility.

All the above terms lead to increased vorticity in the flow. However, in situations most generally encountered by fishes in natural habitats, the vorticity equation simplifies to the following form:

$$\frac{\overrightarrow{D\omega}}{Dt} = v\Delta^2\omega + \frac{\Delta\rho \times \Delta p}{\rho^2} + \vec{\omega}.\Delta\vec{U}$$

2

Biomechanics of Rheotactic Behaviour in Fishes

R.W. Blake* and K.H.S. Chan

INTRODUCTION

Benthic fish experience water currents caused by tides, wave action and stream flow. Orientation to flow and associated behaviour is termed rheotaxis (Frankel and Gunn, 1961). Avoiding displacement allows fish to remain in their habitats for feeding, finding mates and protection. Adaptations associated with station-holding on the bottom that maximize current induced slipping and lift-off involve morphology (e.g., overall body form; frictional devices: frictional pads, fin spines, odontodes; adhesive structures: oral suckers, thoracic suction discs; high density; reduced or absent swimbladder) and behaviour (e.g., burying, heading into the stream direction, adhesion through marginal fins, active fin beating, negative lift production from paired fins) (e.g., Hora, 1922a, b; Keenleyside, 1962; Arnold, 1969; Arnold and Weihs, 1978; Matthews, 1985; Webb, 1989; Macdonnell and Blake, 1990; Blake, 2006). Arnold (1974) gives a comprehensive review of rheotropism in fishes including rheotaxis *per se*, optomotor responses and ecological aspects. Arnold and Weihs (1978) give a thorough analysis of the mechanics and behaviour of rheotaxis in plaice *Pleuronectes platessa* which appears to be designed for

Authors' address: Department of Zoology, University of British Columbia, Vancover, British Columbia, V6T 1Z4, Canada.

*Corresponding author: E-mail: blake@zoology.ubc.ca

minimum drag when heading into a current and maximizes lift-off speeds through behavioural adaptations associated with posture and marginal fin movements. Webb (1989) showed that compressed and depressed forms (*P. platessa* and *Raja clavata*, respectively) are characterized by high lift, low drag designs and behaviours that reduce lift. Benthic Atlantic salmon parr *Salmo salar* place their pectoral and anal fins into the interstices between gravel particles to increase friction and produce negative lift from their pectoral fins (Arnold *et al.*, 1991).

Older studies compared observed slipping speeds with theoretical values (based on technical equivalents); because of the practical difficulties involved in directly measuring the drag of fish attached to a surface (Arnold and Weihs, 1978; Webb, 1989). Blake (2006) directly measured the drag of hill stream catfishes. Direct measurement of drag or lift allows the unknown force to be inferred. In a related study, Blake *et al.* (2007) determine the energy cost of rheotactic behaviour in the loricariid *Pterygoplichthys* spp., showing a stable mass specific metabolic rate over a range of current speeds. The energy cost of station holding by rheotactic behaviour is low relative to that of holding position in the free stream in subcarangiform swimmers (Blake *et al.*, 2007).

We briefly review the basic physics of displacement by slipping and lift-off and give a simple dimensional model of the scaling of these speeds with size and in relation to current speeds. The significance of slipping and lift-off speeds is discussed using two case studies: freshwater stream tropical catfishes and marine flatfishes, focusing on morphological and behavioural adaptations. The adhesive strategies of highly specialized cyprinids and sisorids that inhabit rapidly flowing torrential mountain streams are considered with a focus on suction produced by specialized adhesive discs (oral and thoracic) and the ventral side of marginal fins. The metabolic costs of station-holding in the loricariid *Pterygoplichthys* and *P. platessa* allows for a comparison of the energetics of two rheotactic behavioural stratergies with that of station holding against currents in pelagic and necktonic fish. The mechanism of escape responses (fast-starts) off the substrate in pleuronectids is considered with regard to overcoming adhesive forces. Whilst many benthic fish adhere to the substrate, the pelagic remoras attach to moving hosts (e.g., other fishes, turtles and cetaceans), adhesion mechanics and the behaviour of dolphin hosts in efforts to remove them are discussed. Many rheotactic and non-rheotactic negatively buoyant demersal fish hover and swim close to the substratum where ground effects reduce power requirements in hovering and slow forward swimming. Whilst not rheotaxis *per se*, ground effect is considered in the context of a closely related energy saving behaviour. Finally, a prospectus of possible future studies, including the furtherance of direct experimental force measurements, field studies on stream hydrology and ecology (to establish the relevance of laboratory measurements to the field), and behavioural and biomechanical studies on the inhabitants of fast flowing torrential stream fish fauna, is suggested.

The physics of slipping and lifting

Benthic fish experience forces due to ocean currents, tides and flow in streams and rivers that may cause slippage and lift-off into the water column. Fig. 2.1 gives a simple schematic representation of the forces acting on a fish on the bottom as a function of current speed. Slippage occurs when the drag force ($D=0.5\rho_w AU^2C_D$ where ρ_w, A, U and C_D are the water density, frontally projected area of the fish, water velocity and drag coefficient respectively) generated by the current exceeds the frictional forces between the fish and the substrate (Fig. 2.1). Lifting forces ($L=0.5\rho_w A_p U^2 C_L$ where A_p is the planform area and C_L is lift coefficient) may be positive (away from the substratum) or negative (towards the substratum). Positive lift acts against the submerged weight ($W_o = mg$ where m and g are submerged body mass and acceleration due to gravity respectively) of the fish and negative lift acts in the opposite manner. Body form and other morphological adaptations (e.g., fish density, frictional plates, odontodes, fin spines) and behaviour (e.g., orienting into the flow direction, employing fin and/or body movements to hold position) affect drag and lift.

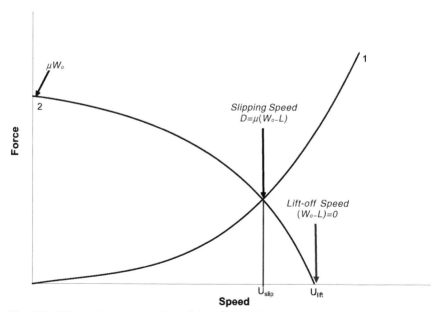

Fig. 2.1 Schematic representation of drag (curve 1) and frictional force (curve 2) with U_{slip} and U_{lift}.

The drag and lift coefficients can be determined from $C_D = 2D(\rho_w A U^2)^{-1}$ and $C_L = 2W_o(\rho_w A_p U^2) - C_D \mu^{-1}$ where μ is the coefficient of static friction. From theses equations, the predicted slipping speed (U_{slip}), lift-off speed (U_{lift}) and their ratio (U_{slip}/U_{lift}) can be calculated as $U_{slip} = \sqrt{\mu W_o (0.5\rho_w A C_D)^{-1}}$,

$U_{lift} = \sqrt{W_o(0.5\rho_w A_p C_{Lmax})^{-1}}$ and $U_{lift}/U_{slip} = \sqrt{C_D(\mu C_L)^{-1}}$ where C_{Lmax} is the maximum lift coefficient. Forms of low drag coefficients are characterized by a high slipping speed. However, low drag coefficients are associated with high lift coefficients and a low lift-off speed. Overall body forms that maximize U_{slip} will minimize U_{lift} and vice versa. Broadly, morphological and behavioural adaptations associated with station-holding in benthic fishes fall into two categories: forms with relatively high drag coefficients that use behaviour and morphological frictional devices to increase slipping speeds (e.g., certain loricariid catfishes) and forms with low drag coefficients and high lift coefficients that use behaviour to reduce lift (e.g., flatfishes) (Arnold and Weihs, 1978; Webb, 1989; Blake, 2006).

Scaling of slipping and lift-off speeds

There are no detailed studies (intra or interspecific) of the scaling of slipping and lift-off speeds in relation to size, perhaps because of the practical difficulties involved in making measurements on large fishes. However, a simple dimensional analysis allows for some understanding of the determinants of size specific values of U_{slip} and U_{lift}. The slipping speed is

$U_{slip} = \sqrt{\mu W_o (0.5\rho_w A C_D)^{-1}}$. Assuming that μ scales isometrically with body length and ρ_w and C_D are not a function of length l [ρ_w is a fixed property and C_D varies little with Reynolds number $R_e = lU/\nu$ (where ν is the kinematic viscosity of water) for $R_e > 10^4$; Blake, 2006], $U_{slip} \alpha (W_o A^{-1})^{0.5}$. Since $W_o \alpha l^3$ and $A \alpha l^2$, $U_{slip} \alpha l^{0.5}$. Therefore, for two geometrically similar fish (A and B) where fish B is twice the length of fish A, the slipping speed of B will be x1.4 that of A. The observed live slipping speeds of two size categories of *Pterygoplichthys* spp. ($\bar{l} \approx 6$ cm (Blake, 2006) and $\bar{l} \approx 35$ cm (Blake *et al.*, 2007)) are 40 cm s^{-1} and 100 cm s^{-1} respectively. The size differential between the two groups is roughly x6 implying that the slipping speed of the larger fish should exceed that of the smaller ones by a factor of 2.5 and this is the case (Blake *et al.*, 2007). Fig. 2.2 shows that experimentally determined values of U_{slip} are not significantly different from those predicted for the loricariids *Otocinclus, Chaetostoma, Hypostoma* and *Farlowella* and the gyrinocheilid *Gyrinocheilus aymonieri* ($P > 0.05$); U_{slip} scales with $l^{0.5}$. Arnold and Weihs (1978) calculated U_{slip} for freshly killed plaice ranging in length from 10 to 50 cm ($U_{slip} \approx 13$ and 31 cm s^{-1} respectively).

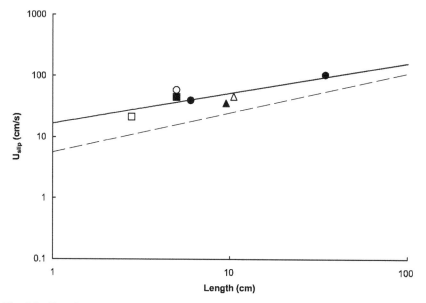

Fig. 2.2 Experimental live U_{slip} (dark line) for *Otocinclus* (empty square), *Pterygoplichthys* (filled circle), *Chaetostoma* (solid square), *Gyrinocheilus* aymonieri (empty circle), *Hypostoma* (filled triangle) and *Farlowella* (empty triangle) and the theoretical curve (broken line). Data from Blake (2006).

Their calculated value for the larger fish is x2.4 that of the smaller, in good agreement with our simple dimensional model which predicts a factor of 2.2. Fig. 2.3 shows submerged weight versus calculated U_{slip} for plaice (curve 2) and a fit to measured U_{slip} values for dead fish (curve 1). The calculated U_{slip} for plaice is within 95% confidence interval of measured U_{slip} values for dead fish. Similar dimensional arguments based on geometric similarity show that U_{lift} also scales as $l^{0.5}$. The calculated U_{lift} for freshly killed plaice ($l = 10$ and 50 cm) differed by a factor of x2.4 (18 and 44 cm s^{-1} respectively; Arnold and Weihs, 1978; Table 3) for $C_L = 1$ (Fig. 2.3; curve 3), close to the predicted value of x2.2. Values of U_{lift} are also shown for $C_L = 1.4$ and 1.8 which also scale as $l^{0.5}$ (curves 4 and 5).

The practical utility of the simple geometric scaling approach can be employed to assess the significance of U_{slip} and U_{lift} over a large size range. The order Heterosomata is a homogeneous group of essentially geometrically similar fishes. Fig. 2.4 shows total length versus weight for sole *Solea*, turbot *Lepidorhombus*, plaice *Pleuronectes* and halibut *Hippoglossus* and a fitted curve for all of the data for $10 < l > 300$ cm referenced to corresponding values of U_{slip} and U_{lift} which allows for reasonable predictions of the relative current speeds necessary for passive slipping and lift-off into the water column for juvenile and adult heterosomatids. The scope of heterosomatids to utilize locomotor

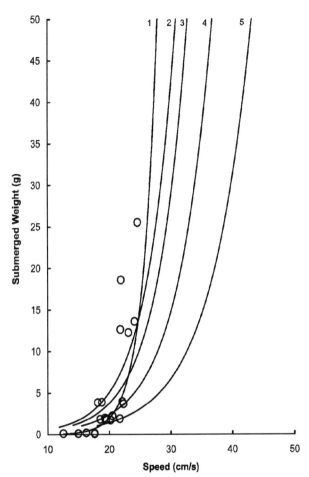

Fig. 2.3 Statistical fit (exponential) for observed slipping speeds (empty circles; Arnold and Weihs, 1978) based on experimental data (curve 1), theoretical U_{slip} (curve 2) and U_{lift} for C_L = 1.0, 1.4 and 1.8 (curve 3, 4 and 5 respectively) for plaice.

strategies (e.g., burst and glide swimming (Weihs, 1973), tidal stream transport (Weihs, 1978) and their capacity to swim in the water column in the absence of tides and currents (Harden Jones, 1977)) is a function of U_{lift}.

Plaice, flounder *Paralichthys* and sole utilize tidal stream transport for diel feeding migrations (Tyler, 1971; Raffaelli *et al.*, 1990), predator avoidance (Ansell and Gibson, 1990) and seasonal off-shore migrations to deeper waters (Rickey, 1995) by resting on the seabed when the tide is acting against them and swimming when it is favourable (Harden Jones, 1977; Greer Walker *et al.*, 1978). Adult flounder travel on the order of

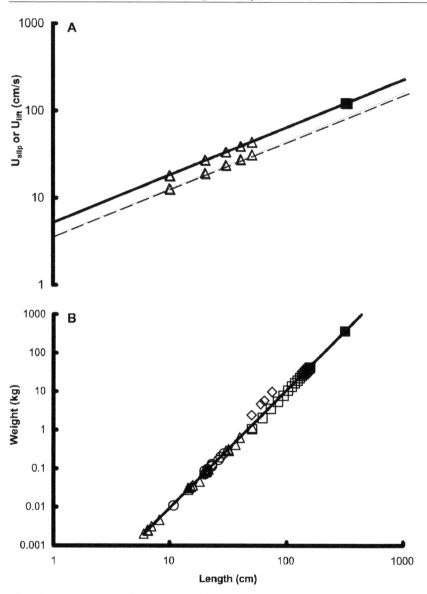

Fig. 2.4 **A:** Predicted U_{slip} (broken line) and U_{lift} (dark line) for plaice (empty triangle) and halibut (solid square). **B:** Weight versus length for halibut (empty square), plaice (empty triangle), turbot (empty diamond) and sole (empty circle). The solid square represents a large halibut of 365 kg. Data from Myrhe (1974), Arnold and Weihs (1978) and Coull *et al.*, *(1989)*.

10^2–10^3 m with each tide in this manner (Wirjoatmodjo and Pitcher, 1984). Weihs (1978) gives a hydromechanical model of tidal-stream transport that describes the energetic significance of the process by comparing its energy requirement with that of steady swimming at a constant velocity. The model predicts energy savings per unit distance traveled of over 90% and 40% for juvenile and adult plaice and sole respectively. Metcalfe and Arnold (1990) and Arnold and Metcalfe (1995) confirm that tidal-stream transport is a key mechanism for plaice swimming in the fast-tidal streams of the European continental shelf. Arnold and Weihs (1978) note that the speed of bottom currents in the southern North Sea is sufficient to allow plaice ranging in length from 10–50 cm to lift off and utilize tidal stream transport.

The largest fish shown on Fig. 2.4 is a Pacific halibut *Hippoglossus stenolepis* (365 kg, corresponding to a length of 3.15 m) which is about x6.3 that of the largest plaice considered by Arnold and Weihs (1978). The simple dimensional geometric scaling model predicts that U_{lift} for this halibut should be 1.10 m s^{-1}. For a body density of 1070 kg m^{-3}, its submerged weight would be 25.6 kg ($W_o = 251$ N). Assuming $C_L = 1$ and that the ocular planform area scales as for plaice ($A_p = 0.3l^2 + 0.17$; Arnold and Weihs, 1978) $A_p = 0.24$ m^2 and U_{lift} is about 1.3 m s^{-1}. Large halibut in the Strait of Juan de Fuca off the coast of British Columbia experience peak current speeds of this order (Huggett *et al.*, 1976). However, the simple analysis above does not take account of any behavioural adaptations that would increase U_{lift}.

When resting on the bottom and swimming freely, flatfish are adapted for minimum drag (Arnold and Weihs, 1978). Lighthill (in Harden Jones, 1977) suggested that plaice would only need to swim at relative water velocities of 10–20 cm s^{-1} to obtain the lift necessary to swim in the water column. Indeed, it is likely that most flatfishes can generate the required lift to raise their submerged weight by active swimming. Once in the water column, negative buoyancy could facilitate energy saving by burst and glide swimming (Weihs, 1973). Under these circumstances, high density might be regarded as an adaptation for both pelagic and benthic phases of life (Arnold and Weihs, 1978).

Slipping and lifting in catfishes and flatfishes

Blake (2006) documents the biomechanical, behavioural and morphological adaptations affecting the drag and lift of tropical freshwater stream dwelling catfishes. For five loricariid genera and one gyrinocheilid, values of the fineness ratios (total length/maximal height; 6.7–9.0), flattening index (maximum body width/maximal height; 0.9–2.0) and lengthening index (distance from within the rostrum to maximum height/total length; 0.17–0.22) are similar to the optimum values for technical bodies (e.g., smooth blisters attached to surfaces; Hoerner, 1965; Hoerner and Borst, 1975) of low drag (10, 2 and 0.3 for fineness ratio, flattening index and lengthening index respectively). In general, hill-stream catfishes maximize slipping speed through high densities (1060–1150 kg m⁻³), high rheotactic oral suction pressures (26–173 Pa) and high frictional coefficients ($\mu = 0.12$–1.2 on a smooth surface). In loricariids, the ventral frictional pad provides a weight bearing surface that increases friction. The ventral frictional pads and the pectoral fin spines are covered by odontodes (Fig. 2.5). Some forms (e.g., *Chaetostoma*) hold their pectoral fins at a negative angle of attack, such that the fins act as negative lift hydrofoils, pushing the fish against the substratum. This is the basis of station-holding in Atlantic salmon parr (Arnold *et al.*, 1991).

Drag coefficients for loricariids and gyrinochelids for $2\times10^4 < R_e > 5\times10^4$ are about 0.25 (based on frontally projected area). These values are about double that estimated for plaice over a similar R_e range (0.08–0.16; Arnold and Weihs, 1978). Forms characterized by specialized frictional (e.g., odontodes, pectoral fin spines) and adhesive (e.g., oral sucker, ventral adhesive disc surrounding the mouth or the thoracic region) adaptations that maximize U_{slip} (e.g., certain Loricariidae, Sisoridae, Cyprinidae, Gyrinocheilidae) are unlikely to experience flows corresponding to U_{lift}. For plaice oriented in the upstream direction on the bottom the lift is of the order of x10–20 drag (Arnold and Weihs, 1978). However, the high lift forces are mitigated through behaviour. At low current velocities, plaice turn upstream by movements of the fins and tail. As velocities increase, the amplitude and frequency of respiratory movements increase and the fish adopts a "clamped-down posture" with marginal fins pressed against the bottom. With increasing current speed, an "arched-back posture" appears, together with vigorous "posterior fin beating" (wave forms that pass back along the marginal fins; Arnold, 1969). Arnold and Weihs (1978) argue that the arched-back response reduces the vertical pressure differential across the fish, thus reducing lift.

Lift coefficients in loricariids are low relative to forms such as plaice. In some, the lift to drag ratio is high with a low slipping speed relative to lift-off speed (e.g., *Otocinclus* and *Pterygoplichthys*; U_{lift}/U_{slip} of 2.9 and 2.0 respectively). In others (e.g., *Farlowella*, *Chaetostoma*, *Gyrinocheilus* and *Hypostoma*), the lift to drag ratio is low and slipping speed is high relative to lift-off speed ($U_{lift}/U_{slip} = 1.1$, 1.2, 1.2 and 1.3 respectively). Different degrees of morphological, behavioural and biomechanical adaptations are likely to reflect differences in stream habitat. Unfortunately, nothing is known regarding stream hydrology, activity levels, depth distribution and the swimming behaviours of loricariids in their natural habitat.

The difference between measured slipping speeds of live fish (U_{live}) and dead fish (U_{dead}) is an indication of the relative importance of frictional forces versus oral suction for station-holding in loricariids. Forms relying principally on frictional devices to prevent slipping for station-holding (e.g., *Otocinclus* and *Hypostomus*, Fig. 2.5) show low values of U_{live}-U_{dead}. Conversely, those placing greater reliance on oral suction (e.g., *Pterygoplichthys*) are characterized by large values of U_{live}-U_{dead}. Values of U_{live}-U_{dead} are consistent with measured oral suction pressure; low values are found for *Otocinclus* and *Hypostomus*, and high values for *Pterygoplichthys* and *Chaetostoma* (Blake, 2006).

In loricariids, the respiratory gap remains open under all flow conditions (Gradwell, 1971), preventing the formation of a complete seal by the oral sucker lips (open sucker; Fig. 2.5). The Gyrinocheilidae are characterized by a closed oral sucker where adhesion is affected by a negative (suction) pressure induced by respiratory currents. This arrangement is more effective in producing adhesion than the open sucker. However, suction pressures produced by hill stream catfishes are much lower than those of certain intertidal fishes that possess a powerful closed-vacuum sucking discs derived from the pelvic fins (e.g., *Liparis montagui* and *Cyclopterus lumpus*; Gibson, 1969) which are of the order of 10^4 Pa. These

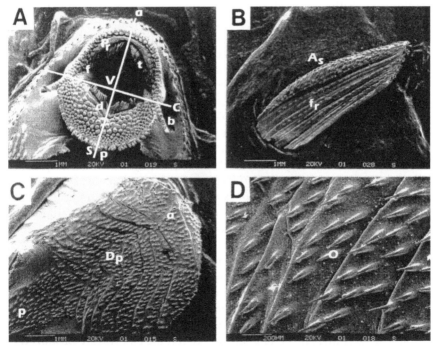

Fig. 2.5 Scanning electron micrographs of *Otocinclus* sp.: oral sucker (A), pectoral fin anterior spine (B), ventral frictional pad (C) and odontodes and dermal plates on the ventral frictional pad (D): anterior (a), posterior (p), cross-sectional axis (c), sagittal axis (s), teeth (t), oral valve (v), respiratory fissure (f), maxillary barbel (b), anterior pectoral fin spine (A_s), fin rays (f_r), dermal plate (D_p), odontodes (O), dentary (d) and premaxillae (P_r).

high suction pressures may be required to resist displacement forces caused by oscillatory flows (unsteady inertial added mass forces) that depend on fluid acceleration and the mass of water (added mass) entrained by the fish. Both open and closed oral suckers can function in respiration, feeding and adhesion. In the intertidal Gobiidae, Gobiesocidae, Liparidae and Cyclopteridae, the oral sucker (derived from pelvic fins) does not play a role in respiration and feeding. This "decoupling" allows the oral sucker to perform optimally in adhesion without functional design trade-offs associated with feeding and respiration.

Adhesion by torrential hill stream fishes

Benthic fish inhabiting the glacier fed mountain streams of the Himalayas and other mountainous areas resist high velocity turbulent currents by the use of highly specialized ventral adhesive discs either surrounding the mouth (cyprinids) or the thoracic region (Sisoridae) (Hora, 1922a, 1930). Fig. 2.6 shows a side and ventral view of *Balitora lineolata* which is dorso-ventrally compressed with a flat ventral surface. The median fins skirt the margins of the body (pectoral and pelvic fins over the anterior and posterior half respectively) giving a similar arrangement to that produced by the dorsal and anal fins of flatfishes. Although there is a superficial resemblance between the peripheral marginal fin skirts of *Balitora* and pleuronectids, the physical basis of adhesion is different. In *Balitora*, an adhesive organ functions together with the marginal fins (R.W. Blake, unpublished observation). In the sisorid *Glyptothorax pectinopterus*, a bilateral adhesive organ is situated posterior to the mouth between the opercular opening and the base of the pectoral fin. Thin, longitudinal ridges and grooves run roughly parallel to the body axis. The ridges are surmounted by keratinized epidermal spines with curved or sucker like tips (Singh and Agrawal, 1991). Hora (1922a) and Bhatia (1950) suggest that the adhesive organ in *G. pectinopterus* functions through friction between the spines and the substratum. However, Saxena and Chandy (1966) propose that a vacuum is generated in the longitudinal grooves through muscular action.

The skin of the outer rays of the pectoral and pelvic fins of *Pseudocheneis sulcatus* are thrown into a series of alternating grooves and ridges. The ridges are covered by spines and mucous pores which are not present in the grooves. The mucous causes a weak adhesion that prepares the substratum for adhesion of the fins (Das and Nag, 2004). Singh and Agrawal (1991) and Das and Nag (2004) suggest that adhesion is affected by a reduced pressure produced by the contraction of musculature attached to the ridges and grooves and that the spines augment this adhesion by anchoring to the substratum. There is no experimental evidence regarding the role of the pectoral and pelvic fins in adhesion of *P. sulcatus*. However, it is likely that they supplement the action

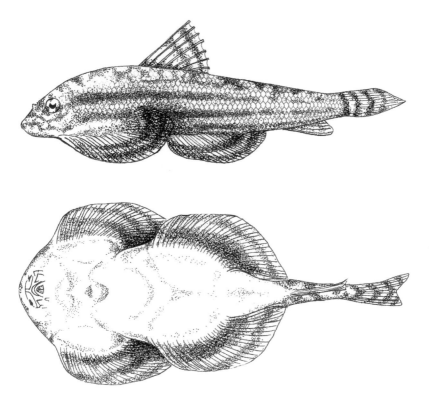

Fig. 2.6 A side and ventral view of the hillstream fish *Balitora lineolata*. Body length: 4.5 cm.

of the thoracic adhesive disc in adhering the fish to the substratum. When swimming in the free stream against the current, the fish is propelled by its expanded paired fins (Das and Nag, 2004). *G. pectinopterus* (Sisoridae) and *Balitora* sp. also swim in a similar manner.

Fishes that climb up waterfalls separating freshwater pools and streams have locomotor, behavioural and morphological adaptations which are akin to those of torrential stream fishes. For example, to reach adult habitats, juvenile (< 3 cm long) Hawaiian gobies (*Lentipes concolor* and *Awaous guamensis*) use "powerbursts" of axial undulation to climb waterfalls up to 350 m in height (Schoenfuss and Blob, 2003). Powerbursts in the two species are characterized by a single rapid adduction of the pectoral fins and high amplitude bodily undulation in a single rapid short burst of movement that may involve hybridized terrestrial and aquatic propulsive mechanisms (Schoenfuss and Blob, 2003). In contrast, *Sicyopterus stimpsoni* exits the water outside of direct flow and moves up vertical surfaces employing terrestrial propulsive mechanisms involving the alternate attachment of oral and pelvic suckers to the surface (Schoenfuss and Blob, 2003).

To date, there are no experimental measurements of the adhesive forces generated by torrential stream fishes. It is likely that behavioural responses (e.g., presenting a profile of minimum drag to the flow) and morphological adaptations (e.g., ventral surface of the adhesive marginal fins, thoracic suction discs) maximize slipping speeds. In addition, completely separate structures for different functions (decoupling: feeding, respiration and adhesion) allows for optimal performance in station-holding.

Metabolic cost of station holding

Whilst the basic physics of behavioural and morphological adaptations associated with station-holding are well understood, little is known about the energetics of the process. The energy cost of station-holding will likely reflect the extent of active rheotactic behavioural responses to flow velocity leading up to U_{slip} and U_{lift}. Forms that exhibit morphological adaptations (e.g., frictional plates, odontodes, oral sucker) to maximize U_{slip} show little reliance on behaviour and are characterized by a more or less constant mass specific active metabolic rate M (Blake *et al.*, 2007). In contrast, when behavioural mechanisms are important in maximizing U_{lift} (e.g., pleuronectids: low drag, high lift), M reflects the energy cost of rheotactic behaviours associated with maximizing U_{lift}.

Like many other loricariids, *Pterygoplichthys* show two behavioural responses to increased velocity: orienting upstream (commonly a passive process) and oral sucker adhesion. In some forms (e.g., *Otocinclus*), a third stage of fin extension occurs at higher velocities (Macdonnell and Blake, 1990). This limited activity is consistent with the flat form of the versus velocity curve for *Pterygoplichthys*. Blake *et al.*, (2007) measured the resting and active mass specific metabolic rate of *Pterygoplichthys* spp. and the resting metabolic rate is in the range of many subcarangiform swimmers (Fig. 2.7). The active metabolic rate decreases slightly with increasing water velocity relative to the resting value, in contrast to the logarithmic increase shown by free stream swimming fish (Fig. 2.7). This implies that the energetic cost of station-holding in *Pterygoplichthys* (and likely other benthic catfishes) is low compared to forms that hold station by swimming against the current. Intermittent burst swimming occurs when the fish begins to slip back at velocities of about 1.1 m s^{-1} which is associated with a rapid rise in M (Blake *et al.*, 2007). The slight negative slope of the oxygen consumption versus water velocity curve for *Pterygoplichthys* may reflect a lower rate and energy cost of respiratory pumping at higher flow velocities.

With increasing current speed, plaice turn their head upstream by movements of the fins and tail. The amplitude and frequency of respiratory movements increase in the "clamped-down posture" which is subsequently followed by an "arched-back posture" with vigorous fin beating (Arnold,

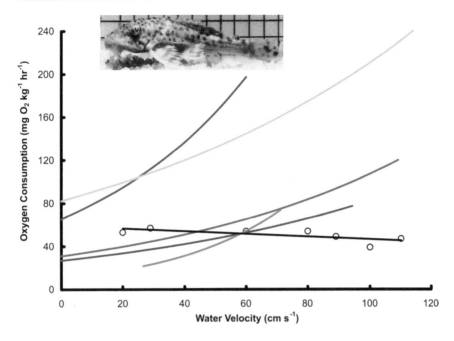

Fig. 2.7 Relationship between mass specific oxygen consumption rate and water velocity for rainbow trout *Oncorhynchus mykiss* (green line), river sockeye salmon *O. nerka* (red line), coho salmon *O. kisutch* Walbaum (purple line), sockeye salmon *O. nerka* (orange line) and Atlantic cod *Gadus morhua* (blue line). Points for *Pterygoplichthys* spp. (inset) are given by open circles which were used to generate a regression (black line). Data from Blake *et al.*, (2007).

1969). These rheotactic behaviours (associated with maximizing U_{lift}) are more complex and intense than those of loricariids where lift-off speeds are high. It is likely that hill stream fish (e.g., Cyprinidae, Siluridae, Gyrinochelidae and Loricariidae) do not naturally experience flows corresponding to their lift-off speeds. As far as we are aware, there are no metabolic measurements of the energy cost associated with the behavioural responses *per se* of forms that maximize U_{lift}. However, it seems likely there would be a significant cost associated with the vigorous posterior fin beating response. Priede and Holliday (1980) measured the swimming metabolic rate of plaice in a tilting tunnel respirometer. Resting metabolic rate is lower than that predicted from extrapolation of the active metabolic rate curve to zero velocity. The difference between the energy expenditure at lift-off (i.e., lowest speed of active swimming) and at rest is an indirect measure of the energetic cost of station-holding behaviour. Plaice lift off at swimming speeds of about 0.5 body lengths per second when the metabolic rate rises sharply to about three times the resting value. A corollary of this is that it is uneconomical for a plaice to swim at speeds below 0.5 body lengths per second (Priede and Holliday, 1980).

Both plaice and *Pterygoplichthys* are characterized by a small aerobic scope (Priede and Holliday, 1980; Blake *et al.*, 2007 respectively) relative to many non-benthic fish. It is likely that this is characteristic of rheotactic benthic fish regardless of their rheotactic adaptations.

Fast-starts off the bottom

In addition to adaptations for station-holding and moving off the bottom in diel and seasonal migrations that may involve locomotor strategies (e.g., tidal-stream transport, burst and glide swimming), rheotactic fishes must avoid and escape from their predators. Pleuronectids avoid predators by burying which involves quick and repetitive movements of the dorsal and anal fins agented by fast "twitch-like" muscle fibres (Gilly and Aladjem, 1987). Escape from piscivorous predatory strikes involves C-type fast-starts (Domenici and Blake, 1997) off the bottom which consists of two stages. In stage 1, the fish lifts the anterior part of its body upward; in stage 2, the fish begins to move forward (Webb, 1981; Brainerd *et al.*, 1997). Lifting off the head causes water to rapidly flow into the space that forms between the fish and the bottom causing a suction (low pressure; Stefan adhesion) under the fish that could resist its movement. Brainerd *et al.*, (1997) showed that an opercular jet attenuates the effects of Stefan adhesion. In addition adhesion may also be reduced by employing the median fins to prop up the body by increasing the initial distance between the fish and the bottom. Webb (1981) measured the fast-start performance of the flatfish *Citharichthys stigmaeus* in the water column and on the wire grid placed at various distances above the bottom and found higher accelerations for fast-starts off the grid. He found no significant difference in performance between starts from the grid on the bottom and starts from the grid above the bottom. Brainerd *et al.*, (1997) point out that fish starting from the grid above the bottom would not experience Stefan adhesion and would, therefore, be expected to have better fast-start performance relative to starting from the grid on the bottom. However, there was no significant difference between the two circumstances. They attribute this to the use of the median fins to prop the body up off the bottom and opercular jet effects.

It is likely that when leaving the ocean bottom to swim in the water column for purposes other than escaping predators the motions involved would be less vigorous (small accelerations) than those associated with fast-starts and, consequently, Stefan adhesion forces will be small.

Remoras: attachment to marine hosts

Echeneids (remoras) attach to marine hosts (e.g., elasmobranches, teleosts, cetaceans and sea turtles; Fertl and Landry, 1999) by means of a long spineless dorsal fin that acts as a laminated adhesive suction disc (O'Toole, 2002). Muscles erect or depress the paired laminae. When erect, the laminae create a sub-ambient pressure chamber allowing attachment to their hosts (Fulcher and Motta, 2006). Evidently, the anterior region of the disk generates most of the sub-ambient pressure differential. The depressor muscles likely assist in detachment by lowering the laminae and releasing the seal. This effect is augmented by the pressure acting on the remora due to its host's forward swimming speed.

Behavioural data for the eight species of remoras (Echeneidae) suggests that "hitchhiking" behaviour developed from general schooling to attaching to hosts in the pelagic environment (O'Toole, 2002). In some species, the remora-host association may be described as mutualistic benefiting the remora (protection from predation, transportation, access to food sources) and the host (removal of ectoparasitic copepods). In other species, there is an absence of strongly defined host specificity implying a commensal association (O'Toole, 2002). However, Weihs *et al.* (2007) suggest that remoras have effectively become hydrodynamic parasites on certain dolphin species due to increased energy cost of swimming associated with their presence and skin irritation (Ridgway and Carder, 1990). The aerial maneuvers of spinner dolphins (*Stenella longirostris*) may be used to remove remoras (Hester *et al.*, 1963) with the combined vertical and tangential velocity of the fall and spin generating enough force to dislodge them (Fish *et al.*, 2006; Weihs *et al.*, 2007).

Ground effects in near bottom swimming

When current velocities equal or exceed U_{slip} and U_{lift}, many rheotactic fish swim close to the bottom holding position against the flow (e.g., Loriicaridae, Pleuronectidae). In addition, many demersal negatively buoyant fish hover and swim close to the bottom (e.g., Selachii, Callionymidae). Under these circumstances, there is a reduction in the power required for hovering and slow forward swimming due to ground effect. Ground effect reduces the power required to support the submerged weight of the fish (induced power) relative to that out of ground effect reducing total power consumption.

Blake (1979a, 1983) determined the thrust and power in hovering for the callionymid *Synchiropus picturatus* at various heights above the substrate (expressed as the height above the bottom divided by the span of a pectoral fin). Values of power required to hover at various heights above the ground relative to that out of ground effect are shown in Fig. 2.8. The power required to hover decreases rapidly as the fish comes into ground effect and vanishes for a height index greater than 3. Typical values of the height index for *S. picturatus* are between 0.25 (about 0.5–1 cm above the bottom) and 0.5, corresponding to power savings of the order of 30–60%. Induced power savings decrease with increasing swimming velocity and the power to overcome

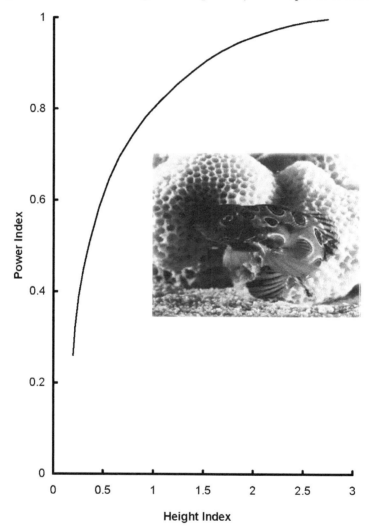

Fig. 2.8 Total power versus height for the mandarin fish *Synchiropus picturatus* (inset). The power required to hover out of ground effect divided by that in ground effect (power index) is plotted against the height index (height above the substrate divided by the span of a pectoral fin). Based on Blake (1979).

body drag increases rapidly, giving a U-shaped total power versus speed curve (Blake, 1979b). Lighthill (1979) gives a detailed mathematical model of the process.

Prospectus

Most flume based laboratory studies of benthic station-holding in fish assess behaviour in relation to observed slipping and lift-off speeds which are compared to theoretically predicted values (e.g., Arnold and Weihs, 1978; Webb, 1989). Calculation of the theoretical speeds requires the selection of values for the drag and lift coefficients of the fish based on those for roughly equivalent technical bodies attached to smooth surfaces. Blake (2006) directly measured the drag force (allowing the lift coefficient to be inferred) acting on catfishes attached to a surface. Further improvement in the apparatus (a biaxial force transducer in the force measuring apparatus) would allow for the simultaneous direct measurement of both the drag and the lift force.

The basic mechanical, morphological and behavioural responses of a number of freshwater and marine rheotactic fish are well understood. The characteristics of station-holding in benthic fish (e.g., compressed or depressed form, high density and weight in water, reduced or no swim bladder, ventral frictional pads and odontodes, adhesive disc and oral suckers, negative lift production from pectoral fins) can work in concert to maximize station-holding ability. However, benthic fish must leave the substratum to travel relatively large distances (e.g., find new habitats, resources, mates) and more frequently shorter distances (e.g., to breathe air). These requirements are associated with behavioural adaptations and energy saving locomotor strategies (e.g., tidal stream transport). Assessment of the overall behavioural ecology of marine and freshwater rheotactic fish with regard to the possible competing demands of station-holding versus swimming in the water column is required.

Field measurements of stream hydrology and the physical characteristics of substrata in relation to distribution, behaviour and swimming activity in rheotactic fish will allow for a better understanding of the relevance of laboratory measurements of behaviour and mechanics. Gordon *et al.* (1992) give a good guide to the assessment and measurement of the relevant hydrological and ecological variables. The same approach could also provide for an understanding of aspects of life history in these fish with regard to size specific habitat selection; for example, the hydrological characteristics of breeding *versus* feeding sites.

Detailed studies of the structure and ultrastructure of the adhesive apparatus of hill stream cyprinid and sisorid fishes that inhabit torrential streams characterized by high flows (greater than 1 m s^{-1}) are available.

However, there are very few descriptions of the behaviour of these fishes or mechanistic accounts of the significance of the structural aspects of their adhesive apparatus. In addition, there are no measurements of slipping or lift-off speeds in these species. Experimental studies of these issues would be rewarding.

Further investigation of the association between remoras and their hosts are warranted. In particular, with regard to remora attachment location sites in relation to the hydrodynamic forces that they and their hosts experience. In addition, the possibility of drafting (force transfer among individuals without physical contact between them) for remoras freely swimming close to their host (as may occur between young dolphin calves and their mothers, Weihs, 2004) could be explored.

List of abbreviations

l	body length
m	submerged body mass
A	frontally projected area of the fish
A_p	planform area
W_o	submerged weight
U	water velocity
U_{slip}	slipping speed
U_{lift}	lift-off speed
D	drag force
L	lift force
C_D	drag coefficient
C_L	lift coefficient
M	mass specific active metabolic rate
γ	acceleration due to gravity
ρ_w	water density
v	kinematic viscosity of water
μ	coefficient of static friction
R_e	Reynolds number

ACKNOWLEDGEMENTS

We would like to thank Ms. T. Mark for drawing Fig. 2.6 and the Natural and Sciences Engineering Research Council of Canada for financial support to R.W.B.

REFERENCES

Ansell, A.D. and R.N. Gibson. 1990. Patterns of feeding and movement of juvenile flatfishes on an open sandy beach. In: *Trophic Relationships in the Marine Environment*, M. Barnes and R.N. Gibson (Eds.). Aberdeen University Press, Scotland, pp. 191–207.

Arnold, G.P. 1969. The reactions of the place (*Pleuronectes platessa* L.) to water currents. *Journal of Experimental Biology* 51: 681–697.

Arnold, G.P. 1974. Rheotropism in fishes. *Biological Reviews* 49: 515–576.

Arnold, G.P. and J.D. Metcalfe. 1995. Seasonal migrations of plaice (*Pleuronectes platessa*) through the Dover Strait. *Marine Biology* 127: 151–160.

Arnold, G.P. and D. Weihs. 1978. The hydrodynamics of rheotaxis in the plaice (*Pleuronectes platessa* L.). *Journal of Experimental Biology* 75: 147–169.

Arnold, G.P., P.W. Webb and B.H. Holford. 1991. The role of the pectoral fins in station-holding of Atlantic salmon parr (*Salmo salar* L.). *Journal of Experimental Biology* 156: 625–629.

Bhatia, B. 1950. Adaptive modification in a hillstream catfish *Glyptothorax telchitta* (Hamilton). *Proceedings of the National Institute of Science of India* 16: 271–285.

Blake, R.W. 1979a. The energetics of hovering in the Mandarin fish (*Synchropus picturatus*). *Journal of Experimental Biology* 82: 25–33.

Blake, R.W. 1979b. The swimming of the mandarin fish (*Synchropus picturatus, Callionyiidae: Teleostei*). *Journal of the Marine Biological Association of the United Kingdom* 59: 421–428.

Blake, R.W. 1983. Hovering performance of a negatively buoyant fish. *Canadian Journal of Zoology* 61: 2629–2630.

Blake, R.W. 2006. Biomechanics of rheotaxis in six teleost genera. *Canadian Journal of Zoology* 84: 1173–1186.

Blake, R.W., P.Y.L. Kwok and K.H.S. Chan. 2007. The energetics of rheotactic behaviour in *Pterygoplichthys* spp. *Journal of Fish Biology* 71: 623–627.

Brainerd, E.L., N.P. Bejamin and F.E. Fish. 1997. Opercular jetting during fast-starts by flatfishes. *Journal of Experimental Biology* 200: 1179–1188.

Coull, K.A., A.S. Jermyn, A.W. Newton, G.I. Henderson and W.B. Hall. 1989. Length-weight relationships for 88 species of fish encountered in the North Atlantic. *Scottish Fisheries Research Report* 43: 1–80.

Das, D., and T.C. Nag. 2004. Adhesion by paired pectoral and pelvic fins in a mountain-stream catfish, *Pseudocheneis sulcatus* (Sisoridae). *Environmental Biology of Fishes* 71: 1–5.

Domenici, P. and R.W. Blake. 1997. The kinematics and performance of fish fast-start swimming. *Journal of Experimental Biology* 200: 1165–1178.

Eschmeyer, W.N., E.S. Herald and H. Hammann. 1983. *A Field Guide to Pacific Coast Fishes of North America*. Houghton Mifflin Co., Boston, Massachusetts.

Ellis, T., B.R. Howell and R.N. Hughes. 1997. The cryptic responses of hatchery-reared sole to a natural sand substratum. *Journal of Fish Biology* 51: 389–401.

Fertl, D. and A.M. Landry. 1999. Sharksucker (*Echeneis naucrates*) on a bottlenose dolphin (*Tursiops truncates*) and a review of other cetacean-remora associations. *Marine Mammal Science* 15: 859–863.

Fish, F.E., A.J. Nicastro and D. Weihs. 2006. Dynamics of the aerial maneuvers of spinner dolphins. *Journal of Experimental Biology* 209: 590–598.

Frankel, G.S. and D.L. Gunn. 1961. *The Orientation of Animals*. Dover Publications Incorporated, New York.

Fulcher, B.A. and P.J. Motta. 2006. Suction disk performance of echeneid fishes. *Canadian Journal of Zoology* 84: 42–50.

Gibson, R.N. 1969. Powers of adhesions in *Liparis montagui* (Donovan) and other shore fish. *Journal of Experimental Marine Biology and Ecology* 3: 179–190.

Gilly, W.F. and E. Aladjem. 1987. Physiological properties of three muscle fibre types controlling dorsal fin movements in a flatfish, *Citharichthys sordidus*. *Journal of Muscle Research and Cell Motility* 8: 407–417.

Gordon, N.D., T.A. McMahon and B.L. Finlayson. 1992. *Stream hydrology: an introduction for ecologists*. John Wiley & Sons, New York.

Gradwell, N. 1971. A muscular oral valve unique in fishes. *Canadian Journal of Zoology* 49: 837–839.

Greer Walker, M., F.R. Harden Jones and G.P. Arnold. 1978. Movement of plaice (*Pleuronectes platessa*) tracked in the open sea. *Journal du Conseil* 38: 58–86.

Harden Jones, F.R. 1977. Performance and behaviour on migration. In: *Fisheries Mathematics*, J.H. Steele (Ed.). Academic Press, New York, pp. 145–170.

Haugen, C.W. 1990. The California halibut, *Paralichthys californicus*, resource and fisheries. *Fishery Bulletin* 174, the Resources Agency, Department of Fisheries and Game, State of California, pp. 1–475.

Hester, F.J., J.R. Hunter and R.R. Whitney. 1963. Jumping and spinning behavior in the spinner porpoise. *Journal of Mammalogy* 44: 586–588.

Hoerner, S.F. 1965. *Fluid-dynamic drag*. Hoerner Fluid Dynamics, New York.

Hoerner, S.F. and H.V. Borst. 1975. *Fluid-dynamic lift*. Hoerner Fluid Dynamics, New York.

Hora, S.L. 1922a. Structural modifications of fishes of mountain torrents. *Records of the Indian Museum* 24: 46–58.

Hora, S.L. 1922b. The modification of the swimbladder in hill-stream fishes. *Journal of the Asiatic Society of Bengal* 18: 5–7.

Hora, S.L. 1930. Ecology, bionomics and evolution of torrential fauna with special reference to the organs of attachments. *Philosophical Transactions of the Royal Society of London* B 218: 171–282.

Hugett, W.S., J.F. Bath and A. Douglas. 1976. *Data of current observations 15. Juan de Fuca Strait 1973*. Institute of Ocean Sciences, Patricia Bay, Victoria, B.C.

Keenleyside, M.H.A. 1962. Skin-diving observation of Atlantic salmon and brook trout in the Miramichi River, New Brunswick. *Journal of the Fisheries Research Board of Canada* 19: 625–634.

Lighthill, J. 1979. A simple fluid-flow model of ground effect on hovering. *Journal of Fluid Mechanics* 93: 781–797.

Macdonnell, A.J. and R.W. Blake. 1990. Rheotaxis of *Otocinclus* sp. (Teleostei: Loricariidae). *Canadian Journal of Zoology* 68: 599–601.

Matthews, W.J. 1985. Critical current speeds and microhabitats of the benthic fishes *Percina roanoka* and *Etheostoma flabellare*. *Environmental Biology of Fishes* 12: 303–308.

Metcalfe, J.D. and G.P. Arnold. 1990. The energetics of migration by selective tidal stream transport: an analysis for plaice tracked in the southern North Sea. *Journal of the Marine Biological Association of the United Kingdom* 70: 149–162.

Myrhe, R.J. 1974. Minimum size and age and optimum age at entry of Pacific halibut. *International Pacific Halibut Commission: Scientific Report*. No. 55.

O'Toole, B. 2002. Phylogeny of the species of the superfamily Echeneoidea (Perciformes: Carangoidei: Echeneidae, Rachycentridae, and Coryphaenidae), with an interpretation of echeneid hitchhiking behaviour. *Canadian Journal of Zoology* 80: 596–623.

Priede, I.G., and F.G.T. Holliday. 1980. The use of a new tilting tunnel respirometer to investigate some aspects of metabolism and swimming activity of the plaice (*Pleuronectes platessa* L.). *Journal of Experimental Biology* 85: 295–309.

Raffaelli, D., H. Richner, R. Summers and S. Northcott. 1990. Tidal migrations in the founder (*Platichthys flesus*). *Marine Behaviour and Physiology* 16: 249–260.

Rickey, M.H. 1995. Maturity, spawning, and seasonal movement of arrowtooth flounder, *Atheresthes stomias*, off Washington. *Fishery Bulletin* 93: 127–138.

Ridgway, S.H. and D.A. Carder. 1990. Tactile sensitivity, somatosensory responses, skin vibrations, and the skin surface ridges of the bottlenose-dolphin, *Tursiops truncatus*.

In: *Sensory Abilities of Cetaceans*, J.A. Thomas and R.A. Kastelein (Eds.). Plenum Press, New York, pp. 163–179.

Saxena, S.C. and M. Chandy. 1966. Adhesive apparatus in certain hillstream fishes. *Journal of Zoology* 148: 315–340.

Schoenfuss, H.L. and R.W. Blob. 2003. Kinematics of waterfall climbing in Hawaiian freshwater fishes (Gobiidae): vertical propulsion at the aquatic-terrestrial interface. *Journal of Zoology* 261: 191–205.

Singh, N. and N.K. Agarwal. 1991. SEM surface structure of the adhesive organ of the hillstream fish *Glyptothorax pectinopterus* (Teleostei: Sisoridae) from the Garhwal Hills. *Functional Development and Morphology* 1: 11–13.

Tyler, A.V. 1971. Surges of winter flounder, *Pseudopleuronectes americanus*, into the intertidal zone. *Journal of the Fisheries and Research Board of Canada* 28: 1727–1732.

Webb, P.W. 1981. The effect of the bottom on the fast start of flatfish *Citharichthys stigmaeus*. *Fishery Bulletin of Fish Wildlife Service of the United States* 79: 271–276.

Webb, P.W. 1989. Station-holding by three species of benthic fishes. *Journal of Experimental Biology* 145: 303–320.

Weihs, D. 1973. Mechanically efficient swimming techniques for fish with negative buoyancy. *Journal of Marine Research* 31: 194–209.

Weihs, D. 1978. Tidal stream transport as an efficient method for migration. *Journal du Conseil* 38: 92–99.

Weihs, D. 2004. The hydrodynamics of dolphin drafting. *Journal of Biology* 3: 8–16.

Weihs, D., F.E. Fish and A.J. Nicastro. 2007. Mechanics of remora removal by dolphin spinning. *Marine Mammal Science* 23: 707–714.

Wirjoatmodjo, S. and T.J. Pitcher. 1984. Flounders follow the tides to feed: evidence from ultrasonic tracking in an estuary. *Estuary and Coastal Shelf Science* 19: 231–241.

3

Fish Guidance and Passage at Barriers

Theodore Castro-Santos* and Alex Haro

INTRODUCTION

Habitat fragmentation resulting from human activities is a major factor contributing to reductions in biodiversity and species abundance worldwide. When movements are restricted, subpopulations become isolated, leading to reduced breeding opportunities, inbreeding depression, and interruption of key life stages. This problem is particularly ubiquitous in riverine ecosystems, where dams, water diversions, culverts, and other structures create barriers to movements of aquatic organisms. Because these organisms are unable to leave their streams to bypass obstacles, and because the bifurcating structure of stream networks precludes the use of alternate pathways to needed habitats, this type of fragmentation can quickly lead to loss of species richness, and compromised ecosystem function (Morita and Yamamoto, 2002; Morita and Yokota, 2002).

Impeded fish passage can have consequences that extend beyond the organismal level. For example, an extensive literature documents the current and past importance of marine-derived nutrients (MDN's) to forest ecosystems in the Pacific Northwest of North America (Naiman *et al.*, 2002). Adult salmon

Authors' address: S.O. Conte Anadromous Fish Research Center, USGS-Biological Resources Discipline, P.O. Box 796, One Migratory Way, Turners Falls, MA 01376. (413)863-3838, USA.
Corresponding author: E-mail: tcastrosantos@usgs.gov

act as a vector transporting nitrogen, phosphorus, and other nutrients into otherwise low-productivity systems. Each year, millions of salmon return to these rivers to spawn and die. Their carcases become an important nutrient source, not only for riverine organisms, but for the surrounding flora and fauna as well. Other anadromous species are iteroparous (spawning and migrating multiple times); among these species the nutrients transported out of the riverine systems by juveniles may exceed what is deposited upstream by adults, resulting in net nutrient removal (Nislow *et al.*, 2004). Ultimately, the nutrient flux is dependent on location- and life-stage-dependent growth and survival, and the consequences have important implications for landscape ecology and fisheries management (Moore and Schindler, 2004; Scheuerell *et al.*, 2005).

Migratory fish play another key role in riverine ecosystems, which is to transport weaker-swimming organisms upstream. An interesting case in point are various species of freshwater mussels. These normally sessile organisms begin life as glochidia, a larval stage that is specially adapted to latch onto the gills of passing fish, dropping off and settling after some period of time or when certain environmental conditions are met (Haag and Warren, 1998). Glochidia are often host-specific, having adapted to attach to a particular fish species—often diadromous or potamodromous— for upstream transport (Haag and Warren, 1998; Freeman *et al.*, 2003; Rashleigh and DeAngelis, 2007). These mussels play an important role in aquatic ecosystem health: growing in dense beds, they filter tremendous volumes of water, removing phyto- and zooplankton from their environment and so playing an important role in water quality and nutrient cycling. In the absense of freely-moving fish populations many species of mussel are unable to recolonize upstream habitats, and freshwater mussels are now among the most endangered group of animals in North America (Richter *et al.*, 1997).

These cascading ecological consequences of barriers to movement in riverine ecosystems are subtle, though, compared to the dramatic losses of anadromous populations. The extensive fragmentation of riverine habitat that has occurred, especially throughout the Northern Hemisphere, has extirpated these and other freshwater migrants by preventing access to key spawning and nursery habitat. These losses—and similar losses occuring worldwide—have significant economic and cultural costs as well, impacting both freshwater and marine fisheries important to subsistence, recreational, and commercial interests. For these reasons, restoration of the river continuum constitutes a major challenge to conservation biologists, managers, and civil engineers. The profile of this goal has been raised by its inclusion as a priority in the recently enacted European Union Water Framework Directive (Bloch, 1999; European Commission, 2000; Howe and White, 2002).

Of course, fragmentation and obstacles to movement can be a natural and important part of riverine ecosystems. Less permanent obstacles can also occur, such as those created by beaver dams and obstructions from large woody debris. These obstacles can delimit boundaries of some species while allowing others through: whether persistent or dynamic, barriers can contribute to biological diversity, with local endemics evolving in protected habitats, and both behavioral and morphological adaptations arising for traversing them (Northcote *et al.*, 1970; Kelso *et al.*, 1981; Ghalambor *et al.*, 2003).

Diversity arising in protected habitats can take tens of thousands of years to evolve, however, and high levels of endemism are typically associated with riverine habitats that do not undergo frequent fragmentation and reconnection (Griffiths, 2006). Anthropogenic barriers, by contrast, have arisen suddenly, recently, and in vast numbers (World Commision on Dams, 2000; U.S. Army Corps of Engineers, 2001). These barriers are effectively permanent, and the resulting fragmentation and elimination of key habitats is not mitigated for by local adaptation. Habitats that have undergone anthropogenic fragmentation tend to have depauperate fauna and elevated extinction rates (Morita and Yamamoto, 2002; Letcher *et al.*, 2007).

The challenge of fish passage, then, is to mitigate for the effects of these barriers by providing passage routes that are both safe and effective. Ideally, this means that passage structures should be 'transparent' to movement, i.e., movements through the obstructed reach of river are no different than would be expected in the absence of any barrier. Such transparency cannot be measured simply by quantifying presence or absence of species from a particular habitat, but must also consider continuous behavioral processes, such as the frequency and timing of movement, use of habitat for cover and feeding, vulnerability to predation, etc. In addition, energetic costs and other physiological stressors should be avoided; in particular there should be no added delays to migration resulting from the structure or obstacle. In short, the objective of fish passage is to neutralize the fitness consequences of anthropogenic barriers to movement (Castro-Santos *et al.*, 2008).

Achieving this objective requires a holistic approach: one that considers innate behaviors, physiology, and capabilities—not just of one species, but of the entire native fauna to which the structure poses a barrier. Although the importance of physiological capabilities to fish passage has long been recognized, the focus has historically been on a limited number of species, and behavioral adaptations for (or obstacles to) traversing barriers have at best been relegated to an ancillary role in design criteria (Powers *et al.*, 1985; Orsborn, 1987; Bell, 1991; Clay, 1995).

Historical Overview

Swimming capacity and behavior have been studied in the context of fish passage for at least 120 years (Day, 1887), however not until the work of Denil (1909, 1937) was capacity explicitly linked to fishway design. The primary focus of these and subsequent studies has been upstream passage. Early work relied on field observations of swimming fish, focusing either on swimming against known flow velocities (Denil, 1909, 1937), deriving maximum attainable swim speeds from leaping behavior (Stringham, 1924), and even attaching tachometers to fishing reels to record run speed of angled fish (Gero, 1952).

None of these methods provided satisfactorily reproducible results, however, and there was a clear need to develop devices for studying swimming performance and energetics in a controlled laboratory setting. A number of different swim chambers were developed by Fry and Hart (1948), Blazka *et al.* (1960), Harden-Jones (1963) and others (see Beamish 1978 for an overview). Probably the most popular model, and the one that set the standard for swimming performance studies during he latter half or the 20th century was developed by Brett (1964). This consisted of a closed-circuit tube with flow driven by a centrifugal pump capable of generating velocities up to 1.1 m·s^{-1} and a transparent chamber for holding and monitoring the fish. Flow straighteners ensured that large-scale eddies were absent and flow velocities were uniform across the cross section of the swim chamber. Using this device, Brett and his successors were able to study fish swimming performance and energetics under highly controlled conditions.

Even more important than his swimming chamber were Brett's contributions to the study of swimming capacity. In particular, he introduced two metrics of swimming ability and associated methods for their quantification: the first and most widely repeated measure is U_{crit}, or critical swimming speed, which originated as an estimate of the maximum speed that fish can sustain indefinitely (U_{ms}, Brett *et al.*, 1958; Brett, 1964). This was done by gradually incrementing flow velocity every 60 minutes until the fish fatigued, with the assumption that a speed that could be maintained for 60 minutes could be maintained indefinitely. Brett offered caveats to this approach, suggesting a true measure of U_{ms} would have required timesteps greater than 200 minutes. The 60 minute threshold was established for convenience of laboratory measurement, and although Brett felt this method produced a reasonable estimate of U_{ms} the shorter timesteps meant that $U_{crit} >$ U_{ms}. Moreover, subsequent researchers have further reduced the timestep and otherwise modified this approach such that U_{crit} is best viewed as a comparative performance index: a valuable experimental technique, but yielding data that are often not readily comparable among studies (Hammer, 1995; Jain *et al.*, 1997).

The second metric Brett developed was a fixed-velocity protocol for estimating the relationship between swim speed and fatigue time at speeds $> U_{ms}$. From this approach, and comparing his data with those of Bainbridge (1960), Brett identified 2 modes of unsustainable swim speeds: a slower, 'prolonged' mode, where bursting and coasting allowed fish to use both aerobic and anaerobic muscle groups to maximize stamina; and a final 'burst' or 'sprint' mode, powered exclusively by anaerobic muscle groups, and consequently resulting in fatigue in the shortest time. Each of these two modes is described by a unique log-linear relationship:

$$LogT_m = a_m + b_m U_s \qquad (1)$$

Where fatigue time (T) within each mode m is determined by swim speed (U_s, typically measured in body lengths·s^{-1}), adjusted by regression coefficients a_m and b_m.

Brett's early work focused on juvenile salmonids (Brett *et al.*, 1958; Brett, 1964). This reflects a primary motivation, which was to identify protective flows for migrating salmon smolts. Hydroelectric development, especially on the Columbia River and its tributaries meant that during their downstream migration juveniles would have to be guided around turbine intakes and into safer alternative passage routes. Brett's measurements of endurance were intended to identify flow velocities that these fish could resist long enough to avoid becoming entrained into the turbines.

These methods were also applied to adults, however, where swimming performance data were needed to establish criteria for construction of fish ladders and other upstream passage structures. By extrapolating Equation 1 to very short fatigue times it was possible to estimate maximum burst speeds, an important criteria for fishway designs (Bainbridge, 1960). Brett and his colleagues were also interested in migratory bioenergetics, and respirometry formed a core application of his swim chamber (Brett, 1962, 1965, 1967).

A common characteristic of the devices used by Brett and his contemporaries was that fish were confined to swim chambers of limited size. This imposed limitations on both the size and types of fish that could be swum in these chambers. It also meant that researchers had to decide whether to include data from fish that performed erratically, which reduced both the precision and accuracy of the tests.

A more generic method for estimating swimming performance was developed by Wardle (1975). Wardle's approach was even more fundamentally physiological than Brett's: using isolated plugs of muscle, he measured maximum twitch velocity. From this, he estimated minimum contraction time and by association maximum tailbeat frequency. Because the relationship between tailbeat frequency and swim speed is often linear (Bainbridge, 1958; Hunter and Zweifel, 1971), Wardle (1975) proposed to use these data to estimate maximum sprint swimming ability for several marine teleost species without confining them to swim chambers. Encouragingly, Wardle's estimates of maximum swimming ability were consistent with those of Brett and Bainbridge, i.e., maximum sprint speeds for 20 cm long fish should be about 10–15 L s^{-1}.

The work of Brett (1965, 1973) and Wardle (1975) was to form the core biological basis of fishway design. The apparent generality of their results lead Zhou (1982) to develop models of swimming performance that could be applied to a broad range of species. These models were predicated on the idea that fatigue resulted from depletion of glycogen stores, and that the rate of this depletion was determined by maximum respiration rate and power requirements of swimming at maximum speeds. These principles were then further developed by Beach (1984) to produce deterministic models that engineers could use to identify maximum velocities within fishways, as well as distances between resting pools. These models were so popular that they were quickly incorporated into fishway design manuals throughout North America and Europe, where they continue in use today.

Unfortunately, the enthusiasm with which these models were accepted appears to have been premature. Castro-Santos and Haro (2006) itemized several flaws in the models themselves; but even more troubling are the questions that have arisen regarding the underlying methodologies. Reviews by Hammer (1995) and Plaut (2001) both raised concerns regarding the relevance of forced swimming tests to field situations; and swim speeds of free-swimming fish are often well in excess of theoretical maxima (Dow, 1962; Weaver, 1963; Castro-Santos, 2005). This may arise in part because fish adjust behaviors and kinematics in the presence of rapid flows in ways that allow them to increase their performance; these adjustments may not be possible in a confined swim chamber causing underestimates of actual swimming potential (Peake, 2004; Peake and Farrell, 2005; Tudorache *et al.*, 2008)

Given that established methods for estimating swimming ability consistently underestimate capacity, one might expect that fish passage structures designed from those estimates should still perform well. Unfortunately the opposite seems to be true. Non-salmonid species in particular seem to struggle with existing designs. A series of papers by Bunt *et al.* (1999, 2000, 2001) showed that fishways deployed to pass several non-salmonid species failed in many cases to pass even half of the fish that entered. Even worse performance was described by Sullivan (2004). Using high-power passive integrated transponders (PIT tags; Castro-Santos *et al.*, 1996) Sullivan measured passage success rates of American shad (*Alosa sapidissima*) ascending a pool-and-weir fishway ranging from 3–17%. These results were particularly troubling, given that this same fishway had been used elsewhere to set design standards for American shad and other clupeid species (Rideout *et al.*, 1985; Larinier and Travade, 2002).

Even salmonid species, for which many of these fishway designs were originally developed, often do not pass as well as expected. A survey of fishways in Northern Spain found that only 58% successfully passed target species (Elvira *et al.*, 1998); low entry rates have been identified in Swedish rivers (Rivinoja *et al.*, 2001); and there is evidence that cumulative effects of poor passage may even be a problem for salmon on the Columbia River (Dauble and Mueller, 2000; Naughton *et al.*, 2005), where many of the most common fishway types were developed (Collins and Elling, 1960).

One possible reason for this reduced performance may be a lack of appropriate hydraulic conditions to stimulate upstream movements. Orsborn (1987) expressed concern that overly-conservative fishway velocities could reduce passage performance. There appears to be a tradeoff between conditions that stimulate movement and those that result in fatigue; although the latter has been studied extensively, the former remains poorly characterized (Powers *et al.*, 1985; Bunt, 2001; Castro-Santos, 2004, 2005). Fish passage technology would greatly benefit from more research on the associations between hydraulics and other environmental stimuli and rheotaxis in migratory fish.

Although upstream passage is necessary for anadromous species to reach spawning habitat, the provision of safe and effective downstream passage at obstacles is equally important to fisheries conservation. In the case of hydroelectric facilities the dangers of downstream passage are obvious: typically, the bulk of the flow at these facilities passes through the turbines. Although survival of juveniles passing through turbines is sometimes surprisingly high (as much as 94% through some turbine types; Muir *et al.*, 2001; Skalski *et al.*, 2002), this is not typically the case. Moreover, there is an inverse relationship between body size and survival, so catadromous species like eel (*Anguilla spp.*) and iteroparous species that emmigrate as adults can incur mortality rates in excess of 50% at each site (Bell and Kynard, 1985; McCleave, 2001). In addition to the dangers posed by hydroelectric plants, water withdrawal structures for irrigation or cooling can entrain large numbers of fish, particularly during their larval and juvenile stages. Even where passage and survival are good at individual facilities, if fish have to pass multiple structures the cumulative mortality can be devastating, and provision of safe downstream passage routes is one of the primary challenges of fisheries conservation.

Migratory delay, poor attraction, fallback and stalling within fishways (for upstream passage), and both immediate and delayed passage-induced mortality (for downstream passage) are common problems identified for a range of fishway types and taxa. Effects of these barriers accumulate on a watershed scale, to the detriment of both local and migratory populations. These comprise the primary issues that fish passage biologists and engineers need to address, and all point to the need for greater incorporation of behavioral data in the development of fishway design.

Fishway Performance

Two distinct but related deficiencies in available data have contributed to the poorer-than-expected performance of existing upstream fish passage structures. First, data on swimming performance are typically conducted under very controlled conditions, where researchers take pains to minimize turbulence and eliminate entrained air. While this produces consistent results, the conditions are very different from fishways, where turbulence and entrained air are endemic. Second, as mentioned above, the apparati used to study swimming performance have typically constrained fish within test structures of limited size, preventing them from exhibiting behaviors that might affect their ability or motivation to pass real-world structures. In the following sections, we describe some behaviors associated with riverine migrations and consider how they might influence passage success.

Upstream passage

Swimming performance and distance maximization

The interaction between behavior and physiological swimming capacity provides an instructive example of the importance of behavior to fish passage and locomotion generally. The shape of the swim speed-fatigue time relationship dictates that the distance fish will be able to ascend against a given flow velocity will depend greatly on a) whether they select prolonged or sprint mode, and b) on the speed at which they actually swim within a given mode. When the speed of flow (U_f) is subracted from swim speed, (producing groundspeed), the distance a fish can traverse relative to the ground (D_g) is described by

$$D_g = (U_s - U_f) \times e^{a_m + b_m U_s} \qquad (2)$$

Where fatigue time is calculated following Equation 1. The response follows a dome-shaped surface, with clear and distinct distance-maximizing optima associated with prolonged and sprint modes, respectively (Fig. 3.1; Bainbridge, 1960; Castro-Santos, 2005). These optima correspond to groundspeeds equal to the negative inverse of the slope of the swim speed-fatigue time relationship (i.e., the optimal groundspeed within a given mode, $U_{g opt_m} = -b_m^{-1}$, Equation 1). Within a given mode this groundspeed remains fixed, regardless of flow velocity, and fish that swim either too fast or too slow incur costs of reduced distance of ascent (Fig. 3.1; Castro-Santos, 2005, 2006). Another way to think of this is the optimal swim speed within a given mode ($U_{s opt_m}$) is always equal to the speed of flow plus $U_{g opt_m}$, i.e., $U_{s opt_m} = U_f - b_m^{-1}$.

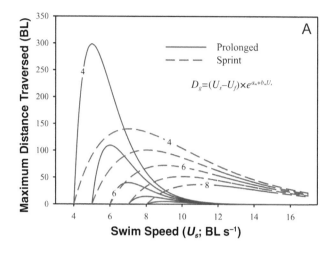

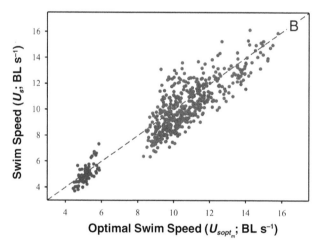

Fig. 3.1 Relationship between swim speed, fatigue time, and flow velocity on distance of ascent through velocity barriers. Data are from American shad (*Alosa sapidissima*; Castro-Santos, 2005) and represent prolonged (blue) and sprint modes (red). (A): Theoretical maximum distance of ascent , from Equation 2 against flow velocities of 4 – 8 BL·s⁻¹ (U_f ; contours) and at swim speeds of 4–17 BL·s⁻¹ (see text for regression coefficients). Note the strong nonlinearity and clear distance-maximizing optima; speeds greater or less than these optima result in reduced distance of ascent. (B): Extent to which American shad approximated distance-maximizing groundspeeds ($U_{s\,opt_m}$, dashed line) for prolonged (blue) and sprint modes (red). Points represent actual swim speeds of 584 American shad swimming against flow velocities of 3.4 – 4.9 BL·s⁻¹ (prolonged mode) and 5.3 – 12.7 BL·s⁻¹ (sprint mode). Note that although most fish swam within 1 BL·s⁻¹ of $U_{s\,opt_m}$ in each mode, mean swim speeds were slightly lower than $U_{s\,opt_m}$ and considerable variability occurred. Consequences of this are shown in Fig. 3.2.

Variation in the ability to approximate the distance-maximizing optimum swim speeds occurs among species, and to a lesser degree among individuals within species (Fig. 3.1b). This highlights one of the greatest weaknesses of Beach's (1984) equations, which was the assumption that fish would swim at a fixed speed, regardless of the speed of flow, and regardless of species. By ignoring behavioral strategies, these models included systematic error that have led to overly-conservative fish passage designs, and very probably reduced overall fishway efficiency (Orsborn, 1987).

Even when founded on sound biological principles and data, deterministic models are intrinsically unrealistic, because they assume fixed behaviors. If we assume, for example, that fish will swim at their distance maximizing speed, then this will only tell us whether a barrier is passable in theory, it does not provide any estimate of what proportion of the population is likely to pass a challenging but passable barrier. Proportion passing, however, is the metric of greatest value to managers and engineers. To meet this need, some researchers have performed empirical studies of passage performance, where distance of ascent was measured directly (Weaver, 1963; Haro *et al.*, 2004). These studies were performed in large, open-channel flumes with regulated flow velocities up to 4.8 m·s^{-1}, and ascents were performed volitionally by unrestrained fish. An example of the data provided by this approach is modeled in Fig. 3.2.

In addition to empirical measurements of distance of ascent through velocity barriers, these apparati also allowed for both the estimation of swim speed-fatigue time relationships for free-swimming fish and the extent to which swim speeds are optimized for distance maximization (Castro-Santos, 2005). Figs. 3.1 and 3.2 were derived from the same data (American shad; N = 584; U_f = 1.5 –4.5 m·s^{-1}; estimated regression coefficients from Equation 2 are a_p = 10.7, b_p = –1.0 , a_S = 6.2 , b_S = –0.33 with m = P and S for prolonged and sprint mode, respectively), yet close examination shows that most individuals of this species failed to achieve their maximum potential distance of ascent. Consider a 30 m (72 BL) long culvert through which fish must pass in order to reach essential habitat. At flow velocities U_f ≤ 6 BL s^{-1} a deterministic model that assumed distance maximizing swim speeds (-b_m^{-1} = 1.0 and 3.0 BL s^{-1} for prolonged and sprint modes, respectively) predicts 100% passage through such an obstacle. The same model predicts 100% failure, however, at distances >52 and 38 body lengths for U_f = 7 and 8 BL s^{-1}, respectively. This is not a realistic depiction of passability. Instead, the proportion of fish passing such a barrier declines continuously with distance and flow velocity (Fig. 3.2).

The fact that failure is not discrete, but rather a continuous function reflects the underlying physiological and, just as importantly, behavioral variability among individuals. Of the six species studied by Haro *et al.* (2004) and Castro-Santos (2005), American shad came closest to approximating their distance maximizing swim speeds. The reduced performance shown in

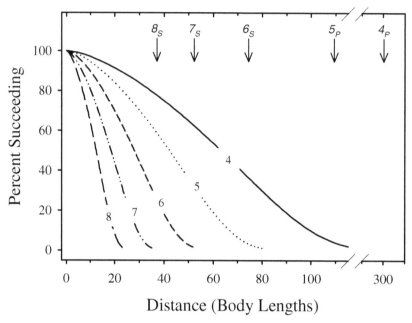

Fig. 3.2 Output of two models predicting distance of ascent attained by American shad (*Alosa sapidissima*) swimming against currents of 4 – 8 BL·s⁻¹ (arrows and contours). Arrows indicate theoretical maximum distance of ascent predicted from the swim speed-fatigue time relationship (Fig. 3.1 and equation 2; subscripted here with *P* or *S* to denote prolonged or sprint mode, respectively); distance-maximizing swim speeds at each flow velocity should result in 100% passage up to the corresponding arrow and 0% passage thereafter. The contours are output from a regression fit to actual distance ascended up an open-channel flume and show that the percent of fish attaining a given distance declines gradually rather than abruptly as the predicted maximum distance of ascent is approached. The range of failure distances described using this second approach arises from variability in swimming capacity and behavior (Fig. 3.1b; Haro *et al.*, 2004; Castro-Santos, 2006).

Fig. 3.2 primarily reflects individual deviation from the distance maximizing optimum (Fig. 3.1b). Other species deviated systematically from the optima, and performance of these species was correspondingly lower than for American shad. Conversely, some of the shad performed better than was theoretically possible, given the swim speed-fatigue time relationship. This occurs in nature because of variability in swimming performance: the theoretical maxima mentioned earlier were for fish of average capacity only; more capable fish can swim further—when under genetic control this variability can lead to population structuring and ultimately to speciation.

Because the behavioral variability was quantified, it is possible to reconstruct the empirical performance curves shown in Fig. 3.2 from the theoretical swim speed-fatigue time relationship. This is important, because the empirical models were developed under uniform, regulated flow conditions. In the field, hydraulic conditions like these are rarely encountered. Instead, velocities vary continuously, and a need remains to be able to predict passage through real-life barriers. By quantifying behavioral variability, it becomes possible to construct an individual-based stochastic model of a population of fish attempting to ascend a velocity barrier. Castro-Santos (2006) demonstrates how this information can be coupled with hydraulic modelling to estimate proportion of a population passing a barrier with non-uniform velocities. As mentioned above, these studies—like most previous studies—were done under controlled conditions with minimal turbulence. Until more work is available quantifying the effects of turbulence on swimming performance these models should be used with caution.

Motivation and guidance

A fundamental assumption underlying technical fishway designs is that fish posess an innate drive to swim upstream. In the case of salmon this is well-justified: notoriously philopatric, salmon may literally swim themselves to death in an effort to pass obstacles and reach their natal streams (Rand and Hinch, 1998). Other species requiring passage, however, are much less philopatric, and an exploration of the necessary and appropriate stimuli to elicit upstream migratory behaviors may shed light on interspecific variability in fishway performance.

Odor and rheotaxis are probably the best-documented cues used in upstream navigation. Although the relative importance at earlier life stages is still debated, it seems clear that among migratory salmonids, imprinting occurs at least during the smolt (outmigrant) phase. In a series of experiments involving both blocking of olfactory apparatus and introduction of artificial odors during the smolt phase, Hasler (reviewed in Hasler, 1971) was able to manipulate homing behaviors, providing clear evidence that olfactory cues recorded as juveniles played a major role in navigation and orientation in upstream adult migrants. Whether the odors associated with a natal stream provide sufficient stimulus for homing, or whether smolts lay down an olfactory map during their outmigration remains open to conjecture. It appears, however, that the more sophisticated map is at least feasible: one can easily imagine benefits accruing from greater navigational accuracy, especially as scents associated with individual streams become diluted by successive confluences (Fig. 3.3).

If philopatry were perfect and invariate, though, colonization of new habitats would be impossible. Straying and invasive behaviors are well-documented, and so it is clear that olfaction is not the only factor affecting migratory navigation. Many other factors have been proposed, including magnetic navigation, use of polarized light and other visual cues, dead reckoning, and the presence of a spatial map (Quinn, 1991; Dittman and Quinn, 1996; Friedland et al., 2001). Although the ability to detect various

A   B   C   D

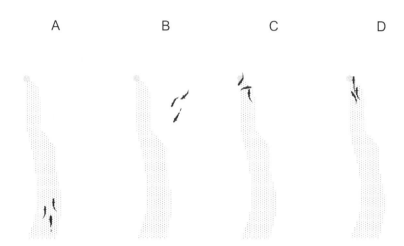

Fig. 3.3 Route selection based on simple "traffic rules", (river flow from top to bottom): A. In presence of home river odor (stippled area), swim upstream; B. If odor cue is absent, turn around and go back downstream until odor is detected again. C. Increase rate of turning until source (gray circle) is identified. D. Cease migrating and wait for appropriate conditions for spawning. Alternatively, the odor cue or source could be thought of as appropriate spawning habitat conditions, flow conditions, etc.

cues has been demonstrated, the ability to orient to flows, or rheotaxis, consistently emerges as a dominant factor during freshwater migrations.

Rheotaxis itself is a result of the coupling of at least two sensory systems: the mechanosensory lateral line and the visual systems. The lateral line system allows fish to detect and respond to changes in water velocity in their immediate vicinity (Coombs and Montgomery, 1999; Mogdans and Bleckmann, 2001; Mogdans *et al.*, 2003). Using this, they are able to detect velocity gradients and direction, and some rheotaxis is possible even among blinded individuals (Gray, 1937). Recent advances in our understanding of the mechanics of the lateral line system have generated considerable enthusiasm, including the idea that velocity gradients and turbulence alone may provide sufficient orienting cues for guiding fish at dams (Coutant, 2001; Goodwin *et al.*, 2006). This, however, ignores the well-established role of vision in rheotaxis; indeed, until recently vision was believed to be the only sensory modality driving rheotaxis (Lyon, 1904; Gray, 1937; Arnold, 1974), and although the lateral line has now been demonstrated to be sufficient for some rheotactic behavior, a major role for vision in is still generally recognized (Montgomery *et al.*, 1997; Liao, 2006).

The importance of rheotaxis in upstream fish passage cannot be overstated. The success of any upstream fish passage device depends on three groups of behaviors: guidance to the fishway entrance; entry rate, or attraction into the fishway; and passage rate through the fishway once entered.

None of these components is sufficient on its own, and measures of passage efficiency must include all three. Moreover, each of these three components is mediated by a suite of behaviors interacting with the physical characteristics of the site.

Optimizing guidance of fish and entry rates into a fishway can be challenging, especially at hydroelectric facilities. This is not surprising when one considers that the volume of competing flow from the turbines typically greatly exceeds the flow issuing from fishways (Fig. 3.4). Fishway engineers use two primary strategies to minimize this problem. The first strategy is to site the fishway entrance as close as possible to the source of the competing flow (Powers *et al.*, 1985). The second strategy is to supplement the flow at the fishway entrance. This increases the ratio of flow at the fishway entrance to turbine flow without increasing the total flow passed through fishway itself. In combination, these strategies are supposed to allow the fish to, (a) approach the site by orienting to the main flow of the river (b) encounter the hydraulic signature of the supplemented fishway entrance flow from some distance downstream, and follow this signature to the fishway entrance (behavioral, or rheotactic guidance) and (c) readily enter the fishway structure once it is encountered. Other strategies for optimizing guidance and entry rates might include modifying characteristics (depth, width, velocity) of the fishway entrance, and placing the entrance along a wall or similar boundary, which will improve the ability of fish to orient using visual cues.

In the absence of philopatry, flow direction and velocity may serve as a sufficient and appropriate stimulus for upstream migration. Other factors may still play important roles, however, depending on the context of the migratory behavior. For example, olfaction may still play an important role if it helps fish to identify good quality habitat upstream, either for spawning, feeding, or other activities.

Olfaction is certainly implicated in the identification and attraction of conspecifics. This fact has been put to use in the control of the invasive sea lamprey (*Petromyzon marinus*) in the Laurentian Great Lakes. With the opening of navigation routes, sea lamprey were able to colonize new habitats and prey upon/parasitize the native fish fauna. The effects of this invasion have been particularly damaging to lake trout (*Salvelinus namaycush*) populations, and ongoing efforts to control lamprey include extensive application of lampricide poison and construction of migratory barriers to exclude lamprey from spawning habitat (Farmer and Beamish, 1973; Hunn and Youngs, 1980; Porto *et al.*, 1999; Madenjian *et al.*, 2002). Both of these control methods have drawbacks, both biologically and politically, and so there has been considerable interest in alternative methods of control. In a suite of papers, Li *et al.* (2003a, b; Li, 2005) showed that male lamprey produce a pheromone that potential mates follow upstream, even at very low concentrations. This fact has been

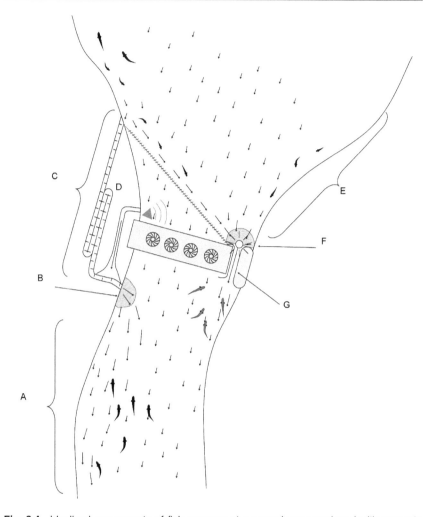

Fig. 3.4 Idealized components of fish passage at a powerhouse equipped with separate up- and downstream fishways. Length and direction of arrows indicate velocity and direction of current. Upstream migrants are guided to the right bank of the river (facing downstream; 'Guidance Zone', A), orienting to hydraulic cues from the fishway (C), supplemental attraction flow (D), and stream morphology. Once they are able to detect the fishway entrance ('Detection Zone', B), they must be attracted to enter the structure (entry rate, time-dependent). Once in the upstream fishway ('Passage Zone', C), passage is a function of capacity, behavior, and motivation; performance can be measured as proportion passing per unit time. Downstream migrants are likewise guided to the fishway entrance using hydraulic cues set up using a louver array ('Guidance Zone', E—this could also represent a screen or other mechanical exclusion). Fish that pass through the screen are behaviorally deflected from the turbines using sound, bubbles, etc. Once in the Detection Zone (F), entry rates are enhanced with a light (but for some species this can be a deterrent!) and an acceleration reducer (bell-mouth entrance). Fish are swept downstream (possibly into schools of waiting predators) through a downstream fishway (G).

used to design effective traps to remove adult lamprey from spawning streams without resorting to the less desirable control methods.

Motivating factors may be intrinsic, such as the extent of development of gonads and gametes, or extrinsic, such as habitat quality or population density of conspecifics. Because of the diversity of contexts that occur, it is unreasonable to expect simple rules to predict migratory motivation. Instead, the life-history of each organism is likely to influence behavior. Thus migratory movements, whether up- or downstream, might be considered the outcome of priming mechanisms and releasing/inhibiting cues (Fig. 3.5). Priming mechanisms include the intrinsic hormonal and developmental processes that occur, linked perhaps to photoperiod, temperature, and nutrition. Releasing or inhibiting cues may include discharge, temperature (or perhaps change in temperature) flow velocity, turbulence, and air entrainment, odor, behavior of conspecifics, and outside disturbances. A fish migrating up a river continually processes these various cues, with motivation being stimulated or inhibited, depending on inputs.

Much of the preceding paragraph is conjecture; good data on proximate behavioral cues are rare, and our understanding of underlying mechanisms is very limited. Fish passage technology would benefit from research improving our understanding of priming, releasing, and inhibiting mechanisms, both intrinsic and extrinsic. In the absence of data, careful consideration of life-history traits might help managers predict passage performance, but such predictions should always be viewed as tenuous at best.

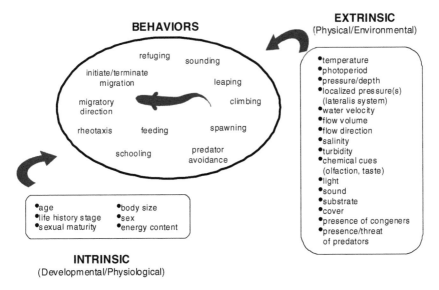

Fig. 3.5 Schematic of intrinsic and extrinsic factors affecting fish passage behavior.

Improved understanding of orienting and rheotactic cues would also be helpful in identifying and avoiding disorienting hydraulic conditions. Although difficult to prove as a phenomenon, some researchers have reported seeing fish swimming or holding in currents when it seems they should be able to find and use obvious cues to locate the upstream passage (Hinch *et al.* 1996; Haro and Kynard, 1997). Are these fish indeed actively migrating but trapped by confusing currents, or could it be that they are selecting one condition as a means of avoiding conditions elsewhere? Without a clearer understanding of cues and responses, questions like these will remain enigmatic.

Some behaviors that are adpative for ascending obstacles in nature can be problematic in the context of technical fishways. For example, Stuart (1962, 1964) described hydraulic conditions that triggered leaping behavior among salmonids. Specifically, leaping appears to be an involuntary response to standing waves below falls. This was further developed by Powers *et al.* (1985), who saw leaping as the most obvious, cost-effective and energetically-efficient way to pass salmon above dams. While this may be true in many situations, it turns out to be more problematic in technical fishways, especially at high dams. The primary problem with leaping behavior of salmonids is that it tends to be poorly oriented (Lauritzen *et al.*, 2005). Thus salmon stimulated to leap up fishways often either strike baffles and other fishway components, or else leap out of the structure altogether. Fishways on the Columbia River, where this problem was first documented, are now configured to discourage leaping, preferring instead for salmon to swim through submerged orifices or slots (Jonas *et al.*, 2004).

Despite the intriguing possibilities stimulated by Stuart's work, relatively few studies exist that relate swimming behaviors to hydraulic phenomena. As flow and kinematic visualization techniques improve, this line of research may hold some interesting solutions. For example, Liao *et al.* (2003a, b) found that trout were able to actively synchronize their swimming kinematics to vortices shed by a cylinder, seemingly realizing significant energetic savings in the process. Others have documented sheltering behavior, where fish take advantage of low-velocity zones (Webb, 1993, 1998). There appear to be limits, either kinematic or behavioral, to the extent to which these strategies can be applied, however. Here again, more information on behavioral and kinematic responses to hydraulic conditions would be helpful.

In addition to the swimming behaviors described above, some fish have special adaptations that enable them to traverse otherwise limiting obstacles. Upon freshwater entry, juvenile eel (*Anguilla* spp.) are able to ascend nearly vertical wetted surfaces. Even larger eels are adept at ascending pegboards and bristle-brush matrices, which make passage provisions for these species relatively simple and cost-effective. Many other fish species are capable of similar climbing feats (Legault, 1988; Ryan, 1991; Baker and Boubee, 2006;

Blob *et al.*, 2006); combined with air breathing this adaptation can permit passage past seemingly intimidating obstacles, and has important implications for conservation of native species, but also for exclusion of invasives.

Downstream passage

Earlier we stated that the goal of the fish passage bioengineer is to develop structures and strategies for passing fish that make the obstacle effectively transparent to movement. This goal applies in both the up- and the downstream direction. Ideally, one might hope to construct a fishway that functioned equally well in both directions; in reality, though, separate structures are usually designed for each (Fig. 3.4). This is true primarily because of differing hydraulic requirements of up- and downstream passage. While upstream passage must address both attraction and passability, downstream passage is mostly just a matter of attraction: high flow velocities in downstream bypasses convey fish passively once they enter the structures. Also, in order to make upstream fishways as economical to operate as possible, size and discharge flowing into the upstream end of these structures are minimized.

As with upstream passage, guidance and attraction to downstream passage routes is largely a matter of providing adequate orienting cues, which may require more discharge than is called for in the typical upstream fishway design. But just what are the appropriate orienting cues for downstream passage, and once a bypass is encountered, what governs the decision to enter? As with upstream migration, hormonal changes appear to serve as priming mechanisms for downstream migration. Synchronized by photoperiod and modulated by both temperature and growth, these hormonal shifts prepare the smolts both behaviorally and physiologically for migration to the marine environment (Zydlewski and McCormick, 1999; Zydlewski *et al.*, 2005).

For at least some diadromous species, temperature, changing discharge, and other environmental cues appear to serve as a releasing mechanism for initiating downstream migration. It is important to recognize that this behavior is ephemeral; some of the same factors that serve as releasing mechanisms can act to reverse both the physiological and behavioral adaptations for seawater entry, including negative rheotaxis, or downstream movement. Other stressors, such as handling, injury, and even migratory delay, are also thought to interrupt or even terminate migrations (Leggett, 2004; Bernard *et al.*, 1999; McCormick *et al.*, 1999). Thus, passage rates through safe routes need to be maximized: to protect the fish from mechanical injury, certainly, but also to minimize delays, stressors, and other risks that may be associated with residency in forebays and reservoirs above dams. Loss of migratory adaptations, including behavior, needs to be considered among these risks.

Because use of downstream fishways is almost entirely a question of guidance and attraction (Fig. 3.4), it is the locomotory behavior of the fish, rather than their swimming ability, that is of greatest concern. Fortunately for some species (Ferguson *et al.*, 1998; Kynard *et al.*, 2003), part of the suite of behaviors associated with downstream migration is a preference for the upper water column. Because turbines tend to draw from deep in the water column, sufficient and appropriate orienting cues for these species can be easily created using a surface bypass structure, such as a log sluice or spillbay. Other species are not so fortunate, preferring instead to pass obstacles deep in the water column. Here, competing flows through the turbines make it much more difficult for fish to find and use safe passage routes. Worse still, many of these bathyphilic species migrate at large size and advanced age. This is significant, both because they are at far greater risk than small fish to turbine strike injury, and because mortality of these individuals has a greater effect on the population as a whole.

The principles governing exclusion from undesirable routes is essentially the converse of attraction: keep fish away from the intakes, and if they do approach them, make them unattractive and keep velocities low so as to reduce both the rate and likelihood of passage. One way to keep fish out of structures is to exclude them—either mechanically, using screens and similar physically impassible devices; or behaviorally, using illumination, bubbles, or sounds that fish are unwilling to pass (Fig. 3.4). Mechanical exclusion is problematic, in part because entrainment can still occur, and fish that become impinged on the intake screens rarely survive (Boubée *et al.*, 2002; Swanson *et al.*, 2005). This leaves behavioral exclusion via diversion and repulsion as the two most viable means of excluding fish from turbines and other industrial intakes.

Conveniently, accelerated passage through a favorable route also achieves the desired result of exclusion from unfavorable routes. This then, is the primary challenge of fishway engineers: to attract fish into the fishway while minimizing exposure to unfavorable routes. Done effectively, this approach will simultaneously minimize delays and hazards of the obstruction, and so increase its overall transparency.

Early recommendations for entrance conditions were similar to both upstream and downstream fishways: place the structures as near as possible to the dominant flow in the system. Every site is unique, of course, and this approach is not always successful. Often passage rates are reduced because fish are unwilling to enter structures once they have located them. Recognizing this, Kynard and O'Leary (1993) sought to bypass the voluntary decision to enter downstream fishways by administering electrical shocks to migrants that were holding just upstream of a bypass entrance. Momentarily immobilized, the fish were then passively swept downstream into the bypass. No studies were done on subsequent survival, but this solution is interesting because it involved deactivating a behavior that was reducing entry rate, rather than modifying the structure to improve attraction or using a mechanical conveyance (e.g., a crowding screen to force fish to enter the bypass).

Although mechanical exclusion has been used at some locations (McMichael *et al.*, 2004), it is often the solution of last resort. Recently, there has been movement toward a more holistic approach: alter the bypass entrance conditions to make them more attractive to the migrants. This may be done, for example, with lighting or with physical/hydraulic alterations. Louver arrays employing so-called 'sweeping flow' have been used to seemingly good effect, guiding fish past dangerous routes and toward desirable routes of passage (Kynard and Buerkett, 1997; Kynard and Horgan, 2001; Scruton *et al.*, 2002; Swanson *et al.*, 2004, 2005). Also, entry rates can be increased by reducing the rate of flow acceleration over the weir. Evidently, rapid accelerations associated with sharp-crested weirs are repulsive to some species of fish, and when these are modified by gradually accelerating flow using, say, a bell-mouth entrance (Robertson and Crowe, 1997), fish enter and pass downstream bypasses at increased rates (Haro *et al.*, 1998; Kemp *et al.*, 2005). This technology has enabled the Ice Harbor Dam on the Snake River to reduce the amount of flow passing over the spillways by 35%, realizing the same survival as through existing spillgates. Applied basin-wide, this technology may enable the hydrosystem to realize savings of $10's of millions per year (Noah Adams, USGS Columbia River Research Laboratory, *pers. comm.*).

Reliability of savings like these could be increased if the effectiveness of fishway entrances were quantified in ways that allow scientists and engineers to estimate time to locate and pass them. Exposure and passage time through alternate (undesirable) routes is also important, both for the perspective they offer regarding specific sites, but also because all these data together describe the conditions that are most conducive to expedited passage.

One intriguing development in downstream passage technology is the use of eulerian-lagrangian models (Goodwin *et al.*, 2006). These are essentially individual-based models that overlay a set of behavioral rules on a three-dimensional hydraulic model, generated using computational fluid dynamics (CFD). Existing models operate on the assumption that rheotaxis is primarily modulated through the lateral line, and that turbulence and shear are sufficient and appropriate orienting cues to guide and pass fish through downstream bypasses. Under some conditions these models appear to effectively predict passage route selection. At the moment they are not predicated on the location-entry rate models we describe here (owing, it seems, to lack of data), and other biological aspects, such as the role of vision and conspecific interactions are also lacking. Nevertheless, the approach holds promise and is a good example of applied behavioral bioengineering, with behavioral outcomes predicted from measurable physical parameters.

The application of individual-based modeling to fish passage scenarios holds considerable promise, both for understanding behavioral rules and for improved structure design. Field calibration, coupled with rigorous, theory-based experimentation, will help bioengineers assess and refine their assumptions regarding guidance and entrance efficiency. At the same time, this approach will help to further prioritize experimental and observational studies, ultimately with the goal of identifying and quantifying guiding and releasing cues that result in expedited passage.

CONCLUSION

Fish passage bioengineering is the process of developing technical solutions within the boundaries defined by site-specific hydraulic and physical characteristics on the one hand, and by biological capabilities on the other. Application of these biological aspects of fish passage to fishway design requires active integration of both physiological and behavioral components. Although conservative estimates of swimming performance are widely available, swimming capacity alone is a poor predictor of passage performance. Instead, passage performance is the product of locomotory behavior generally, including guidance, attraction, and ascent or descent through the fishway.

Upstream fishways must provide the necessary cues to motivate the fish to continue their ascent (and discourage descent) up complex hydraulic structures. This ascent should be as rapid as possible, while incurring minimal stress. If fish cannot complete the transit of the entire length of a fishway without reaching their physiological or behavioral limits, resting areas must be incorporated that allow recovery from fatigue, but without loss of motivation.

Downstream passage is governed by similar processes, but guidance and attraction dominate. Physiological determinants have more to do with the ability to locate and use downstream fishways than with swimming ability, and here again behavioral responses may be as important as physiological capacities.

The challenge of fish passage is therefore both physiological and behavioral, and neither component is sufficient on its own to produce consistent and acceptable fishway performance. To date, many of the basic data needed to relate behavioral responses associated with fish passage to physical stimuli are rudimentary or lacking. If we hope to reverse the trend of fragmentation of riverine habitats through improved fish passage, then there is a clear and urgent need to advance and integrate knowledge on the interactions between the physical environment, physiology, and behavior.

REFERENCES

Arnold, G.P. 1974. Rheotropism in fishes. *Biological Reviews*. 49: 515–576.

Bainbridge, R. 1958. The speed of swimming of fish as related to size and to the frequency and amplitude of the tail beat. *Journal of Experimental Biology* 35: 109–133.

Bainbridge, R. 1960. Speed and stamina in three fish. *Journal of Experimental Biology* 37: 129–153.

Baker, C.F. and J.A.T. Boubee. 2006. Upstream passage of inanga *Galaxias maculatus* and redfin bullies *Gobiomorphus huttoni* over artificial ramps. *Journal of Fish Biology* 69: 668–681.

Beach, M.H. 1984. *Fish pass design—criteria for the design and approval of fish passes and other structures to facilitate the passage of migratory fish in rivers*. Ministry of Agriculture, Fisheries, and Food; Directorate of Fisheries Research; Fisheries Research Technical Report. Lowestoft. pp. 1–46

Beamish, F.W.H. 1978. Swimming capacity. In: *Fish Physiology*, W.S. Hoar and D.J. Randall(Eds.). Academic Press, London. Vol. 7: Locomotion. pp. 101–187.

Bell, C.E. and B. Kynard. 1985. Mortality of adult American shad passing through a 17-megawatt Kaplan turbine at a low-head hydroelectric dam. *North American Journal of Fisheries Management* 5: 33–38.

Bell, M.C. 1991. *Fisheries Handbook of Engineering Requirements and Biological Criteria*. U.S. Army Corps of Engineers. Portland, OR.

Bernard, D.R., J.J. Hasbrouck and S.J. Fleischman. 1999. Handling-induced delay and downstream movement of adult chinook salmon in rivers. *Fisheries Research* 44: 37–46.

Blazka, P., M. Volf and M. Cepela. 1960. A new type of respirometer for the determination of the metabolism of fish in an active state. *Physiol.Bohemoslov* 9: 553–558.

Blob, R.W., R. Rai, M.L. Julius and H.L. Schoenfuss. 2006. Functional diversity in extreme environments: effects of locomotor style and substrate texture on the waterfall-climbing performance of Hawaiian gobiid fishes. *Journal of Zoology* 268: 315–324.

Bloch, H. 1999. The European Union Water Framework Directive, taking European water policy into the next millennium. *Water Science and Technology* 40: 67–71.

Boubee, J., B. Chisnall, E. Watene, E. Williams and D. Roper. 2002. Enhancement and management of eel fisheries affected by hydroelectric dams in New Zealand. In: *Biology, Management, and Protection of Catadromous Eels*, D.A. Dixon, (Ed.). *American Fisheries Society Symposium* 33, Bethesda, MD, pp. 191–205.

Brett, J.R. 1962. Some considerations in the study of respiratory metabolism in fish, particularly salmon. *Journal of the Fisheries Research Board of Canada* 19: 1025–1038.

Brett, J.R. 1964. The respiratory metabolism and swimming performance of young sockeye salmon. *Journal of the Fisheries Research Board of Canada* 21: 1183–1226.

Brett, J.R. 1965. The relations of size to the rate of oxygen consumption and sustained swimming speeds of sockeye salmon (*Oncorhynchus nerka*). *Journal of the Fisheries Research Board of Canada* 22: 1491–1501.

Brett, J.R. 1967. Swimming performance of sockeye salmon in relation to fatigue time and temperature. *Journal of the Fisheries Research Board of Canada* 24: 1731–1741.

Brett, J.R. 1973. Energy Expenditure of Sockeye Salmon, *Oncorhynchus nerka*, During Sustained Performance. *Journal of the Fisheries Research Board of Canada* 30: 1799–1809.

Brett, J.R., M. Hollands and D.F. Alderdice. 1958. The effect of temperature on the cruising speed of young sockeye and coho salmon. *Journal of the Fisheries Research Board of Canada* 15: 587–605.

Bunt, C.M. 2001. Fishway entrance modifications enhance fish attraction. *Fisheries Management and Ecology* 8: 95–105.

Bunt, C.M., S.J. Cooke and R.S. McKinley. 2000. Assessment of the Dunnville fishway for passage of walleyes from Lake Erie to the Grand River, Ontario. *Journal of Great Lakes Research* 26: 482–488.

Bunt, C.M., C. Katopodis and R.S. McKinley. 1999. Attraction and Passage Efficiency of White Suckers and Smallmouth Bass by Two Denil Fishways. *North American Journal of Fisheries Management* 19: 793–803.

Bunt, C.M., B.T. van Poorten and L. Wong. 2001. Denil fishway utilization patterns and passage of several warmwater species relative to seasonal, thermal and hydraulic dynamics. *Ecology of Freshwater Fish* 10: 212–219.

Castro-Santos, T. 2004. Quantifying the combined effects of attempt rate and swimming capacity on passage through velocity barriers. *Canadian Journal of Fisheries and Aquatic Sciences* 61: 1602–1615.

Castro-Santos, T. 2005. Optimal swim speeds for traversing velocity barriers: an analysis of volitional high-speed swimming behavior of migratory fishes. *Journal of Experimental Biology* 208: 421–432.

Castro-Santos, T. 2006. Modeling the effect of varying swim speeds on fish passage through velocity barriers. *Transactions of the American Fisheries Society* 135: 1230–1237.

Castro-Santos, T., A. Cotel and P.W. Webb. 2009. Fishway evaluations for better bioengineering — an integrative approach. Pp. 557–575 *in: Challenges for diadromous fishes in a dynamic global environment*, A.J. Haro, K.L. Smith, R.A. Rulifson, C.M. Moffit, R.J. Klauda, M.J. Dadswell, R.A. Cunjak, J.E. Cooper, K.L. Beal, and T.S. Avery (Eds.). American Fisheries Society Symposium 69, Bethesda, MD.

Castro-Santos, T. and A. Haro. 2006. Biomechanics and fisheries conservation. In: *Fish Physiology* R.E. Shadwick and G.V. Lauder, (Eds.). Academic Press, New York, Vol. 23: Fish Biomechanics. pp. 469–523.

Castro-Santos, T., A. Haro and S. Walk. 1996. A passive integrated transponder (PIT) tagging system for monitoring fishways. *Fisheries Research* 28: 253–261.

Clay, C.H. 1995. *Design of Fishways and Other Fish Facilities*, 2nd Edition. Lewis Publishers. Boca Raton.

Collins, G.B. and C.H. Elling. 1960. *Fishway Research at the Fisheries-Engineering Research Laboratory*. U.S. Fish and Wildlife Service Circular 98. Washington D.C.

Coombs, S. and J.C. Montgomery. 1999. The enigmatic lateral line system. In: *Comparative Hearing: Fish and Amphibians*, R.R. Fay and A.N. Popper (Eds.). Springer Verlag, New York, pp. 319–361.

Coutant, C.C. 2001. Turbulent attraction flows for guiding juvenile salmonids at dams. *American Fisheries Society Symposium* 26: 57–77.

Dauble, D.D. and R.P. Mueller. 2000. Difficulties in estimating survival for adult chinook salmon in the Columbia and Snake rivers. *Fisheries* 25: 24–34.

Day, F. 1887. *British and Irish salmonidae*. Williams and Norgate, London.

Denil, G. 1909. Les eschelles a poissons et leru application aux barrages des Meuse et d'Ourthe. *Annales des travaux publics de Belgique* 2: 1–152.

Denil, G. 1937. La mecanique du poisson de riviere: les capacites mecaniques de la truite et du saumon. *Annales des travaux publics de Belgique* 38: 412–423.

Dittman, A. and T. Quinn. 1996. Homing in pacific salmon: mechanisms and ecological basis. *Journal of Experimental Biology* 199: 83–91.

Dow, R.L. 1962. Swimming speed of river herring *Pomolobus pseudoharengus* (Wilson). *Journal du Conseil, Conseil Permanent International Pour L'Exploration de la Mer* 27: 77–80.

Elvira, B., G.G. Nicola and A. Almodovar. 1998. A Catalogue of Fish Passes at Dams in Spain. In: *Fish Migration and Fish Bypasses*, M. Jungwirth, S. Schmutz and S. Weiss (Eds.). Fishing News Books, Cambridge, pp. 203–207.

European Commission. 2000. *Directive 2000/60/EC of the European Pariliament and of the council of 23 October 2000 establishing a framework for Community action in the field of water policy*. Official Journal of the European Community. 43 (L327): 1–72.

Farmer, G.J. and F.W.H. Beamish. 1973. Sea lamprey (*Petromyzon marinus*) predation on freshwater teleosts. *Journal of the Fisheries Research Board of Canada* 30: 601–605.

Ferguson, J.M., T.P. Poe and T.J. Carlson. 1998. Surface-oriented bypass systems for juvenile salmonids on the Columbia River, USA. In: *Fish Migration and Fish Bypasses*. M. Jungwirth, S. Schmutz, and S. Weiss (Eds.). Fishing News Books, Cambridge, pp. 281–299.

Freeman, M.C., C.M. Pringle, E.A.Greathouse and B.J. Freeman. 2003. Ecosystem-level consequences of migratory faunal depletion caused by dams. *Biodiversity, Status, and Conservation of the World's Shads* 35: 255–266.

Friedland, K.D., R.V. Walker, N.D. Davis, K.W. Myers, G.W. Boehlert, S. Urawa and Y. Ueno. 2001. Open-ocean orientation and return migration routes of chum salmon based on temperature data from data storage tags. *Marine Ecology-Progress Series* 216: 235–252.

Fry, F.E.J. and J.S. Hart. 1948. Crusing speed of goldfish in relation to temperature. *Journal of the Fisheries Research Board of Canada* 7: 169–175.

Gero, D.R. 1952. The hydrodynamic aspects of fish propulsion. *American Museum Novitates* 1601: 1–32.

Ghalambor, C.K., J.A. Walker and D.N. Reznick. 2003. Multi-trait selection, adaptation, and constraints on the evolution of burst swimming performance. *Integrative and Comparative Biology* 43: 431–438.

Goodwin, R.A., J.M. Nestler, J.J. Anderson, L.J. Weber and D.P. Loucks. 2006. Forecasting 3-D fish movement behavior using a Eulerian-Lagrangian-agent method (ELAM). *Ecological Modelling* 192: 197–223.

Gray, J. 1937. Pseudo-rheotropism in fishes. *Journal of Experimental Biology* 14: 95–103.

Griffiths, D. 2006. Pattern and process in the ecological biogeography of European freshwater fish. *Journal of Animal Ecology* 75: 734–751.

Haag, W.R. and M.L. Warren. 1998. Role of ecological factors and reproductive strategies in structuring freshwater mussel communities. *Canadian Journal of Fisheries and Aquatic Sciences* 55: 297–306.

Hammer, C. 1995. Fatigue and exercise tests with fish. *Comparative Biochemistry and Physiology* A112: 1–20.

Harden-Jones, F.R. 1963. The reaction of fish to moving backgrounds. *Journal of Experimental Biology* 40: 437–446.

Haro, A., T. Castro-Santos, J. Noreika and M. Odeh. 2004. Swimming performance of upstream migrant fishes in open-channel flow: a new approach to predicting passage through velocity barriers. *Canadian Journal of Fisheries and Aquatic Sciences* 61: 1590–1601.

Haro, A. and B. Kynard. 1997. Video evaluation of passage efficiency of American shad and sea lamprey in a modified Ice Harbor fishway. *North American Journal of Fisheries Management* 17: 981–987.

Haro, A., M. Odeh, J. Noreika and T. Castro-Santos. 1998. Effect of water acceleration on downstream migratory behavior and passage of Atlantic salmon smolts and juvenile American shad at surface bypasses. *Transactions of the American Fisheries Society* 127: 118–127.

Hasler, A.D. 1971. Orientation and fish migration. In: *Environmental Relations and Behavior*, W.S. Hoar and D.J. Randall(Eds.). Academic Press, New York, pp. 429–510.

Hinch, S.G., R.E. Diewert, T.J. Lissimore, A.M.J. Prince, M.C. Healey and M.A. Henderson. 1996. Use of electromyogram telemetry to assess difficult passage areas for river-migrating adult sockeye salmon. *Transactions of the American Fisheries Society* 125: 253–260.

Howe, J. and I. White. 2002. The potential implications of the European Union Water Framework Directive on domestic planning systems: A UK case study. *European Planning Studies* 10: 1027–1038.

Hunn, J.B. and W.D. Youngs. 1980. Role of physical barriers in the control of sea lamprey (*Petromyzon marinus*). *Canadian Journal of Fisheries and Aquatic Sciences* 37: 2118–2122.

Hunter, J.H. and J.R. Zweifel. 1971. Swimming speed, tail beat frequency, tial beat amplitude, and size in jack mackerel, *Trachurus symmetricus*, and other fishes. *Fishery Bulletin* 69: 253–266.

Jain, K.E., J.C. Hamilton and A.P. Farrell. 1997. Use of a ramp velocity test to measure critical swimming speed in rainbow trout (*Oncorhynchus mykiss*). *Comparative Biochemistry and Physiology* 117A: 441–444.

Jonas, M.R., J.T. Dalen, S.T. Jones and P.L. Madson. 2004. *Evaluation of the John Day Dam south fish ladder modification 2003*. U.S. Army Corps of Engineers CENWP-OP-SRF. Cascade Locks, Oregon.

Kelso, B.W., T.G. Northcote and C.F. Wehrhahn. 1981. Genetic and environmental aspects of the response to water current by rainbow-trout (*Salmo gairdneri*) originating from inlet and outlet streams of 2 lakes. *Canadian Journal of Zoology* 59: 2177–2185.

Kemp, P.S., M.H. Gessel and J.G. Williams. 2005. Fine-scale behavioral responses of Pacific salmonid smolts as they encounter divergence and acceleration of flow. *Transactions of the American Fisheries Society* 134: 390–398.

Kynard, B. and C. Buerkett. 1997. Passage and behavior of adult American shad in an experimental Louver bypass system. *North American Journal of Fisheries Management* 17: 734–742.

Kynard, B. and M. Horgan. 2001. Guidance of yearling shortnose and pallid sturgeon using vertical bar rack and louver arrays. *North American Journal of Fisheries Management* 21: 561–570.

Kynard, B. and J. O'Leary. 1993. Evaluation of a bypass system for spent American shad at Holyoke Dam, Massachusetts. *North American Journal of Fisheries Management* 13: 782–789.

Kynard, B., M. Horgan and E. Theiss. 2003. Spatial distribution and jumping of juvenile shads in the Connecticut River, Massachusetts during seaward migration. *Jornal of Ichthyology* 43 (Supplement 2): S228–S236.

Larinier, M. and F. Travade. 2002. The design of fishways for shad. *Bulletin Francais de la Peche et de la Pisciculture* 364 (Supplement): 135–146.

Lauritzen, D.V., F. Hertel and M.S. Gordon. 2005. A kinematic examination of wild sockeye salmon jumping up natural waterfalls. *Journal of Fish Biology* 67: 1010–1020.

Legault, A. 1988. The dam clearing of eel by climbing study in Sevre Niortaise. *Bulletin Francais de la Peche et de la Pisciculture* 308: 1–10.

Leggett, W.C. 2004. The American shad, with special reference to its migration and population dynamics in the Connecticut River. In: *The Connecticut River Ecological Study (1965–1973) Revisited: Ecology of the Lower Connecticut River 1973–2003*, P.M. Jacobson, D.A. Dixon, W.C. Leggett, B.C. Marcy, Jr. and R.R. Massengill (Eds.). American Fisheries Society Monograph 9, Bethesda, MD, pp. 181–238.

Letcher, B.H., K.H. Nislow, J.A. Coombs, M.J. O'Donnell and T.L. Dubreuil. 2007. Population response to habitat fragmentation in a stream-dwelling brook trout population. *PLoS ONE* 2(e1139): 1–11.

Li, W.M. 2005. Potential multiple functions of a male sea lamprey pheromone. *Chemical Senses* 30: i307–i308.

Li, W.M., A.P. Scott, M.J. Siefkes, S.S. Yun and B. Zielinski. 2003a. A male pheromone in the sea lamprey (*Petromyzon marinus*): an overview. *Fish Physiology and Biochemistry* 28: 259–262.

Li, W.M., M.J. Siefkes, A.P. Scott and J.H. Teeter. 2003b. Sex pheromone communication in the sea lamprey: Implications for integrated management. *Journal of Great Lakes Research* 29: 85–94.

Liao, J.C. 2006. The role of the lateral line and vision on body kinematics and hydrodynamic preference of rainbow trout in turbulent flow. *Journal of Experimental Biology* 209: 4077–4090.

Liao, J.C., D.N. Beal, G.V. Lauder and M.S. Triantafyllou. 2003a. Fish exploiting vortices decrease muscle activity. *Science* 302: 1566–1569.

Liao, J.C., D.N. Beal, G.V. Lauder and M.S. Triantafyllou. 2003b. The Karman gait: novel body kinematics of rainbow trout swimming in a vortex street. *Journal of Experimental Biology* 206: 1059–1073.

Lyon, E.P. 1904. On rheotropism. I. Rheotropism in fishes. *American Journal of Physiology* 12: 149–161.

Madenjian, C.P., G.L. Fahnenstiel, T.H. Johengen, T.F. Nalepa, H.A. Vanderploeg, G.W. Fleischer, P.J. Schneeberger, D.M. Benjamin, E.B. Smith, J.R. Bence, E.S. Rutherford, D. S. Lavis, D.M. Robertson, D.J. Jude and M.P. Ebener. 2002. Dynamics of the Lake Michigan food web, 1970-2000. *Canadian Journal of Fisheries and Aquatic Sciences* 59: 736–753.

McCleave, J.D. 2001. Simulation of the impact of dams and fishing weirs on reproductive potential of silver-phase American eels in the Kennebec River basin, Maine. *North American Journal of Fisheries Management* 21: 592–605.

McCormick, S.D., R.A. Cunjak, B. Dempson, M.F. O'Dea and J.B. Carey. 1999. Temperature-related loss of smolt characteristics in Atlantic salmon (*Salmo salar*) in the wild. *Canadian Journal of Fisheries and Aquatic Sciences* 56: 1649–1658.

McMichael, G.A., J.A. Vucelick, C.S. Abernethy and D.A. Neitzel. 2004. Comparing fish screen performance to physical design criteria. *Fisheries* 29: 10–16.

Mogdans, J. and H. Bleckmann. 2001. The mechanosensory lateral line of jawed fishes. In: *Sensory Biology of Jawed Fishes: New Insights*, B.G. Kapoor and T.J. Hara (Eds.). Science Publishers, Inc., Enfield(NH),USA, pp. 181–213.

Mogdans, J., J. Engelmann, W. Hanke and S. Krother. 2003. The fish lateral line: How to detect hydrodynamic stimuli. In: *Sensors and Sensing in Biology and Engineering*, F.G. Barth, G. Friedrich, J.A.C. Humphrey, and T.W. Secomb, (Eds.). Springer Verlag, New York, pp. 173–185.

Montgomery, J.C., C.F. Baker and A.G. Carton. 1997. The lateral line can mediate rheotaxis in fish. *Nature* (London)389: 960–963.

Moore, J.W. and D.E. Schindler. 2004. Nutrient export from freshwater ecosystems by anadromous sockeye salmon (*Oncorhynchus nerka*). *Canadian Journal of Fisheries and Aquatic Sciences* 61: 1582–1589.

Morita, K. and S. Yamamoto. 2002. Effects of habitat fragmentation by damming on the persistence of stream-dwelling charr populations. *Conservation Biology* 16: 1318–1323.

Morita, K. and A. Yokota. 2002. Population viability of stream-resident salmonids after habitat fragmentation: a case study with white-spotted charr (*Salvelinus leucomaenis*) by an individual based model. *Ecological Modelling* 155: 85–94.

Muir, W.D., S.G. Smith, J.G. Williams and B.P. Sandford. 2001. Survival of juvenile salmonids passing through bypass systems, turbines, and spillways with and without flow deflectors at Snake River dams. *North American Journal of Fisheries Management* 21: 135–146.

Naiman, R.J., R.E. Bilby, D.E. Schindler and J.M. Helfield. 2002. Pacific salmon, nutrients, and the dynamics of freshwater and riparian ecosystems. *Ecosystems* 5: 399–417.

Naughton, G.P., C.C. Caudill, M.L. Keefer, T.C. Bjornn, L.C. Stuehrenberg and C.A. Peery. 2005. Late-season mortality during migration of radio-tagged adult sockeye salmon (*Oncorhynchus nerka*) in the Columbia River. *Canadian Journal of Fisheries and Aquatic Sciences* 62: 30–47.

Nislow, K., J. Armstrong and S. McKelvey. 2004. Phosphorus flux due to Atlantic salmon (*Salmo salar*) in an oligotrophic upland stream: effects of management and demography. *Canadian Journal of Fisheries and Aquatic Sciences* 61: 2401–2410.

Northcote, T.G., S.N. Williscr and H. Tsuyuki. 1970. Meristic and lactate dehydrogenase genotype differences in stream populations of rainbow trout below and above a waterfall. *Journal of the Fisheries Research Board of Canada* 27: 1987.

Orsborn, J.F. 1987. Fishways—historical assessment of design practices. *American Fisheries Society Symposium* 1: 122–130.

Peake, S. 2004. An evaluation of the use of critical swimming speed for determination of culvert water velocity criteria for smallmouth bass. *Transactions of the American Fisheries Society* 133: 1472–1479.

Peake, S.J. and A.P. Farrell. 2005. Postexercise physiology and repeat performance behaviour of free-swimming smallmouth bass in an experimental raceway. *Physiological and Biochemical Zoology* 78: 801–807.

Plaut, I. 2001. Critical swimming speed: its ecological relevance. *Comparative Biochemistry and Physiology.* AB131: 41–50.

Porto, L.M., R.L. McLaughlin and D.L.G. Noakes. 1999. Low-head barrier dams restrict the movements of fishes in two Lake Ontario streams. *North American Journal of Fisheries Management* 19: 1028–1036.

Powers, P.D., J.F. Orsborn, T.W. Bumstead, S. Klinger-Kingsley and W.C. Mih. 1985. *Fishways—An assessment of their development and design.* Bonneville Power Administration, US Department of Energy.

Quinn, T.P. 1991. Models of Pacific salmon orientation and navigation on the open ocean. *Journal of Theoretical Biology* 150: 539–545, 1991.

Rand, P.S. and S.G. Hinch. 1998. Swim speeds and energy use of upriver-migrating sockeye salmon (*Oncorhynchus nerka*)—simulating metabolic power and assessing risk of energy depletion. *Canadian Journal of Fisheries and Aquatic Sciences* 55: 1832–1841.

Rashleigh, B. and D.L. DeAngelis. 2007. Conditions for coexistence of freshwater mussel species via partitioning of fish host resources. *Ecological Modelling* 201: 171–178.

Richter, B.D., D.P. Braun, M.A. Mendelson and L.L. Master. 1997. Threats to imperiled freshwater fauna. *Conservation Biology* 11: 1081–1093.

Rideout, S.G., L.M. Thorpe and L.M. Cameron. 1985. Passage of American shad in an Ice Harbor style fishladder after flow pattern modifications. In: *Procedings of the Symposium on Small Hydropower and Fisheries*, F.W. Olson, R.G. White, and R.H. Hamre (Eds.). American Fisheries Society, Bethesda, MD, pp. 251–256.

Rivinoja, P., S. McKinnell and H. Lundqvist. 2001. Hindrances to upstream migration of Atlantic salmon (*Salmo salar*) in a northern Swedish river caused by a hydroelectric power-station. *Regulated Rivers-Research and Management* 17: 101–115.

Robertson, J.A. and C.T. Crowe. 1997. *Engineering Fluid Mechanics.* John Wiley & Sons, New York. 6th Edition.

Ryan, P.A. 1991. The success of the Gobiidae in tropical pacific insular streams. *New Zealand Journal of Zoology* 18: 25–30.

Scheuerell, M.D., P.S. Levin, R.W. Zabel, J.G. Williams and B.L. Sanderson. 2005. A new perspective on the importance of marine-derived nutrients to threatened stocks of Pacific salmon (*Oncorhynchus* spp.). *Canadian Journal of Fisheries and Aquatic Sciences* 62: 961–964.

Scruton, D.A., R.S. McKinley, N. Kouwen, W. Eddy and R.K. Booth. 2002. Use of telemetry and hydraulic modeling to evaluate and improve fish guidance efficiency at a louver and bypass system for downstream-migrating Atlantic salmon (*Salmo salar*) smolts and kelts. *Hydrobiologia* 483: 83–94.

Skalski, J.R., R. Townsend, J. Lady, A.E. Giorgi, J.R. Stevenson and R.D. McDonald. 2002. Estimating route-specific passage and survival probabilities at a hydroelectric project from smolt radiotelemetry studies. *Canadian Journal of Fisheries and Aquatic Sciences* 59: 1385–1393.

Stringham, E. 1924. The maximum speed of fresh-water fishes. *American Naturalist* 18: 156–161.

Stuart, T.A. 1962. *The leaping behavior of salmon and trout at falls and obstructions.* Department of Agriculture and Fisheries for Scotland, Freshwater and Salmon Fisheries Research, His Majesty's Stationery Office. Edinburgh.

Stuart, T.A. 1964. Biophysical aspects of leaping behaviour in salmon and trout. *Annals of Applied Biology* 53: 503–505.

Sullivan, T.J. 2004. *Evaluation of the Turners Falls fishway complex and potential improvements for passing adult American shad.* M.S. Thesis. University of Massachusetts, Amherst.

Swanson, C., P.S. Young and J.J. Cech. 2004. Swimming in two-vector flows: Performance and behavior of juvenile Chinook salmon near a simulated screened water diversion. *Transactions of the American Fisheries Society* 133: 265–278.

Swanson, C., P.S. Young and J.J. Cech. 2005. Close encounters with a fish screen: Integrating physiological and behavioral results to protect endangered species in exploited ecosystems. *Transactions of the American Fisheries Society* 134: 1111–1123.

Tudorache, C., P. Viaenen, R. Blust and G. De Boeck. 2007. Longer flumes increase critical swimming speeds by increasing burst and glide swimming duration in carp (*Cyprinus carpio*, L.). *Journal of Fish Biology* 71: 1630–1638.

U.S. Army Corps of Engineers. 2001. *U.S. National Inventory of Dams.* Alexandria, VA (USA).

Wardle, C.S. 1975. Limit of fish swimming speed. *Nature* (London) 255: 725–727.

Weaver, C.R. 1963. Influence of water velocity upon orientation and performance of adult migrating salmonids. *Fishery Bulletin* 63: 97–121.

Webb, P. W. 1993. The effect of solid and porous channel walls on steady swimming of steelhead trout *Oncorhynchus mykiss*. *Journal of Experimental Biology* 178: 97–108.

Webb, P.W. 1998. Entrainment by river chub *Nocomis micropogon* and smallmouth bass *Micropterus dolomieu* on cylinders. *Journal of Experimental Biology* 201: 2403–2412.

World Commision on Dams. 2000. *Dams and development: a new framework for decision-making.* Earthscan Publications, Ltd. London.

Zhou, Y. 1982. The swimming behaviour of fish in towed gears; a reexamination of the principles. *Working Paper, Deptartment of Agriculture and Fisheries, Scotland* 4: 1–55.

Zydlewski, G.B., A. Haro and S.D. McCormick. 2005. Evidence for cumulative temperature as an initiating and terminating factor in downstream migratory behavior of Atlantic salmon (*Salmo salar*) smolts. *Canadian Journal of Fisheries and Aquatic Sciences* 62: 68–78.

Zydlewski, G.B. and S.D. McCormick. 1999. The role of thyroid hormones in the transmission of environmental cues used for downstream migration of Atlantic salmon smolts. *American Zoologist* 39: A123.

4

Swimming Strategies for Energy Economy

Frank E. Fish

INTRODUCTION

The laws of physics are rigid and fixed. Animals, therefore, are restricted in their use of available energy in accordance to the constraints of mechanics and thermodynamics. Evolutionary success dictates that a large proportion of the available energy reserves be allocated to reproductive effort despite the demand by other energy consuming functions. Therefore, mechanisms that reduce energy costs for non-reproductive functions relative to total energy reserves have an adaptive benefit for individuals (Fausch, 1984). As the laws of physics are inflexible and the available energy limited, animals have found ways to exploit these laws for their own benefit.

Locomotion is an energy demanding activity. Swimming by fishes permits them to seek out new energy resources, but comes at a cost to transport the body mass over a distance. For movement in water, the kinetic energy is transferred from the movement of the body to the fluid medium, both to replace momentum losses associated with viscosity and induced energy losses associated with this replacement. This energy loss from the animal is due to the high density and viscosity of the water in conjunction with the shape and texture of the body surface, and the animal's propulsive movements (Webb,

Author's address: Department of Biology, West Chester University, West Chester, PA 19383 USA. E-mail: ffish@wcupa.edu

1988). The rate of energy loss, sometimes considered the "drag" power, P_D, for a body relative to a water flow can be determined according to the equation:

$$P_D = 0.5 \, \rho \, S \, C_D \, U^3 \qquad (1)$$

where ρ is the density of the water (freshwater, 1000 kg/m^3; salt water, 1025 kg/m^3), S is the wetted surface area of the body, U is the velocity, and C_D is the dimensionless drag coefficient (Webb, 1975; Vogel, 1994). While arguments of energy saving by maintaining laminar versus turbulent flow conditions in the boundary layer have occurred (Webb, 1975; Fish, 2006a), body shape has a greater influence in determination of the drag and energy consumption while moving through water. Drag power is minimized primarily by streamlining the shape of the body and the appendages. However, even with the most optimized of streamlined forms, advantageous geometry of the surface texture, and control of flow around the body, drag can only be minimized and never eliminated. The consequences for a fish are that energy must be allocated to overcome power loss, this being actively generated by the fish as "thrust" power. The propulsive movements of the swimming fish further draw upon available energy reserves. Propulsive movements of the body and appendages are expected to increase C_D by a factor of 3–5 (Lighthill, 1971).

How can swimming animals gain an energetic advantage when subjected to the unrelenting demands associated with the drag power? As P_D is dependent on the cube of U, the velocity becomes the important factor in determining the rate of energy expended in swimming. For some fishes, energetic gain can be maximized relative to locomotor costs by adopting a sit-and-wait or ambush predator strategy, reducing translocation to very short periods. In this strategy, energy cost=0 while U=0. By choosing among velocities of the flow, a swimming fish or a fish in a current can similarly reduce the energy cost of locomotion. Fishes use advantageous flow conditions as a behavioral strategy to reduce drag.

This review will examine behavioral strategies to minimize drag and energy expenditure. The primary thrust of most research on swimming fishes has been concerned with the interaction of morphology, muscle physiology, and hydrodynamics. The carefully executed experiments for these examinations constrained fish to swim against constant flows in linear flow tanks or annular swimming rings without interference with the solid walls of the test chamber (Webb, 1975). The results of these studies have benefited understanding of physiological mechanisms, but have been more limited in an ecological context. Steady rectilinear swimming may be more the exception than the rule. By using alternative means of swimming, fishes "cheat" in order to reduce costs due to swimming and maximize net energy resources. Fishes, therefore, use a variety of behavioral mechanisms to take advantage of local flow conditions.

Optimal speed and gaits

As the mechanical power due to drag of a fish increases with the cube of the velocity (equation 2), the metabolic cost, representing the power input, also increases curvilinearly with increasing swimming velocity according to the power function (i.e., $y = ax^b$) with an exponent $b>1$ (Brett, 1964, 1965; Webb, 1975; Videler and Nolet, 1990; Videler, 1993). The intersection of the curve, expressed by this power function, and the tangent that passes through the origin indicates the minimum cost of transport (CT) (Fig. 4.1; Tucker, 1975; Videler, 1993).

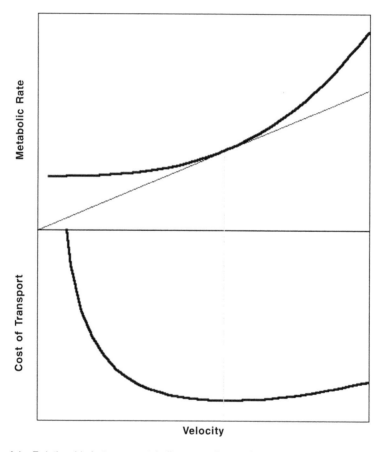

Velocity

Fig. 4.1   Relationship between metabolic rate and cost of transport (CT) of a swimming fish and velocity. The optimal swimming speed, where the greatest energy efficiency is realized, occurs at a velocity where the tangent line intersects with the curve of the metabolic rate. This corresponds to the velocity of the minimum cost of transport, which is indicated by the vertical line.

CT is defined as the metabolic energy required to transport a unit mass a unit distance. When plotted against velocity, CT displays a U-shaped curve (Brett, 1965; Videler and Nolet, 1990; Videler, 1993). At low swimming speeds, the metabolic cost is high relative to locomotor costs and the total costs per unit distance are high (Webb, 1975). At high swimming speeds, CT is high due to the disproportional rise in drag. The minimum CT corresponds to the velocity where the distance traveled for the energy available is maximized (i.e., maximum range).

The optimal swimming speed should coincide with the speed of minimum CT. Weihs (1973) predicted a theoretical optimal cruising speed of approximately one body length/s. An ultrasonic telemetry study on migrating sockeye salmon (*Oncorhynchus nerka*) showed that estimated ground speed corresponded to a mean 1.0 lengths/s in open waters (Quinn, 1988). The optimal speed predicted by Weihs (1973) was considered low when compared with the maximum endurance speeds of fishes. A review of CT based on metabolic studies showed a range of optimal swimming speeds of 0.7 to 5.8 lengths/s for a variety of fishes (Videler, 1993). Scombrids cruise at about 0.3 to 2.2 lengths/s (Magnuson, 1978). Dewar and Graham (1994a) found the optimal swimming speed of 2.0 lengths/s for yellowfin tuna (*Thunnus albacares*) measured in a water tunnel to be close to the mean swimming speed of yellowfin tuna tracked at sea (Holland *et al.*, 1990).

Changes in optimal swimming speed can be exploited by using different gaits. Changes in gait are considered to minimize energy costs when moving at a particular speed or increase stability (Hoyt and Taylor, 1981; Arreola and Westneat, 1996; Hove *et al.*, 2001; Alexander, 2003). Many fishes swim with various kinematics patterns of combinations of median and paired fins (Webb, 1994, 1998b). As defined by Alexander (1989), a gait is "a pattern of locomotion characteristic of a limited range of speeds described by quantities of which one or more change discontinuously at transitions to other gaits." While gaits are typically associated with terrestrial locomotion, the description of gaits is also applicable to swimming (Webb, 1994). The swimming gaits of fish are median-paired fin (MPF) gaits and body-caudal fin (BCF) gaits (Webb, 1998b). MPF gaits involve undulatory or oscillatory movements of the paired (e.g., pectoral) or median (e.g., dorsal, anal) fins. These fins can be used singly or in combination. MPF swimmers hold the body rigid. BCF gaits involve undulations of the body and caudal fin, or oscillations confined to the caudal fin (Breder, 1926; Webb, 1975, 1994, 1998b; Lindsey, 1978; Hale *et al.*, 2006). MPF gaits are generally associated with relatively slow swimming, particularly used for precise maneuvering and stability control. BCF gaits involve generally more rapid swimming for cruising and high-speed accelerations (e.g., burst swimming, fast-starts).

Gait transition with changing swimming speed is common in fish. As the power requirement for swimming increases with the cube of the speed (equation 1), a given gait that is powered with a restricted muscle mass can only generate sufficient power efficiently over a small range of speeds (Webb, 1994). Therefore, varied gaits are necessitated to swim over an extended performance range. The sequence of gait changes is MPF swimming, burst and coast BCF swimming (see below) powered by red muscle, steady BCF swimming with red muscle, burst and coast BCF swimming with white muscle, and steady BCF swimming with white muscle (Alexander, 1989; Webb, 1994; Hale *et al.*, 2006). For MPF swimmers, the pectoral fins are beat synchronously at low speeds and alternately at high speeds or vice versa (Hove *et al.*, 2001; Hale *et al.*, 2006).

MPF swimming is used in hovering and station holding (i.e., zero ground speed) (Webb, 1994, 1998b). Sustained low-speed swimming also employs MPF gaits. At higher swimming speeds, MPF gaits are abandon in favor of BCF swimming, with the exception of labriform fishes that use almost exclusively MPF swimming which is supplemented with BCF swimming (Drucker and Jensen, 1996a, b; Mussi *et al.*, 2002). The MPF gaits use a small mass of propulsive appendicular muscle to power the fins. In transition to BCF swimming, power is generated from the myotomal muscle mass. At low BCF swimming speeds, power is generated by red muscle, which comprises a small proportion of the entire myotomal muscle mass (Greer-Walker and Pull, 1975; Jayne and Lauder, 1995; Coughlin and Rome, 1996; Knower *et al.*, 1999). However with increasing speed, a greater proportion of the musculature is recruited in the form of pink and white muscle fibers (Rome *et al.*, 1990, 1993).

The energy cost of swimming is low for MPF swimming compared to BCF gaits (Korsmeyer *et al.*, 2002; Cannas *et al.*, 2006). CT decreases with increasing speed using MPF swimming until gait transition. Above the gait transition speed, CT increases with the shift to BCF swimming. The difference in energy cost between gaits is likely due to the relatively smaller muscle mass required to swim at low speeds and reduced drag due to a rigid body. Gait transition is necessary as the forces (e.g., drag, acceleration reaction, lift) involved in the production of thrust for each of the gaits is maximal for only a limited range of speeds (Vogel, 1994; Webb, 1994; Sfakiotakis *et al.*, 1999). It has been assumed that gait transition is associated with a change from aerobic to anaerobic swimming (Drucker and Jensen, 1996, Cannas *et al.*, 2006). Increases in speed require shifting to different muscle masses to take advantage of different optimum muscle shortening velocities (Drucker and Jensen, 1996).

Burst and Coast Swimming

The use of intermittent locomotion has been demonstrated to reduce the energy cost of locomotion. Burst and coast swimming is a two-phase periodic behavior (Weihs and Webb, 1983). Alternating bouts of active swimming movements (burst phase) with passive coasts or glides (coast phase) conserves on the energy by reducing the amount of muscular effort over a prolonged time period (Fig. 4.2). The burst phase accelerates the body forward, while in the coast phase, the body decelerates. Burst and coast swimming is a behavioral strategy that exploits the lower drag of a rigid, non-flexing animal during the coast compared to when it is actively swimming (Weihs and Webb, 1983). Many fish swim using alternating bursts and coasts. Fish that use this behavior have a body morphology within a range of fineness ratio (body length/maximum girth) of 4.0 and 6.5 (Blake, 1983). The hydrodynamic drag of a flexing fish body is three to five times higher than when the fish is straight and coasting at the same speed (Lighthill, 1971; Webb, 1975). Therefore, the advantage of this swimming strategy would be realized by BCF swimmers using subcarangiform and carangiform swimming. As MPF swimmers such as labriform fishes (e.g., wrasse) and to a lesser extent thick-bodied thunniform swimmers (e.g., tuna) already maintain a rigid body, little advantage would be gained by adopting burst and coast swimming.

Weihs (1974) developed a theoretical model that estimates that efficiency on the initial and final speeds and the difference between the drags associated with active swimming and coasting. The optimal condition

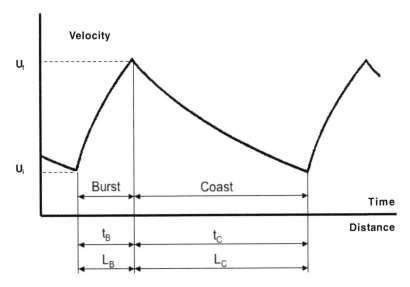

Fig. 4.2 Burst and coast swimming. During the burst phase, there is a rapid increase in velocity from a low initial velocity (U_i) to a final velocity (U_f). The burst is short in duration (t_B) and propels the fish a relatively short distance (L_B). During the coast phase, the fish decelerates over a long time (t_C)and is transported a relatively long distance (L_C). Figure based on Weihs (1974).

Water Surface

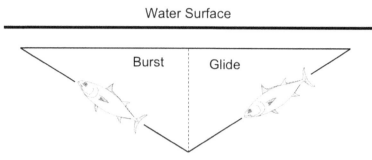

Fig. 4.3 Hypothetical path of negatively buoyant fish during burst and glide swimming. The fish sinks to a lower depth using a passive glide. During the burst phase, the fish actively swims to power itself to a higher depth. Figure based on Weihs (1973a).

was obtained at low average speeds, but for gadids, energetic advantage also accrues when using high-speed anaerobically powered sprints (Videler and Weihs, 1982). Weihs (1974) estimated that an energy savings of over 50% was possible by alternating accelerated motion with powerless glides. Burst-and-coast swimming becomes more economical as the animal's size or speed increases (Videler and Weihs, 1982). In addition, burst time is reduced by a fish trailing between and behind other fish in a school, resulting in a 29% energy savings compared to burst-and-coast swimming by solitary fish (Fish *et al.*, 1991).

A variant of the burst and coast behavior was applied to negatively buoyant fish as swim and glide (Weihs, 1973a, 1984). Energy is conserved by gliding downward with no propulsive motions and then regaining altitude by active swimming (Fig. 4.3; Weihs, 1973a; Weihs and Webb, 1983). Weihs (1973a) calculated an energy savings of 20% for a kawakawa using this technique to traverse the same horizontal distance in 7% more time. Energy savings for other tuna were estimated at 7% for skipjack tuna and 16% for albacore (Magnuson, 1978). Increased energy savings are possible if the angle between the plane of the surface and downward trajectory of the fish on the glide is kept small (Weihs, 1973a).

Schooling

Energy economy for swimming by fish has been suggested as a possible advantage to schooling. Increased energy savings would be advantageous by permitting faster swimming speeds during foraging or increasing range particularly for migration. The pattern of the school is hypothesized to be dependent on the undulatory movements of the fish, which produce a series of counter-rotating vortices in its wake as a thrust-type vortex street (reverse Kármán vortex street) (Belyayev and Zuyev, 1969; Weihs, 1972a, 1973b).

The vorticity in thrust production is necessary to transport momentum from the propulsor into the fluid. Each vortex is formed from the reversal in direction of the propulsor, requiring a reversal in circulation of the bound vortex. At the end of each half-stroke, the bound vortex is shed from the propulsor and as the propulsor is reaccelerated in the opposite direction a new bound vortex is formed with a reversed circulation (Vogel, 1994). The shed vortex is known as the "stopping vortex." The oscillating motion of the propulsor thus produces two parallel trails of staggered vortices perpendicular to the plane of oscillation and with opposite circulations. The direction of the circulation is oriented so that the tangential velocity is parallel to the trails and directed posteriorly between the trails and anteriorly on the outside of the trails.

The thrust-type vortex street is a two-dimensional representation of the flow field shed from an oscillating propulsor. In three dimensions, the stopping vortices are actually connected by tip vortices, forming a folded chain of vortex rings (Videler, 1993; Vogel, 1994). Such a flow field is essential to the generation of thrust in swimming fish. Although the vorticity convected into the wake represents an energy loss, the vortex street has possibilities for reduction in energy costs by schooling fish (Weihs, 1973b, 1975). The thrust-type vortex system reduces the drag on individuals positioned parallel and lateral to the street. Thus, a fish swimming diagonally behind another achieves a low relative velocity and high energetic advantage. Due to the rotation of the vortices, a fish following directly behind another will experience a higher relative velocity and would have to expend a greater amount of energy (Weihs, 1973b).

A three-dimensional, inviscid flow model by Weihs determined the optimal configuration between fish in a school for energy conservation (Weihs, 1973b, 1975). The model considered the structure of an infinite array of identical fish swimming in an oncoming flow. The fish were organized in discrete layers and in evenly spaced rows so that fish in the trailing row were staggered and centered between two fish in the leading row. The three-dimensional arrangement of fish in the theoretical school is like a crystal lattice (Weihs, 1975). The stationary position of the fish in the flow is maintained by the oscillatory sideways propulsive motions of the fish from its body and caudal fin. The propulsive motions produce the vortex wake in which the rate of change of momentum in the water is equal and opposite to the thrust, which opposes the total drag on the body. The equality of thrust and drag is maintained as the fish swims at constant velocity.

Considering only one layer of the theoretical school, the model predicted that the relative velocity directly behind a fish would be high, whereas the relative velocity would be lowest outside the vortex street (Breder, 1965; Belyayev and Zuyev, 1969; Weihs, 1972a, 1973b, 1975). Because the vortex wake takes time to fully develop and then dissipate further downstream, the

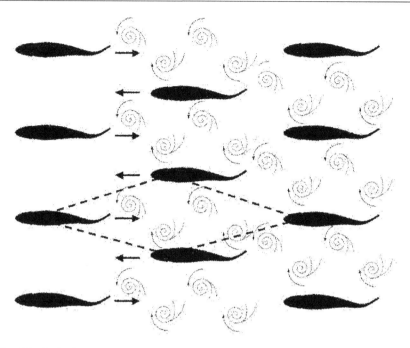

Fig. 4.4 Optimal arrangement for maximum energy savings of a fish school swimming in a horizontal layer. Vortex patterns in the wake of the fish are illustrated. Arrows show direction of induced flow relative to vortices. The diamond configuration of fish is shown by the dashed lines. Based on illustration from Weihs (1973b).

optimal configuration is a diamond or shallow rhombus pattern with a leading fish, two fish in the second row, and a fish in the third row (Fig. 4.4). The angles within the pattern are 30° and 150° (Weihs, 1973b).

The first row of fish swimming into undisturbed water will have the same relative and absolute velocities. Fish in the second row experience a relative velocity 40–50% of the free stream velocity and a reduction of the force generated for swimming by a factor of 4–6 (Weihs, 1973b, 1975). However, the decrease in relative velocity is not maintained with each successive row due to destructive interference. The vortex wakes of two successive rows will cancel because the vortices from each row are in line and have opposite vorticity when the lateral distances between adjacent fish is twice the width of the vortex street. The third row thus encounters undisturbed flow and incurs no reduction in relative velocity and drag. The reduced relative velocity occurs with alternate rows. Integrated over the entire formation, the school will have only a 50% savings in energy of the second row (Weihs, 1973b).

Individual fish in alternate rows not experiencing reduced drag from the interaction of the vortex wake of the previous row may still contract a benefit from effects due to lateral spacing. As spacing decreases there is a channeling effect so that the force produced by a fish in a row may be twice that of a single fish (Weihs, 1975). The channeling effect is not added to the energy savings of the row encountering the effects of the vortex wake. The two effects may not be superimposed, because the trailing row has a reduced relative velocity, which decreases lateral interactions.

In addition, Weihs (1973b, 1975) suggested that tip vortices from the pectoral fins could be exploited for lift by trailing fish. Tip vortices are produced from the movement of fluid caused by the pressure difference over a three-dimensional foil generating lift (Vogel, 1994; Fish, 1999). A foil that is directed into a flow with a positive angle of attack (i.e., angle to the incipient flow) will have a relatively faster flow on the upper surface (suction side) than the lower surface (pressure side). The difference in velocity translates into a pressure difference due to the Bernoulli effect, which generates lift. Fluid moves around the foil tip from the pressure side (high pressure) to the suction side (low pressure). The induced flow around the tip creates the tip vortex. The tip vortex encounters the main flow over the foil and is sheared. As the tip vortex is shed from the foil, it leaves a vortex trail. The vortex is positioned slightly inboard of the tip.

Fish such as tuna are negatively buoyant and maintain trim from lift generated by the pectoral fins (Magnuson, 1978). The spin of tip vortices from leading fish would provide an upwash lateral and posterior of the pectoral fin tips. Trailing fish would benefit by being in the tip vortex wake. This is analogous to the mechanism used by birds flying in V-formations, which for the trailing individual reduces induced drag, increases lift, and reduces the energetic cost of locomotion (Fish, 1999).

The three-dimensional complexity of fish schools has made data collection to validate the model difficult. Schools of 20–30 individuals of saithe (*Pollachius virens*), herring (*Clupea harengus*) and cod (*Gadus morhua*) were tested in a circular tank (Partridge and Pitcher, 1979). Although the fish generated vortices in their wake as predicted by Weihs, the schools were not organized according to the model. In all cases the fish exhibited non-random spacing but did not mimic the optimal configuration predicted by Weihs' model. Trailing fish often swam with their snout ahead of the tail of the leading fish. Despite the lack of data supporting energy reductions by trailing fish, the observed lateral separation was about 0.9 body lengths and could still provide a 35% reduction in energy.

There are few examples of fish that swim in the hydrodynamically advantageous arrangement predicted by Weihs (1975) (Pitcher and Partridge, 1979). Golden shiners (*Notemigonus crysoleucas*) were not observed to swim in the theoretical arrangement when in groups of 30 individuals (Boyd and Parsons, 1998). However, Fish *et al.* (1991) found that groups of three golden shiners did swim in positions that would benefit a trailing individual.

Observations on scombrid and salmonid fish do support the configuration predicted from the model. Jack mackerel (*Trachurus symmetricus*) and pink salmon (*Oncorhynchus gorbuscha*) swam in formations, which approximated the diamond shape (Breder, 1976; Keenleeyside and Dupuis, 1988). Estimates of lateral distance by scombrids show relatively narrow spacings of less than one body length in which propulsive force is increased. Pacific mackerel (*Scomber japonicus*) and bluefin tuna (*Thynnus thynnus*) in schools had lateral distances of 0.4–0.6 body lengths (Belyayev and Zuyev, 1969; van Olst and Hunter, 1970; Partridge *et al.*, 1983).

Indirect measures of energy economy of fish schools have been based on the kinematics of propulsive movements. Maximum duration of fish swimming in schools is 2–6 times longer than for a single fish (Belyayev and Zuyev, 1969), which suggests energy economy and reduced metabolic effort. As fish increase tail beat frequency with increasing swimming speed and thus experience an increased drag, the frequency reflects the relative energy expenditure (Parrish and Kroen, 1988). Tail beat frequency was demonstrated to be lower for some individuals of Pacific mackerel (*Scomber japonicus*) when schooling than when swimming alone (Fields, 1990). The tail beat frequency was higher for sea bass (*Dicentrarchus labrax*) swimming at the front of a school than for fish at the rear of the school (Herskin and Steffensen, 1998). Individual fish at the rear of the school displayed frequencies that were 9–14% lower than fish swimming in the front. Svendsen *et al.* (2003) swam schools of eight roach (*Rutilus rutilus*) at velocities of 2, 3 and 4 body lengths/ s. The roach showed consistent intra-school positional preferences. Tail beat frequencies of trailing fish were 7.3–11.9% lower compared to roach in leading positions.

In addition to the use of vortices from leading fish in a school, the reduced swimming effort of trailing fish has been hypothesized to be due to shed mucus from leading fish (Breder, 1976; Herskin and Steffensen, 1998). The addition of dilute solutions of long-chain polymers into flow is well established as a means of drag reduction (Rosen and Cornford, 1971; Bushnell and Moore, 1991). Drag reduction by introduced polymers is possible under the following conditions: (1) turbulent or pulsed laminar flow in the boundary layer, (2) the polymer is linear and soluble, (3) the polymer has a molecular weight of 50,000 or more and (4) the density and viscosity of the fluid from the surface outwards must be constant (Hoyt, 1975; Daniel, 1981). The mucus secreted by fish over the body surface is considered to meet these conditions. Mucus is a combination of mucopolysaccharides, nucleic acids, proteins, and surfactants in the form of lipids, phospholipids and lipoproteins (Bushnell and Moore, 1991). Reductions in drag of 50–65% were observed with fish mucus (Rosen and Cornford, 1971; Hoyt, 1975; Daniel, 1981). While these drag reducing properties were measured for the mucus of several species of fish, Parrish and Kroen (1988) found no advantage for the mucus of silversides (*Menidia menidia*).

Oxygen consumption by fish schools was reported to be significantly lower than the collective consumption of an equivalent number of solitary fish (Belyayev and Zuyev, 1969; Parker, 1973). However, this effect has been attributed more to group effect than to formation swimming. Parker (1973) argued that the decrease in metabolic rate of schooling fish compared to individuals tested in isolation resulted from a "calming effect" by the group. Schooling and nonschooling species both exhibited the effect when swimming in a rotating annular tank. However, nonschooling *Micropterus salmoides* swam in a single file formation and demonstrated no discernable reduction in metabolic cost between solitary and grouped fish. Obligate schoolers such as *Dorosoma cepedianum* and *Mugil cephalus*, however, showed 42–58% reductions in metabolic rate when swimming in schools as opposed to solitary swimming. The reduction of tail beat frequency for trailing sea bass over a range of swimming speeds (see above) was correlated with a reduction in oxygen consumption corresponding of 9–23% (Herskin and Steffensen, 1998). To remove the possibility of group effect, three fish were tested as a school in a water current of 0.07 m/s (Abrahams and Colgan, 1985). Oxygen consumption was measured for the entire school and on individuals separated from the other two by a clear partition. The partition allowed the fish to maintain visual contact without experiencing flow distortion. A 13% reduction in oxygen consumption was found for the school compared to the sum for the individual fish. However, only schools of large individuals (approximately 60 mm in length) demonstrated measurable energy savings. The small diameter (50 mm) of the test chamber may have introduced errors due to blocking and wall effects.

The possibility that fish swimming in schools reduces the energy cost of swimming assumes that all the fish in the school are swimming continuously to use a constant vortex pattern. Fish such as scombrids beat their tails continuously when swimming; however, many fish swim intermittently using a burst-and-coast strategy (Weihs, 1974; Weihs and Webb, 1983; Fish *et al.*, 1991). In burst-and-coast swimming, the fish alternates bouts of active accelerated swimming with passive glides or coasts. The fish realizes an energy savings by decreasing its drag during the coast phase. Lower total energy expenditure to travel a given distance thus is achieved when compared to a fish, which is constantly undulating over the same distance.

When three fish were examined swimming in the formation predicted by Weihs (1973b, 1975) for energy savings, the lead fish had equivalent mean times for bursting and coasting with solitary fish; whereas, the trailing fish demonstrated a reduced burst time and an increased coast time (Fish *et al.*, 1991). Coast time for trailing fish was 58 and 115% greater than coast time for leading and solitary fish, respectively. Coast and burst phases of trailing fish were nearly equal in duration. An energy reduction of 29 and 21% was estimated for the trailing fish relative to solitary fish swimming continuously and leading fish, respectively. Although the combination of formation swimming and burst-and-coast strategies allows for increased energy savings, the simultaneous use of both behaviors reduces the effectiveness of either strategy alone. The interaction of the two strategies would negate attainment of the optimal configuration, because as the trailing fish coasts it moves backward relative to the leading fish. The trailing fish must then accelerate to return to its original position. In addition, coasting by a leading fish does not generate the vortex pattern exploited by trailing fish. Such conditions may explain the deviation in natural formations from the model (Partridge and Pitcher, 1979; Fish *et al.*, 1991).

Position within a school relates to competing strategies associated with responses to stimulus detection and energetics. Location at the periphery of a school will allow individuals to more readily detect predators and sources of food. However in this position, the fish are more exposed to predators and leading fish can incur higher energy requirements to swim. Fish in the middle of the school are less likely a direct target for predators and can experience a reduction in drag when swimming in the wake of leading fishes. However, the energetic price for trailing in a school can be influenced by a delay in food procurement and a reduction in oxygen availability (McFarland and Moss, 1967; Domenici *et al.*, 2002).

Hitchhiking

Hitchhiking is a mechanism to reduce locomotor costs by direct physical attachment to another animal in motion. As the hitchhiker is passively towed along, it saves considerable amounts of energy that it would have to expend by muscular contraction to swim. For this behavior to be of benefit to the hitchhiker without being a detriment to the other animal, the hitchhiker must be much smaller in body size compared to the host animal.

The remora or sharksucker (Echeneididae) include eight species that rely upon hitchhiking. Theses fishes attach onto much larger fish, turtles, and whales (Fertl and Landry, 1999; Guerrero-Ruiz and Urbán, 2000; O'Toole, 2002; Brunnschweiler and Sazima, 2006; Sazima and Grossman, 2006). Even larger sharksuckers (*Echeneis naucrates*) can be a host for juveniles (Brunnschweiler and Sazima, 2006). Attachment onto large mobile hosts minimizes the energy expenditure to locomote by remoras. When attached to a moving object or in a current, remoras use the water flow for respiration, switching from active branchial ventilation of the gills to passive ram gill ventilation (Parin, 1968; Steffensen and Lomholt, 1983). Oxygen consumption is 3.7–5.7% lower for ram gill ventilation compared to branchial pumping. As the cost of gill ventilation can be 5–28% of the standard metabolism in fish (Cameron and Cech, 1970; Edwards, 1971), the shift to ram gill ventilation can significantly reduce the total energy cost of the remora. The association with larger hosts provides the remora, in addition to a reduction in locomotor costs, with feeding opportunities. Remoras will forage on food scraps left by their host, the host's feces and vomit, parasites on the host's body, and plankton from ram-feeding while attached to the host (Sazima *et al.*, 1999, 2003; O'Toole, 2002; Sazima and Grossman, 2006).

Remoras attach onto their larger host with a suction disk (Fig. 4.5). The sucking disk of remoras is a modified dorsal fin with slat-like transverse ridges, which are modified spines (Moyle and Cech, 1988). Muscles erect or depress the spines on the disk. During suction, the spines are erected creating a sub-ambient chamber and a pressure differential for suction (Fulcher and Motta, 2006). The suction pressure generated by *E. naucrates* has been recorded up to –103 kPa. The pressure difference is highest when the sharksucker is attached to a smooth surface, but decreases to –47 kPa on sharkskin. However, more force is required to dislodge the sharksucker from rough sharkskin

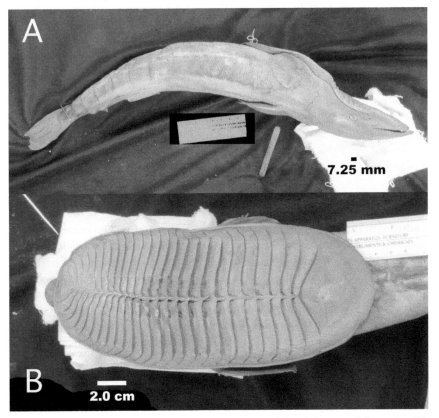

Fig. 4.5 A photograph of a remora showing a lateral view (A) and close-up of the sucker (B).

compared to a smooth surface as friction is increased to the sucker spines (Fulcher and Motta, 2006). Townsend (1915) found that 0.61 and 0.67 m long specimens of *E. naucrates* could lift pails of water weighing 93.4 and 107.9 N, respectively.

The association can negatively impact the remora's host. Remoras are considered a hydrodynamic parasite as they disrupt the flow of water over the body of their host and potentially increase the host's drag (Moyle and Cech, 1988). In addition, the suction developed by the disk and presence of the disk spines can act as an irritant on the skin of the host. Schwartz (1992) observed an open sore on a sheepshead (*Archosargus probatocephalus*) due to the attachment of a remora. Remoras would be particularly irritating to a highly sensitive and naked skin, such as in dolphins (Fish *et al.*, 2006). The spinning aerial leaps that are executed by spinner dolphins (*Stenella longirostris*) were suggested to be a mechanism in the removal of remoras (Hester *et al.*, 1963; Fish *et al.*, 2006). The spinning may be a behavioral response to remora attachment necessitated by an inability to remove parasites by use of traditional mechanisms of biting, combing, scratching and rubbing (Gunter, 1953). The high rotation rates could aid in peeling away the suction disk from the dolphin's body and the orientation of the dolphin's body during re-entry into the water could produce enough hydrodynamic force to shear off the unwanted remoras (Fish *et al.*, 2006; Weihs *et al.*, 2007). Sharksuckers appear to irritate sharks and induce rotational and non-rotational behavioral patterns. Jumping by blacktip sharks (*Carcharhinus limbatus*) was proposed as a means of dislodging attached remoras (Ritter, 2002; Ritter and Brunschweiler, 2003).

Drafting

A fish can take advantage of the thrust produced by a larger animal by swimming in a hydrodynamically favorable region. The close proximity of the two bodies induces an interaction of their pressure and flow fields. The presence of the neighboring bodies causes the flow between them to increase, which produces a pressure drop (Bernoulli effect). The pressure drop creates a suction force resulting in a mutual attraction that exerts an equal and opposite force on each animal (Kelly, 1959; Weihs, 2004). Due to the difference in masses of the animals, the smaller fish is pulled along with the larger. A smaller individual can swim with reduced energy costs at the expense of the larger. This free-riding behavior is associated with drafting or slipstreaming, and bow-riding (Lang, 1966; Fish, 1999; Weihs, 2004).

Drafting uses the flow field along the sides of a large animal to "pull" the smaller fish forward, whereas, bow-riding uses the high pressure generated in the anterior region of a large individual to "push" a smaller fish forward without physical contact. This latter mechanism is often observed by dolphins swimming in front of ships (Fish and Hui, 1991). The pilotfish (*Naucrates*) swims in the flow field in front and along the sides of larger fish like sharks (Parin, 1968; Magnuson and Gooding, 1971; Clark and Nelson, 1997). Although a reduction in energy expenditure by pilotfish has been ascribed to swimming in the boundary layer of the shark, such is not the case. The boundary layer is generally too thin relative to the size of the pilotfish. It is rather the interaction of the mutual pressure fields of the fishes that permits drafting. Trout (*Oncorhynchus mykiss*) are able to hold station in front of a cylinder in a water current (Liao *et al.*, 2003b). The trout displays reduced tail-beat amplitudes and body wave speeds, indicating low energy costs. This effect is due to the same mechanism as bow wave riding.

A variant of drafting is the use by animals of the wake from obstructions in a flow stream (Webb, 1998a). By swimming behind an object shedding a vortex wake, an animal could extract energy from the flow (Fig. 4.6). The flow behind a rigid, bluff (non-streamlined) body is non-steady. The pressure on the downstream side of the bluff body is low forming a suction and drawing fluid back toward the body. This reversal in flow direction leads to instabilities in the boundary flow adjacent to the body and causes flow separation from its surface. Flow separates alternately from each side of the body producing two staggered rows of vortices, which are shed into the wake. All the vortices in one row rotate in the same direction, but opposite to that of the other row. This flow pattern around a bluff body is a drag-type of vortex street or Kármán vortex street (Fig. 4.6). The vortex pattern is stable for a long distance downstream if the distance between successive vortices on the same side is 3.56 times the distance between the two rows (Vogel, 1994). Due to the direction of rotation in the vortex street, the flow in the center of the street is directed anteriorly. By swimming between the parallel vortex rows of the Kármán vortex street, the cost of locomotion can be reduced (Fish, 1999). The trailing fish will experience a reduced relative velocity compared to the free stream velocity. As drag is proportional to the square of the velocity, the fish will expend a reduction in energy to maintain position between the bluff body.

Fishes are able to use vortices shed by objects in flowing water and radically alter their locomotor kinematics to maintain station well downstream of any suction region (Webb, 1998; Liao *et al.*, 2003a, b; Liao 2004; Beal *et al.*, 2006). The fish can extract sufficient energy from the oncoming vortices to develop thrust and overcome the drag on the body (Beal *et al.*, 2006). Fish coordinate the swimming motions with the alternating vortices of the Kármán street. This locomotor gait, termed the Kármán gait, was observed in fishes swimming 3–6 fish body lengths downstream of a cylindrical object in the Kármán vortex street shed by this object.

Trout (*Oncorhynchus mykiss*) swimming in a Kármán street alter the body kinematics. Large lateral oscillations of the center of mass occur as the body

Fig. 4.6 Karman vortex gait. The fish is swimming directly behind a bluff body, which is alternately shed vortices. The fish stays between the two rows of vortices.

is buffeted from side to side in the vortex street (Fig. 4.6). The amplitude of lateral body movement and body curvature are both much greater than during free-stream locomotion (Liao *et al.* 2003b). Tail-beat frequency is lower than when swimming in the free stream and matches the frequency of vortex shedding by the cylinder (Liao *et al.*, 2003b).

Trout can maintain position in the central downstream flow between vortices with very little activity of body musculature (Liao *et al.*, 2003a). This reduction in muscle activity when swimming in a Kármán vortex street strongly suggests that fishes are experiencing a considerable energy savings compared to free-stream locomotion. In effect, the fish can capture energy from vortices generated by the environment. The strength of this mechanism was demonstrated when a dead fish was towed behind a cylinder (Liao, 2004; Beal *et al.*, 2006). With no muscular effort, the dead fish "swam" against the water current and moved forward passively into the suction region behind the cylinder.

Naturally formed vortices are generated by obstructions in a flow (e.g., stones, tree branches) and interactions with surface interfaces (i.e., river bottom, stream banks). For fish holding station, the vortices need to be predictably shed (Webb, 1998; Liao *et al.*, 2003a, b). Unsteady perturbations will require greater energy allocation for stability.

Vorticity control

The production of vorticity is a consequence of the propulsive motions of animals in a fluid. As a shearing action is imposed on a separating flow, it induces a rotation in the form of a vortex. A vortex functions to transport momentum into the fluid, which can be used in the generation of thrust (Rayner, 1985, 1995). However, the vorticity will demand the expenditure of energy and lower propulsive efficiency. Thus, there has been interest in using vorticity to enhance propulsive performance.

Rosen (1959) developed the vortex peg hypothesis. He observed cross flows around the dorsal and ventral surfaces of a sprinting fish. The cross flows produced vortices on alternate sides of the body. Rosen postulated that the fish actively produced the vortices to act as quasi-static pegs. The fish could push off the vortices to affect propulsion. Rosen (1959) concluded that the drag on the fish was near zero. However, the velocity of flow adjacent to the fish from the vorticity would be higher than the free stream velocity, resulting in higher drag (Webb, 1975).

More recently, the idea that vorticity generated along the body or leading edge of the tail of a swimming fish upstream of the trailing edge of the propulsive caudal fin could enhance thrust production and increase efficiency (Ahlborn *et al.*, 1991; Triantafyllou *et al.*, 1993, 1996; Gopalkrishnan *et al.*, 1994; Anderson, 1996; Anderson *et al.*, 1998; Bandyopadhyay and Donnelly, 1998; Barrett *et al.*, 1999; Wolfgang *et al.*, 1999). Ahlborn *et al.* (1991) developed the vortex excitation/ destruction model. In the model, a starting vortex is produced before being acted upon as the fin is quickly reversed. This action produces new vortices on the opposite side of the fin, which gain strength at the expense of the primary vortex. Higher power is achieved (Ahlborn *et al.*, 1991). This mechanism is particularly applicable to starts from rest.

Using digital particle image velocimetry (DPIV), Anderson (1996) and Wolfgang *et al.* (1999) were able to demonstrate the development of vorticity along the sides of an undulating fish. This vorticity was developed in a manner similar to flow along an undulating plate (Wu, 1971a). The bound vorticity was conducted toward the trailing edge of the caudal fin. The bound vortices combined as they were being shed into the wake to produce an amplified vortex. The next set of vortices shed into the wake had the opposite rotation. This produced a pair of counter-rotating vortices and a thrust jet. Continuous vortex shedding produces a wake with the thrust type reverse Karman vortex street (Weihs, 1972; McCutchen, 1977; Müller *et al.*, 1997; Wolfgang *et al.*, 1999; Lauder, 2000). The interaction of vorticity generated along the body and shed at the caudal fin conformed to the mechanism discussed by Gopalkrishnan *et al.* (1994). A similar pattern of vorticity was observed for fish executing a turn (Anderson, 1996; Wolfgang *et al.*, 1999). It was postulated that this mechanism of propulsion was dependent on active control involving coordination of the body undulation and caudal fin motion (Wolfgang *et al.*, 1999).

Use of vorticity generated along the body to enhance wake structure is limited in high-performance thunniform swimmers. These animals are relatively stiff anteriorly, have nearly circular or elliptical cross-sections, and display extreme narrow-necking (Webb, 1975; Lindsey, 1978). These morphological features should not promote the production and conduction of bound vorticity along the body. However, the heaving and pitching motions of the relatively stiff, high-aspect ratio caudal propulsor could produce leading-edge vortices, which would impact the wake structure (Anderson *et al.*, 1998). The development of leading edge vortices from dynamic stall has been experimentally demonstrated to produce high-lift forces (Ellington, 1995; Dickinson, 1996; Ellington *et al.*, 1996; Dickinson *et al.*, 1999). The leading edge vortex would interact with the trailing edge vortex to produce thrust. This vorticity control is the principle mechanism by which high efficiency is achieved (Anderson, 1996; Anderson *et al.*, 1998). Anderson *et al.* (1998) demonstrated propulsive efficiencies of over 85% for a rigid flapping foil.

Experiments with the biologically-inspired RoboTuna showed that under a particular set of kinematic conditions the swimming robot could reduce its drag in excess of 70% compared to the same body towed straight and rigid (Barrett *et al.*, 1999). The conditions, which produced this reduction in drag, deviated from the kinematics found in living thunniform swimmers. In particular, the lateral excursion of the caudal propulsor was 12% of body length (Barrett *et al.*, 1999), whereas the typical excursion of the caudal propulsor in animals is approximately 20% of body length (Fierstine and Walters, 1968; Webb, 1975; Lindsey, 1978; Fish *et al.*, 1988; Fish, 1993b, 1998a; Dewar and Graham, 1994b; Gibb *et al.*, 1999). Deviation from the optimal settings of kinematic parameters for RoboTuna resulted in a drag augmentation of 300% (Barrett *et al.*, 1999).

Large energy gains by vortex control can only be realized when the vortices are from energy sources external of the fish (e.g., schooling). A fish that is not subjected to external energy sources and the water immediately around it are a closed system with a finite amount of energy. The internally generated mechanical energy required for thrust production cannot be enhanced beyond the energy available in the system. Vorticity shed anterior of the caudal fin is essentially wasted energy. Capture of this wasted vorticity can reduce the total wasted energy and increase propulsive efficiency (Lighthill 1970; Wu, 1971b; Webb, 1975; Weihs, 1989; Barrett *et al.*, 1999). However, this increase in efficiency may only be marginal. In the case of vorticity shed from anterior fins (e.g., dorsal fin), a proportion of the additional thrust generated by the fin would be used to overcome the drag of the fin. Removal of the anterior fin, although affecting stability, may not adversely affect propulsive performance. Indeed, during steady propulsion, BCF swimmers will collapse highly flexible fins to reduce the lateral body profile and drag (Webb, 1977).

Aerial behavior

The high energy demand of movement in water can be reduced by leaving the water and entering an aerial phase. This behavior is performed by penguins and the fastest mammalian swimmers (Hui, 1987; Reidman, 1990;

Fish and Hui, 1991) in a series of rhythmic leaps, referred to as porpoising (Fish and Hui, 1991). At high speeds, the energy required to leap a given distance is considered less than the energy to swim at the water surface (Fig. 4.7; Weihs, 2002). Although they do not porpoise, flying fish (Exocoetidae) have the ability to launch themselves from the water and glide in the air (Aleyev, 1977; Fish, 1990; Azuma, 1992). Gliding has been considered as a behavior that decreases the energetic expenditure of locomotion with a low-drag aerial phase (Shoulejkin, 1929; Rayner, 1981, 1986; Davenport, 1994). Another possibility is that gliding is used to transport flying fish from food-poor to food-rich patches in the ocean (Davenport, 1994).

Flying fish possess wing-like surfaces to generate lift for flight. The wings are composed of elongate pectoral and pelvic fins, with the pectoral fins being the primary wing (Breder, 1930; Fish, 1990; Davenport, 1992). The morphological differences in wing structure were expressed by Breder (1930) as based on the airplane design at the time. A monoplane type was exemplified by the genus *Exocoetus*, which possessed a single set of long narrow pectoral fins, and the biplane type by *Cypselurus* with expanded pectoral and pelvic fins. These aerodynamic configurations displayed differences in flight performance for the two genera (Fish, 1990).

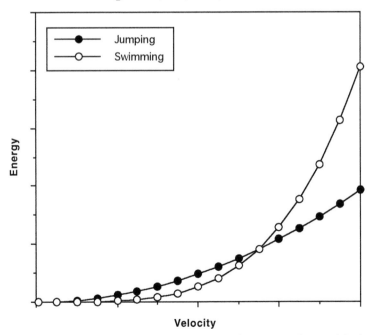

Fig. 4.7 The energies required for swimming close to the water surface and for jumping according to the model proposed by Au and Weihs (1980). The crossover velocity (i.e., where it becomes more economical to jump a given distance than to swim) occurs at the intersection of the two curves.

Take-offs vary with the genus of flying fish. *Cypselurus* uses a "taxiing glide" at the water surface for the initiation of flight (Breder, 1929; Hubbs, 1933, 1936; Hertel, 1966). The ventral surface of these fish is flat to act as a planing surface as the fish emerges from the water. During the taxi, the elongate lower or hypocaudal lobe of the caudal fin continues to remain in the water and generate thrust, increasing speed for take-off from 10 to 16–20 m/s (Shoulejkin, 1929; Mills, 1936a; Edgerton and Breder, 1941; Franzisket, 1965; Aleyev, 1977). The average taxi distance is 9 m (Hubbs, 1933). Once the trunk of the fish clears the water, the pectoral fins are opened and set at a small positive angle of attack to generate lift (Hubbs, 1933; Aleyev, 1977). The pectoral fins are opened lifting the tail from the water as the fish becomes airborne (Hubbs, 1933). Alternatively, initiation of the flight of *Exocoetus* is characterized by emergence from the water at an angle up to 45° (Shadbolt, 1908; Abel, 1926; Hubbs, 1933; Edgerton and Breder, 1941; Hertel, 1966).

The glide path of flying fish is regarded as relatively flat (Seitz, 1891; Breder, 1930, 1937). Maximum height of the glide is 6–7 m above the water (Ahlborn, 1897; Aleyev, 1977). Glide distances are generally greater for *Cyselurus* compared to *Exocoetus* and other flying fish genera that use a short taxi (Hubbs, 1933; 1936). Typical flights of 15 to 92 m for *Cypselurus* were reported. It was claimed that when flying with the wind flying fish could traverse 400 m in a single glide (Hubbs, 1918; Aleyev, 1977; Davenport, 1994). Total flight distance is extended by successive glides (Dahl, 1891; Breder, 1929; Hubbs, 1933, 1936; Forbes, 1936; Loeb, 1936, Mills, 1936a, b). At the end of a single flight, the flying fish will lower its hypocaudal lobe of the caudal fin into the water and accelerate the body to produce enough thrust for another glide. Successive glides, totaling as many as 12, can increase flight time and greatly increase the maximum flight distance (Hankin, 1920; Breder, 1929; Aleyev, 1977).

The reduction in drag in an aerial flight could minimize the energy expenditure in swimming over the same distance, particularly if swimming were to occur close to the water surface (Hertel, 1966). However, the increased drag incurred by wings and accelerated swimming during the taxi may limit any substantial energy savings. While energetic arguments at this time cannot be validated, a more probable use of an aerial phase by flying fish is the assertion that gliding is an anti-predator mechanism (Gill, 1905; Hubbs, 1933). In contrast to simple ballistic jumps, the aerial trajectory of a flying fish makes the position of reentry unpredictable to any predator confined to the water (Davenport, 1994).

Aerial behavior is also exemplified by the leaps of other fishes, including needlefishes (Belonidae), halfbeaks (Hemiramphidae), silversides (Atherinidae), and salmon and trout (Salmonidae) (Gudger, 1944; Gunter, 1953; Stuart, 1962). Leaps by salmonids are used to negotiate obstacles and waterfalls during upstream migrations (Osborne, 1961; Stuart, 1962; Lauritzen *et al.*, 2005). To make such jumps, the fish must accelerate quickly to a high take-off speed and expend considerable energy. A higher jump requires a higher take-off speed (Lauritzen *et al.*, 2005). However, a fish can be aided by using the upward tangential flow of the vortex produced by the water falling into a pool at the base of a waterfall (Stuart, 1962).

Currents and microhabitats

Fish will often migrate in areas where constant current or tidal currents are present (Weihs and Webb, 1983). When migrating downstream, the fish benefits by swimming with the current. The relative velocity between the fish and the water flow is reduced with a concomitant decrease in drag. Selective tidal stream transport by fish has been shown theoretically to provide a means of energy conservation (Weihs, 1978). When tidal current speeds are high, the fish can save 50% or more of its energy by swimming in midwater in the same direction of the current and resting on the bottom when the tide is moving in the opposite direction (Weihs, 1978; Weihs and Webb, 1983).

In a variable current, such as in estuaries and coastal areas, the energetic cost of swimming is minimized if these variations are ignored and a constant speed is maintained (Trump and Leggett, 1980). The cost of swimming for migrating plaice (*Pleuronectes platessa*) using tidal stream transport is about 20% less than the cost for continuous swimming (Metcalfe *et al.*, 1990). Even greater energy saving can be realized by these flatfish if burst-and-coast swimming is employed. An effective strategy used by flatfish is to rest on the bottom when the tidal current is flowing in a direction that is unfavorable for the fish (Weihs and Webb, 1983). The fish then swims when the tide is flowing in favorable direction. Even when the fish is swimming against the current reverse-flow vortices may be used to reduce the energy cost of swimming (Hinch and Rand, 2000). Such vortices would be more prevalent in river and stream habitats, where interactions between the water and solid interfaces would more readily shed vorticity into the flow.

Swimming near a solid surface, such as a river bottom or bank, can be advantageous when moving against a current. Despite the flow, the water at the interface with the solid has the same velocity as the solid (Vogel, 1994). Thus a hard river bottom will have a zero velocity as will the water adhering to its surface. This is the "no slip" condition. The layers of fluid adjacent to the interface are sheared due to viscous effects between the successive layers and the flow in the outer stream. This effect results in a velocity gradient within the fluid with the velocity varying with distance from the surface (Vogel, 1994). This layer of sheared flow is the boundary layer with a velocity gradient varying from zero at the interface to 99% of the free stream velocity. Fish swimming near the interface and in the boundary layer experience a lower velocity relative to the current in the open water.

A number of behaviors described as "flow refuging" (Webb, 1998) minimizes or avoids the energy cost of swimming by taking advantage of low-flow microhabitats. By swimming in pools, behind physical obstructions, or within the boundary flow against a solid substrate, a fish takes advantage of the lower flow compared to swimming in the main stream or current (Feldmeth and Jenkins, 1973; Fausch, 1984; Webb, 1993a, 1998). Osborne (1961) argued that migratory fish would seek low-velocity water close to the bottom or shoreline. Rainbow trout (*Oncorhynchus mykiss*) and rosyside dace (*Clinostomus funduloides*) typically swim in microhabitats with velocities substantially lower than all the velocities available, avoiding energetically expensive microhabitats (Facey and Grossman, 1990, 1992). Osborne (1961) proposed that salmon migrating upstream could take advantage of irregularities in the bottom contour to reduce energy costs.

The flow patterns due to microhabitats are used to reduce energy costs in station holding. Mottled sculpin (*Cottus bairdi*) are a relatively inactive benthic fish (Grossman and Freeman, 1987). Sculpins have a strong preference for zero or low velocity microhabitats (Facey and Grossman, 1992). In a flow, the sculpin holds position by clinging to the substrate. The enlarged pectoral fins generate a downward force due to the water flowing over them. Similarly, plaice (*Pleuronectes platessa*) utilize changes in body and fin posture for station holding on the bottom and avoid active swimming in a flow (Arnold, 1969; Arnold and Weihs, 1978; Webb, 1989, 2002). Cod (*Gadus morhua*) and plaice are able to take advantage of substratum ripples to find areas of retarded flow to station-hold (Gerstner, 1998; Gerstner and Webb, 1998). With its compressed body form, the benthic plaice can be sheltered by bottom ripples with heights of 4% of total fish length or wavelengths of four times total fish length (Gerstner and Webb, 1998). Fusiform fishes, such as cod and cyprinids, require that bottom ripples have heights of 6–8% of total length (Gerstner, 1998; Webb, 2006).

In shallow water, the optimal position in a current may not always be close to a solid substrate, such as the bank of a stream. Fish swimming in deeper water experience less wave drag. Wave drag is the additional resistance from gravitational forces in the production of surface waves when moving at or near the air-water interface. Kinetic energy from the animal motion is lost as it is changed to potential energy in the formation of waves (Vogel, 1994). The wave drag can reach a maximum of five times frictional drag (Hertel, 1966). Maximum wave drag occurs when the body is just submerged at a relative depth of 0.5 of the maximum body diameter and negated at a depth of three times maximum body diameter (Hoerner, 1965; Hertel, 1966). Huges (2004) generated a wave drag model that accurately predicted that large chinook salmon (*Oncorhynchus tshawytscha*) swam further from the bank to avoid wave drag. In addition, the model predicted that smaller sockeye salmon (*Oncorhynchus nerka*) should swim relatively closer to the bank to minimize migration costs.

Despite the energetic advantage of swimming or station holding in a flow near an interface with a solid body, there is a trade-off in remaining in a low velocity microhabitat. Under such flow conditions, the availability of prey items may be reduced. This is particularly the case in drift feeding, where the rate of prey availability is determined by the flow (Fausch, 1984; Facey and Grossman, 1992). Feeding in fast flowing currents may then facilitate a net gain in energy. Juvenile smallmouth bass (*Micropterus dolomieu*) increase the rate of net energy gain by moving into shallow-fast microhabitats (Sabo *et al.*, 1996). Stream salmonids were observed to take up positions in moderate current speeds that were close to faster flowing water (Everest and Chapman, 1972). This position was proposed to maximize the quantity of prey delivery, while minimizing the energy cost to maintain station.

CONCLUDING STATEMENT

The ability to control internal energy expenditures and exploit external energy sources permits fishes to reduce their energy for swimming (Table 4.1). While various behaviors for energy economy have been recognized, it has been mainly the mechanistic basis of the interaction between the fish and the water, which has been explored. These behaviors and their apparent energy savings have not been adequately investigated with respect to the muscle dynamics, total energy budget, or ecology. Are muscles tuned for particular species to operate efficiently with an energy saving behavior? If energy is a limited commodity for wild stocks, what are the savings as a proportion of total energy demands? Are energy savings by a particular behavior substantial enough that individuals not effectively using them are subject to reduced fitness or elimination by natural selection?

A uniform conceptual framework is lacking. While motion and theoretical studies have dominated the discipline (Weihs, 1973b, 1974; Lighthill, 1975; Webb, 1975), experiments using metabolic and muscle measures have been more limited (Belyayev and Zuyev, 1969; Liao *et al.*, 2003). Laboratory studies need to be performed in which metabolic measurements and electromyography are integrated with controlled flow tank experiments over a range of swimming speeds. Furthermore, field energetics need to be correlated with swimming behaviors.

Finally, work in this area has application to the development of nautical technologies that enhance increased efficiency and energy economy (Bandyopadhyay and Donnelly, 1997; Fish, 2006b; Fish and Lauder, 2006). The construction of biorobotic models provides a new direction for the development of engineered systems with improved propulsive performance

Table 4.1 Behavior and energy sources for energy economy by swimming fish.

Behavior	Energy Source	Mechanism of energy savings
Optimal swimming speed	Internal	Partitioning of metabolic and propulsive energy consumption
Gait transition	Internal	Gearing of muscle power
Burst and coast	Internal	Reduced muscular power and drag by maintaining a straight body
Swim and glide	External	Conversion of potential energy to kinetic energy and reduced muscular power and drag by maintaining a straight body
Schooling	External	Induced current from leading fishes reverses flow direction and reduces relative velocity on trailing fish
Hitchhiking	External	Direct attachment to a moving body
Drafting	External	Reduced relative velocity around drafting animal
Bow riding	External	High pressure behind fish from trailing body
Vorticity control	Internal	Constructive interference between vortices shed from anterior body and caudal fin of same fish
Aerial behavior	External	Change in medium reduces drag
Currents	External	Reduced relative velocity due to mass flow
Microhabitat	External	Induced current or vortices from physical obstructions

(Triantafyllou *et al.*, 1993; Barrett *et al.*, 1999; Bandyopadhyay, 2005; Long *et al.*, 2006; Lauder *et al.*, 2007; Pfeifer *et al.*, 2007; Bandyopadhyay *et al.*, 2008; Techet, 2008). The morphology and energy conservation strategies of fish can serve as model systems to design autonomous aquatic vehicles to function in remote, hostile environments for extended deployments. As the need for greater efficiency and energy economy are desired in engineered systems, imaginative solutions from nature can serve as the inspiration for new technologies (Fish, 2006b).

REFERENCES

Abel, O. 1926. Beobachtungen an Flugfischen im mexikanischen Golf. *Aus Natur und Museum, Frankfurt* 56: 129–136.

Abrahams, M.V. and P.W. Colgan. 1985. Risk of predation, hydrodynamic efficiency and their influence on social structure. *Environmental Biology of Fishes* 13: 195–202.

Ahlborn, B., D.G. Harper, R.W. Blake, D. Ahlborn and M. Cam. 1991. Fish without footprints. *Journal of Theoretical Biology* 148: 521–533.

Ahlborn, F. 1897. Der Flug der Fische. *Zoologische Jahrbuecher* (Syst.) 9: 329–338.

Alexander, R. McN. 1989. Optimization and gaits in the locomotion of vertebrates. *Physiological Reviews* 69: 1199–1227.

Alexander, R. McN. 2003. *Principles of Animal Locomotion.* Princeton University Press, Princeton.

Aleyev, Y.G. 1977. *Nekton,* W. Junk, The Hague.

Anderson, J.M. 1996. *Vorticity control for efficient propulsion.* Ph.D. Dissertation. Massachusetts Institute of Technology, Cambridge, MA.

Anderson, J.M., K. Streitlien, D.S. Barrett and M.S. Triantafyllou. 1998. Oscillating foils of high propulsive efficiency. *Journal of Fluid Mechanics* 360: 41–72.

Arnold, G.P. 1969. The reactions of plaice (*Pleuronectes platessa*) to water currents. *Journal of Experimental Biology* 51: 681–697.

Arnold, G.P. and D. Weihs. 1978. The hydrodynamics of rheotaxis in the plaice (*Pleuronectes platessa* L.). *Journal of Experimental Biology* 75: 147–169.

Arreola, V.I. and M.W. Westneat. 1996. Mechanics of propulsion by multiple fins: kinematics of aquatic locomotion in the burrfish (*Chilomycterus schoepfi*). *Proceedings of the Royal Society of London* B263: 1689–1696.

Azuma, A. 1992. *The Biokinetics of Flying and Swimming.* Springer-Verlag: Tokyo.

Bandyopadhyay, P.R. 2005. Trends in biorobotic autonomous undersea vehicles. *IEEE Journal of Oceanic Engineering* 29: 1–32.

Bandyopadhyay, P.R. and M.J. Donnelly. 1997. The swimming hydrodynamics of a pair of flapping foils attached to a rigid body. In: *Tenth International Symposium on Unmanned Untethered Submersible Technology: Proceedings of the Special Seesion on Bio-Engineering Research Related to Autonomous Underwater Vehicles Sp. Ses. Bio-Eng Res. Related to Autonomous Underwater Vehicles*, Autonomous Undersea Systems Institute, Lee, NH, pp. 27–43.

Bandyopadhyay, P.R., D.N. Beal and A. Menozzi. 2008. Biorobotic insights into how animals swim. *Journal of Experimental Biology* 211: 206–214.

Barrett, D.S., M.S. Triantafyllou, D.K.P. Yue, M.A. Grosenbaugh and M.J. Wolfgang. 1999. Drag reduction in fish-like locomotion. *Journal of Fluid Mechanics* 392: 183–212.

Beal, D.N., F.S. Hover, M.S. Triantafyllou, J.C. Liao and G.V. Lauder. 2006. Passive propulsion in vortex wakes. *Journal of Fluid Mechanics* 549: 385–402.

Belyayev, V.V. and G.V. Zuyev. 1969. Hydrodynamic hypothesis of school formation in fishes. *Problems of Ichthyology* 9: 578–584.

Blake, R.W. 1983. *Fish Locomotion.* Cambridge University Press, Cambridge.

Boyd, G.L. and G.R. Parsons. 1998. Swimming performance and behavior of Golden shiner, *Notemigonus crysoleucas*, while schooling. *Copeia* 1998: 467–471.

Breder, C.M., Jr. 1926. The locomotion of fishes. *Zoologica* 4: 159–297.

Breder, C.M., Jr. 1929. Field observations on flying fishes: a suggestion of methods. *Zoologica* 9: 295–312.

Breder, C.M., Jr. 1930. On structural specialization of flying fishes from the standpoint of aerodynamics. *Copeia* 1930: 114–121.

Breder, C.M., Jr. 1965. Vortices and fish schools. *Zoologica* 50: 97–114.

Breder, C.M., Jr. 1976. Fish schools as operational structures. *Fisheries Bulletin* 74: 471–502.

Brett, J.R. 1964. The respiratory metabolism and swimming performance of young sockeye salmon. *Journal of the Fisheries Research Board of Canada* 21: 1183–1226.

Brett, J.R. 1965. The swimming energetics of salmon. *Scientific American* 213: 80–85.

Brunnschweiler, J.M. 2005. Water-escape velocities in jumping blacktip sharks. *Journal of the Royal Society Interface* 2: 389–391.

Brunnschweiler, J.M. and I. Sazima. 2006. A new and unexpected host for the sharksucker (*Echeneis naucrates*) with a brief review of the echeneid-host interactions. *JMBA2-Biodiversity Records* (on-line).

Bushnell, D.M. and K.J. Moore. 1991. Drag reduction in nature. *Annual Review of Fluid Mechanics* 23: 65–79.

Cameron, J.N. and J.J. Cech. 1970. Notes on the energy cost of gill ventilation in teleosts. *Comparative Biochemistry and Physiology* 34: 447–455.

Cannas, M., J. Schaefer, P. Domenici and J.F. Steffensen. 2006. Gait transition and oxygen consumption in swimming striped surfperch *Embiotoca lateralis* Agassiz. *Journal of Fish Biology* 69: 1612–1625.

Carlson, J.K., K.J. Goldman and C.G. Lowe. 2004. Metabolism, energetic demand, and endothermy. In: *Biology of Sharks and Their Relatives*, J.C. Carrier, J.A. Musick and M.R. Heithaus (Eds.). CRC Press, Boca Raton, pp. 203–224.

Clark, E. and D.R. Nelson. 1997. Young whale sharks, *Rhincodon typus*, feeding on a copepod bloom near La Paz, Mexico. *Environmental Biology of Fishes* 50: 63–73.

Coughlin, D.J. and L.C. Rome. 1996. The roles of pink and red muscle in powering steady swimming in scup, *Stenotomus chrysops*. *American Zoologist* 36: 666–677.

Cressey, R.F. and E.A. Lachner. 1970. The parasitic copepod and life history of diskfishes (Echeneidae). *Copeia* 1970: 310–318.

Dahl, F. 1891. Die Bewegung der fliegenden Fische durch die Luft. *Zoologische Jahrbuecher* (Sst.) 5: 679–688.

Daniel, T.L. 1981. Fish mucus: *In situ* measurements of polymer drag reduction. *Biological Bulletin* 160: 376–382.

Davenport, J. 1992.Wing-loading, stability and morphometric relationships in flying fish (Exocoetidae) from the North-Eastern Atlantic. *Journal of the Marine Biological Association of the United Kingdom* 72: 25–39.

Davenport, J. 1994. How and why do flying fish fly? *Reviews in Fish Biology and Fisheries* 4: 184–214.

Dewar, H. and J.B. Graham. 1994a. Studies of tropical tuna swimming performance in a large water tunnel. I. Energetics. *Journal of Experimental Biology* 192: 13–31.

Dewar, H. and J.B. Graham. 1994b. Studies of tropical tuna swimming performance in a large water tunnel. III. Kinematics. *Journal of Experimental Biology* 192: 45–59.

Dickinson, M.H. 1996. Unsteady mechanisms of force generation in aquatic and aerial locomotion. *American Zoologist* 36: 537–554.

Dickinson, M.H., F.O. Lehmann and S.P. Sane. 1999. Wing rotation and the aerodynamic basis of insect flight. *Science* 284: 1954–1960.

Domenici, P., R.S. Ferrari, J.F. Steffensen and R.S. Batty. 2002. The effects of progressive hypoxia on school structure and dynamics in Atlantic herring *Clupea harengus*. *Proceeding of the Royal Society of London* B269: 2103–2111.

Drucker, E.G. and J.S. Jensen. 1996a. Pectoral fin locomotion in striped surfperch. I. Kinematics effects of swimming speed and body size. *Journal of Experimental Biology* 199: 2235–2242.

Drucker, E.G. and J.S. Jensen. 1996b. Pectoral fin locomotion in striped surfperch. II. Scaling swimming kinematics and performance as a gait transition. *Journal of Experimental Biology* 199: 2243–2252.

Edgerton, H.E. and C.M. Breder. 1941. High-speed photographs of flying fishes in flight. *Zoologica* 26: 311–314.

Edwards, R.R. 1971. An assessment of the energy cost of gill ventilation in the plaice, *Pleuronectes platessa*. *Comparative Biochemistry and Physiology* A40: 391–398.

Ellington, C.P. 1995. Unsteady aerodynamics of insect flight. In: *Biological Fluid Dynamics*, C.P. Ellington and T. J. Pedley (Eds.). The Company of Biologists: Cambridge, pp. 109–129.

Ellington, C.P., C. van den Berg, A.P. Willmott and A.L.R. Thomas. 1996. Leading-edge vortices in insect flight. *Nature* (London) 384: 626–630.

Everest, F.H. and D.W. Chapman. 1972. Habitat selection and spatial interaction by juvenile Chinook salmon and steelhead trout in two Idaho streams. *Journal of the Fisheries Research Board of Canada* 29: 91–100.

Facey, D.E. and G.D. Grossman. 1990. The metabolic cost of maintaining position for four North American stream fishes: effects of season and velocity. *Physiological Zoology* 63: 757–776.

Facey, D.E. and G.D. Grossman. 1992. The relationship between water velocity, energetic costs, and microhabitat use in four North American stream fishes. *Hydrobiologia* 239: 1–6.

Fausch, K.D. 1984. Profitable stream positions for salmonids: relating specific growth rate to net energy gain. *Canadian Journal of Zoology* 62: 441–451.

Fertl, D. and Jr. A.M. Landry. 1999. Sharksucker (*Echeneis naucrates*) on a bottlenose dolphin (*Tursiops truncatus*) and a review of other cetacean-remora associations. *Marine Mammal Science* 15: 859–863.

Feldmeth, C.R. and T.M. Jenkins. 1973. An estimate of energy expenditure by rainbow trout (*Salmo gairdneri*) in a small mountain stream. *Journal of the Fisheries Research Board of Canada* 30: 1755–1759.

Fields, P.A. 1990. Decreased swimming effort in groups of Pacific mackerel (*Scomber japonicus*). *American Zoologist* 30: 134A.

Fierstine, H.L. and V. Walters. 1968. Studies of locomotion and anatomy of scombrid fishes. *Memoirs of the Southern California Academy of Sciences* 6: 1–31.

Fish, F.E. 1990. Wing design and scaling of flying fish with regard to flight performance. *Journal of Zoology, London* 221: 391–403.

Fish, F.E. 1999. Energetics of swimming and flying in formation. *Comments on Theoretical Biology* 5: 283–304.

Fish. F.E. 2006a. Drag reduction by dolphins: myths and reality as applied to engineered designs. *Bioinspiration and Biomimetics* 1: R17–R25.

Fish, F.E. 2006b. Limits of nature and advances of technology: What dooes biomimetics have to offer to aquatic robots? *Applied Bionics and Biomechanics* 3: 49–60.

Fish, F.E. and C.A. Hui. 1991. Dolphin swimming—a review. *Mammal Review* 21: 181–195.

Fish, F.E. and G.V. Lauder. 2006. Passive and active flow control by swimming fishes and mammals. *Annual Review of Fluid Mechanics* 38: 193–224.

Fish, F.E., S. Innes and K. Ronald. 1988. Kinematics and estimated thrust production of swimming harp and ringed seals. *Journal of Experimental Biology* 137: 157–173.

Fish, F.E., J.F. Fegely and C.J. Xanthopoulos. 1991. Burst-and-coast swimming in schooling fish (*Notemigonus crysoleucas*) with implications for energy economy. *Comparative Biochemistry and Physiology* 100A: 633–637.

Fish, F.E., A.J. Nicastro and D. Weihs. 2006. Dynamics of the aerial maneuvers of spinner dolphins. *Journal of Experimental Biology* 209: 590–598.

Forbes, A. 1936. Flying fish. *Science* 83: 261–262.

Fulcher, B.A. and P.J. Motta. 2006. Suction disk performance of echeneid fishes. *Canadian Journal of Zoology* 84: 42–50.

Gerstner, C.L. 1998. Use of substratum ripples for flow refuging by Atlantic cod, *Gadus morhua*. *Environmental Biology of Fishes* 55: 455–460.

Gerstner, C.L. and P. W. Webb. 1998. The station-holding by plaice *Pleuronectes platessa* on artificial substratum ripples. *Canadian Journal of Zoology* 76: 260–268.

Gibb, A.C., K.A. Dickson and G.V. Lauder. 1999. Tail kinematics of the chub mackerel *Scomber japonicus*: testing the homocercal tail model of fish propulsion. *Journal of Experimental Biology* 202: 2433–2447.

Gill, T. 1905. Flying fishes and their habits. *Annual Report of the Board of Regents of the Smithsonian Institution* 1904: 495–515.

Gopalkrishnan, R., M.S. Triantafyllou, G.S. Triantafyllou and D. Barrett. 1994. Active vorticity control in a shear flow using a flapping foil. *Journal of Fluid Mechanics* 274: 1–21.

Greer-Walker, M. and G.A. Pull. 1975. A survey of red and white muscle in marine fish. *Journal of Fish Biology* 7: 295–300.

Grossman, G.D. and M.C. Freeman. 1987. Microhabitat use in a stream fish assemblage. *Journal of Zoology, London* 212: 151–176.

Gudger, E.W. 1944. Fish that play "leapfrog." *American Naturalist* 78: 451–463.

Guerrero-Ruis, M. and J.R. Urbán. 2000. First report of remoras on killer whales (*Orcinus orca*) in the Gulf of California, Mexico. *Aquatic Mammals* 26.2: 148–150.

Gunter, G. 1953. Observations on fish turning flips over a line. *Copeia* 1953: 188–190.

Hale, M.E., R.D. Day, D.H. Thorsen and M.W. Westneat. 2006. Pectoral fin coordination and gait transitions in steadily swimming juvenile reef fishes. *Journal of Experimental Biology* 209: 3708–3718.

Hankin, E.H. 1920. Observations on the flight of flying-fishes. *Proceedings of the Zoological Society of London* 32: 467–474.

Herskin, J. and J.F. Steffensen. 1998. Energy savings in sea bass swimming in a school: measurements of tail beat frequency and oxygen consumption at different swimming speeds. *Journal of Fish Biology* 53: 366–376.

Hertel H. 1966. *Structure, Form, Movement.* Reinhold, New York.

Hester, F.J., J.R. Hunter and R.R. Whitney. 1963. Jumping and spinning behavior in the spinner porpoise. *Journal of Mammalogy* 44: 586–588.

Hinch, S.G. and P.S. Rand. 2000. Optimal swimming speeds and forward-assisted propulsion: energy-conserving behaviours of upriver-migrating adult salmon. *Canadian Journal of Fisheries and Aquatic Sciences* 57: 2470–2478.

Hoerner S.F. 1965. *Fluid-Dynamic Drag.* Published by author, Brick Town, New Jersey.

Holland, K.N., R.W. Brill and R.K.C. Chang. 1990. Horizontal and vertical movements of yellowfin and bigeye tuna associated with fish aggregating devices. *Fishery Bulletin* 88: 493–507.

Hove, J.R., L.M. O'Bryan, M.S. Gordon, P.W. Webb and D. Weihs. 2001. Boxfishes (Teleostei: Ostraciidae) as a model system for fishes swimming with many fins: kinematics. *Journal of Experimental Biology* 204: 1459–1471.

Hoyt, D.F. and C.R. Taylor. 1981. Gait and energetics of locomotion in horses. *Nature* 292: 239–240.

Hoyt, J. W. 1975. Hydrodynamic drag reduction due to fish slimes, In: *Swimming and Flying in Nature,* T.Y. Wu, C.J. Brokaw, and C. Brennen (Eds.). Plenum Press, New York, Vol. 2, pp. 653–672.

Hubbs, C.L. 1918. The flight of the California flying fish. *Copeia* 1918: 85–88.

Hubbs, C.L. 1933. Observations on the flight of fishes, with a statistical study of the flight of the Cypselurinae and remarks on the evolution of the flight of fishes. *Papers of the Michigan Academy of Sciences* 22: 641–660.

Hubbs, C.L. 1936. Further observations and statistics on the flight of fishes. *Papers of the Michigan Academy of Sciences* 22: 641–660.

Hughes, N.F. 2004. The wave-drag hypothesis: an explanation for size-based lateral segregation during the upstream migration of salmonids. *Canadian Journal of Fisheries and Aquatic Sciences* 61: 103–109.

Hui, C.A. 1987. The porpoising of penguins: an energy-conserving behavior for respiratory ventilation? *Canadian Journal of Zoology* 65: 209–211.

Jayne, B.C. and G.V. Lauder. 1995. Red muscle motor patterns during steady swimming in largemouth bass: effects of speed and correlations with axial kinematics. *Journal of Experimental Biology* 198: 1575–1587.

Keenleyside, M.H.A. and H.M.C. Dupuis. 1988. Courtship and spawning competition in pink salmon (*Oncorhynchus gorbuscha*). *Canadian Journal of Zoology* 66: 262–265.

Kelly, H.R. 1959. A two -body problem in the echelon-formation swimming of porpoise. U.S. Naval Ordinance Test Station, China Lake, CA, Technical Note 40606-1.

Knower, T., R.E. Shadwick, S.L. Katz, J.B. Graham and C.S. Wardle. 1999. Red muscle activation patterns in yellowfin (*Thunnus albacares*) and skipjack (*Katsuwonus pelamis*) tunas during steady swimming. *Journal of Experimental Biology* 202: 2127–2138.

Korsmeyer, K.E., J.F. Steffensen and J.H. Herskin. 2002. Energetics of median and paired fin swimming, body and caudal fin swimming, and gait transition in parrotfish (*Scarus schlegeli*) and triggerfish (*Rhinecanthus aculeatus*). *Journal of Experimental Biology* 205: 1253–1263.

Lang, T.G. 1966. Hydrodynamic analysis of cetacean performance. In: *Whales, Dolphins and Porpoises,* K.S. Norris (Ed.). University of California Press, Berkley. pp. 410–432.

Lauder, G.V. 2000. Function of the caudal fin during locomotion in fishes: kinematics, flow visualization, and evolutionary patterns. *American Zoologist* 40: 101–122.

Lauder, G.V., E.J. Anderson, J. Tangorra and P.G.A. Madden. 2007. Fish biorobitics: kinematics and hydrodynamics of self-propulsion. *Journal of Experimental Biology* 210: 2767–2780.

Lauritzen, D.V., F. Hertel and M.S. Gordon. 2005. A kinematic examination of wild sockeye salmon jumping up natural waterfalls. *Journal of Fish Biology* 67: 1010–1020.

Liao, J. 2004. Neuromuscular control of trout swimming in a vortex street: implications for energy economy during the Kármán gait. *Journal of Experimental Biology* 207: 3495–3506.

Liao, J., D.N. Beal, G.V. Lauder and M.S. Triantafyllou. 2003a. Fish exploiting vortices decrease muscle activity. *Science* 302: 1566–1569.

Liao, J., D.N. Beal, G.V. Lauder and M.S. Triantafyllou. 2003b. The Kármán gait: novel body kinematics of rainbow trout swimming in a vortex street. *Journal of Experimental Biology* 206: 1059–1073.

Lighthill, J. 1970. Aquatic animal propulsion of high hydromechanical efficiency. *Journal of Fluid Mechanics* 44: 265–301.

Lighthill, J. 1971. Large-amplitude elongate-body theory of fish locomotion. *Proceedings of the Royal Society* B179: 125–138.

Lighthill, J. 1975. *Mathematical Biofluiddynamics*. Society for Industrial and Applied Mathamatics, Philadelphia.

Lindsey, C.C. 1978. Form, function, and locomotory habits in fish. In: *Fish Physiology:* W.S. Hoar and D. J. Randall (Eds.). Academic Press, New York, Vol. 7: *Locomotion*, pp. 1–100.

Loeb, L.B. 1936. The "flight" of flying fish. *Science* 83: 260–261.

Long, J.H., Jr., J. Schumacher and N. Livingston. 2006. Four flippers or two? Tetrapodal swimming with an aquatic robot. *Bioinspiration and Biomimetics* 1: 20–29.

Magnuson, J.J. 1978. Locomotion by scombrid fishes: hydrodynamics, morphology and behaviour. In: *Fish Physiology*, W.S. Hoar and D.J. Randall (Eds.). Academic Press, London, Vol. 7, pp. 239–313.

Magnuson, J.J. and R.M. Gooding. 1971. Color patterns of pilotfish (*Naucrates ductor*) and their possible significance. *Copeia* 1971: 314–316.

McCutchen, C.W. 1977. Froude propulsive efficiency of a small fish, measured by wake visualization. In: *Scale Effects in Animal Locomotion*, T.J. Pedley (Ed.). Academic Press, London, pp. 339–363.

McFarland, W.N. and S.A. Moss. 1967. Internal behavior in fish schools. *Science* 156: 260–262.

Metcalfe, J.D., G.P. Arnold and P.W. Webb. 1990. The energetics of migration by selective tidal stream transport: an analysis for plaice tracked in the southern North Sea. *Journal of the Marine Biological Association United Kingdom* 70: 149–162.

Mills, C.A. 1936a. Source of propulsive power used by flying fish. *Science* 83: 80.

Mills, C.A. 1936b. Propulsive power used by flying fish. *Science* 83: 262.

Moyle, P.B. and J.J. Cech, Jr. 1988. *Fishes: An Introduction to Ichthyology*. Prentice Hall, Englewood Cliffs, NJ.

Müller, U.K., B.L.E. van den Heuvel, E.J. Stamhuis and J.J. Videler. 1997. Fishfoot prints: morphology and energetics of the wake behind a continuously swimming mullet (*Chelon labrosus* Risso). *Journal of Experimental Biology* 200: 2893–2906.

Mussi, M., A.P. Summers and P. Domenici. 2002. Gait transition speed, pectoral fin-beat frequency and amplitude in *Cymatogaster aggregata*, *Embiotoca lateralis* and *Damalichthyes vacca*. *Journal of Fish Biology* 61: 1282–1293.

Osborne, M.F.M. 1961. The hydrodynamic performance of migratory salmon. *Journal of Experimental Biology* 38: 365–390.

O'Toole, B. 2002. Phylogeny of the species of the superfamily Echeneoidea (Perciformes: Carangoidei: Echeneidae, Rachycentridae, and Coryphaenidae), with an interpretation of echeneid hitchhiking behaviour. *Canadian Journal of Zoology* 80: 596–623.

Parin, N.V. 1970. *Ichtyofauna of the Epipelagic Zone.* Israel Program for Scientific Translations, Jerusalem.

Parker, F.R., Jr. 1973. Reduced metabolic rates in fishes as a result of induced schooling. *Transactions of the American Fisheries Society* 1973: 125–131.

Partridge, B.L. and T.J. Pitcher. 1979. Evidence against a hydrodynamic function for fish schools. *Nature* (London) 279: 418–419.

Partridge, B.L., J. Johansson and J. Kalish. 1983. The structure of schools of giant bluefin tuna in Cape Cod Bay. *Environmental Biology of Fishes* 9: 253–262.

Parrish, J.K. and W.K. Kroen. 1988. Sloughed mucus and drag-reduction in a school of Atlantic silversides, *Menidia menidia. Marine Biology* 97: 165–169.

Pfeiffer, R., M. Lungararella and F. Iida. 2007. Self-organization, embodiment, and biologically inspired robotics. *Science* 318: 1088–1093.

Quinn, T.P. 1988. Estimated swimming speeds of migrating adult sockeye salmon. *Canadian Journal of Zoology* 66: 2160–2163.

Rayner, J.M.V. 1981. Flight adaptations in vertebrates. *Symposia, Zoological Society of London* 48: 137–172.

Rayner, J.M.V. 1985. Vorticity and propulsion mechanics in swimming and flying animals. In: *Konstruktionsprinzipen lebender und ausgestorbener Reptilien,* J. Riess and E. Frey (Eds.). Universität Tubingen, Germany, pp. 89–118.

Rayner, J.M.V. 1986. Pleuston: animals which move in water and air. *Endeavour* 10: 58–64.

Rayner, J.M.V. 1995. Dynamics of the vortex wakes of flying and swimming vertebrates. In: *Biological Fluid Dynamics,* C.P. Ellington and T.J. Pedley (Eds.). Society for Experimental Biology, Cambridge, pp. 131–155.

Reidman, M. 1990. *Seals, Sea Lions, and Walruses.* University of California Press, Los Angles.

Ritter, E.K. 2002. Analysis of sharksucker, *Echeneis naucrates,* induced behavior patterns in the blacktip shark, *Carcharhinus limbatus. Environmental Biology of Fishes* 65: 111–115.

Ritter, E.K. and A.J. Godknecht. 2000. Agonistic displays in the blacktip shark (*Carcharhinus limbatus*). *Copeia* 2000: 282–284.

Ritter, E.K. and J.M. Brunschweiler. 2003. Do sharksuckers, *Echeneis naucrates,* induce jump behaviour in blacktip sharks, *Carcharhinus limbatus? Marine and Freshwater Behaviour and Physiology* 36: 111–113.

Rome, L.C., R.P. Funke and R. McN. Alexander. 1990. The influence of temperature on muscle velocity and sustained swimming performance in swimming carp. *Journal of Experimental Biology* 154: 163–178.

Rome, L.C., D. Swank and D. Corda. 1993. How fish power swimming. *Science* 261: 340–343.

Rosen, M.W. 1959. Water flow about a swimming fish. *U.S. Naval Ordinance Test Station Technical Publication* 2298: 1–96.

Rosen, M.W. and N.E. Cornford. 1971. Fluid friction of fish slimes. *Nature* (London) 334: 49–51.

Sabo, M.J., D.J. Orth and E.J. Pert. 1996. Effect of stream microhabitat characteristics on rate of net energy gain by juvenile smallmouth bass, *Micropterus dolomieu. Environmental Biology of Fishes* 46: 393–403.

Sazima, I. and A. Grossman. 2006. Turtle riders: remoras on marine turtles in southwest Atlantic. *Neotropical Ichthyology* 4: 123–126.

Sazima, I., C. Sazima and J.M. Silva, Jr. 2003. The cetacean offal connection: feces and vomits of spinner dolphins as a food source for reef fishes. *Bulletin of Marine Science* 72: 151–160.

Sazima, I., R.L. Moura and M.C.M. Rodrigues. 1999. Juvenile sharksucker, *Echeneis naucrates* (Echeneidae), acting as a station-based cleaner fish. *Cybium* 23: 377–380.

Schwartz, F.J. 1992. Effects of the sharksucker, *Echeneis naucrates*, family Echeneididae, on captive sheepshead, *Archosargus probatocephalus*. *Journal Elisha Mitchell Scientific Society* 108: 55–56.

Sfakiotakis, M., D.M. Lane and J.B.C. Davies. 1999. Review of fish swimming modes for aquatic locomotion. *Journal of Ocean Engineering* 24: 237–252.

Shadbolt, L. 1908. On the flying fish. *Aeronautical Journal* 12: 111–114.

Shoulejkin, W. 1929. Airdynamics of the flying fish. *Internationale Revue der gesamten Hydrobiologie und Hydrographie* 22: 102–110.

Steffensen, J.F. and J.P. Lomholt. 1983. Energetic cost of active branchial ventilation in the sharksucker, *Echeneis naucrates*. *Journal of Experimental Biology* 103: 185–192.

Stuart, T.A. 1962. The leaping behaviour of salmon and trout at falls and obstructions. *Freshwater and Salmon Fisheries Research* 28: 1–46.

Svendsen, J.C., J. Skov, M. Bildsoe and J.F. Steffensen. 2003. Intra-school positional preference and reduced tail beat frequency in trailing positions in schooling roach under experimental conditions. *Journal of Fish Biology* 62: 834–846.

Techet, A.H. 2008. Propulsive performance of biologically inspired flapping foils at high Reynolds numbers. *Journal of Experimental Biology* 211: 274–279.

Townsend, C.H. 1915. The power of the shark-sucker's disk. *Zoological Society Bulletin* 18: 1281–1283.

Triantafyllou, G.S., M.S. Triantafyllou and M.S. Gosenbaugh. 1993. Optimal thrust development in oscillating foils with application to fish propulsion. *Journal of Fluids and Structures* 7: 205–224.

Triantafyllou, M. S., D.S. Barrett, D.K.P. Yue, J.M. Anderson, M.A. Grosenbaugh, K. Streitien and G. S. Triantafyllou. 1996. A new paradigm of propulsion and maneuvering for marine vehicles. *SNAME Transactions* 104: 81–100.

Trump, C.L. and W.C. Leggett. 1980. Optimum swimming speeds in fish: the problem of currents. *Canadian Journal of Fisheries and Aquatic Sciences* 37: 1086–1092.

Tucker V.A. 1975. The energetic cost of moving about. *American Scientist* 63: 413–419.

van Olst, J.C. and J.R. Hunter. 1970. Some aspects of the organization of fish schools. *Journal of the Fisheries Research Board of Canada* 27: 1225–1238.

Videler J.J. 1993. *Fish Swimming*. Chapman and Hall, London.

Videler, J.J. and D. Weihs. 1982. Energetic advantages of burst-and-coast swimming of fish at high speed. *Journal of Experimental Biology* 97: 169–178.

Videler J.J. and B.A. Nolet. 1990. Cost of swimming measured at optimum speed: Scale effects, differences between swimming styles, taxonomic groups and submerged and surface swimming. *Comparative Biochemistry and Physiology* 97A: 91–99.

Vogel, S. 1994. *Life in Moving Fluids*. Princeton University Press, Princeton.

Webb, P.W. 1975. Hydrodynamics and energetics of fish propulsion. *Bulletin of the Fisheries Research Board of Canada* 190: 1–158.

Webb, P.W. 1977. Effects of median-fin amputation on fast-start performance of rainbow trout (*Salmo gairdneri*). *Journal of Experimental Biology* 68: 123–135.

Webb, P.W. 1988. Simple physical principles and vertebrate aquatic locomotion. *American Zoologist* 28: 709–725.

Webb, P.W. 1989. Station holding by three species of benthic fishes. *Journal of Experimental Biology* 145: 303–320.

Webb, P.W. 1993a. The effect of solid and porous channel walls on steady swimming of steelhead trout *Oncorhynchus mykiss*. *Journal of Experimental Biology* 178: 97–108.

Webb, P.W. 1994. The biology of fish swimming. In: *Mechanics and Physiology of Animal Swimming*. L. Maddock, Q. Bone, and J.M.V. Rayner (Eds.). Cambridge University Press, Cambridge, pp. 45–62.

Webb, P.W. 1998a. Entrainment by river chub *Nocomis micropogon* and smallmouth bass *Micropterus dolomieu* on cylinders. *Journal of Experimental Biology* 201: 2403–2412.

Webb, P.W, 1998b. Swimming. In: *The Physiology of Fishes*, D.H. Evans (Ed.). CRC, Boca Raton, pp. 3–24.

Webb, P.W. 2002. Kinematics of plaice, *Pleuronectes platessa*, and cod, *Gadus morhua*, swimming near the bottom. *Journal of Experimental Biology* 205: 2125–2134.

Webb, P.W. 2006. Use of fine-scale current refuges by fishes in a temperate warm-water stream. *Canadian Journal of Zoology* 84: 1071–1078.

Webb, P.W., P.T. Kostecki and E.D. Stevens. 1984. The effect of size and swimming speed on locomotor kinematics of rainbow trout. *Journal of Experimental Biology* 109: 77–95.

Weihs, D. 1972. Semi-infinite vortex trails, and their relation to oscillating airfoils. *Journal of Fluid Mechanics* 54: 679–690.

Weihs, D. 1973a. Mechanically efficient swimming techniques for fish with negative buoyancy. *Journal of Marine Research* 31: 194–209.

Weihs, D. 1973b. Hydromechanics of fish schooling. *Nature* (London) 241: 290–291.

Weihs, D. 1974. Energetic advantages of burst swimming of fish. *Journal of Theoretical Biology* 48: 215–229.

Weihs, D. 1975. Some hydrodynamical aspects of fish schooling. In: *Swimming and Flying in Nature*, T.Y. Wu, C.J. Brokaw and C. Brennen (Eds.). Plenum Press, New York, Vol. 2, pp. 703–718.

Weihs, D. 1978. Tidal stream transport as an efficient method for migration. *Journal du Conseil International pour l'Exploration de la Mer* 38: 92–99.

Weihs, D. 1984. Bioenergetic considerations in fish migration. In: *Mechanisms of Migration in Fishes*, J.D. McCleave, G.P. Arnold, J.J. Dodson and W.H. Neill (Eds.). Plenum Press, New York, pp. 487–506.

Weihs, D. 1989. Design features and mechanics of axial locomotion in fish. *American Zoologist* 29: 151–160.

Weihs, D. 2002. Dynamics of dolphin porpoising revisited. *Integrative and Comparative Biology* 42: 1071–1078.

Weihs, D. 2004. The hydrodynamics of dolphin drafting. *Journal of Biology* 3: 8.1–8.16.

Weihs, D. and P. W. Webb. 1983. Optimization of locomotion. In: *Fish Biomechanics*, P.W. Webb and D. Weihs (Eds.). Praeger, New York, pp. 339–371.

Weihs, D., F.E. Fish and A. J. Nicastro. 2007. Mechanics of dolphin spinning for remora removal. *Marine Mammal Science* 23: 707–714.

Wolfgang, M.J., J.M. Anderson, M.A. Grosenbaugh, D.K.P. Yue and M.S. Triantafyllou. 1999. Near-body flow dynamics in swimming fish. *Journal of Experimental Biology* 202: 2303–2327.

Wu, T.Y. 1971a. Hydrodynamics of swimming propulsion. Part 1. Swimming of a two-dimensional flexible plate at variable forward speeds in an inviscid fluid. *Journal of Fluid Mechanics* 46: 337–355.

Wu, T.Y. 1971b. Hydrodynamics of swimming propulsion. Part 3. Swimming and optimal movements of slender fish with side fins. *Journal of Fluid Mechanics* 46: 545–568.

5

Escape Responses in Fish: Kinematics, Performance and Behavior

Paolo Domenici

INTRODUCTION

The escape response is the main defense mechanism used by most fish species when facing a predator attack. Escape responses have been studied in a variety of species, revealing a high variability in kinematics and behavior. Typically, fish respond to a startle stimulation with a unilateral contraction of their axial muscle (called a "C-start") in a direction away from the threat (Domenici and Blake, 1997). This contraction (stage 1), can be followed by a further contraction on the opposite side of the body (stage 2) (Fig. 5.1). During stages 1 and 2, high accelerations and turning rates are produced. Little is known about the fish's swimming behavior beyond these two stages. Weihs (1973) and Webb (1976) defined this period as "stage 3", during which fish may show variable behavior, i.e., continue swimming, glide or brake. Escape responses are usually triggered by one of a pair of large reticulospinal cells (the Mauthner cells, M-cells) and several other related neurons found in the hindbrain (Eaton and Hackett, 1984). The Mauthner cells receive stimulation from a number of ipsilateral sensory inputs, typically visual and/or mechanoacoustic. As a result of the

Authors' address: CNR-IAMC Località Sa Mardini 09072 Torregrande (Or) Italy.
E-mail: paolo.domenici@iamc.cnr.it

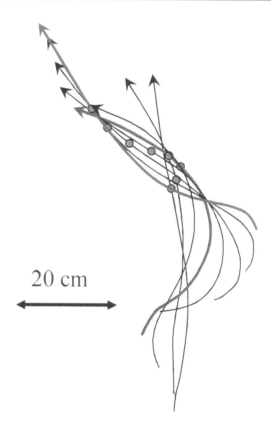

20 cm

Fig. 5.1 Escape response in dogfish (*Squalus acanthias*). Midline and centre of mass (red circles) of the fish at 40·ms intervals from the onset of the response. Head is indicated by the arrow. Red and blue lines indicate the end of stage 1 and stage 2, respectively. (Based on Domenici *et al.*, 2004 adapted with permission of the Journal of Experiemental Biology).

firing of a M-cell, the contralateral axial musculature as well as other motor components (e.g., pectoral and dorsal fins, opercula) are activated, while the opposite M-cell is inhibited (Eaton and Hackett, 1984). A large body of evidence now demonstrates the existence of many "variants" of the typical escape response described above. For example, escape responses may consist of stage 1 only (Domenici and Blake, 1997), certain species may exhibit an S-start rather than a C-start, where contractions on both sides of the body occur during stage 1 (Hale, 2002), and responses in which Mauthner cells are not activated have been observed (Eaton *et al.*, 1984; Hale, 2000 and Kohashi and Oda, 2008).

In addition, while escape responses are a common reaction of fish when under threat, fish may also show other startle responses, such as withdrawal (head retraction) and freezing responses (immobility response, Godin *et al.*,

1997) (Fig. 5.2a). Head retraction has been observed in elongated fishes as a response to head stimulation, although it can occur also as a result of tail stimulation (in *Erpetoichthys calabaricus*, Bierman *et al.*, 2004). Head retraction usually involves no propulsive phase and it is typical of elongated species such as *Anguilla rostrata, Petromyzon marinus*, living in structurally complex environments, in which predators are avoided by retracting the head back into a burrow rather than by swimming away from the threat (Ward and Azizi, 2004). In certain species, both withdrawal and escape can be observed (e.g., *Lepidoseren paradoxical*, Ward and Azizi, 2004; gunnels *Pholis laeta*, P. Domenici personal observations) (see Ward and Azizi, 2004, for a insightful analysis of the convergent evolution of head retraction). Withdrawal responses are also thought to be mediated by the M-cells (Currie and Carlsen, 1987). The Mauthner cells of the lamprey, unlike those of teleosts, are excited by both the ipselateral and contralateral sensory innervation (8th nerve), and lack reciprocal inhibition, so that both cells fire at the same time (Rovainen, 1979). There appears to be a relationship between the morphology of the axon cap and the type of startle behavior exhibited (Meyers *et al.*, 1998). Elongated fishes with no or reduced axon caps tend to exhibit withdrawal behaviors, while elongated fish with axon caps perform escape responses (Meyers *et al.*, 1998). However, there are exceptions, like *Erpetoichthys calabaricus* which has an axon cap but performs withdrawal behavour (Bierman *et al.*, 2004).

Freezing responses have been observed particularly as a response to chemical odours (Brown and Smith 1998; Brown and Cowan 2000; Brown *et al.*, 2001), and in species that tend to rely primarily on alternative anti-predator strategies such crypsis (e.g., flatfish) or body armours (see Godin 1997 for a review on immobility as an anti-predator response). When startled, sole (*Solea solea*) may show a freezing response as an alternative to escape responses (Ellis *et al.*, 1997), which is usually accompanied by a decrease in ventilation (Cannas *et al.*, 2007). Interestingly, for a given species, the same individuals may exhibit freezing or escape responses depending on their seasonal color, associated with different degrees of crypsis. Orange-throat darters *(Etheostoma spectabile)* switch to fleeing during the breeding season when they are conspicuously colored, as opposed to freezing as they do outside the breeding season, when they are cryptically colored (Radabaugh, 1989). Freezing responses have also been observed in many non-cryptic species such as seabass, grey mullet, and salmon (Hawkins *et al.*, 2004, Shingles *et al.*, 2005; Cooke *et al.*, 2003). In both cryptic and non-cryptic species, freezing as an anti-predator response is often accompanied by behavioral and/or physiological phenomena, such as changes in ventilation (a decrease or an increase, depending on the species) and heart rate (Hawkins *et al.*, 2004, Shingles *et al.*, 2005; Cooke *et al.*, 2003), erection of fins and spines (Huntingford *et al.*, 1994), and subtle body postures (Godin, 1997).

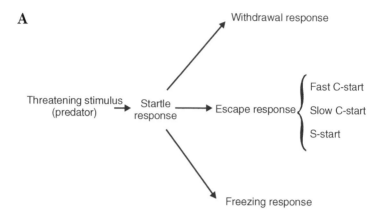

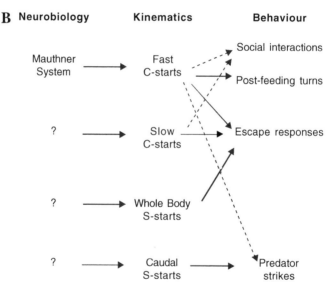

Fig. 5.2 A: Diagram illustrating various types of startle responses. **B:** Diagram illustrating various types of fast-start behaviors (both escapes and strikes) and their relative kinematics and neural control. Largely based on Domenici and Blake (1997) and successive modifications by Hale (2002) and Wohl and Schuster (2007) adapted with permission of the Journal of Experimental Biology. Broken arrows departing from the (Mauthner-cell mediated) fast C-start category indicate that no intracellular recording was made in order to ascertain that M-cells are involved in social interactions and predator strikes, while M-cells are known to be involved in post-feeding turns (Canfield and Rose, 1993) and escape responses (Eaton and Hackett, 1984). In addition, it is possible that the social interactions observed by Fernald (1975) correspond to slow (as indicated by the broken arrow), rather than fast C-starts, given the low turning rate observed (ranging 200–2000 degrees/s in 7 cm long fish, compared with turning rates around 4000 degrees/s in fast C-starts of similar sized fish, Domenici 2001). Question marks indicate that the neural commands driving slow C-starts (as in Domenici and Batty, 1994, 1997), whole body and caudal S-starts (as in Schriefer and Hale, 2004) are unknown (see text for details).

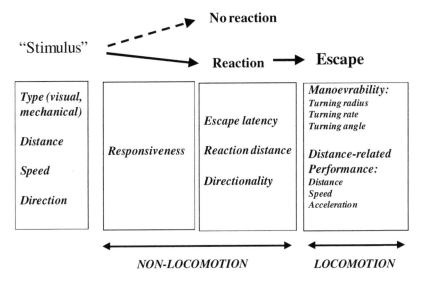

Fig. 5.3 The phases of an escape response, from stimulation to escape and relative variables. Variables are divided into non-locomotion and locomotion ones. Largely based on Domenici *et al.*, 2007.

This review focuses solely on escape responses. Fig. 5.2b shows a diagram illustrating the relationship between neural control, kinematics and various fast-start behaviors. Three main types of escape responses can be observed, i.e., slow and fast C-starts, and S-starts. Work done on locomotor kinematics and performance of escape responses is considered, as well as on non-locomotor variables related to the behavioral and sensory performance of fish, since all of these variables are an integral part of escape performance (Fig. 5.3). The ecological significance of escape responses in relation to various aspects such as size effects, functional morphology and habitat types is discussed. This review is mainly focused on adult fish. Burst and escape swimming in larvae is discussed in Chapter 11, by Fischer and Leis. In addition, various specific aspects of the escape response in flatfish, such as interactions with the substratum, are presented in the Chapter 2, by Blake and Chan. Another purpose of this chapter is to outline the variability present in escape responses, both at the intraspecific and interspecific level, and to argue that this variability is ecologically relevant in terms of the fish's antipredator strategies.

Escape Performance—more than just locomotion

Escape responses invariably imply a burst motion that allows fish to dart away from a threat. While locomotion clearly plays a fundamental role in determining escape performance, work on predator-prey encounters has highlighted the importance of non-locomotor variables (e.g., reaction distance and responsiveness) for the ability of fish to escape from predator attacks (Webb, 1986, Scharf *et al.*, 2003; Walker *et al.*, 2005, Fuiman *et al.*, 2006). In this review, therefore, all of these variables are considered as part of escape performance, and some of the causes of their variability will be discussed. Environmental effects are discussed in the Chapter 9, by Wilson *et al.*.

Figure 5.3 shows the various steps and relative characteristics of escape responses in fish. Fish can be startled by a threatening stimulus, with certain characteristics of speed, distance, direction and size. In laboratory settings, stimuli can be set up to be limited to only one sensory channel, usually the visual or the mechanical channel. However, natural threatening stimuli such as an approaching predator, may startle prey through a number of sensory channels. Stimulus characteristics may affect a number of escape response variables, and therefore they will be considered here when relevant in discussing prey behavior. Once a threatening stimulus, such as a predator, has made an attack, fish may or may not respond. This can be scored as the fish's responsiveness (Fig. 5.3). If the startled fish exhibits an escape response, its reaction distance can be measured as the distance between the stimulus and the prey at the time of response. If the stimulus is sudden such as a burst of sound or a flash of light, a latency can be calculated as the time between stimulus onset and the first visible movement of the fish. Fish will usually contract their body into a C-shape opposite to the stimulus (away responses *sensu* Blaxter *et al.*, 1981) as a result of the activation of the Mauthner cell ipsilateral to the stimulus. However, occasionally, this bend is oriented towards the stimulus (i.e., in towards responses). The ratio of away responses/total responses can be scored as directionality (Blaxter *et al.*, 1981, Domenici and Blake, 1993a). These variables are related to the sensory performance of fish and they will be discussed here since they are an integral part of the escape response, both functionally and in terms of affecting the outcome of the interaction. Locomotor performance during an escape response can be assessed using a number of variables related to distance-time (propulsive) performance (i.e., distance covered, speed and acceleration) and to manoeuvrability (i.e., turning radius, turning rate and turning angle).

Responsiveness

Responsiveness is a key variable for escape responses, because a fish that does not respond to a predator attack is most likely a dead fish. Work on predator-prey encounters by Fuiman *et al.* (2006) clearly shows that one of the main factors affecting fish vulnerability is responsiveness. Fuiman *et al.* (2006) show that in red drum larvae, responsiveness to predator attacks was the primary determinant of success in predation trials, accounting for 86% of the variation in escape potential. When a fish does not respond to an approaching stimulus, this phenomenon may be equivalent to having a relatively high response threshold, such that approaching predators might never cause that threshold to be reached. In predator-prey encounters staged in a laboratory setting, Webb (1982) found that fathead minnow responded to most attacks by trout, bass and rock bass, but only rarely to attacks by pike (28%). When escape responses were observed as a reaction to attacks by pike, these responses were at a much shorter reaction distance than responses to other predator species. Webb (1982) suggests that the rounded body section of pike results in a high prey response threshold, and therefore prey may fail to respond before they are captured.

In laboratory experiments using artificial stimulation, low responsiveness may imply that the stimulation used is below escape response threshold. Decreased responsiveness compared to a control situation was suggested to imply increased vulnerability to predation (Lefrançois and Domenici, 2006). Variation in responsiveness can be due to environmental factors (see Wilson *et al.*, Chapter 9). Responsiveness can also be affected by habituation. Work by Eaton *et al.* (1977) shows that responsiveness in larval zebrafish decreases in repetitive trials with 5 seconds between stimulations by means of mechanical vibration. Responsiveness could then be restored in the following ways: (a) if a rest period (6 minutes) was given, (b) if stimulus intensity was increased and (c) if fish were injected with strychnine, thereby presumably decreasing their response threshold. In addition to increasing stimulus intensity, changing other characteristics of the stimulus may have the same effect. Schleidt *et al.* (1983) showed that responsiveness in angelfish (*Pterophyllum eimekei*) was restored after habituation, once the direction of the predator model was reversed. The ecological significance of this phenomenon can be in maintaining a high responsiveness for novel threats, while habituation implies decreasing responsiveness for predictable, and seemingly not dangerous, repeated stimulations (Schleidt *et al.*, 1983).

Latency

If fish respond to a threat, an escape latency (in the rest of the chapter referred to as "latency" for simplification) can be calculated between the stimulation and the visible response of the fish. While a physiological latency must occur in all cases, due to the conduction time between the sensory system and the control neurons, and between control neuron and the musculature, measurements of behavioral latency can be accomplished only when a sudden stimulation is provided. In these cases, the time between stimulus onset and the visible response is taken as the latency, with the assumption that the onset of the sudden stimulus is what triggers the response. As long as the stimulation is short lived and does not increase in intensity with time, this assumption is quite reasonable. On the other hand, latencies cannot be measured when a fish is startled by an approaching object, since one cannot determine by external observation the precise instant at which the threshold for response is reached, and therefore the time interval between stimulus and response cannot be calculated. Therefore, while latencies can be easily determined in a laboratory setting where sudden stimulation can be provided, determining latencies in natural situations is often unrealistic because an attacking predator usually corresponds to a stimulus which increases in strength as it approaches the prey.

Latencies are the sum of a number of temporal components (Eaton *et al.*, 1981; Turesson and Domenici, 2007). Turesson and Domenici (2007) defined these components as T_{mc} (time from stimulus onset to Mauthner cell firing), T_{emg} (time between Mauthner cell firing and muscle activation, i.e., the electromyographic signal) and T_m (motor time, i.e., the time between muscle activation and the first visible movement of the fish). Typical latencies for Mauthner-mediated responses to mechanical stimuli are of the order of 10–20 ms (Eaton and Hackett, 1984). In 10 cm goldfish startled by mechanical stimulation, T_{mc} was of the order of 7 ms , although it was quite variable and could be as short as 2 ms (Eaton *et al.*, 1981). T_{emg} was 2.2 ms on average, while T_m was much longer, about 6 ms (Eaton *et al.*, 1981). These relatively short latencies to mechanical stimulation result from the direct neural connection of the octavolateralis nerve and the lateral line nerve to the Mauthner cells (Eaton and Hackett, 1984). Short latencies of the order of 10–20 ms were found in various species startled with mechanical stimulations (e.g., Eaton and Hackett 1984; Lefrancois *et al.*, 2005; Turesson and Domenici, 2007). On the other hand, measuring latency to visual stimulation is challenging because fish do not readily respond to sudden visual stimuli like flashes. Where latencies could be measured (e.g., Batty 1989), visual stimulation resulted in relatively long latencies, of the order of 100 ms, possibly as a result of a longer neural pathway between the optic nerve and the Mauthner cells, due to intermediate processing by the optic tectum (Zottoli *et al.*, 1987). Latencies can also vary depending on the intensity of the stimulation. Eaton *et al.* (1984) showed that latency decreased from 15 to 10 ms when doubling the relative intensity of the mechanical stimulation. Similarly, Blaxter *et al.* (1981) found that latencies in herring responding to sound stimuli (80 Hz) increased from about 25 ms when the stimulus intensity was well above threshold, to 40 ms when intensity was near threshold.

Theoretically, latency may be expected to increase with fish size, since two components of latency, T_{emg} and T_m, are predicted to increase with body size and may not be fully compensated with faster signal propagation (Turesson and Domenici, 2007). A size-dependent relationship can be expected for T_m on the theoretical grounds that muscle power is proportional to area, while the inertia of the body and entrained water that resists acceleration is proportional to body volume (Daniel and Webb, 1987; Turesson and Domenici, 2007). However, using a range of fish size from 6 to 29 cm, Turreson and Domenici (2007) found no effect of size on latency. There could be various reasons for this finding. One could be that there are compensatory mechanisms, for example in the time interval from stimulus to Mauthner cell excitation (T_{mc}), since it is possible that large fish show higher sensory acuity, compensating for any disadvantages in T_{emg} and T_m. It is also possible that slight differences in latencies are too small to be detected with the temporal resolution used, although this was relatively high (500 Hz, i.e., 2 ms between frames). It is unlikely that temporal differences below 2 ms would result in ecologically relevant scaling effects, therefore this work suggests that latency is not an important factor in the scaling of escape success in fish.

Latencies may also depend on the type of neural command involved. As shown by Eaton *et al.* (1984), Hale (2000) and Kohashi and Oda 2008), escape responses are not always triggered by M-cells, but may also be triggered by other reticulospinal neurons. Work by Eaton *et al.* (1984) shows that non-Mauthner-cell responses have longer latencies (20–40 ms) than Mauthner-cell responses (10–15 ms) in goldfish. Domenici and Batty (1994, 1997) show that fish may use two relatively distinct latencies, with separate peaks (at around 20 and 100 ms) and attributed this to different neural circuits used (e.g., Mauthner and non-Mauthner cells). See heading "Types of escape responses" below.

Reaction distance

Fish do not escape as soon as they perceive a danger, such as a predator, since escaping is a costly activity and it might also imply missing feeding opportunities (Ydenberg and Dill, 1986). In addition, escaping at the last possible instance may serve as a swerving maneuver similar to the "matador" strategy suggested by Blaxter and Fuiman (1990). The reaction distance of fish to an approaching threat is usually calculated as the distance between the threat and the fish at the time of the escape response. Reaction distance was shown to increase with the speed and size of the approaching object (Dill, 1974a). Dill (1974a) showed that fish escape once a specific threshold of looming rate (the rate of change of the angle subtended by the predator frontal profile as seen by the prey) is reached. This threshold is called ALT (Apparent Looming Threshold), where "apparent" stands for the fact that external observations can only tell us the "apparent response", not the actual (larger) distance at which the sensory system is triggered. Figure 5.4 shows the eye of a prey being attacked, and the visual angles subtended by a predator of size S at distances D_0 and D_1. The rate of change of this visual

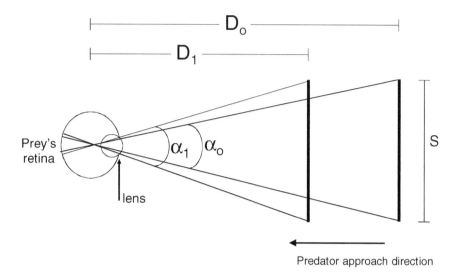

Predator approach direction

Fig. 5.4 The angle subtended by the predator's maximum profile on the prey's retina. S, the size of the predator profile, D_0 and D_1 are subsequent distances at which the predator is perceived by the prey, α_0 and α_1 are subsequent angles subtended on the prey's retina. Based on Dill (1974a) and Domenici (2002a) adapted by permission of the publisher (Taylor & Francis Ltd).

angle ($d\alpha/dt$; i.e., from D_0 to D_1) can be calculated as (Dill, 1974a; Domenici 2002a):

$$\frac{d\alpha}{dt} = \frac{4U_{pred}S}{4D^2 + S^2} \tag{1}$$

where U_{pred} is predator speed, S is the maximum frontal profile of the predator, measured as the mean of the maximum depth and maximum width (Webb, 1981), D is the distance from the prey's eye to the point of the predator's maximum profile (Fig. 5.4). Prey react at a reaction distance RD, once $d\alpha/dt$ reaches a certain threshold (ALT). Substituting ALT for $d\alpha/dt$ and RD for D, and solving for RD, it can be determined that:

$$RD = \sqrt{\frac{U_{pred}S}{ALT} - \frac{S^2}{4}} \tag{2}$$

This equation shows that reaction distance increases with the speed of the predator and with its profile. Empirical data (Dill, 1974a; Webb, 1982) are in general agreement with this model. Predators attacking at high speed elicit an early prey response (i.e., large RD), and so do predators with relatively large body profiles. Recent work (Meager et al., 2006) however, shows that fish startled by a predator model showed similar reaction distances when two predator speeds were compared, i.e., at each predator speed ALT was different. More work is therefore needed in order to test the hypothesis that ALT is independent of predator speed. So far, predator models have been used which approach the prey at a constant speed (Dill, 1974a; Meager *et al.*, 2006). The possibility that predator acceleration may also have an effect on RD should be explored.

ALT is relatively constant for a given prey, therefore prey may have developed a general mechanism that allows them to assess an approaching danger simply based on its size (frontal profile) and speed. This implies that predators with a narrow frontal profile, and with their maximum depth positioned posteriorly, may be able to get quite near their prey, before the prey responds. There is indeed evidence that such predators (e.g., pike) enjoy a relatively high capture success (and short prey RDs) when compared to predators with larger profiles (Webb, 1986). On this basis, it is possible that predators may vary their attack speed depending on the prey' ALT, as modeled by Domenici (2002a). Theoretical work suggests that a predator with a maximum profile up front should attack at maximum speed, while a predator with its maximum depth positioned far back, should approach prey slowly to avoid an early response (Domenici, 2002a). While ALT was shown to be relatively constant for different predators, prey may be less sensitive to attacks by predators with a narrow profile, which can therefore get closer to the prey than predicted (Webb, 1982). In addition, ALT can vary with the experience of the prey (Dill, 1974b).

Prey size may also affect reaction distance. Several theoretical considerations support this idea (Domenici, 2002a). Size is known to affect the perceptual and motor performance of fish, their vulnerability to predation and their energy requirements (Godin, 1997). Experimental work has however provided contradictory results. In studies on larvae, variation in ALT (and therefore reaction distance) can mainly be attributed to ontogenetic development of the sensory and motor systems (Webb, 1981 and Fuiman, 1993). Studies on adult fish report both an increase (Grant and Noakes, 1987; Paglianti and Domenici, 2006) and a decrease (Abrahams 1995) of reaction distance with prey size. However, Abrahams found a size effect only on armoured prey (brook sticklebacks, *Culea inconstans*) and no effect on non-armoured prey (yellow perch, *Perca flavescens*), and therefore it is possible that the scaling of reaction distance was also affected by the armour, i.e., large armoured prey may be less susceptible to predation. Theoretical considerations (Domenici, 2002a) which take into account the size-independency of speed reached within a fixed time, are in agreement with experimental work showing a large reaction distance in large prey (on staghorn sculpins; Paglianti and Domenici, 2006). However, since experience can affect ALT (Dill, 1974 b), in natural situations the scaling of ALT may vary. Reaction distance was found to be size independent in two species of fish tested in nature (De lucia *et al.*, in prep). Other local factors may modulate

ALT in nature, such as the presence of refuges (Godin, 1997), the activity in which fish are engaged (e.g. feeding vs. non feeding, Godin 1997), and human disturbance. Recent work shows that reaction distances vary in different areas inside vs. outside a marine reserve (De lucia *et al.*, in prep).

Directionality

Directionality is defined as the direction of the C-bend made by the fish body when performing a C-start escape response, relative to the stimulus (Blaxter *et al.*, 1981). Responses can be classified as "away" or "towards", corresponding to a muscle contraction in a contralateral or ipsilateral direction relative to the stimulus, respectively. Directionality can therefore be used as a proxy for determining which Mauthner cell is triggered in a given escape response. "Away" responses are the most common case given that the M-cell circuitry implies that stimuli from one side trigger the ipsilateral M-cell, and this M-cell then triggers the muscle contraction on the contralateral side (Eaton and Hackett, 1984). Blaxter and Batty (1985) found that fish showed an increase in the proportion of "away" responses during development in herring startled by mechanical stimulus. The older larvae made a significant proportion of away responses only once the head lateral line (and its coupling with the bulla) was fully developed. For adult fish, away responses are usually more frequent than towards responses. However, directionality may be modulated by various intrinsic and external factors. Domenici and Blake (1993a) found that the proportion of "away" responses was highest when fish were nearly perpendicular to the stimulus, and fell off gradually when the stimulus was at a small angle, to the extent that there were almost as many away as towards responses when the fish was in the 0–30° and the 150–180° sectors of orientation at the time of stimulation (i.e., stimulus head on or tail on). Canfield and Eaton (1990) have shown that swimbladder acoustic pressure transduction triggers Mauthner-mediated responses, and suggest that the detection of particle motion ensures that fish turn away from the acoustic stimulation. Although mechanical sensitivity is maintained for 360° around the fish (Hawkins and Horner, 1981), left–right discrimination decreases when the stimulus is more in line with the longitudinal axis of the fish, because of limits in angular discrimination between two sound sources (Schuijf, 1975). In addition, for stimuli directed to the front or the back of the fish, directionality may be of little biological significance. Similarly, for close-up stimuli, it is possible that directionality may be affected by where along the fish the stimulus is delivered (e.g., a side stimulus near the head vs one near the tail).

The type of stimulus also appears to have an effect on directionality. Using three types of stimuli on herring larvae, visual stimuli yielded the most directional responses, i.e., the highest proportion of away responses, compared with tactile and sound stimuli (Blaxter and Batty, 1985; Yin and

Blaxter, 1987; Batty, 1989). Recent work (Banet *et al.*, in prep.) shows that directionality is also affected by the vicinity of the wall. When fish were positioned within about <0.4 body lengths from the wall, they produced more towards responses (i.e., swimming away from the wall) than away responses. Schooling was also found to affect directionality, i.e., schooling fish show a higher proportion of away responses than solitary fish, possibly because of the sensory integration of the stimulus and the neighbours' responses (Domenici and Batty, 1997).

In addition to these external factors related to the stimulus or the environment around the fish, the swimming phase during which stimulation is provided may affect whether the response is away or towards. Blaxter and Batty (1987) found that gliding fish were more likely to show away responses, while swimming fish showed a high proportion of away responses only if they were swimming in a phase in which the contralateral muscle was already contracted. Similarly, Sillar and Roberts (1988) found that M-cells in *Xenopus* larvae were gated during swimming, to prevent simultaneous contractions of muscle on both sides of the body. However, Jayne and Lauder (1993) found that escape responses could be triggered during any phase of the swimming cycle and Domenici and Batty (1994) found that swimming phase had no effect on directionality. These results are in agreement with the finding that activating the M-cell overrides the swimming motor output to produce an output appropriate for escape regardless of the phase of swimming at which the M-cell was activated (Svoboda and Fetcho, 1996).

The M-cell and related reticulospinal neurons appear therefore to be quite flexible in their functioning, since various intrinsic (gliding phase) and external (proximity to wall, neighbours in a school, type of stimulus) factors may modulate the side which will contract depending on the situation. Recent work by Wohl and Schuster (2007) shows that archerfish stimulated by the fall of a prey onto the water surface, dart towards the prey using a response that resembles, kinematically, an escape response, but that is directed towards the stimulus. Similarly, work by Moulton and Dixon (1967) shows that conditioning goldfish to a light stimulus which anticipates a food supply, could result in triggering "towards" startle responses in the direction of the stimulus source. Clearly, there is need for further work on the mechanisms underlying the flexibility of the escape system in terms of its directional response.

Escape Trajectories

Given that fish usually contract their musculature away from the stimulus, the final swimming trajectory at the end of stage 2 also tends to be away from the stimulus. Previous work has shown that there is however a certain variability in escape trajectories (Eaton and Emberley, 1991; Domenici and Blake, 1993a;

Foreman and Eaton, 1993). This variability may be an adaptive property of the system, making escapes relatively unpredictable, since repeating the same strategy over and over could be detrimental if predators with learning abilities were involved. The use of circular statistics introduced for analysing escape trajectories (ETs) has permitted to establish patterns previously undetected by linear analysis of stimulus angles vs escape angle x-y plots (Domenici and Blake, 1993a) Escape trajectories (ETs) are defined as the swimming trajectories taken by the fish at the end of stage 2, in relation to the pre-startle direction of the stimulus, thus potentially spanning 360° (Domenici and Blake, 1993a; Domenici, 2002b). From a theoretical point of view, the optimal ETs were shown to lie within the sector 90–180° away from the line of attack of the predator (Weihs and Webb, 1984; Domenici, 2002a). Specific optimal angles of ET depend on the relative speed of predator and prey (Weihs and Webb, 1984; Domenici, 2002a; Fig. 5.5). Experimental data show that indeed fish use the 90–180° sector of ETs. While there is a relatively high variability, work on C-starts by Domenici and Blake (1993a) and Domenici and Batty (1997) shows that there may be two specific preferred ETs (130° and 180°, Fig. 5.6), independent of fish size, although the functional bases for these trajectories is not clear. These two ETs (180° and 130°) are observed in away responses (the majority of the responses), while towards responses are quite scattered around 360°. Domenici and Blake (1993a) suggest that these ETs may correspond to maximizing the distance away from the stimulus and escaping at the furthest limits of the fish's discrimination zone (defined as the angular sector of orientation which results in non-random escape directions),

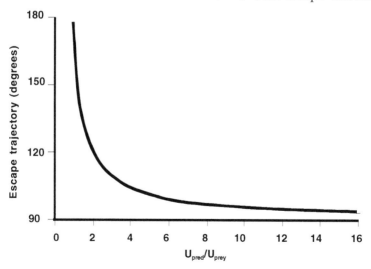

Fig. 5.5   The relationship between optimal ET and the ratio between the speed of the predator (U_{pred}) and that of the prey (U_{prey}) (based on Domenici, 2002a) adapted with permission of the publisher (Taylor & Francis Ltd).

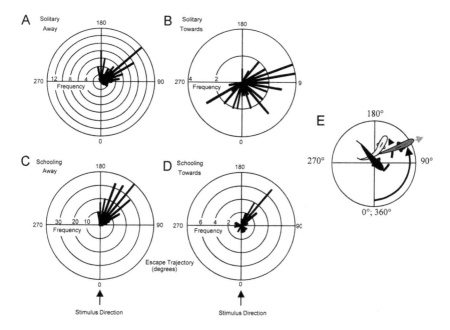

Fig. 5.6 Circular frequency distributions of escape trajectories in response to sound stimulus, plotted with respect to direction of the stimulus (at 0°). Separate distributions are shown for away and towards responses of both solitary and schooling herring *Clupea harengus*. Responses with stimulus to left or right of a fish are plotted as if stimulus were always on right; Frequency interval is 10°. A Solitary away responses; B solitary towards responses; C schooling away responses; D schooling towards responses. Away responses show a bimodal distrlbution in solitary fish and a unimodal distribution in schooling fish. The distribution of towards responses is mainly directed towards the stimulus in solitary fish, while is directed away from the stimulus in schooling fish (based on Domenici and Batty, 1997 With kind permission of Springer Science+Business Media. © Springer-Verlag 1997). E. An escape response drawn within a circular statistics reference as in A-D. Initial position of the fish before stimulation (black silhouette), the fish at the end of stage 1 (stippled fish) and at the end of stage 2 (grey fish). The stimulus direction is indicated by the arrow at the bottom. Broken curved arrow indicates stage 1 angle, solid curve arrow indicates escape trajectory, and grey arrow indicates the swimming direction of the fish at the end of stage 2. Fish escape trajectories can potentially span throughout the whole circle (0–360°) (Domenici, 2002b).

respectively. Schooling herring, however, did not show such a bimodal pattern, but show more homogenous trajectories in both away and towards responses, possibly due to interactions among neighbors (Domenici and Batty, 1997; Fig. 5.6). Little is known however on the ETs of S-starts, although these responses tend to be triggered by tail stimulation and result in relatively little body turning (Hale, 2002; Schriefer and Hale, 2004), which is in line with observations on C-starts when triggered caudally (Domenici and Blake, 1993a), although C-starts can exhibit a wider range (spanning about 180°) of turning angles (i.e., the angle accomplished by the anterior portion of the fish during stage 1).

While variability in ETs has been advocated as contributing to the necessary unpredictability of prey response, a relevant question is whether fish may be unable to escape with more precise trajectories relative to the stimulation, e.g., at 180° only, because of the high speed of the response. Work by Wohl and Schuster (2007) on strikes that are kinematically similar to escape responses, show that these feeding responses are quite precisely aimed at the prey, i.e., the sensory motor system appears to be capable of producing fast "escape-like" responses with high directional precision. Therefore, the variability around ETs is likely to be a functional characteristic that has evolved for high unpredictability in the response while using escape angles that are within the optimal range (90–180°) as shown by theoretical calculations (Weihs and Webb, 1984; Domenici, 2002a).

Maneuverability

Maneuverability is a fundamental component of the ability of fish to avoid predation. In particular, small fish would be captured by their larger predator in most cases, if they were escaping along a straight path, both because of the predictability of the path and because larger fish are faster if performance is measured along a straight line. Swerving, therefore, can be useful to prey because it increases the unpredictability of the swimming behavior, and because small fish are generally more maneuverable than their larger predators (see heading "size effects: a synthesis" below). Maneuverability has been measured using a number of variables, all of which are important depending on the type of interaction. The turning radius can be measured as the approximately circular path of the centre of mass during an escape (reviewed in Domenici and Blake, 1997). The turning radius is therefore a measure of how sharp a turn can be. Sharp turns can be advantageous to prey for swerving away from the predators' path (Howland, 1974; Webb, 1976). In escape responses, turning radius was shown to be independent of speed, and to increase proportionally with fish length (Domenici and Blake, 1997). Turning radius is therefore usually measured in body lengths. Typical values are about 0.1–0.15 Lengths. Fish from structurally complex habitats (CH fish) tend to have tighter turning radii than fish from open (pelagic) habitats (OH fish) (Domenici, 2003). For example, angelfish and knifefish have a turning radius of 0.06 and 0.04 Lengths, respectively, while the turning radius of tuna is 0.4 Lengths (Blake *et al.*, 1995; Domenici and Blake, 1997). These differences in turning radii may be related to the differences in the flexibility of the body. While CH fish are usually quite flexible (Aleev, 1977), the flexibility of open habitat fish (OH fish) is usually limited since body rigidity is advantageous for minimizing recoil, and therefore drag, during continuous swimming (Webb, 1984a; Domenici, 2003).

While the inverse relationship between body rigidity and turning radius is relatively intuitive, there are some exceptions that need to be considered. Boxfishes, for example, have a rigid body, yet they can turn with a turning radius that approaches zero (Walker, 2000). While these turns were not measured during escape responses, but during maneuvers powered by pectoral fins, this result points out that it is important to also consider the whole body of the animal and not just the path of its center of mass. Walker (2000) suggests that turning space (the minimum radius of the circle required to turn) may be a relevant measure of maneuverability. The turning space of boxfish is relatively large when compared with values from more flexible fish (Walker, 2000). Unfortunately there are very few values of turning space in the literature, and none on escaping fish.

The maximum turning angle (the angle between the head of the fish at the onset of the escape esponse and at the end of stage 1, Domenici and Blake, 1997) accomplished within a muscular contraction (half a tail beat, i.e., stage 1 in an escape response) can be a relevant measure of maneuverability because it gives an indication of the changes in direction a fish can accomplish within a relatively short time. Since escape trajectories are usually directed in the sector 90–180° away from the threat (see above), and predators tend to attack from the side (Webb and Skadsen, 1980), turning angles tend to be quite large, since fish often need to escape using >90° angle. In order to measure maximum turning angles, maneuvers in response to stimuli from all directions should be used in order to gather a wide range of angular responses. Turning angles are largely determined by stimulus angle and tend to be large when the stimulus is nearly frontal (Eaton and Emberley 1991; Domenici and Blake, 1993a). Mean values are not an informative measure, since they depend on the average stimulus angle used (Domenici and Blake, 1997). If the stimulus is presented randomly, the mean turning angle is likely to be approximately 90° (i.e., the mean value of a uniform distribution between 0 and 180°). Rigidity may also limit the angle (and the curvature) a fish can accomplish within a tail beat, i.e., the stage 1 of an escape response. Pelagic fish such as tuna, which are relatively rigid, can accomplish a 180° turn only by using more than half a tail beat (2 to 2.5 tailbeats, Blake *et al.*, 1995) (3 tailbeats in yellowtail *Seriola*, Webb and Keyes, 1981). Brainerd and Patek (1998) showed that the maximum body curvature with the first half-tail beat (i.e., stage 1) of escape swimming in four different species of tetraodontiform fishes was related to the number of functional inter-vertebral joints. Unfortunately, there are very few values of the maximum turning angle fish can perform. Domenici and Blake (1993b) suggest that in angelfish, maximum turning angles may decrease with fish size. However, no scaling trend is apparent when various species are considered (Domenici, 2001).

The rate at which a given turning angle is accomplished is defined as the turning rate (Domenici, 2001). Turning rate is presumably of fundamental importance since it measures how fast a fish can change its direction of motion. Recent work on predator-prey encounters show that turning rate is one of the key factors determining escape success (Walker *et al.*, 2005). Here we are mainly considering the turning rate measured as the rate of change of the angle made by the anterior portion of the fish (snout to centre of mass) during the first contraction of an escape response, i.e., stage 1 (Eaton *et al.*, 1977; Domenici, 2001; Hale *et al.*, 2002). However, turning rate has also been measured as the angular velocity of the centre of mass (Fish, 1999; Domenici, 2001). These two measures should be relatively similar in order of magnitude (Domenici, 2001). Turning rate should be related to the amount of power available for the muscular contraction, and inversely related to the body rigidity and drag forces experienced during the turn (Domenici, 2001). Turning rates decrease with size (Fig. 5.7). This is expected since the minimum contraction time increases with size (Wardle, 1975). However, the decrease of turning rate with size is steeper than expected based on minimum muscular contraction time data (Domenici, 2001). Domenici (2001) suggest that this may be because of other additional factors, such as the forces experienced in the water (minimum contraction times are calculated from experiments in vitro; Wardle, 1975) which affect the turning rate of large fish to a larger extent, since any drag force and acceleration reaction force is proportional to L^2 and L^3, respectively (Daniel and Webb, 1987). Overall, the data show that small fish turn faster. For example, a 90° turn is accomplished in about 20 ms in a 10 cm fish, and in about 100 ms in a 30 cm fish. In addition, turning rates tend to be lower than expected in OH fish with relatively low flexibility around the center of mass, such as tuna (Fig. 5.7). Turning rates have also been used to characterize the type of response (see below), since relatively distinct turning rates were observed during fast responses, slow responses and routine turns (Domenici and Batty, 1997; Domenici *et al.*, 2004; Meager *et al.*, 2006). Recently, stimulus direction was also shown to affect turning rates. Larger, faster escape turns result from visually stimulating the retina with rostral cues whereas shallower, slower turns occur after stimulating the retina with caudal cues (Canfield, 2006).

Propulsive performance

The escape response consists of an extreme acceleration event, of the order of >10 *g*. Most studies have investigated escape responses as accelerations from a still position. However, escape responses can also be triggered during routine swimming phases (Blaxter *et al.*, 1987; Jayne and Lauder, 1993; Domenici and Batty, 1994). In all cases, fish produce a peak of acceleration during stage 1, which may be followed by a second peak if stage 2 is present.

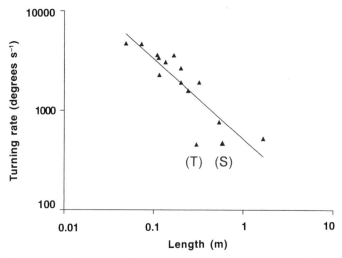

Fig. 5.7 The relationship between fast-start average turning rates (TR) and length (L) in fish. Modified based on McKenzie *et al.* (2007). Log TR = -0.81 log L +2.7 (r^2=0.76; P<0.0001). (T) and (S) indicate tuna (*Thunnus albacares*) and dogfish (*Squalus acanthias*), respectively. The relatively low turning rate of tuna may be related to its body rigidity.

Therefore, the timing of the propulsive power is related to the kinematics of the tailbeats (i.e., one acceleration peak per stage, see Fig. 5.8). The speed achieved by the end of stage 2 is usually in excess of 10 body lengths s^{-1}, depending on species and size (summarized in Domenici and Blake, 1997). Small fish, such as 5 cm angelfish (*Pterophyllum eimekei*), can reach speeds of around 30 body lengths s^{-1} (Domenici and Blake, 1993b). Maximum speed is a relevant measure of performance particularly in chase situations (Guinet *et al.*, 2007). However, in short-lived predator-prey encounters, distance and speed achieved within a relatively short (fixed) time (Webb, 1976; Domenici and Blake, 1993b) are relevant indicators of locomotor performance leading to escape success (Walker *et al.*, 2005). Therefore, an important issue is the choice of an experimental time interval to analyse, which is relevant for predator-prey interactions. This can be determined by analysing actual predator-prey encounters. Work by Walker *et al.*, (2005) suggest that the relevant time interval for measuring fast-start performance in guppies attacked by pike cichlid is from the beginning of the escape until the end of stage 2. It is likely, though, that the relevant time interval will vary with the predator attack strategy and the type of habitat (e.g., structural complexity), because the higher the structural complexity, the higher the probability that the prey finds a refuge and therefore, in such cases, predator-prey encounters are likely to be short-lived.

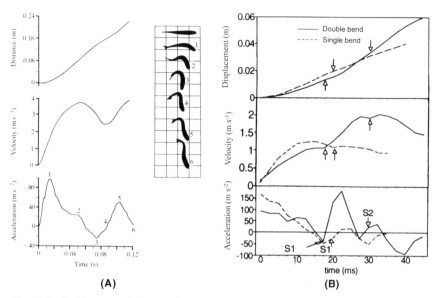

Fig. 5.8 A: The propulsive performance of an escape response in rainbow trout (*Oncorhynchus mykiss- syn. Salmo gairdneri*). Numbers on the tracings correspond to those on the acceleration plot. Position 3 corresponds to the end of stage 1 and the beginning of stage 2. Two peaks in acceleration are shown (one per stage). (From Harper and Blake, 1990 Reproduced with permission) **B:** Propulsive performance in knifefish (*Xenomystus nigri*) escape responses with (continuous line, single bend responses) and without (dotted line, double bend response) stage 2. S1 and S2 indicate the end of stage 1 and stage 2, respectively. Two high acceleration peaks (around 100–150 m s⁻²) are present in each stage in double bend responses, while single bend responses show only one peak in stage 1, and much lower acceleration (near zero) after the end of stage 1 (From Kasapi *et al.*, 1993 Reproduced with permission).

Both maximum burst speed (Wardle, 1975; Videler, 1993) and speed measured within a fast start (Domenici, 2001) increases with fish size. This is because minimum muscle contraction time increases with fish size (Wardle, 1975). Accordingly, the duration of stage 1 for any given angle of turn also increases with size, (Domenici, 2001). While maximum speed increases with fish length, so does the time to reach such a speed (Webb, 1976). However, when measured within a fixed time, speed was size independent (Webb, 1976; Domenici and Blake, 1993b). This suggests that acceleration should also be size independent. However, theoretical arguments predict that, at the relatively high Reynolds number in which fish live, acceleration should decrease with fish size (Daniel and Webb, 1987). This prediction is based on the fact that the thrust (proportional to L^2) used to "maneuver a resistance", i.e., the inertia of the fish's body and added mass (proportional to L^3), resists acceleration. While a tendency towards a decrease is present once fish and cetaceans are considered together (Domenici, 2001), data on fish do not show any particular trend.

This may be because: (1) as suggested by Webb and Johnsrude (1988), acceleration might not decrease with size because of summation of muscle twitches in larger fish, allowing them to perform with higher acceleration than predicted, (2) acceleration data, being a second derivative, is very sensitive to variation in methodology, such as framing rates and image size (Walker, 1998; Harper and Blake, 1989). Acceleration data gathered in different species, using different video set ups, may therefore be subject to large variations and subsequent noise in the data, thereby decreasing the power for finding any specific trend. For this reason, while acceleration is often claimed as an important variable in escape responses (Domenici and Blake, 1997; Walker *et al.*, 2005), it is desirable to accompany it with other, less noisy, measurements, such as distance covered and speed within a given time.

Variability in escape respsonses

Are escape responses stereotypic?

Fish escape responses are sometimes referred to as stereotypic / stereotyped (Eaton *et al.*, 1984; Swanson and Gibb, 2004; Higham *et al.*, 2005; Langerhans *et al.*, 2005; Weiss *et al.*, 2006). There is, however, a large body of evidence demonstrating that a number of escape variables show a high degree of variation, both within and across species. For example, escape angles can vary from about 0 degrees to over 180° (Eaton and Hackett, 1984; Domenici and Blake, 1991, 1997) and latencies can vary from 10 to over 100 ms (Domenici and Batty, 1997). In addition, a number of kinematic types have been described in addition to the most common biphasic C-starts (response with unilateral muscle contractions during stage 1, followed by a contralateral contraction during stage 2). Among these are S-starts, in which the muscle on both sides of the body is contracted during stage 1 (Hale, 2002), and single bend responses, described in a number of species, in which no stage 2 is observed (Domenici and Blake, 1997; Lefrançois *et al.*, 2005). Whether these categories can be considered discrete behaviors with alternative neural commands, or the extremes of a continuum, they nevertheless challenge the notion of stereotypy in escape responses.

In order to assess the degree of stereotypy in escape responses relative to other stereotypic behavours, I have taken some examples of escape variables from the literature, and compared their variability with other behaviors (including motion and sound production) which were claimed to be stereotypic. The examples include "time variables" (such as duration of a behavioral act), "form variables" (such as body angles) and "rate variables" (e.g., turning rate and sound frequency). While these examples include behaviors that do not necessarily involve body motion, they nevertheless represent a sample of stereotypic behavior with which escape responses can be compared in terms of their variability.

Here, I have considered the coefficient of variation (CV= SD /mean) of escape behaviors of adult fishes and of "stereotypic behaviors" of various animals taken from the literature. CV should be inversely related to stereotypy, i.e., behaviors with a high CV show low stereotypy. For the purpose of comparison, I have taken the following examples from the literature:

(a) Duration and form (body angle) of the bow movements of various races of canids, claimed to be highly stereotypic (Bekoff, 1977).
(b) The boatwhistle advertisement calls recorded from male toadfish (Barimo and Fine, 1998), which were claimed to be highly stereotypic, with frequency (a "rate" variable) showing a coefficient of variation averaging less than 1%, and duration with a coefficient of variation of 8%.
(c) The electric organ discharge by mormyrid fish, claimed to be stereotypic in their transient burst called "scallops" (Carlson and Hopkins, 2004). Data on the CV of the total duration of the behavior and the maximum EOD rate (in EOD/s, i.e., a rate variable) were used (data provided by B. Carlson).
(d) The duration of bites aimed at fish and squid by the nurse shark *Ginglymostoma cirratum*, which is claimed to be an inertial suction feeder with stereotypic prey capture kinematics (Matott *et al.*, 2005).

Figure 5.9 shows CV for the various behaviors taken from the literature, and escape responses. All behaviors claimed to be stereotypic show CVs between 0–20%, and in most cases (10 out of 14) less than 10%. Escape responses, on the other hand, rarely show CV<20% (7 out of 29 cases). In many cases, CVs for escape responses exceed 40%. The mean CV for the "stereotypic behaviors" taken from the literature is 11.2 ± 1.7 (Mean ± SE), while that for the escape response is 40.7 ± 5.6. The CVs of escape behaviors were compared with those of the "stereotypic" behaviors, using only one datum per species (or race, in the case of the canid work by Bekoff, 1977), i.e., either a duration variable (latency in the case of escape responses) or a form variable (stage 1 angle in the case of escape responses). Significant differences were found in both cases [duration variables; Mann Whitney test; P<0.0005; mean=69.84 ± 10.16, N=12 (escape responses) and mean=13.46 ± 2.58, N=7 (stereotypic behaviors); Form variables; Mann Whitney test; P<0.002; mean=31.23 ± 3.33, N=10 (escape responses) and mean=8.52 ± 1.16, N=4 (stereotypic behaviors)]. No test was carried out on rate variables because there were only 2 rate variables for the stereotypic responses. Nevertheless, the mean CVs were 9.7 ± 9.1 (N=2) for stereotypic responses, and 26.7 ± 3.0 (N= 13) for escape turning rates. It is also interesting to notice that in those studies in which routine turns

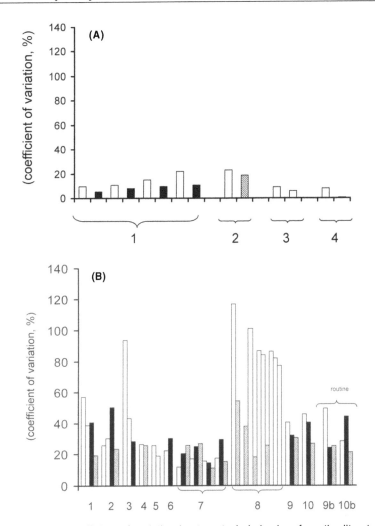

Fig. 5.9   A: The coefficient of variation in stereotypic behaviors from the literature (1, Bekoff, 1977; 2, Barimo and Fine, 1998; 3, Matott *et al.*, 2005; 4, Carlson and Hopkins, 2004). Open bars (time variables), shaded bars (form variable), hatched bars (rate variables).
B: The coefficient of variation in escape response from the literature (1, based on Lefrançois and Domenici 2006, unreported data; 2, based on Lefrançois *et al.*, 2005; unreported data; 3, Based on Domenici and Batty 1997, unreported data; 4, Tytell and Lauder, 2002; 5, Gamperl *et al.*, 1991 (only control values); 6, Domenici and Blake, 1991 (pooled data); 7, Hale *et al.*, 2002; 8, Eaton *et al.*, 1977 (Only species in which at least 3 responses were recorded are plotted. Only biphasic responses.); 9, Domenici *et al.* (2004) (pooled data), 10, Meager *et al.*, 2006 (pooled data of fast and slow predator model, only clear water treatment). 9b and 10b indicate routine turns from Domenici *et al.* (2004) and Meager *et al.* (2006), respectively. Shaded bars indicate temporal variables (latency or s1 duration), open bars indicate form variables (s1 angle) and hatched bars indicate turning rate.

(spontaneous turns not elicited by stimulation) were also considered (Domenici *et al.*, 2004; Meager *et al.*, 2006), the CVs (of turning angle and turning rates) in routine turns are of the same order of magnitude as those for escape response in their respective species (Fig. 5.9). Hence, spontaneous turning behavior accomplished by fish when completely undisturbed appears to be as stereotypic as escape behavior.

From this comparison, it appears that the CVs of a number of key variables of escape responses are generally higher than that observed in behaviors claimed to be stereotypic. In addition, CVs from different escape response studies show a high variation. This can be ascribed in part to different methodologies. Studies that kept certain external variables fixed (such as stimulus orientation, Hale *et al.*, 2002), tend to show lower CVs in stage 1 angle and duration, since the escape angle is in large part determined by stimulus orientation (Eaton and Hackett, 1984; Domenici and Blake, 1993a). In terms of variability, it cannot be excluded that previous work may have pooled escape responses that belonged to different "types", thereby increasing the variability of any given variable (i.e., their CV). For this purpose, for work in which escapes were divided into escape types, I considered only cases in which pooled data of various escape types were given (e.g., Domenici and Blake, 1991). These pooled data were used for calculation of CV, in order to use escape behavior as a single category. In the following section, I will show that CVs indeed decrease if discrete behaviors are considered. I will argue, in the section "types of escape responses" below, that a source of variation contributing to the relatively high CV observed is the existence of a number of types of escape responses, which may correspond to distinct behavioral responses possibly related to different neural commands.

A further issue related to variability, at the inter-individual level that would deserve more attention, is that of repeatability. Repeatability is a measure of the ability of an individual to achieve the same performance levels in tests carried out at different times. Repeatability is a fundamental measure for assessing the reliability of performance in terms of its fitness value (Arnold, 1983; Huey and Dunham, 1987; Dohm, 2002, Claireaux *et al.*, 2007). While work on the repeatability of swimming performance in adult fish has been carried out in relation to a number of performance variables [e.g., endurance swimming (Ucrit), Kolok (1992), Gregory and Wood (1998), Nelson *et al.* (1996), Claireaux *et al.* (2007) and sprint performance, Reidy *et al.* (2000), Martinez *et al.* (2002), Nelson and Claireaux (2005), Claireaux *et al.* (2007), Nelson *et al.* (2008), the repeatablity of escape performance has been little studied. Langerhans *et al.* (2004) and Gibson and Johnston (1995) found that escape speed is repeatable. However, very little is known about other traits such as escape angles and latencies. Working on short-term repeatability (i.e., with 5 trials separated by 40 minutes beween trials) in the escape performance of red drum larvae (*Sciaenops ocellatus*), Fuiman and Cowan

(2003) found that most escape variables showed significant repeatability (i.e., in responsiveness, latency to visual stimulus, distance covered, response duration, speed, but not in latency to acoustic stimulus). However, mean CVs for these variables were relatively high (ranging from 14.5% to 90.6%) and comparable to those reported in Fig. 5.9B. In addition, Fuiman and Cowan (2003) found that intra-individual variability was smaller than inter-individual variability. In conclusion, while more work is needed on inter and intra-individual variability in escape performance, it appears that escape responses are not as "stereotypical" when compared to other stereoptypic behaviours found in the literature, and that intra-individual variability might be lower (i.e., higher repeatability) than inter-individual variability.

Types of escape responses

C- and S-starts

While the most common escape response consists of a unilateral muscular contraction, which bends fish into a "C" shape (a C-start), recent work on elongated species (*Esox*) shows that escape responses can also consist of a bilateral muscular contraction (anteriorly on one side and posteriorly on the opposite side), which bends the fish into an "S" shape during stage 1 (S-starts, Hale, 2002; Fig. 5.10). Previous work has also suggested the presence of S-start escapes in other species (e.g., carp, Spierts and van Leeuwen, 1999). However, as argued by Domenici and Blake (1997) and Jayne and Lauder (1993), an "S" bend may be observed kinematically even if the muscle is contracted unilaterally. This is the case in which the backward bend of the tail may be due to the resistance of the water, particularly in elongated fish. Therefore, recordings of muscle contraction activity are necessary in order to characterize S-starts unequivocally as in Hale (2002). Hale suggests two alternative pathways through which S-start escapes may be generated, i.e., via reticulospinal neurons (as in C-starts), or via a combination of reticulospinal neurons and local spinal cord circuits. More recent work suggests that S-start escapes (whole-body S-starts) may be mediated by a simple system of descending reticulospinal input to spinal neurons as opposed to the S-start strikes (caudal S-starts), which may involve more complex neural circuit and greater stage 1 modulation (Schriefer and Hale, 2004). To complicate matters, bilateral activity has also been observed in escape responses that are kinematically described as C-starts (Foreman and Eaton, 1993; Westneat *et al.*, 1998; Tytell and Lauder, 2002). In this specific case, these authors suggest that bilateral activity may serve as a mechanism for increasing stiffness during the C-start.

Thus far, "true" S-start responses (differing from C-starts both in terms of kinematics and muscle activation) have been observed in species (such as

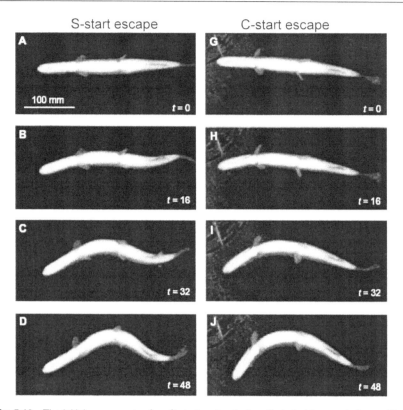

Fig. 5.10 The initial movements of an S-start and a shallow C-start of the muskellunge (*Esox masquinongy*). (from Hale, 2002) reproduced with permission of the Journal of Experimental Biology. (A–D) S-start behavior of the muskellunge. The fish bends into an S shape early in the behavior, 16–32 ms after initiation of movement (B,C). The subsequent return flip of the tail is not shown. (G-J) C-start behavior of the muskellunge. A comparison of S-start images B and C with C-start images H and I demonstrates the difference in caudal bending between these two response types. Time (*t*) is shown in milliseconds. Scale bar, 100 mm.

Esox masquinongy) that also show S-start predatory attacks. It is therefore possible that the flexibility of the system allows these fish to show both behaviors in different situations (escapes and attacks). The question remains as to whether fish that do not show S-start attacks may also show S-start escapes. In terms of variability, it is possible that previous work on certain species may have considered S- and C- start escapes together. This would increase their variability and consequently their CV. High flexibility of fast start kinematics is also shown by predators. While predator strikes are commonly "S" starts, recent work shows that C-starts can also be used by predators, both during predator strikes (Wohl and Schuster, 2007), as well as in the post-feeding turns observed by Canfield and Rose (1993) (Fig. 5.2b).

Single bend and double bend responses

There is evidence from both kinematics and muscle EMG recordings, that some C-start escape responses may lack a contralateral contraction (stage 2) after stage 1 (Foreman and Eaton, 1993; Domenici and Blake, 1997). These responses were defined "single bend" as opposed to "double bend" in which a contralateral contraction (stage 2) is present (Domenici and Blake, 1991, 1997). The absence of stage 2 is apparent from the turning rate *vs.* time traces (Fig. 5.11). Various authors have shown that these two escape types produce different performance levels (Domenici and Blake, 1991; Kasapi *et al.*, 1993; Lefrançois *et al.*, 2005; Lefrançois and Domenici, 2006). The acceleration produced after stage 1 in single-bend responses suggests that thrust production may be due to passive elastic elements (such as skin, collagen fibres and elasticity in the muscle itself), as suggested by Kasapi *et al.* (1993) (Fig. 5.8B). It is unknown whether single bend and double bend are triggered by different neural commands. It is possible that they are both triggered by the M-cells since their main differences are in stage 2 performance, and M-cell are thought to control mainly stage 1 (Eaton *et al.*, 2001). Nevertheless, regardless of whether single bend and double bend are distinct behaviors or extremes of a continuum (in which stage muscle activity may include a range of EMG signals from strong to weak to absent), the presence of such diverse patterns is a source of variability. For example, CV of stage 1 duration in angelfish from Domenici and Blake (1991) drops to 13 (single bend) and 19.5 (double bend) compared to 22.4 when all escape responses are treated together.

Slow and fast turning rates

Work on a number of species (herring *Clupea harengus*, Domenici and Batty, 1997; dogfish *Squalus acanthias*, Domenici *et al.*, 2004; cod *Gadus morhua*, Meager *et al.*, 2006), suggests that escape responses may show distinct turning rates, defined as the angular velocity of the anterior part of the fish during stage 1. Turning rate is a kinematic indicator of muscular contraction speed. Work on these species suggests that escape responses show distinct types of kinematics based on turning rates, rather than a continuum. On the basis of turning rates, escape responses can be classified as fast or slow responses (Domenici and Blake, 1997). Domenici and Batty (1997) and Domenici *et al.* (2004) show different stage 1 angle—stage 1 duration relationship (implying different turning rates) for these two types of responses (Fig. 5.12).

   In addition, slow responses show turning rates that are higher than those recorded during spontaneous activity ("routine turns"; Domenici *et al.*, 2004; Meager *et al.*, 2006; Fig. 5.12). Domenici and Batty (1997) suggest that these slow responses can be considered a type of escape response that

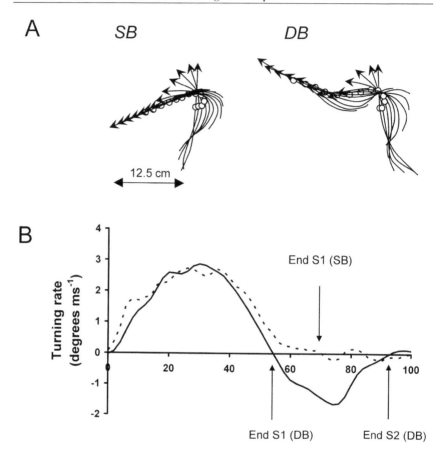

Fig. 5.11 A: Tracings of single bend (SB) and double bend (DB) escape responses in *Liza aurata*. Midlines (curved arrows) and centre of mass (circles) of the fish are shown at 10 ms frame intervals from the frame preceding the onset of the response. While the double bend responses show a reversal in the direction of turning of the head (at frame six) and the fish return flip is complete, in the single bend response, the head does not show a reversal of turning direction, and the tail straightens as the fish goes into a glide after frame six.
B: Turning rate of single (dotted line) and double bend responses (continuous line) shown in A. While the turning rate of the SB response oscillates around zero after the end of stage 1, in the DB response the turning rate after the end of stage 1 shows relatively high turning rate in the opposite direction (i.e., stage 2). (From Lefrançois *et al.,* 2005).

may be triggered by different neural circuits than fast responses. Nissanov *et al.,* (1990) found that escape responses triggered by electrical stimulation of single Mauthner cells (M-cells) showed a slower head turning rate than sensory-evoked responses, where both M-cells and parallel reticulospinal circuits were triggered. In addition, because specific electrical stimulation of the M-cell results in weaker and less variable responses than sensory-

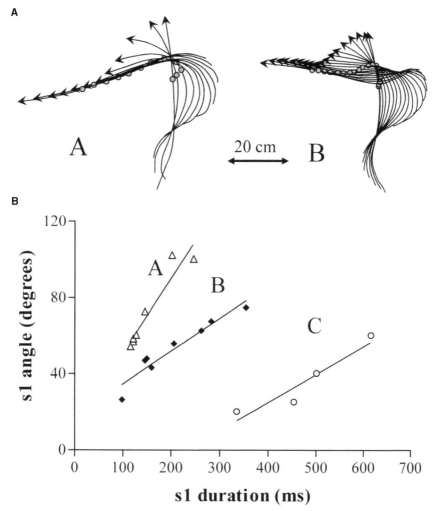

Fig. 5.12 A: Midline and centre of mass (red circles) of escaping dogfish (*Squalus acanthias*) at 40·ms intervals from the onset of the response. Head is indicated by the arrow. A and B correspond to a fast and a slow response, respectively. Note the longer time (i.e., higher number of frames) taken by fish in B in order to achieve a similar turning angle as A.
B: The relationship between stage 1 angle and stage 1 duration in fast responses (open triangles), slow responses (filled diamonds) and routine turns (open circles). (A) $Y=0.40+10.8X$, $P<0.001$, $r_2=0.91$, $N=7$; (B) $Y=0.17+17.4X$, $P<0.001$, $r_2=0.92$, $N=8$. (C) $Y=0.15-34X$, $P<0.05$, $r_2=0.90$, $N=4$. (From Domenici *et al.*, 2004) reproduced with permission of the Journal of Experimental Biology.

evoked ones, it appears that M-cell alone cannot reproduce all forms of escape (Nissanov *et al.*, 1990). Ablation experiments on larval zebrafish by Liu and Fetcho (1999) aimed at partitioning the contribution of M-cell from that of parallel reticulospinal neurons, i.e., the Mauthner-like segmental homologs MiD2cm and MiD3cm. Together with the M-cell, these form the Mauthner array. Ablating the M-cell resulted in escape responses with slower turning rate compared to intact fish only for tail directed stimuli, while no difference was found for head directed stimuli. On the other hand, ablating the whole Mauthner array resulted in decreased performance (slower turning rate) for both tail and head directed stimuli (Liu and Fetcho, 1999). Hence, a combination of different neural commands and stimulus directions may result in responses with different strengths. While it is not known to what extent these situations apply to the intact fish, it is plausible that intact fish may employ fast and slow responses associated with different neural commands, since non-Mauthner cell responses have also been observed in intact fish (Eaton *et al.*, 1984; Kohashi and Oda, 2008). In addition, the two response types may reflect different muscle contraction speeds and activation patterns. Domenici and Batty (1997) show that a higher proportion of slow responses occurred at larger stimulus distance. Similarly, Meager *et al.* (2006) show that slow responses occurred most often when the visual stimulus was presented in turbid conditions. Therefore, the type ("strength") of the stimulation may in large part determine the strength of the response. The idea that escape behavior is not maximized as an 'all or nothing' maneuver is supported by numerous studies (e.g., Webb, 1982, 1986; Domenici and Batty, 1994, 1997; Domenici *et al.*, 2004). Therefore, it seems that fish can employ a "two-gear" escape system with which they respond to startling stimuli. Such a system allows them to react using different response intensities and, therefore, energetic costs, perhaps depending on the degree of the threat perceived. As suggested by Domenici *et al.* (2004), this system does not appear to be graded, possibly as a result of neuromuscular design features. Escape responses with different turning rates were found in both teleosts with Mauthner cells (e.g., herring, Domenici and Batty 1994, 1997; cod, Meager *et al.*, 2006) and in chondrichthyans (such as *Squalus achantias*) which possess Mauthern cell only at the embyionic stage (Stefanelli, 1980). Therefore, this two-gear escape system may be a common feature of the escape responses of many fish, including teleosts and chondrichthyans, regardless of the presence of the Mauthner system.

The presence of different escape types is a source of high variability in the temporal and angular components of stage 1. For example, in dogfish, when all escape responses are taken together, CVs are 40.5, 32.2 and 30.5 for stage 1 duration, stage 1 angle and turning rate, respectively. However, when response types are considered separately, CVs for fast and slow responses are 32 and 41 (stage 1 duration), 28.8 and 28.9 (stage 1 angle), 6.7 and 14.4 (turning rate). This does not necessarily imply that each response type can be considered stereotypic, since their variability remains relatively high. I suggest that escape responses are intrinsically variable, and can consist of various response types which may correspond to different neural commands. This variability in escape "strength" may be ecologically meaningful, since responding with maximum effort to any kind of danger, even low level threats, is an energetically costly strategy, both in terms of the cost of high speed swimming and of potential loss of feeding opportunities. On the other hand, not responding at all once a sub-maximal danger is approaching, may also be costly in terms of survival.

Different latencies

Along with the different turning rates observed in the escape responses of various species, distinct latencies have also been observed (Domenici and Batty, 1994, 1997; Lefrancois *et al.*, 2005; Turesson *et al.*, 2009; Fig. 5.13). Short latencies (10–50 ms) are in line with expectations based on the giant neuron (M-cell) control. However, most long latencies (>50 ms) may be controlled by alternative neural (reticulospinal) circuits, since previous work suggests that non M-cell responses show longer latencies than M-cell responses (Eaton *et al.*, 1984; Kohashi and Oda, 2008). In addition, similarly to turning rates, ablation experiments show that fish with intact parallel reticulospinal neurons (Mauthner array, i.e., M-cell plus homologue reticulospinal neurons) produce short latencies as a response to stimuli from all directions (head and tail), while fish with ablated Mauthner array produce mainly long latencies (Liu and Fetcho, 1999). Specific ablation of M-cells affect increase latencies only in responses to tail directed stimulation (Liu and Fetcho, 1999), suggesting that stimulus direction modulates the effect of neural activation on response latency.

There is a relationship between these latency types and the turning rates in intact fish, i.e., all short latency responses also show high turning rates (Domenici and Batty, 1997). However, long latency responses may include both high and low turning rates (Domenici and Batty, 1997; Fig. 5.13B and inset). It is therefore possible that long latency/slow turning rate escapes are non M-cell responses, while fast turning rates responses (either with short or long latencies) may be M-cell responses.Occasionally, fast turning responses occur with long latency, possibly because of a delay in the stimulus threshold triggering a fast response, or to temporal inhibitions in the firing of Mauthner cells. These are hypotheses that need to be tested using M-cell recording.

Just as for low turning rates, long latencies occur more frequently at longer stimulus distance (Domenici and Batty, 1997). In addition, schooling increases the proportion of these long latencies. Domenici and Batty (1997) hypothesize that schooling may raise the threshold for initiation of fast escape responses, causing longer latencies and slower responses which are more appropriate in their directionality and reduce the possibility of collisions

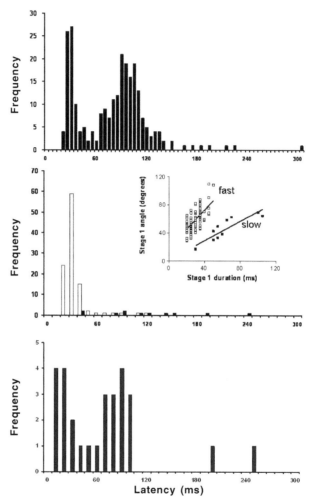

Fig. 5.13   A: Frequency distribution of escape latency in responses to a sound stimulus in schooling herring (*Clupea harengus*). From Domenici and Batty (1994). Two peaks are present, at around 30 ms and a broad peak around 100 ms.
B: Frequency distribution of escape latency in responses to a sound stimulus in solitary herring (*Clupea harengus*). From Domenici and Batty (1997) Reproduced with kind permission of Springer Science+Business Media. © Springer-Verlag 1997. A main peak is present, at around 30 ms. Various responses with latencies > 50 ms are also present. Inset shows the relationship between stage 1 angle and stage 1 duration. Two distinct regressions are shown, i.e. fast responses (with high turning rates) and slow responses (with slow turning rates). All slow responses show long latencies (>50 ms; shaded bars in the frequency distribution graph), while most fast responses show short latencies (<50 ms; open bars in the frequency distribution graph).
C: Frequency distribution of escape latency in responses to mechanical stimulation in black goby (*Gobius niger*). Adapted with permission of the Journal of Experimental Biology, from Turesson et al., 2009. Two peaks in latency, similar to those found in schooling herring, are shown.

with neighbors. Fig. 5.13 shows that long latencies are more common in schooling than in solitary herring. It cannot be excluded that some of the long latency responses observed in schooling herring may be due to responses to neighbors, rather than to the artificial sound stimulation provided experimentally (Domenici and Batty, 1994). Nevertheless, short and long latencies are not specific to gregarious species such as herring, since they are also found in non-gregarious species (e.g., black gobies; Fig. 5.13). Similarly to the various escape types discussed above, the presence of different latencies is a source of high variability, leading to a relatively high CV.

It is also likely that a source of variability may be the stimulus type, which is not always standardized nor reported, in terms of distance and direction. Certainly, varying stimulus orientation can affect turning angles (Domenici and Blake, 1993a), and varying stimulus distance can affect latency and turning rates (Domenici and Batty, 1997). Further work on the variability of escape response triggered by standardized stimuli would be necessary in order to partition the variability due to different stimulus characteristics, from intrinsic escape variability.

Size effects. A synthesis

Scaling of the kinematics and performance of fast start is of great ecological interest, since it can provide testable hypotheses for the scaling of escape success, a fundamental component of predator-prey relationships. The scaling of a number of kinematic variables in fast starts is discussed in detail in Domenici (2001). Here, I will provide a synthesis (focusing on adult fish) including recent findings, and point out open questions.

Table 5.1 provides a summary of the scaling of fast start variables, including both behavioral and kinematic variables. These size effects may be relevant both for the scaling of prey responses (behavioral and kinematic variables), and for the scaling of fast-start performance in terms of predators (large) *vs.* prey (small) (only kinematic variables) in adult fishes. Ontogenetic studies including larval and juvenile fish may show different scaling than that indicated in Table 5.1, especially because of morphological changes associated with ontogeny. For example, while acceleration appears to be size independent in adult fish (Table 5.1; Webb, 1976; Domenici and Blake 1993b), it may increase with size in relation to changes in morphology resulting from ontogeny (Hale, 1999; Wakeling *et al.*, 1999).

Regarding the scaling of kinematic performance within the context of predator-prey encounters, it is clear from Table 5.1 that small fish (i.e., prey) show superior performance compared to large fish (i.e., predators) in most aspects of unsteady maneuvers, i.e., they show superior or equal fast-start kinematic performance level (in maneuverability and distance-time performance, respectively) when compared to large fish, except for burst

Table 5.1 Size effects on behavioral and kinematic variables in escape responses of adult fishes. ⌐ and ⌐ represent a size-related increase and a decrease, respectively.

Variable type	Variable	Size effect	Performance level	Reference
Escape behavior	Responsiveness	=	=	(a)
	Latency	=	=	(b)
	Prey Reaction distance	↗*	↗*	(c) (a)
	Directionality	*Not tested*		
	Escape trajectories	=	=	(a) (d)
Kinematics (maneuverability)	Turning radius	↗	↘	(e) (f)
	Maximum turning angles	↘ =	↘ =	(g) § (f)
	Turning rate	↘	↘	(f)
Kinematics (Distance-time variables)	Fast start duration	↗	#	(e) (g)
	Distance covered during fast start	↗	#	(e) (g)
	Distance covered within a fixed time	=	=	(e) (g)
	Speed within fast start	↗	#	(e) (f)
	Speed within fast start	=		(g)
	Speed within a fixed time	=	=	(e) (g)
	Burst speed	↗	↗	(h) $
	Acceleration	=	=	(e) (g) (f)

(a) Paglianti and Domenici (2006); (b) Turesson and Domenici (2007); (c) Grant and Noakes (1987); (d) Domenici and Blake (1993a); (e) Webb (1976); (f) Domenici (2001); (g) Domenici and Blake (1993b); (h) Videler (1993).

§ - only for double bend responses.

$ - for swimming performance beyond fast start stages.

* - Abrahams (1985) found the opposite trend on armoured species and no trend for non-armoured prey (see text).

- Longer distance covered and higher speed within a fast start in large fish do not necessarily imply a higher performance from an ecological perspective, since fast start duration also increases with fish size.

swimming which includes swimming performance beyond the duration of stage 1 and 2 of a fast start. However, piscivorous predators regularly catch their prey (Chapter 6 by Rice and Hale). This may be related to a number of considerations. One of the key issues to be considered here is the scaling of acceleration, burst swimming and maneuverability. Experimental results suggest that acceleration is size-independent. Consequently, speed within a fixed (and relatively short) time, is also size independent (Webb, 1976; Domenici and Blake, 1993b). However, once the escape response in underway, any further linear progression (i.e., a chase) would be to the advantage of the predator, because, if time is available, large fish (predators) can reach a higher burst speed than their (smaller) prey (burst swimming, Table 5.1). Prey can try to counteract this problem by swerving, which they can do at a faster turning rate and smaller radius than their predators (Domenici, 2001; Table 5.1). In addition, predators may catch their prey by surprise, i.e., if attacking from a hiding place.

Regarding the scaling of prey response, a fundamental component is that of reaction distance. If predators manage to get relatively close to the prey before lunging, then prey might not be able to react in time, and any potential advantage in unsteady locomotion they have (e.g., in maneuverability) is of no use. Reaction distance was indeed found to be a fundamental component of escape success in a number of studies (Webb, 1986; Scharf et al., 2003; Walker et al., 2005). Within this context, the body size of the prey and its relative vulnerability also need to be taken into account. A lot of work has focused on considering the center of mass of the prey as the point of interest for evaluating escape performance (reviewed in Domenici and Blake, 1997). This simplification relates to the fact that this point is the one at which predators usually aim, and it approximates the point upon which forces act (Webb, 1978). While this simplification may be necessary in biomechanical studies, ecologically relevant studies focusing on scaling need to take into account the different body sizes of prey. Therefore, while a small prey that moves 5 cm away from its initial position in 50 ms may end up outside a predator's reach, a large prey that travels the same distance within the same time frame, may have its tail end within predator's reach (Domenici, 2002a; Paglianti and Domenici, 2006). Paglianti and Domenici found that large prey have longer reaction distances than small prey, and suggest that this may be a mechanism for compensating for the large body area that could be bit by a predator. A further consideration (also discussed below) is that the type of predator-prey encounters may depend largely on the fish's behavior and the habitat in which it lives. For example, predator-prey encounters may be based more on long chases in open water habitats, while they are usually brief encounters in structurally complex habitats that offer plenty of refuge opportunities. In open habitats, burst and endurance swimming performances rather than performance in single fast starts may be important factors determining the outcome of the encounter, and these are higher in larger fish (Guinet et al., 2007).

The outcome of predator prey encounters therefore depends on an arms race where certain factors may be to the advantage of the prey and others to the advantage of the predator. Predators can minimize prey reaction distances (especially in complex habitats where predators could attack from hiding places). In complex habitats, if not successful with the first attack, predators may loose their prey since prey can swerve and use their higher maneuverability, eventually finding a refuge to avoid predation. In an open habitat without refuges, predators could eventually catch their prey by using chases, since they have a higher swimming performance in terms of burst and endurance swimming (Videler, 1993).

Performance, habitat type and functional morphology

Fish body form is the result of adaptations for various functions (Alexander, 1974; Domenici, 2003). These are not restricted to swimming, but also conspicuousness, foraging, defense against predators, and sensory performance (Domenici, 2003). Similarly, swimming performance is not only the result of fish body shape, but is also affected by a number of other "internal" factors (the design box, *sensu* Domenici, 2003, Fig. 5.14), such as anatomical, physiological and biochemical ones (e.g., gill area, skeletal properties, flexibility, skin properties, enzyme activity, proportion of red and white muscle, muscle fiber properties etc). In addition, body shape may have important effects on various other factors that are not directly related to swimming performance, such as feeding (how close a predator can get near a prey; Dill, 1974a; Webb, 1986) and vulnerability (i.e., deep body form against gape limited predation; Bronmark and Miner, 1992) . Therefore, fish shape and swimming performance are not directly linked in a cause-effect relationship (Domenici, 2003). This makes it challenging for any given "design" factor to be a reliable predictor of swimming performance. For example, a variety of body shapes were found to show similar escape performance (Domenici and Blake, 1997; Domenici, 2003). Similarly, knifefish show relatively high fast start performance although they were expected to show low fast start performance based on functional morphology theories, since they lack a caudal fin (Kasapi *et al.,* 1993).

The strength of the relationship between body design and performance may also depend in part on the type of habitat considered. For example, in structurally complex habitats, where many species do not swim continuously, design constraints related to drag minimization are not as severe as in pelagic habitats, but they may depend largely on the specific lifestyle of that species (e.g., living in burrows, or in association with the bottom, among rocks, etc.). Accordingly, fish from complex habitats show a high variety of body shapes, which may be associated with their mode of life (Domenici, 2003). For examples, predators may tend to have a narrow body profile which allows

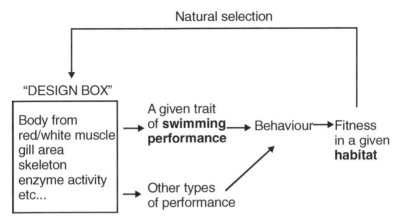

Fig. 5.14 Diagram illustrating the relationships between swimming performance, design and habitat. The 'design box' includes various phenotypic characteristics such as body form, gill area, etc. Each of the design characteristics can affect given traits of swimming performance and other types of performance unrelated to swimming. All these performances have an effect on fitness in a given habitat type, and this effect is modulated by behavior. The level of fitness will determine the extent of natural selective pressure on design traits. From Domenici (2003).

them to get near a prey before the prey reacts (Webb, 1984b), and prey may show deep body forms that may deter gape limited predators (Bronmark and Miner, 1992). On the other hand, most pelagic fish swim continuously (many of them have no choice since they are obligate ram ventilators, e.g., many scombrids) and their body fineness ratio is very close to optimal (Length/ Depth =4.5; Blake, 1983).

Webb (1984a) identified specific features that should result in high performance in one of three swimming performance areas: cruising, acceleration and maneuverability. Webb's scheme suggests that high performance in any one area results in low performance in the other two. This scheme has been amended by Domenici and Blake (1991, 1997). While a trade-off between cruising and unsteady swimming performance was confirmed in various studies (low steady swimming performance in accelerators; Webb, 1988; low maneuverability in cruisers, Blake *et al.,* 1995), the trade off between "maneuverers" and "accelerators" is less straightforward. Typical maneuverers such as angelfish were found to have a high acceleration performance. This is because "maneuver" specialists such as the angelfish swim continuously using their pectoral fins, and they use a decoupled system (axial locomotion) for their high performance fast starts. In decoupled systems, adaptation for one may not necessarily impair the other (Domenici, 2003). Similarly, accelerator specialists such as pike show high maneuverability (Domenici and Blake, 1997). A new scheme largely based on Webb (1984) shows the characteristics of fish with high unsteady swimming performance *vs.* fish with high steady swimming performance (Table 5.2). The characteristics of these two "extremes" clearly trade off. Similar trade-offs were suggested for fin shapes in pectoral fin swimmers, where high aspect ratio fins were associated with flapping at relatively high speeds, while low aspect ratio fins are associated with rowing at lower speeds, but with higher maneuverability (Drucker *et al.*, 2006).

Domenici (2003) proposed an alternative way of predicting swimming performance, based on habitat type (CH, Structurally complex habitats, such as coral reefs and weedy rivers, OH, open habitats, such as the pelagic zone, and IH, habitats of intermediate complexity). The link between habitat type and swimming performance was not suggested as an alternative to the link between design and performance (Arnold, 1983), but simply as a factor to be integrated within the overall scheme that takes design, performance, behavior and habitat use into account. Fish living in any given habitat supposedly have evolved certain swimming performance characteristics that are suited for that habitat (e.g., high endurance for open habitats, high maneuverability for structurally complex habitats). As a result, Domenici (2003) found that the unsteady swimming performance (turning radius and acceleration) of fish from structurally complex habitats (CH fishes) was higher than that of fish from habitat of intermediate structural complexity (IH fishes) and fish from open habitats (OH Fishes). Similarly, the endurance of fish from open habitats was the highest. According to Domenici (2003), swimming patterns are related to habitat type. CH fish tend to swim either slowly or to perform acceleration bursts at high speed, while OH fish tend to swim at intermediate speeds most of the time. IH fishes exploit most of the range of swimming patterns, although on average their swimming performance is lower in any one area than that of specialists. CH fishes use slow maneuvers whether they are piscivorous predators approaching a prey or particulate feeders searching for their food. In addition, CH species exploit swimming speeds at the other end of the spectrum, indicatively between 10 and 25 length/s. These high burst swimming events are used by both predators and prey in short encounters involving high acceleration maneuvers. OH fish, on the other hand, employ intermediate cruising speeds (indicatively between 0.5 and 5 length/s) most of the time while searching for widely dispersed food. Their swimming behavior is in accordance with their high endurance capabilities. They occasionally achieve higher speeds, up to about 15 length/s, when engaged in predator–prey encounters or chases. However, their unsteady swimming performance is the poorest (Domenici, 2003).

Further analysis carried out by Langerhans and Reznick (Chapter 7 in this book) on the same group of species used by Domenici (2003) shows that these species tend to evolve predictable morphologies along with relative swimming performance in different habitat types. Therefore, despite

Table 5.2   Habitat type, morphological features and predicted performance. The features identified are some examples of those that should lead to high performance in one given area, and compromise performance in another area of swimming. Not all features are expected to be present at once in any given species living in a given environment, since body form and swimming performance are not directly linked in a cause-effect relationship (see text for details). (Largely based on Webb, 1984a; Domenici, 2003; Drucker *et al.*, 2006).

Habitat type	*Feature*	*High performance*	*Trading off*
Structurally Complex Habitats	High body flexibility	High acceleration (large amplitudes) and high maneuverability (small radii)	steady swimming because of large recoil, hence high drag
	Large body depth, especially caudally	High acceleration (high thrust)	Steady swimming because of high drag
	Large tail and caudal peduncle	High acceleration (high thrust)	Steady swimming because of high drag
	Lateral insertion of low aspect ratio, large pectoral fins with many degrees of freedom	High maneuverability	Pectoral fins for hydrodynamic lift
	High proportion of white/red muscle	High power during accelerations	Steady swimming (endurance- aerobic swimming)
Open habitats	High body rigidity	Minimize recoil and hence drag in steady swimming	Unsteady maneuvers (low amplitudes and large radii)
	Body depth near 4.5 fineness ratio	Minimize drag in steady swimming	Acceleration (Thrust production)
	Lunate tail and narrow caudal peduncle	Minimize drag in steady swimming	Acceleration (Thrust production)
	Pectoral fins as stiff hydroplanes	Hydrodynamic lift in sustained swimming	Maneuverability
	High proportion of red/white muscle	High endurance (aerobic swimming)	Acceleration (anaerobic swimming)

the complicated nature of the morphology → performance relationships, trends across a diverse group of fishes may be predicted. Recent work using intra-specific comparisons is promising in linking performance with design without other taxonomically confounding factors. For example, Langerhans *et al.* (2004) show that populations of mosquitofish (*Gambusia affinis*) subject to different predator pressures have different morphologies. In line with functional morphology theories, individuals coexisting with piscivorous fish have a larger caudal region and a shallower anterior body/head region, associated with higher fast start speed, compared to individuals from predator-free sites. Crucian carp are an interesting model to study the relationship between body form and function, because their phenotypic plasticity allows them to change shape within a relatively short time (about

10% increase in body depth in two months) when in the presence of predators (Bronmark and Pettersson 1994). In line with functional morphology predictions, Domenici *et al.* (2008) show that these deep-bodied crucian carp have higher unsteady swimming performance (speed, acceleration and turning rates) than shallow-bodied individuals from predator-free sites.

Predator–prey relationships and the ecological significance of escape responses

Work on escape responses can be of great ecological value, since most fish species use escape responses to avoid being preyed upon, and therefore their escape performance can be directly linked to their overall fitness. While locomotion represents an important component of escape performance, it is clear from a number of studies (Webb, 1986; Scharf *et al.*, 2003; Walker *et al.*, 2005; Fuiman *et al.*, 2006) that other "non-locomotor" components such as reaction distance and responsiveness may determine the outcome of predator-prey interactions. Therefore, for escape response studies to be ecologically relevant, they need to use an integrative approach in which all potentially relevant variables (locomotor and "non-locomotor ones) are considered. This approach can allow investigators to make predictions about the effects of various "internal" (e.g., predator and prey sizes) and "external" (e.g., environmental) variables on predator-prey interactions.

In escape response studies, at least three levels of investigation have been adopted, which need to be further integrated. In the majority of studies, escape responses have been observed as a reaction to a standardized stimulus (e.g., Webb, 1976; Langerhans *et al.*, 2004; LeFrançois *et al.*, 2005). These studies can be useful in making predictions about the behavior of the prey, but do not yield any information on the behavior of their predators. For example, while prey size has an effect on reaction distance (Paglianti and Domenici, 2006), it is also possible that it may have an effect on predator attack strategy. Similarly, oxygen level has an effect on escape response (Domenici *et al.*, 2007), but it may also affect the behavior of various species of predators. On the other hand, studies using video observations of predator-prey staged encounters may be more realistic in determining the variables which affect the outcome of escape responses (Webb, 1986; Walker *et al.*, 2005; Gremillet *et al.*, 2006; Enstipp *et al.*, 2007), although they may suffer from the artificial effect of confinement which does not provide predators and prey with all the options available in their natural environment (e.g., hiding places).

The most complex level of investigation is that of observing predator-prey interactions in natural conditions (e.g., Domenici *et al.*, 2000; Fuiman *et al.*, 2002; Guinet *et al.*, 2007). While this type of study is quite challenging and may also suffer from potentially confounding factors related to multiple environmental effects, it offers the opportunity to learn how predators and prey interact and may therefore give indications of what is relevant to focus on when predictions are to be made for any given population. The use of miniature video cameras mounted on predators (Davis *et al.*, 1999; Fuiman *et al.*, 2002; Gremillet *et al.*, 2006) has provided useful information on their attack strategies and success as well as on the behavior of their prey. This kind of information is fundamental in order to link the factors that modulate the behavior of predator and prey (e.g., the locomotor performance) with ecologically relevant variables such as the distribution and the abundance of fish in their natural environment. Similarly, as argued by Domenici (2003), the outcome of predator-prey encounters may obey different rules in habitats with different characteristics. In the pelagic environments, stamina at relatively high speeds (e.g., Guinet *et al.*, 2007) may be the most relevant variable, while in habitats that offer refuges, high unsteady swimming performance and maneuverability may be more important. In addition, predator and prey lifestyles may differ within a given habitat. For example, sympatric predator species may have different attack strategies that range from ambush (e.g., pike) to chasing (e.g., pikeperch, trout) (Webb, 1984a; Turesson and Bronmark, 2004). Therefore, considerations on the species' lifestyle and habitats need to be taken into account when inference on escape or attack success are to be made based on their swimming performance.

CONCLUSIONS

Recent work suggests that escape responses are not stereotypic but show a relatively high variability in performance levels, which may be in part due to distinct response "types" with different strengths. This variability allows fish to adapt their response type to the level of threat they are facing. Integration of variables that are typically "biomechanical" (locomotion), with behavioral ones (i.e., related to the reactivity of fish) is necessary in order to fully assess the escape performance of fish within an ecologically relevant context. Work on these different types of variables needs to be undertaken using a multidisciplinary approach. This makes the escape response a promising model of integrative biology, since it encompasses various fields of study, from physiology and biomechanics to behavior and ecology. Integration of laboratory work, staged arenas and field studies is also necessary, since each level of investigation can be used for testing different hypotheses and is complementary to the other levels of investigation.

ACKNOWLEDGEMENTS

I wish to thank Brian Langerhans and Melina Hale for very useful comments on an earlier version of this chapter, and B. Carlson for providing the coefficient of variation data of the electric organ discharge by mormyrid fish.

REFERENCES

Abrahams, M.V. 1985. The interaction between antipredator behaviour and antipredator morphology: Experiments with fathead minnows and brook sticklebacks. *Canadian Journal of Zoology* 73: 2209–2215.

Aleev, Y.G. 1977. *Nekton*. W. Junk, The Hague.

Alexander, R. McN. 1974 Functional design in fishes. Hutchinson Publishers, London.

Arnold, S.J. 1983. Morphology, performance and fitness. *American Zoologist* 23: 347–361.

Banet, M., G. Serena and P. Domenici. (In Preparation). Wall effects on escape response kinematics in staghorn sculpins.

Barimo, J.F. and M.L. Fine. 1998. Relationship of swim-bladder shape to the directionality pattern of underwater sound in the oyster toadfish. *Canadian Journal of Zoology*. 76: 134–143.

Batty, R.S. 1989. Escape responses of herring larvae to visual stimuli. *Journal of the Marine Biological Association of the United Kingdom*. 69: 647–654.

Bekoff, M. 1977. Social communication in canids: Evidence for the evolution of a stereotyped mammalian display. *Science* 197: 1097–1099.

Bierman, H.S. J.E. Schriefer, S.J. Zottoli and M.E. Hale. 2004. The effects of head and tail stimulation on the withdrawal startle response of the rope fish (*Erpetoichthys calabaricus*) *Journal of Experimental Biology* 207: 3985–3997.

Blake, R.W. 1983. Fish Locomotion. Cambridge University Press, Cambridge.

Blake, R.W., L.M. Chatters and P. Domenici.1995. Turning radius of yellowfin tuna (*Thunnus albacares*) in unsteady swimming manoeuvres. *Journal of Fish Biology* 46: 536–538.

Blaxter J.H.S. and R.S. Batty. 1985. The development of startle responses in herring larvae. *Journal of the Marine Biological Association of the United Kingdom* 65: 737–750.

Blaxter J.H.S. and R.S. Batty. 1987. Comparisons of herring behavior in the light and dark: changes in activity and responses to sound. *Journal of the Marine Biological Association of the United Kingdom* 67: 849–859.

Blaxter, J.H.S. and L.A. Fuiman. 1990. The role of the sensory system of herring larvae in evading predatory fishes. *Journal of the Marine Biological Association of the United Kingdom* 70: 413–427.

Blaxter J.H.S., J.A.B Gray and E.J. Denton. 1981. Sound and startle responses in herring shoals. *Journal of the Marine Biological Association of the United Kingdom* 61: 851–869.

Brainerd, E.L. and S.N. Patek. 1998. Vertebral column morphology, C-start curvature, and the evolution of mechanical defenses in Tetraodontiform fishes. *Copeia* 1998, 971–984.

Brönmark, C. and J.G. Miner, 1992. Predator-induced phenotypical change in body morphology in crucian carp. *Science* 258: 1348–1350.

Brönmark, C. and L.B. Pettersson. 1994. Chemical cues from piscivores induce a change in morphology in crucian carp. *Oikos* 70: 396–402.

Brown, G.E. and R.J.F. Smith 1998. Acquired predator recognition in juvenile rainbow trout (*Oncorhynchus mykiss*): conditioning hatchery-reared fish to recognize chemical cues of a predator. *Canadian Journal of Fisheries and Aquatic Sciences* 55: 611–617.

Brown, G.E. and J. Cowan. 2000. Foraging trade-offs and predator inspection in an Ostariophysan fish: Switching from chemical to visual cues. *Behavior* 137: 181–195

Brown, G.E., J.C. Adrian, T. Patton and D.P. Chivers. 2001. Fathead minnows learn to recognize predator odour when exposed to concentrations of artificial alarm pheromone below their behavioral-response threshold. *Canadian Journal of Zoology* 79: 2239–2245.

Canfield, J.G. 2006. Functional evidence for visuospatial coding in the Mauthner neuron. *Brain Behavior and Evolution*. 67: 188–202.

Canfield, J.G and R.C. Eaton. 1990. Swimbladder acoustic pressure transduction initiates Mauthner-mediated escape. *Nature* (London) 347: 760–762.

Canfield, J.G. and G.J. Rose. 1993. Activation of Mauthner neurons during prey capture. *Journal of Comparative Physiology* A 172: 611–618.

Cannas, M., A. Bayle , O.L.C. Wing, P. Domenici and C. Lefrançois. 2007. Behaviour and physiology of the startle response in common sole (*Solea solea*) exposed to hypoxia. *Comparative Biochemistry and Physiology* A 146: S85.

Carlson, B.A. and C.D. Hopkins. 2004. Stereotyped temporal patterns in electrical communication. *Animal Behaviour* 68: 867–878

Claireaux G., C. Handelsman, E. Standen and J.A. Nelson. 2007. Thermal and temporal stability of swimming performances in the European sea bass. *Physiological and Biochemical Zoology* 80: 186–196.

Cooke S.J., J. Steinmetz, J.F. Degner, E.C. Grant and D.P. Philipp. 2003. Metabolic fright responses of different-sized largemouth bass (*Micropterus salmoides*) to two avian predators show variations in nonlethal energetic costs. Canadian Journal of Zoology 81: 699–709.

Currie, S.N. and R.C. Carlsen. 1987. Functional significance and neural basis of larval lamprey startle behaviour. Journal of Experimental Biology 133: 121–135.

Daniel, T.L. and P.W. Webb, 1987. Physical determinants of locomotion. In: *Comparative Physiology: Life in Water and on Land*. P. Dejours, L. Bolis, C.R. Taylor, E.R. Weibel (Eds.). Liviana Press, Padova, pp. 343–369.

Davis R.W., L.A. Fuiman, T.M. Williams, S.O. Collier, W.P. Hagey , S.B. Kanatous, S. Kohin and M. Horning 1999. Hunting behavior of a marine mammal beneath the Antarctic fast ice *Science* 283: 993–996.

DeLucia, A., H. Turesson, S. Como and P. Domenici. The effect of a Marine Protected Area on fish reaction distance (In Preparation).

Dill, L.M. 1974a. The escape response of the zebra danio (*Brachydanio rerio*). I. The stimulus for escape. *Animal Behaviour* 22: 710–721.

Dill, L.M. 1974b. The escape response of the zebra danio (*Brachydanio rerio*). II. The effect of experience. *Animal Behaviour* 22: 723–730.

Dohm M.R. 2002. Repeatability estimates do not always set an upper limit to heritability. *Functional Ecology* 16: 273–280.

Domenici, P. 2001. Scaling the locomotor performance in predator-prey interactions: from fish to killer whales. *Comparative Biochemistry and Physiology* A. 131: 169–182.

Domenici, P. 2002a. The visually-mediated escape response in fish: predicting prey responsiveness and the locomotor behaviour of predators and prey. *Marine and Freshwater Behavior and Physiology* 35: 87–110.

Domenici, P. 2002b. Escape trajectory, ecological. In: *Encyclopedia of Environmetrics*. Abdel H. El-Shaarawi and Walter W. Piegorsch, (Eds.). John Wiley & Sons, Chichester, Vol. 2, pp. 708–711.

Domenici, P. 2003. Habitat, body design and the swimming performance of fish. In: *Vertebrate Biomechanics and Evolution*, V.L. Bels, J.-P. Gasc and A. Casinos (Eds.). BIOS Scientific Publishers Ltd, Oxford, pp. 137–160.

Domenici, P. and R.S. Batty. 1994. Escape manoeuvres in schooling *Clupea harengus*. *Journal of Fish Biology* 45 (Supplement A): 97–110.

Domenici, P. and Batty, R.S. 1997. The escape behaviour of solitary herring (*Clupea harengus* L.) and comparisons with schooling individuals. *Marine Biology*, 128: 29–38.

Domenici, P. and R.W. Blake. 1991. The kinematics and performance of the escape response in the angelfish, *Pterophyllum eimekei. Journal of Experimental Biology* 156: 187–205

Domenici, P. and R.W. Blake 1993a. Escape trajectories in angelfish (*Pterophyllum eimekei*). *Journal of Experimental Biology* 177: 253–272.

Domenici, P. and R.W. Blake. 1993b. The effect of size on the kinematics and performance of angelfish (*Pterophyllum eimekei*) escape responses. *Canadian Journal of Zoology* 71: 2319–2326.

Domenici, P. and R.W. Blake. 1997. The kinematics and performance of fish fast-start swimming. *Journal of Experimental Biology* 200: 1165–1178.

Domenici, P., R.S. Batty, T. Simila and E. Ogam. 2000. Killer whales feeding on schooling herring using underwater tail-slaps: kinematic analyses of field observations. *Journal of Experimental Biology* 203: 283–294.

Domenici, P., E.M. Standen and R.P. Levine. 2004. Escape manoeuvres in the spiny dogfish (*Squalus acanthias*). *Journal of Experimental Biology* 207: 2339–2349.

Domenici, P., C. Lefrançois and A. Shingles. 2007. The effect of hypoxia on the antipredator behaviour of fish. *Philosophical Transactions of the Royal Society B* 362: 2105–2121

Domenici P., H. Turesson, J. Brodersen and C. Brönmark. 2008. Predator-induced morphology enhances escape locomotion in crucian carp. *Proceedings of the Royal Society B* 275: 195–201.

Drucker, E.G., J.A.Walker and M.W. Westneat 2006. Mechanics of pectoral fin swimming. In: *Fish Biomechanics*, (Eds.). R.E. Shadwick and G.V. Lauder, Academic Press, San Diego, CA: pp. 369–423.

Eaton, R.C., R.A. Bombardieri and O.H. Meyer. 1977. The Mauthner-initiated startle response in teleost fish. *Journal of Experimental Biology* 66: 65–81.

Eaton, R.C., W.A. Lavender and C.M. Wieland. 1981. Identification of Mauthner initiated response patterns in goldfish: evidence from simultaneous cinematography and electrophysiology. *Journal of Comparative Physiology* A 144: 521–531.

Eaton, R.C. and J.T. Hackett. 1984. The role of Mauthner cells in fast-starts involving escape in teleost fish. In: *Neural Mechanisms of Startle Behavior*, R.C. Eaton (Ed.). Plenum Press, New York, pp. 213–266.

Eaton, R.C., J.J. Nissanov and C.M. Wieland. 1984. Differential activation of Mauthner and non-Mauthner startle circuits in the zebrafish-implications for functional substitution. *Journal of Comparative Physiology* A 155: 813–820.

Eaton, R.C. and D.S. Emberley. 1991. How stimulus direction determines the trajectory of the Mauthner-initiated escape response in a teleost fish. *Journal of Experimental Biology* 161: 469–487.

Eaton, R.C., R.K.K. Lee and M.B. Foremam. 2001. The Mauthner cell and other identified neurons of the brainstem escape network of fish. *Progress in Neurobiology* 63: 467– 485.

Ellis T, B.R. Howell and R.N. Hughes. 1997. The cryptic responses of hatchery-reared sole to a natural sand substratum. *Journal of Fish Biology* 51: 389–401.

Enstipp, M.R., D. Gremillet and D.R. Jones 2007. Investigating the functional link between prey abundance and seabird predatory performance *Marine Ecology Progress Series* 331: 267–279.

Fernald, R. D. (1975). Fast body turns in a cichlid fish. *Nature* 258, 228–229.

Fish, F.E. 1999. Performance constraints on the manoeuvrability of flexible and rigid biological systems. *Eleventh International Symposium on Unmanned Untethered Submersible Technology. Autonomous Undersea Systems Institute*, Durham, NH, pp. 394–406.

Foreman, M. B. and R.C. Eaton. 1993. The direction change concept for reticulospinal control of goldfish escape. *Journal of Neuroscience* 13: 4101–4133.

Fuiman, L.A. 1993. Development of predator evasion in Atlantic herring, *Clupea harengus* L. *Animal Behaviour* 45: 1101–1116.

Fuiman L.A, R.W. Davis and T.M. Williams 2002. Behavior of midwater fishes under the Antarctic ice: observations by a predator. *Marine Biology* 140: 815–822.

Fuiman L.A. and J.H. Cowan. 2003. Behaviour and recruitment success in fish larvae: repeatability and covariation of survival skills. *Ecology* 84: 53–67.

Fuiman L.A, K.A. Rose, J.H. Cowan and E.P Smith. 2006 Survival skills required for predator evasion by fish larvae and their relation to laboratory measures of performance *Animal Behaviour* 71: 1389–1399.

Gamperl, A.K., D.L. Schnurr and E.D. Stevens. 1991. Effect of a sprint-training protocol on acceleration performance in rainbow trout *Salmo gairdneri*. *Canadian Journal of Zoology* 69: 578–582.

Gibson, S. and I.A. Johnston. 1995. Scaling relationships, individual variation and the influence of temperature on maximum swimming speed in early settled stages of the turbot *Scophthalmus maximus*. *Marine Biology* 121: 401–408.

Godin, J.G.J. 1997. Evading predators. In: *Behavioural Ecology of Teleost Fishes*, J.G.J. Godin (Ed.). Oxford University Press, Oxford, pp. 191–236.

Grant, J.W.A. and D.L.G. Noakes. 1987. Escape behaviour and use of cover by young-ofthe- year brook trout, *Salvelinus fontinalis*. *Canadian Journal of Fisheries and Aquatic Sciences* 45: 1390–1396.

Gregory R.T. and C.M. Wood. 1998. Individual variation and interrelationships between swimming performance, growth rate, and feeding in juvenile rainbow trout (*Oncorhynchus mykiss*). *Canadian Journal of Fisheries and Aquatic Sciences* 55: 1583–1590.

Gremillet, D., M.R. Enstipp, M. Boudiff and H. Liu. 2006. Do cormorants injure fish without eating them? An underwater video study. *Marine Biology* 148: 1081–1087.

Guinet C., P. Domenici, R. de Stephanis, L. Barrett-Lennard, J. K. B. Ford and P. Verborgh. 2007. Killer whale predation on bluefin tuna: exploring the hypothesis of the endurance-exhaustion technique. *Marine Ecology Progress Series* 347: 111–119.

Hale, M.E. 1999. Locomotor mechanics during early life history: effects of size and ontogeny on fast-start performance of salmonid fishes. *Journal of Experimental Biology* 202: 1465–1479.

Hale M.E. 2000. Startle responses of fish without Mauthner neurons: Escape behavior of the lumpfish (*Cyclopterus lumpus*). *Biological Bulletin* 199: 180–182.

Hale, M.E. 2002. S- and C-start escape responses of the muskellunge (*Esox masquinongy*) require alternative neuromotor mechanisms. *Journal of Experimental Biology* 205: 2005–2016.

Hale, M.E., J.H. Long, M.J. McHenry and M.W. Westneat. 2002. Evolution of behavior and neural control of the fast-start escape response. *Evolution* 56: 993–1007.

Harper, D.G.and R.W. Blake. 1989. On the error involved in high speed film when used to evaluate maximum accelerations of fish. *Canadian Journal of Zoology* 67: 1929–1936.

Harper, D.G and R.W: Blake. 1990. Fast-start performance of rainbow trout *Salmo gairdneri* and northern pike *Esox lucius*. *Journal of Experimental Biology* 150: 321–342.

Hawkins A.D. and K. Horner. 1981. Directional characteristics of primary auditory neurons from the cod ear. In: *Hearing and Sound Communication in Fishes*, W.N. Tavolga, A.N. Popper and R. Fay, (Eds.). Springer-Verlag, New York, pp. 311–327.

Hawkins L.A., J.D. Armstrong and A.E. Magurran. 2004. Predator-induced hyperventilation in wild and hatchery Atlantic salmon fry. *Journal of Fish Biology* 65: 88–100.

Higham, T.E., B. Malas, B.C. Jayne and G.V. Lauder. 2005. Constraints on starting and stopping: behavior compensates for reduced pectoral fin area during braking of the bluegill sunfish *Lepomis macrochirus*. *Journal of Experimental Biology* 208: 4735–4746

Howland, H.C. 1974. Optimal strategies for predator avoidance: the relative importance of speed and manoeuvrability. *Journal of Theoretical Biology* 134: 56–76.

Huey, R.B. and Dunham. A.E. 1987. Repeatability of locomotor performance in natural populations of the lizard *Sceloporus merriami*. *Evolution* 41: 1116–1987.

Huntingford, F.A., P.J. Wright and J.F. Tierney. 1994. Adaptive variation in antipredator behaviour in threespine stickleback. In: *The Evolutionary Biology of the threespine stickleback*. M.A Bell and S.A. Foster (Eds.). Oxford University Press, Oxford, pp. 277–296.

Jayne, B.C. and G.V. Lauder. 1993. Red and white muscle activity and kinematics of the escape response of bluegill sunfish during swimming. *Journal of Comparative Physiology* A173: 495–508.

Kasapi, M., P. Domenici, R.W. Blake and D.G. Harper. 1993. The kinematics and performance of the escape response in the knife fish (*Xenomystus nigri*). *Canadian Journal of Zoology* 71: 189–195.

Kohashi, T. and Y. Oda . 2008. Initiation of Mauthner or Non-Mauthner-Mediated Fast Escape Evoked by Different Modes of Sensory Input. *Journal of Neuroscience* 28 (42): 10641–10653

Kolok, A.S. 1992. The swimming performances of individual largemouth bass, Micropterus salmoides, are repeatable. *Journal of Experimental Biology* 170: 265–270.

Langerhans, R.B., C.A. Layman, M. Shokrollahi and T. DeWitt, 2004. Predator-driven phenotypic diversification in *Gambusia affinis*. *Evolution* 58: 2305–2318.

Langerhans, R.B., C.A. Layman and T.J. DeWitt. 2005. Male genital size reflects a tradeoff between attracting mates and avoiding predators in two live-bearing fish species. *Proceedings of the National Academy of Sciences of the United States of America* 102: 7618–7623.

Lefrançois, C., A. Shingles and P. Domenici. 2005. The effect of hypoxia on locomotor performance and behaviour during escape in *Liza aurata*. *Journal of Fish Biology* 67: 1711–1729.

Lefrançois, C. and P. Domenici. 2006. Locomotor kinematics and behaviour in the escape response of European sea bass, *Dicentrarchus labrax* L., exposed to hypoxia. *Marine Biology* 149: 969–977

Liu K.S. and J.R. Fetcho. 1999. Laser ablations reveal functional relationships of segmental hindbrain neurons in zebrafish. *Neuron.* 23: 325–335.

Martinez, M., H. Guderley, J.A. Nelson, D. Webber and J.D. Dutil. 2002. Once a fast cod, always a fast cod: maintenance of performance hierarchies despite changing food availability in cod (*Gadus morhua*). *Physiological and Biochemical Zoology* 75: 90–100.

Matott, M.P., P.J. Motta and R.E. Hueter. 2005. Modulation in feeding kinematics and motor pattern of the nurse shark *Ginglymostoma cirratum*. *Environmental Biology of Fishes* 74: 163–174.

McKenzie D, M. Hale, and P. Domenici (2007) Locomotion in primitive fishes . In: *Primitive Fishes* (Fish physiology Series, McKenzie D, Brauner C, and Farrell A eds). Academic Press. pp 319–380.

Meager, J.J., P. Domenici, A. Shingles and A.C. Utne-Palm 2006. Escape responses in juvenile Atlantic cod *Gadus morhua* L.: the effects of turbidity and predator speed. *Journal of Experimental Biology* 209: 4174–4184.

Meyers, J.R., E.H. Copanas and S.J. Zottoli. 1998. Comparison of fast startle responses between two elongate bony fish with an anguilliform type of locomotion and the implications for the underlying neuronal basis of escape behavior. *Brain Behavior and Evolution* 52: 7–22

Moulton, J. M. and R.H. Dixon. 1967. Directional hearing in fishes. In: *Marine Bioacoustics,* W. N. Tavolga (Ed.). Pergamon Press, Oxford: pp. 187–203.

Nelson, J.A. and G. Claireaux. 2005. Sprint Swimming Performance of Juvenile European Sea Bass. *Transactions of the American Fisheries Society* 134: 1274–1284.

Nelson, J.A., P.S. Gotwalt, C.A. Simonetti and J.W. Snowgrass. 2008. Environmental correlates, plasticity, and repeatability of differences in performance among Blacknose Dace (*Rhinichthys atratulus*) populations across a gradient of urbanization. *Physiological and Biochemical Zoology* 81: 25–42.

Nelson, J.A., Y. Tang and R.G. Boutilier. 1996. The effects of salinity change on the exercise performance of two Atlantic cod (*Gadus morhua*) populations inhabiting different environments. *Journal of Experimental Biology* 199: 1295–1309.

Nissanov, J., R.C. Eaton and R. DiDomenico. 1990. The motor output of the Mauthner cell, a reticulospinal command neuron. *Brain Research* 517: 88–98.

Paglianti, A. and P. Domenici (2006). The effect of size on the timing of visually mediated escape behaviour in staghorn sculpin *Leptocottus armatus*. *Journal of Fish Biology* 68: 1177–1191.

Radabaugh D.C. 1989. Season color changes and shifting antipredatro tactics in darters. *Journal of Fish Biology* 34: 679–695

Reidy S., S.R. Kerr and J.A. Nelson. 2000. Aerobic and anaerobic swimming performance of individual Atlantic cod. *Journal of Experimental Biology* 203: 347–357.

Rovainen, C.M. 1979. Electrophysiology of vestibulospinal and vestibuloreticulospinal systems in lampreys . *Journal of Neurophysiology* 42: 745–766.

Scharf, F.S., J.A. Buckel, P.A. McGinn and F. Juanes. 2003. Vulnerability of marine forage fishes to piscivory: effects of prey behavior on susceptibility to attack and capture. *Journal of Experimental Marine Biology and Ecology* 294: 41–59.

Schleidt, W.M., M.D. Shalter and T.C. Carawan. 1983. The effect of spatial context on habituation to a predator model. *Journal of Comparative Ethology* 61: 67–70.

Schriefer, J.E. and M.E. Hale, 2004. Strikes and startles of northern pike (*Esox lucius):* a comparison of muscle activity and kinematics between S-start behaviors. *Journal of Experimental Biology* 207: 535–544

Schuijf, A. 1975. Directional hearing in the cod under approximate free field conditions. *Journal of Comparative Physiology* 98: 307–332.

Shingles, A., D.J. McKenzie, G. Claireaux, G. and P. Domenici. 2005. Reflex cardioventilatory responses to hypoxia in the flathead grey mullet (*Mugil cephalus*) and their behavioral modulation by perceived threat of predation and water turbidity. *Physiological and Biochemical Zoology* 78: 744–755.

Sillar, K.T. and A. Roberts. 1988. A neuronal mechanism for sensory gating during locomotion in a vertebrate. *Nature* 331 : 262-265.

Spierts, I.L. and J.L. Van Leeuwen, 1999. Kinematics and muscle dynamics of C- and Sstarts of carp (*Cyprinus carpio* L.). *Journal of Experimental Biology* 202: 393–406.

Stefanelli, A. 1980. I neuroni di Mauthner degli Ittiopsidi. Valutazioni comparative morfologiche e funzionali. *Lincei Memorie Scienze Fisiche Matematiche e Naturali* 16: 1–45.

Svoboda K.R. and J.R. Fetcho 1996. Interactions between the neural networks for escape and swimming in goldfish. *Journal of Neuroscience* 16: 843–852.

Swanson, B.O. and A.C. Gibb 2004. Kinematics of aquatic and terrestrial escape responses in mudskippers. *Journal of Experimental Biology* 207: 4037–4044.

Turesson, H and C. Brönmark. 2004. Foraging behaviour and capture success in perch, pikeperch and pike and the effects of prey density. *Journal of Fish Biology* 65: 363–375.

Turesson, H. and P. Domenici. 2007. Escape latency is size independent in grey mullet. *Journal of Fish Biology* 71: 253–259.

Turesson, H., A. Satta and P. Domenici. 2009. Preparing for escape: Anti-predator posture and fast-start performance in gobies. *Journal of Experimental Biology.* 212: 2925–2933.

Tytell, E.D. and G.V. Lauder. 2002. The C-start escape response of *Polypterus senegalus:* bilateral muscle activity and variation during stage 1 and 2. *Journal of Experimental Biology* 205: 2591–2603.

Videler, J.J. 1993. *Fish Swimming.* Chapman and Hall, London.

Wakeling, J.M., K.M. Kemp and I.A. Johnston. 1999. The biomechanics of fast-starts during ontogeny in the common carp *Cyprinus carpio. Journal of Experimental Biology* 202: 3057–3067.

Walker, J.A. 1998. Estimating velocities and accelerations of animal locomotion: a simulation experiment comparing numerical differentiation algorithms. *Journal of Experimental Biology* 74: 211–266.

Walker, J.A. 2000. Does a rigid body limit maneuverability? *Journal of Experimental Biology* 203: 3391–3396.

Walker, J.A., C.K. Ghalambor, O.L. Griset, D. McKenney and D.N. Reznick. 2005. Do faster start increase the probability of evading predators? *Functional Ecology.* 19: 808– 815.

Ward, A.B. and E. Azizi. 2004. Convergent evolution of the head retraction escape response in elongate fishes and amphibians. *Zoology* 197: 205–217.

Wardle, C.S. 1975. Limits of fish swimming speed. *Nature* (London) 255: 725–727.

Webb, P.W. 1976. The effect of size on fast-start performance of rainbow trout *Salmo gairdneri* and a consideration of piscivorous predator–prey interaction. *Journal of Experimental Biology* 65: 157–177.

Webb, P.W. 1978. Fast-start performance and body form in seven species of teleost fish. *Journal of Experimental Biology* 74: 211–226.

Webb, P.W. 1981. Responses of northern anchovies, *Engraulis mordax,* larvae to predation by a biting planktivore, Amphiprion percula. *Fisheries Bulletin.* 79: 727–735.

Webb, P.W. 1982. Avoidance responses of fathead minnow to strikes by four teleost predators. *Journal of Comparative Physiology* A 147: 371–378.

Webb, P.W. 1984a. Body form, locomotion and foraging in aquatic vertebrates. *American Zoologist* 24: 107–120.

Webb, P.W. 1984b. Body and fin form and strike tactics of four teleost predators attacking fathead minnow (*Pimephales promelas*) prey. *Canadian Journal of Fisheries and Aquatic Sciences* 41: 157–165.

Webb, P. W. 1986. Effect of body form and response threshold on the vulnerability of four species of teleost prey attacked by largemouth bass (*Micropterus salmoides*). *Canadian Journal of Fisheries and Aquatic Sciences* 43: 763–771.

Webb, P.W. 1988. Steady swimming kinematics of tiger musky, an esociform accelerator, and rainbow trout, a generalist cruiser. *Journal of Experimental Biology* 138: 51–69.

Webb, P.W. and R.S. Keyes. 1981. Division of labor between median fins in swimming dolphin (Pisces: Coryphaenidae). *Copeia* 1981: 901–904.

Webb, P.W. and C.L. Johnsrude, 1988. The effect of size on the mechanical properties of the myotomal skeletal system of rainbow trout *Salmo gairdneri. Fish Physiology and Biochemistry* 5: 163–171.

Webb, P.W. and J.M. Skadsen. 1980. Strike tactics of Esox. *Canadian Journal of Zoology* 58: 1462–1469.

Weihs, D. 1973.The mechanism of rapid starting of slender fish. *Biorheology* 10: 343–350.

Weihs, D. and P.W. Webb. 1984. Optimal avoidance and evasion tactics in predator–prey interactions. *Journal of Theoretical Biology* 106: 189–206.

Weiss, S.A., S.J. Zottoli, S.C. Do, D.S. Faber and T. Preuss. 2006. Correlation of C-start behaviors with neural activity recorded from the hindbrain in free-swimming goldfish (*Carassius auratus*). *Journal of Experimental Biology* 209: 4788–4801.

Westneat, M.W., M.E. Hale, M.J. McHenry and J.H. Long. 1998. Mechanics of the faststart: muscle function and the role of intramuscular pressure in the escape behavior of *Amia calva* and *Polypterus palmas. Journal of Experimental Biology* 201: 3041–3055.

Wohl S. and S. Schuster. 2007. The predictive start of hunting archer fish: a flexible and precise motor pattern performed with the kinematics of an escape C-start. *Journal of Experimental Biology* 210: 311–324.

Yin, MC and J.H.S. Blaxter. 1987. Escape speeds of marine fish larvae during early development and starvation. *Marine Biology* 96: 459–468.

Ydenberg, R.C. and L.M. Dill 1986. The economics of fleeing from predators. *Advanced Study of Behaviour* 16: 229–249.

Zottoli, S.J., A.R. Hordes and D.S. Faber. 1987. Localization of optic tectal input to the ventral dendrite of the goldfish Mauthner cell. *Brain Research.* 401: 113–121.

6

Roles of Locomotion in Feeding

Aaron N. Rice[1] and Melina E. Hale[2]

INTRODUCTION

For most fish species, locomotion plays integral roles in the two fundamental phases of energy acquisition: foraging (the act of searching for suitable food items) and feeding (the physical act of food procurement). The vast majority of fishes actively swim to search for food. Some use sustained swimming to migrate to food resources or to track prey over long distances. Many intermittently swim shorter distances searching for food within much smaller spatial areas (e.g., O'Brien *et al.*, 1990). Others use subtle, fine-tuned fin movements to hold station and remain cryptic as prey come to them. Once a food item is identified, locomotion is often essential for successful prey capture. The median or paired fins and/or the body axis may be used to position the body relative to the food item for effective suction feeding, during high acceleration strikes and high-speed chases (e.g., Keast and Webb, 1966; Blake, 1983; Webb, 1984a, c, 1986; Webb and de Buffrénil, 1990; Domenici, 2001; Guinet *et al.*, 2007). Lastly, locomotion may also be used to rapidly turn, brake and reverse body movement after the food item is acquired. Thus, the phases of prey acquisition behavior, from foraging to feeding, require a range of locomotor behaviors and force the

Authors' addresses: [1]Department of Neurobiology and Behavior, Cornell University, Ithaca, NY 14853, USA. E-mail: arice@cornell.edu
[2]Department of Organismal Biology and Anatomy, University of Chicago, Chicago, IL 60637, USA. E-mail: mhale@uchicago.edu

organism to encounter varying fluid mechanical constraints for steady- (i.e., constant swimming) and unsteady-state (accelerating, turning) swimming modes (Daniel, 1984; Webb, 1988). Phases of predation and the type of locomotion typically associated with them are diagrammed in Fig. 6.1.

In this chapter, we review the use of locomotion in feeding behavior of fishes and discuss the roles, demands, and patterns of locomotion as they relate to the different aspects of the feeding cycle. Though often synonymous in the literature, here we distinguish between the terms "foraging" and "feeding", where foraging is a locomotor act of searching for food items and feeding is the specific, integrated mechanical act of locomotion and jaw movement resulting in prey capture (or attempted prey capture). Additionally, the temporal properties of these behaviors differ substantially, often by orders of magnitude: while the search for food can take a substantial amount of time, catching food happens extremely rapidly. By creating distinctions between these two behaviors, it becomes possible to establish further criteria for patterns in which to assign locomotor movements or morphologies. In this review, we present a synthesis of the interactions between locomotor and feeding mechanisms in fishes. In particular, we examine these

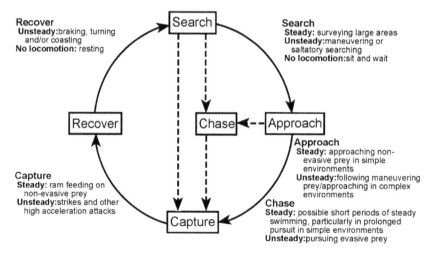

Fig. 6.1 Typical phases of a successful predation cycle (adapted from Webb, 1986). Examples of locomotor demands demonstrate diverse uses of steady and unsteady swimming during a feeding event. The types of swimming a fish may use for predation depend on a number of factors including the phase of the behavior, the movement of its prey and whether it is feeding in a structurally and/or hydrodynamically simple or complex environment. Phases of the predation cycle (in boxes) typically include the search for a food resource, approach to prey once it has been targeted, capture and recovery. Many species also chase their prey between approach and capture. Recovery is often followed by a new search for another prey item. The ordering of phases of is indicated with arrows. Dashed paths indicate alternatives or additions to the four primary events.

interactions from a mechanistic perspective of prey capture and divide the overall feeding behavior into discrete units based on locomotor behavior.

The locomotor performance parameters that may affect feeding are diverse and dependent on many intrinsic factors including body morphology (Keast and Webb, 1966; Werner, 1977), fin morphology (Webb, 1982, 1984c), feeding strategy (Rice and Westneat, 2005; Rice et al., 2007; Rice, 2008) and prey type (Vinyard, 1982); and the demands on locomotor performance may change during different periods of the prey capture event. As reviewed by Higham (2007a), during the immediate prey capture movement acceleration, deceleration, speed, movement accuracy and maneuverability are the central locomotor characteristics involved in a range of feeding strategies (also see Harris, 1937: Webb, 1984c, 1986; Nemeth, 1997; Gerstner, 1999). However, environmental factors such as habitat complexity (e.g., Crowder and Cooper, 1982; Savino and Stein, 1982; Stoner, 1982; Savino and Stein, 1989a, b; James and Heck, 1994; Domenici, 2003) and flow regime (Asaeda et al., 2001; Priyadarshana et al., 2001; Asaeda et al., 2005; Priyadarshana and Asaeda, 2007; Rincón, et al., 2007) can additionally influence predation success as a function of the fish's swimming ability. As discussed by Domenici (Chapter 5) for startles, it is also important to consider other aspects of performance that affect success of prey capture including the latency to initiate a prey capture event and ability to coordinate motor systems (Webb, 1984b).

Prey type and environmental context vary considerably among fish taxa and feeding events. Fishes have evolved to exploit almost all available trophic resources in their environment (e.g., Kotrschal, 1988, Kotrschal and Thomson, 1989). Such a diversity of prey types has required a similar array of locomotor strategies to facilitate successful prey capture. As such, fishes possess a staggering diversity of both feeding and locomotor mechanisms. This review attempts to address the synergy between these seemingly disparate mechanical systems in an attempt to elucidate broader patterns in the evolution of fishes.

The ecological attributes of prey items have important implications for the role of locomotion in the feeding (Fig. 6.2). Two of the main determinants of different feeding strategies and feeding-associated locomotion during all stages of a feeding event—before, during and after prey capture—may be the evasiveness of the prey, and the complexity of the habitat in which the prey lives. With these two major demands, the predator must be able to successfully find, approach, and outmaneuver its prey, while at the same time being able to traverse the physical demands of the habitat in which that prey item lives (Fig. 6.3).

The critical roles of locomotion in fish feeding suggest that many aspects of these behaviors and their associated morphologies are closely linked and have evolved in conjunction with one another (Rice, 2006; Collar et al., 2007; Higham, 2007a). Although there is an extensive body of work on both

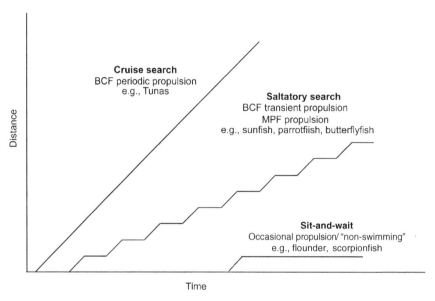

Fig. 6.2   Three types of locomotor modes employed in foraging (adapted from O'Brien *et al.* (1990) and Hart (1997)) and examples of fishes that fit these patterns.

locomotion and feeding in fishes, the relationship between them has received much less attention. We suggest that the interaction of locomotor and feeding systems provides a rich and necessary area for future work.

LOCOMOTOR DEMANDS AND FORAGING

In the broad context of foraging, many locomotor performance parameters become important (Webb, 1984c; Hart, 1997). For example, searching for widely distributed prey may necessitate efficient endurance swimming (Rice, 2008) and station holding to scan for more food items may require improved stability or postural control (Cyrus and Blaber, 1982; O'Brien *et al.*, 1989). However, such locomotor behaviors are often not considered part of the immediate prey capture event, though they directly determine feeding success. Alternative uses of the locomotor morphology in different phases of feeding and in non-feeding locomotor contexts may put competing demands on locomotor morphology. Understanding the complex integration of these varying ethological, biomechanical and ecological demands is important for understanding how these behaviors and associated morphology evolved.

The classical dichotomy for locomotor patterns during foraging is the cruising, or actively-searching, predator and the sit and wait predator (Hart, 1997) (Fig. 6.2). Clearly, these movement patterns associated with foraging represent endpoints in a continuum; one requiring efficient, sustained swimming abilities, and the other relying on some form of crypsis and then an explosive burst of movement to catch the prey. Cruising predators are likely to have more frequent encounters with less mobile prey, while sit-and-wait predators have infrequent encounters with highly evasive prey items (Norton, 1991). A third, intermediate form of foraging is known as saltatory searching (O'Brien *et al.*, 1986, 1989, 1990) (Fig. 6.2). In this searching mode, fishes search for a prey item, pursue and capture it, and come to a complete stop for handling before repeating the cycle (O'Brien *et al.*, 1986, 1990). Such a foraging mode requires energetically efficient stopping and starting (often accomplished with paired fins) (Kendall *et al.*, 2007), and is a common feeding mode for many fishes, especially planktivores (sensu Davis and Birdsong, 1973; Werner, 1977; Hobson and Chess, 1978; Webb, 1984c; O'Brien *et al.*, 1986; Hobson, 1991). These fishes will vary the duration of swimming, or the duration of stopping depending on the risk of predation or the density of suitable food items (O'Brien *et al.*, 1989).

THE ROLE OF BODY SHAPE, LOCOMOTION AND FEEDING

The overall body shape has a profound impact on the locomotor and hydrodynamic properties of fishes. In their seminal paper, Keast and Webb (1966) found that it is often easy to infer the feeding ecology and locomotor habits of a fish simply based on the fish's overall body and fin shapes. In those fishes that frequently engage in sudden accelerations, a body shape that includes characteristics such as a caudal fin with a large surface area, posteriorly placed median fins and a flexible and deep body is often selected for (e.g., Webb, 1982, 1984a, 1998; Rincón, 2007); this morphology is observed in both predators and prey (Priyadarshana and Asaeda, 2007). In one example, the esocids, an elongate body with a high lateral caudal surface area is extremely efficient for explosive acceleration as a sit-and-wait predator (Keast and Webb, 1966; Webb, 1986), but not for active searching (Webb, 1986). The elongation of the body of predators has also been suggested to decrease the reaction distance of the prey during a strike (Webb, 1986; Domenici, 2002).

Different body shapes also help potential prey fishes evading predators. Planktivorous damselfishes display variation in their body shape: those fishes which forage close to the reef have a deeper body, while those that forage in the water column, and need to rapidly retreat to the safety of the reef to evade predators have a much more fusiform body (Davis and Birdsong, 1973; Hobson and Chess, 1978; Webb, 1984c; Hobson, 1991). Alternatively, carp with a deeper body achieve a higher acceleration and velocity than shallow bodied forms, and are more successful at evading predators; this may be due to the larger proportion of muscle mass and higher thrust (Domenici *et al.*, 2008).

Thus, increased body depth may increase the probability of escaping predators, but additionally increase drag during cruising in deeper bodied morphs(Webb, 1982, 1984c, 1998).

Fishes foraging and feeding in high flow regimes often display relatively more fusiform body shapes, and as a result, increase their prey capture success and energy efficiency (Schaefer *et al.*, 1999; Rincón *et al.*, 2007; Langerhans, 2008). For fishes that station-hold in the water column, the energetic demand for this behavior is substantial, with the energetic cost being a direct function of the fish's drag multiplied by the water's current velocity (Schaefer *et al.*, 1999). In stream-dwelling minnows, overall body shape, and the associated biomechanical consequences on locomotion, is the main factor determining feeding success (Rincón *et al.*, 2007). Thus, different locomotor efficiencies during foraging and feeding due to differences in body shape in varying flow regimes and habitat complexity may be a selection pressure or sorting mechanism within microhabitats (Keast and Webb, 1966; Werner, 1977; Schaefer *et al.*, 1999; Asaeda *et al.*, 2005; Priyadarshana and Asaeda, 2007; Rincón *et al.*, 2007).

Fishes which require a lot of maneuvering when foraging and feeding typically possess a shorter deeper body, with anteriorly placed pectoral fins positioned in the middle of the body, and pelvic fins beneath the pectorals (Breder, 1926; Harris, 1937; Webb, 1982; Lauder and Drucker, 2002; Lauder *et al.*, 2002). This body form may assist in maneuvering in unsteady conditions (Webb, 1984c; Rincón *et al.*, 2007), but it significantly increases drag (and thus energy required) during steady swimming (Webb, 1984c). Although it is an important factor for fishes in a wide range of habitats, the effect of overall body shape becomes particularly pronounced for fishes feeding in hydrodynamically or physically complex environments, where this shape ultimately determines feeding success (Keast and Webb, 1966; Savino and Stein, 1982; Rincón *et al.*, 2007).

EFFECTS OF THE PHYSICAL ENVIRONMENT ON LOCOMOTION AND FORAGING

The environment in which the prey lives plays an important but as yet lesser-studied role in the predator's use of locomotion during feeding (but, for example, see Crowder and Cooper, 1982; Asaeda *et al.*, 2001, 2005). Fortunately, the interaction of behavior and environment is demonstrated many times throughout the chapters of this book. Complex structural environments offer very different challenges to prey capture from complex flow environments to open water environments. For example, topographically complex habitats will constrain predator and prey movements and strategies, in terms of the fish's ability to maneuver and encounter prey (Crowder and

Cooper, 1982). In environments of increasing complexity, swimming speeds (Savino and Stein, 1982; Diehl, 1988; Savino and Stein, 1989a; Priyadarshana et al., 2001; Priyadarshana and Asaeda, 2007), and attack rates (Priyadarshana and Asaeda, 2007) decrease substantially. Challenges to feeding with environmental complexity are particularly likely to affect locomotor aspects of feeding as it should particularly affect foraging strategy and positioning prior to the prey capture event.

A major complication of feeding in an aquatic environment faced by all fishes is that the physical medium can move. Fishes are not living in simple, still environments like those of a typical filming tank but with complex structures and/or changing flow. Both conditions would likely increase the use of locomotion during feeding. For example, in the case of structurally complex environments, maneuvering is likely to be important for searching among structures in the environment and for orienting to prey positioning themselves relative to structure. It is typically fishes employing a median-paired fin swimming mode that are specialized to forage in complex habitats (Webb, 1984c).

In the case of complex flows, water velocity can dramatically affect the fish's feeding strategy or feeding success. Fishes feeding in low flow regimes typically have a higher rate of prey capture success than those feeding at higher flow speeds (Flore and Keckeis, 1998, Asaeda et al., 2005). Fishes modulate their swimming patterns to finely adjust their position to compensate for flow-induced movement and to respond to flow-induced movement of the prey. At higher flow velocities, drift-feeding cyprinids will increase their cruising and pursuit speeds, and decrease the approach angle and attack distance relative to the prey (Asaeda et al., 2005). As a consequence of feeding in high flow regimes, the energetic expenditure can increase substantially (Asaeda et al., 2005). Successful prey capture in high flow regimes may also be dependent on body size, as larger individuals within a species typically have higher capture success (Flore and Keckeis, 1998). Schaefer et al. (1999) assert that fishes with a more fusiform body shape are much more efficient at prey capture in higher flow environments than congeners with a more stout body shape. Alternatively, Rincón et al. (2007) suggest that while streamlining of the body reduces energy expenditure for continuous swimming, it decreases efficiency in swimming during prey capture.

In a study of blennies (*Acanthemblemaria spinosa* and *A. aspera*) feeding in still and turbulent flow environments on evasive and non-evasive prey, Clarke and colleagues (2005) found that water turbulence decreased the ability of the blennies to feed on non-evasive prey with a decrease in capture success from 100% to 78%. However, they found that turbulence had the opposite effect for feeding on evasive prey, increasing their prey capture success from 21% to 56%. They suggest that while turbulence makes the movement of non-evasive prey more difficult to predict and thus catch, it also hydrodynamically masks the presence of an approaching predator thus, prey that potentially could evade capture have less time to react.

One of the major functions of the locomotion in fishes is the search for food resources. It has been shown that a considerable portion of most fishes' active periods are involved with foraging and feeding (Webb, 1984c; Hart, 1997). Prey resources are frequently patchy and may be distributed over a broad area, many move with changes in environmental conditions such as light or temperature fluctuations, or may be transiently abundant in distant locations. Feeding on widely distributed or migrating prey requires a mobile searching tactic to forage over large spatial ranges (Hart, 1997) and many of the fishes that take advantage of these resources perform highly-efficient sustained swimming (Webb and de Buffrénil, 1990; Bellwood and Wainwright, 2001; Fulton and Bellwood, 2002; Wainwright *et al.*, 2002; Domenici, 2003; Rice and Westneat, 2005).

SPECIALIZATION AND DUAL USES OF LOCOMOTOR ANATOMIES IN FORAGING

There are diverse adaptations to typically locomotor structures that are associated with foraging. For example, some fish fins have evolved to assist with prey detection. The pelvic fin rays of the hake, *Urophycis chuss*, are innervated by not only spinal (motor) nerves, but also projections from cranial nerve VII (gustatory sense), and can be used to literally taste for benthic or buried prey to localize their position (Bardach and Case, 1965). There are several fishes that use their pectoral fins to physically uncover or contact potential prey items. The goby, *Pterogobius virgo*, uses its pectoral fins and body to dig and excavate sand to expose buried polychaetes, its primary prey item (Choi and Gushima, 2002). The anterior-directed pectoral fin sweeps uncover the polychaete, and the posterior-directed sweep removes the sand from the excavation site. Similarly, parental cichlids (*Cichlasoma* [=*Archocentrus*] *nigrofasciatum*) generate thrust with their pectoral fins to uncover prey items for their fry (Wisenden *et al.*, 1995). The pectoral fin rays of the fourhorn poacher (*Hypsagonus quadricornis*) are used to move shells and rocks in search of hidden invertebrates (Jensen, 2005). The use of fins to search for fast moving prey items, allows the jaws to be strictly devoted to prey capture, rather than being constrained by competing demands of both exposing prey potential items and having to quickly grasp them (Choi and Gushima, 2002; Jensen, 2005).

Several groups of pelagic predators have evolved their axial muscle to serve not only in locomotion but also as a heater organ. Representatives of the lamnid sharks (Carey *et al.*, 1982, Block and Carey, 1985; Wolf *et al.*, 1988) and tunas (Scombridae) (e.g., Carey 1982) as well as, to a lesser extent, swordfish (*Xiphias gladius*) (Carey, 1982; Tullis *et al.*, 1991) use internalized axial red muscle and countercurrent heat exchangers to elevate the temperature of the body (reviewed with cranial endothermy by Block and Finnerty, 1994). It is believed that one of the main functions of endothermy in fishes is to allow fishes to follow migrations of prey (e.g., Carey, 1992; Graham and Dickson, 2001) through significant temperature clines.

SWIMMING DURING THE DIRECT APPROACH TO A FOOD ITEM

Once a fish detects a suitable food item while foraging, it enters the next phase of the prey acquisition process, the approach. This involves both achieving an appropriate distance from the food, approach speed, and body orientation relative to the prey's location (Nyberg, 1971, Rice *et al.*, 2008). Demands on the locomotor system vary markedly among fish depending on their food resources. At two extremes are fishes that feed on highly mobile prey that have well-developed escape responses and those that feed on attached, immobile prey. For the former, locomotion may be more important for achieving a reasonable strike distance without being detected and then rapid, and precisely directed acceleration toward the prey item during strike and pursuit swimming. For the latter, locomotion may help the fish perfectly position itself and move relative to the surface on which the food is found, and assist in detaching the item from the substrate. While some specialized fishes obligately feed on limited prey types, many are generalist feeders who can take advantage of a wide range of food resources but may be less able to exploit any particular one. Figure 6.3 diagrams some typical prey resources for fish, their evasiveness and position in the water column. We will examine differences in locomotor strategy across levels of food evasiveness and resource specialization.

Position and approach speed of the predator is often a function of the disparity in body size between predator and its prey (Webb, 1986; Webb and de Buffrénil, 1990; Domenici, 2001). As the predator's mouth size gets bigger relative to the size of the prey, the distance that the prey can escape in a given amount of time decreases (Webb and de Buffrénil, 1990), and thus large predators typically do not need to quickly maneuver for prey capture. When feeding on relatively non-evasive prey, positioning is frequently performed by the predator to optimize approach or energy demand of capturing prey (Vinyard, 1982). However, for both large and small evasive prey, positioning may rely completely or in part on the movement of the prey item (Webb, 1986). Depending on how food is sensed and the structure of the environment, this stage may involve considerable locomotion toward the food source or none at all. For example, Heyman *et al.* (2001) found

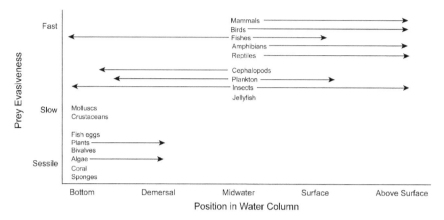

Fig. 6.3 Examples of typical prey resources of fishes, characterized as a function of prey mobility (y-axis) and habitat (x-axis). Although, many taxonomic categories of prey will span a range of levels of mobility, we show a common level of mobility for each here. The speed required to capture the prey and the physical complexity of the habitat will exercise competing demands (speed versus maneuverability) on the locomotor system of the predator.

that whale sharks (*Rhincodon typus*) would approach a gamete cloud from over 40 meters within two minutes of the spawning event. In contrast, cryptic predators frequently wait for the prey to come to them (e.g., Grobecker and Pietsch, 1979; Grobecker, 1983). Grobecker (1983) found that during some feeding events of the cryptic and benthic stonefish (*Synanceia verrucosa*), the only movement observed prior to the strike were eye movements used to track the prey item and subtle elevation of the head.

Most species range between these extremes, with fishes using finely controlled movements of the fins and body axis to position themselves for a successful strike after a food resource has been identified. While controlling body orientation and distance from prey are critical for most species, independent demands specific to prey type, environment and the kinematics of the strike influence this pre-strike behavior. This component of the feeding has not been studied in detail, with the focus of most studies being the movements associated with jaw opening and closing. Here, we discuss examples of pre-strike behavior of predators that target evasive and non-evasive prey; however, there is likely to be considerable diversity in this component of the feeding event that has yet to be explored.

Locomotor movements that precede the prey strike have been most broadly studied in fishes that target evasive prey and pre-strike movement must orient and position the body while also being subtle to avoid detection by the prey. As will be discussed in depth below, these predators often use rapid jaw protrusion and suction feeding with coordinated locomotor movement to capture prey (e.g., Westneat and Wainwright, 1989). Many use fins for steady swimming and maneuvering and are likely to use similar movement patterns in the phases leading up to feeding; however, the kinematics during the transition from foraging to feeding has not been described in detail.

One group of fish that feed on evasive prey that has been widely studied is the Esocidae, which includes pikes, pickerels and muskies. These fish are axial acceleration specialists (e.g., Webb and Skadsen, 1980; Rand and Lauder, 1981; Harper and Blake, 1991; Schriefer and Hale, 2004) and, while they may generate some suction to help draw the prey item into the oral cavity, rely heavily on rapid acceleration of the body to overcome the prey. During a feeding event, these fishes use fine fin movements to approach and orient to the prey (examined in the tiger musky, a hybrid between *Esox lucius* x *E. masquinongy*; Webb and Skadsen, 1980). New and colleagues (2001) suggest that, for the muskellunge (*E. masquinongy*), this minimizes detection by the prey. Once positioned for the strike, esocids demonstrate several different axial kinematic patterns which Webb and Skadsen (1980) termed type A and type B for startles of pike (*Esox*). In type A responses, the fish begins the strike with a straight body bending rapidly into an S-shape and then accelerating toward the prey. In type B responses, the predator holds the S-shaped bend in the body prior to initiating the strike. Type B strikes occurred at a shorter distance to the prey (1.7–7.7 cm) compared to Type A strikes (3.8 to 33.9 cm) which were also found for the pickerel (*Esox niger*) (Rand and Lauder, 1981) and were significantly more successful (95% for type B compared to 42% for type A). Schriefer and Hale (2004) saw a range of behaviors spanning types A and B and suggest that these types are ends of a continuum. However, in some of their trials they found axial bending preceded activity of the white muscle. This suggests that in some trials slow axial bending prior to the strike may be generated by red muscle and a different motor program and represents a different mechanism for generating an S-bend.

The term S-start has in the past been used to describe both strike and startle S-shaped bends. Both behaviors involve an "S"-like body bend that precedes high acceleration axial movement—in the case of escape, away from the predator, and in the case of strike, toward the prey—and may help position the body to maximize acceleration performance. However, there are significant differences between S-starts used in these contexts. The bends of the startle S-start occur along the length of the body and there is considerable angular rotation of the head, even during low curvature S-starts. In contrast, the strike-associated S-start involves primarily caudal bending and the head remains stable (Schriefer and Hale, 2004) (Fig. 6.4). These differences may reflect different constraints on the movement between these contexts. During feeding strikes, S-bends may be smaller and concentrated caudally to minimize detection. The stable position of

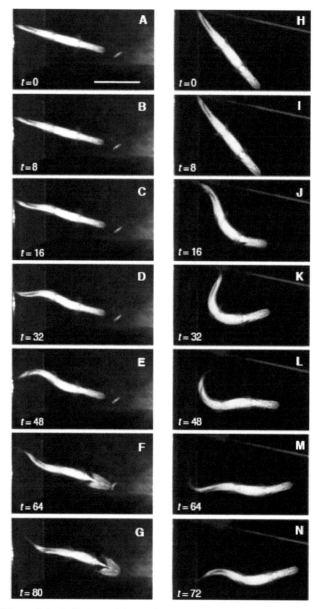

Fig. 6.4 Strike and startle S-starts of the northern pike (*Esox lucius*) (reprinted from Schriefer and Hale, 2004). These representative trials demonstrate several key differences between S-starts used in feeding (A-G) and those used in escape (H-N). For feeding strikes, prior to forward movement, pikes bend their axis into an S-shape in the caudal region of the body without appreciable movement of the head while the S-shaped bend that occurs prior to forward movement in escape occurs along the length of the body with significant angular movement of the head. Scale bar = 10 cm.

the head appears to be important for targeting and timing the strike. This is supported by data from New *et al.* (2001) that demonstrated that in the muskellunge, sensory information from visual orientation is important for successful prey capture. During escape, the initial S-bend may be important for moving local regions of the body away from a predator and for orienting the body for propulsion. Because of differences in the S-start between strikes and startles, we suggest that they are produced by different, although possibly overlapping motor control mechanisms.

C-start behavior, which, like the S-start, is typically associated with startle behavior, also can function in feeding. Archerfish use an unusual approach to feeding in which they shoot jets of water at insects on branches above them (Elshoud and Koomen, 1985; Timmermans, 2000). When the insects fall, the archer fish can determine the trajectory of their fall and capture them after they hit the water (Rossel *et al.*, 2002; Wöhl and Schuster, 2006). The initial movement of the fish to reach the falling insect is a C-shaped turn that orients the fish toward the position where the prey should fall. Wöhl and Schuster (2007) compared the escape C-start to the C-start used in prey capture and found that the kinematics of the two behaviors are comparable: both are rapid responses with high angular acceleration and similar turn distance. They suggest that these responses are driven with the same reticulospinal circuits. One of the features of the Mauthner cell-initiated neural circuit which drives the C-start response is that the latency between the trigger and the initiation of movement is very short (reviewed by Zottoli and Faber, 2000). It would be interesting to know what triggers the archer fish to begin its C-start and how long as well as the length and variability of this latency. As discussed below, C-starts have also been associated with later components of prey capture (Canfield and Rose, 1993) and a broader look at the role of C-starts in feeding, and particularly the Mauthner cell circuit would be valuable.

Among predators feeding on non-evasive food, there are a range of strategies for prey capture, and these strategies can influence pre-strike behavior. The strategies are influenced in large part by the type of food resource used. Non-evasive prey includes food items that are free floating and can be ingested whole, or prey items that are substrate attached and must be dislodged, or are buried and must be dug up to be eaten. Unlike feeding on evasive prey, fishes consuming sessile organisms do not need to avoid detection by the prey item, but may have to deal with other factors outside of the immediate predator-prey dynamic, such as the possibility of competition or of vulnerability to predators. As with feeding on evasive prey, fish feeding on sessile prey requires the fish to position their bodies appropriately in order to feed effectively. One challenge for feeding on non-evasive prey is that the prey can be positioned at any angle, in extremely complex structural environments; thus fishes exploiting this trophic resource need to maximize their maneuverability, this is often accomplished with the pectoral fins (Harris, 1937; Webb, 1984c; Choat, 1991; Gerstner, 1999). Since the pectoral fins facilitate both foraging and feeding, they have to both ensure efficient propulsion during prey localization, and provide the ability to properly orient the fish for the strike (Rice and Westneat, 2005). During the feeding strike on non-evasive prey, this increased maneuverability is often at the expense of reduced speed (Blake, 1983) to ensure precise placement of the jaws for food acquisition.

One of the experimental systems that may be most tractable for examining locomotory patterns during prey capture is the larval zebrafish, which has become a model species for studying motor control. Several recent studies have examined locomotion before and during prey capture (Budick and O'Malley, 2000; Borla et al., 2002; McElligott and O'Malley, 2005) describing a range of feeding strategies being used to feed on paramecia in just the first week or so post-fertilization. Larval zebrafish were found to use slow swimming and a caudal "J-bend" to prepare for the strike (Budick and O'Malley, 2000). Slow swimming allows the fish to approach the prey and the J-bends are used to orient the predator. The J-bends appear unique to the feeding event and are believed to allow the fish to turn gradually possibly to reduce disturbance of the fluid environment (McElligott and O'Malley, 2005). Because these fish are small, around 4 mm, Reynolds numbers are in the intermediate and low range making viscosity a significant factor in movement through fluid. Pike also use a caudal bend prior to the strike (e.g., Webb and Skadsen, 1980; Schriefer and Hale, 2004) but in that case it appears to be used to bend the body for an effective strike rather than in turning, which seems most dependent on the fins. J bends may have other functions in relation to feeding. In several cases, J-bends were also used by the predator to back up and reposition after a failed attack (McElligott and O'Malley, 2005).

Although the nature of locomotor movements prior to prey capture has not been studied in depth and has primarily been investigated in constrained laboratory settings, it is clear that sensory, locomotor and feeding systems must work in concert during this component of feeding to prepare the fish for effective prey capture. In fact, it is possible that this part of the response involves the greatest integration of feedback with relatively little modulation during the rapid prey-capture phase of feeding (Nyberg, 1971). The interaction between the prey capture phase of feeding with prior locomotor movements may provide insight into how sequential phases of movement are coordinated and controlled.

Here, we have examined the pre-strike portion of the feeding event independently from the actual predatory strike. If the strike is rapid, there may not be enough time to integrate feedback from the environment and adjust movement. In these cases, pre-strike positioning is critical for prey capture success and is the last time before prey capture is initiated that the feeding event can be adjusted. However, for other species, in which the prey capture event occurs over a longer time frame, the positioning prior to the prey capture event may be continuous with reorientation during prey capture. Many feeding studies focus on body motions that are concurrent with jaw movements, and we suggest that broadening the period examined to cover pre-strike locomotor behavior would provide important information for understanding the coordination of locomotion between these stages of feeding. In addition, as shown by McElligott and O'Malley (2005), Webb and Skadsen (1980) and others, there are unique kinematic patterns only used during this stage of feeding. This demonstrates that pre-strike movements, can be distinct from routine swimming and maneuvering.

LOCOMOTION DURING PREY CAPTURE

From the initiation of a predatory strike on evasive prey items, two relevant locomotor variables are likely to determine the success of the feeding event. First, predators attempt to minimize the overall duration of the attack, as the longer the encounter lasts, the higher the probability that the prey item will be able to escape safely to cover (Webb, 1986). Thus, the predator will maximally accelerate, and strike as quickly as possible (Daniel, 1984; Webb, 1986; Webb and de Buffrénil, 1990; Hart, 1997; Domenici, 2001). Second, the predator has to be able to outmaneuver its prey, often measured in terms of the fish's turning radius (Webb, 1984c, 1986; Webb and de Buffrénil, 1990; Domenici, 2001). Complex maneuvers involved in prey capture, such as turning, include an energetic cost several times greater than swimming in a constant direction (Boisclair and Tang, 1993; Hughes and Kelly, 1996; Asaeda *et al.*, 2005). However, because this burst swimming is energetically costly (Hart, 1997), predators will often modulate their feeding strike depending on the evasiveness of the prey item (Nyberg, 1971; Vinyard, 1982; Coughlin and Strickler, 1990; Ferry-Graham *et al.*, 2001, 2002). Many fishes feed on both sessile and mobile prey, and are often able to modulate their feeding strike to meet the functional demands of benthic prey capture (Liem, 1979; Nemeth, 1997), or to decrease the energetic cost involved in the strike (Vinyard, 1982).

The fins and body axis take on diverse roles during prey capture. In some species, the fins stabilizes the body while jaws protrude and suction forces carry the food item into the mouth (Rice *et al.*, 2008). In others, axial or fin locomotion is critical to bring the body to the prey bringing the fish to the food. These two general feeding strategies have been called, respectively, suction and ram feeding (e.g., Grubich, 2001). These approaches are not mutually exclusive and most fish use a combination of suction and ram during each prey capture event (e.g., Norton and Brainerd, 1993). The interaction of ram locomotion and suction is interesting and still being examined. It is not a simple relationship. Ram has been shown to augment suction in prey capture of some species (Wainwright, 2001; Higham *et al.*, 2005) while flow generated by the movement of the body through the water has generally been thought to hinder the generation of suction forces (e.g., Nyberg, 1971; Muller and Osse, 1984; Ferry Graham *et al.*, 2003).

Suction feeding involves a relatively consistent pattern of jaw kinematics and muscle activity. First, the mouth opens and buccal cavity expands rapidly to generate a negative pressure inside. This negative pressure draws the prey and water surrounding it into the mouth. Second, the mouth closes and buccal cavity compresses forcing the water inside to be filtered through the now-open operculi and trapping the prey (e.g., Lauder, 1985; Grubich, 2001). Several studies (e.g., Wainwright *et al.*, 1989; Westneat and Wainwright, 1989; Grubich, 2001) have demonstrated that even in diverse actinopterygian fishes, from the basal bowfin (*Amia calva*), to basal teleosts (*Notopterus chitala, Megalops atlanticus*), to Perciformes including extreme forms like the sling-jaw wrasse (*Epibulus insidiator*), the recruitment order of major jaw muscles used in feeding has been conserved although there is variation in the durations of muscle activity and other muscle activity parameters (reviewed by Grubich, 2001). Despite these similarities across taxa, when observing fish feeding there appears to be a diverse array of cranial movements. These data on motor pattern and others on structure of the jaws (Wainwright and Richard, 1995; Wainwright and Shaw, 1999; Wainwright *et al.*, 2000; Westneat, 2004, 2006) indicate the major diversity in jaw movement patterns is due to the diversity of jaw morphology rather than how it is controlled (Wainwright *et al.*, 1989; Wainwright, 2002). An intriguing exception to this is the high variability in both motor activity pattern and feeding kinematics in parrotfishes (Alfaro and Westneat, 1999).

Although the focus of work on prey capture has been jaw movement and motor control particularly during suction feeding, there is a growing body of work analyzing fin and tail movements during this stage of feeding. In fact, there is substantial diversity in the coordination and control of the locomotor apparatus in conjunction with the activity driving the jaws, at least when comparing among fish that use suction feeding. Locomotor patterns during the strike differ depending on the mobility of food item. Beyond the broad comparison of evasive to non-evasive prey, the demands for prey capture within these categories can also differ fundamentally. For example, some non-evasive prey is eaten whole while others must be bitten off the food source in pieces. Some non-evasive prey is free floating while others are substrate associated. For evasive prey, there are differences in how the prey sense the predator and their escape responses. Although the diversity of feeding associated locomotion is immense, here we focus on several of the best-known examples of this diversity: strikes with an extensive ram component, ram filter feeding, primarily suction feeding, and substrate biting.

The strikes of esocids have a large ram component although suction is still a factor. Figure 6.4 shows the movement of a northern pike performing a strike which would be classified as type B under Webb and Skadsen (1980) because the initial S-shaped bend was held prior to the response. The pre-strike S-bend is formed without forward movement of the head (t = 0 to t = 32). During the strike the fish bends toward the opposite direction and the head moves rapidly toward the prey (t = 32 to t = 80). Toward the end of the strike (t = 64), the prey item is seen bending at the middle. It is unclear whether the minnow is initiating an escape response or is bending due to the suction force generated with the clear concurrent buccal expansion. The propulsive strike movements have been called stage 2 as they are similar to stage 2 of the startle response (Webb and Skadsen, 1980; Schriefer and Hale, 2004). Rand and Lauder (1981) found no difference in the rate of lateral tail movement or anal fin movement or of the rostrocaudal propagation of the bending wave down the body during propulsion between A and B type strikes. During this stage a rostral to caudal wave of body bending is propagated along the posterior region of the fish. It is associated with simultaneous activity of the rostral and midbody musculature on both sides of the body. Schriefer and Hale (2004) suggested that this muscle activity may help stiffen the rostral region and maintain its orientation during propulsion as usually rapid axial acceleration is accompanied by lateral movements along the entire body length. Using bilateral muscle activity to stiffen the body has been suggested previously to occur during the fast-start escape response (Diamond, 1971; Foreman and Eaton, 1993; Westneat et al., 1998; Hale et al., 2002; Tytell and Lauder, 2002). It is also possible that bilateral muscle activity is occurring in order to raise the neurocranium during jaw opening as found by Thys (1997) during prey capture of the largemouth bass (*Micropterus salmoides*).

Another behavior that relies heavily on ram but in a very different feeding strategy is ram filter feeding, which occurs when the predator/prey size ratio is large (Webb and de Buffrénil, 1990; Domenici, 2001). During ram filter feeding, steady swimming is used to move the fish through a planktonic food cloud. Water and plankton are taken in through the mouth and water filtered out through the gill slits. Ram filter feeding is characteristic of large sharks: whale sharks (*Rhincodon typus*), basking sharks (*Cetorhinus maximus*) and megamouths (*Megachasma pelagios*) (e.g., Taylor, et al., 1983; Compagno, 1984; Sims, 1999; Heyman et al., 2001). Heyman et al. (2001) observed several variations of ram filter feeding by whale sharks. Whale sharks appear to typically slow their steady swimming rate when they are feeding. In some midwater feeding events they were observed to swim through a plankton cloud holding their mouths open until they reached the other side (Heyman et al., 2001). They also did this at the surface keeping their upper jaw above the water for many seconds while slow swimming through the plankton. When feeding on very dense plankton, they swam very slowly through or stopped in the cloud using repeated suction feeding cycles to bring food into the mouth. Basking sharks do not use suction feeding but rely completely on ram filter feeding (Sims, 1999). Like whale sharks, they decrease the rate of swimming during feeding. Swimming during filter feeding was also more variable than during non-feeding periods and it is suggested that these fish are modulating their swimming speed in association with prey densities in the water (Sims, 1999). This also appears to be the case for the behavior of whale sharks that swim very slowly or stop when feeding in dense plankton. Sims (1999) compared non-feeding and feeding swimming speeds to the optimal swimming speed and optimal filter-feeding speed modeled by Weihs and Webb (1983). While steady, non-feeding locomotor speed was similar to values suggested by the model, swimming speed during feeding was much lower in that study and previous work (Matthews and Parker, 1950, Harden Jones, 1973) than had been predicted. Much of the filter feeding data was taken as the shark swam at the surface and, as discussed by Sims (1999), surface swimming may induce greater drag and cost of swimming than swimming below the surface. This additional drag may account for the slower-than-predicted swimming speeds (Videler, 1993) and result in slower swimming (Priede, 1984).

A number of smaller ray-finned fishes also ram filter feed (Sanderson and Wassersug, 1993). In contrast to the large sharks, swimming during ram filter feeding events in small fish including anchovies (James and Probyn, 1989), mackerel (Pepin *et al.*, 1988), and menhaden (Durbin *et al.*, 1981) is generally faster than is typical for cruising. Sanderson *et al.* (1994), found that this is even true for paddlefish, at least when they are at smaller sizes (36–45 cm). The increase in swimming speed as been suggested to be related to the balance of energy intake with energy expenditure and growth (Durbin and Durbin, 1983, discussed by Sanderson *et al.*, 1994). Additional work on the hydrodynamics and kinematics of ram filter feeding to complement the physiology would be useful for resolving why the pattern of swimming speed during ram filter feeding differ across taxa and teasing apart effects of phylogeny, scaling and feeding strategy.

Many fish do not use extensive axial movement as part of the feeding strike instead using the fins, particularly pectoral fins, and jaws in this last approach to the prey. While the jaw movements of suction feeding have received considerable attention in many species (Lauder, 1982, 1983; Ferry-Graham and Lauder, 2001; Westneat, 2006), particularly perciform centrarchids (Wainwright and Lauder, 1986; Wainwright, 1996; Wainwright and Shaw, 1999; Carroll *et al.*, 2004) and labrids (Wainwright *et al.*, 2000; Ferry-Graham *et al.*, 2001; Ferry-Graham *et al.*, 2002), only recently have the associated movements of the fins begun to draw attention (Nyberg, 1971;

Rice, 2005; Rice and Westneat, 2005; Higham, 2007b; Rice *et al.*, 2008). In centrarchids, Higham described the approach to the prey for both largemouth bass (*Micropterus salmoides*) and bluegill sunfish (*Lepomis macrochirus*) in the 120 ms prior to maximum gape as having a long deceleration phase followed by shorter acceleration phase as the jaws opened to maximum gape (Higham, 2007b). After maximum gape was achieved, fish of both species had a sharp deceleration phase in which the fins were used as a brake to stop forward movement. The pectoral fins are important during this stage of movement in diverse suction feeding species (Breder, 1926; Harris, 1937; Werner, 1977; Westneat and Wainwright, 1989; Higham, 2007b; Rice *et al.*, 2008). Work in centrarchids has identified a number of factors that are important to braking performance including fin size and shape, the rate of protraction, the speed of swimming prior to protraction, and suggests others that have yet to be explored including the performance of fin muscles (discussed by Higham, 2007a, b). Braking by the pectoral fins was augmented by concurrent braking with the anal fin and caudal fin as has been described previously for these species and others (e.g., Geerlink, 1983; Drucker and Lauder, 2003; Rice and Westneat, 2005). Higham (2007b) observed yawing movement during the later stage of deceleration of the largemouth bass and suggests that this is due to the asymmetrical braking by the caudal fin, which has a considerably larger area than the pectoral fins in this species. With increasing ram speed during the feeding event, yaw became more pronounced and it was suggested that this effect may be related to the decreased accuracy during higher speed swimming during feeding. Higham (2007b) did not see yawing in bluegill sunfish which have a much smaller caudal fin compared to their pectoral fins. In a broader analysis across an array of labrid species, Collar and colleagues (2007) indicate that fin morphology and cranial morphology have coevolved to complement each other in different feeding modalities. An important area of research in need of additional data is the coordination of fin movements prior to the strike with the movements during the expansion phase of feeding.

In contrast to the types of feeding above which involve drawing food and water into the mouth through suction and/or ram, some fishes bite their food. An example of this are the parrotfishes which bite coral and algae. Unlike with the other feeding strategies discussed, these fish often feed on food attached to substrate which has a number of important consequences for the strike. One of the main constraints is that the food has to be detached from the substrate. Not only does the fish need to bite to separate the food in its mouth from its attachment, it may also need to pull the food off of the source which involves moving backward away from the substrate. Also unlike ram and suction feeders, the fish cannot move through the location of the prey after it is acquired as the attachment surface remains.

Rice and Westneat (2005) examined the coordination of feeding and locomotion in two species of parrotfishes, *Sparisoma radians* and *Scarus quoyi*, which had previously been shown to differ in bite rate with *Sparisoma*'s rate being lower (Bruggemann *et al.*, 1994). In addition, they differ in food resources typically consumed with *Sparisoma* feeding on sea grass and *Scarus* scraping algae and other potential food from corals (reviewed by Rice and Westneat, 2005). In the laboratory, each species would only reliably eat food presented in a similar way to its natural food resources, tethered lettuce for *Sparisoma* and algae smeared on coral for *Scarus*. This indicates that they are highly specialized for different food resources but also prevented comparison of feeding on comparable food between the species. Both species approached the food source with pectoral fin movements with similar protraction and retraction movement patterns. Despite the difference in food, these species had similar patterns of fin movement after the initial approach. Both used downward protraction of the pectoral fins to brake forward movement during the strike. Braking begins as the mouth is opening prior to contact with the prey and continues as the fish backs away from the target. Figure 6.5a shows data for the relationship between gape and pectoral fin protraction through the strike for *S. quoyi*.

Comparing closely related taxa with a diversity of feeding ecologies can reveal a substantial amount of information about trends in how fish approach food items (Rice and Westneat, 2005; Rice, 2006; Higham, 2007a; Higham *et al.*, 2007; Rice *et al.*, 2008). The carnivorous cheiline wrasses may represent different points along the continuum of predatory behavior on benthic food items. With their varying diets, as these species approach sessile food items, there is substantial variation in body speed and pectoral fin beats during the strike (Rice and Westneat, 2005; Rice *et al.*, 2008). *Cheilinus fasciatus* feeds on sessile or slow moving molluscs, gastropods and crustaceans, and exhibits a slow and steady approach and strikes when in close proximity (Rice *et al.*, 2008). The closely related *Epibulus insidiator* has an extremely fast-moving and highly-protrusible mouth (Westneat and Wainwright, 1989; Westneat, 1991) allowing it to slowly approach food items with a low body velocity, use its pectoral fins to maneuver into optimal strike position, and then execute the strike from a relatively large distance while the body remains stationary (Rice *et al.*, 2008). These two predatory cheilines provide a stark contrast to the explosive feeding strike of the closely related cheiline, *Oxycheilinus digrammus*, which, even when feeding on sessile prey items, displays a very rapid attack with few short duration fin movements (Ferry-Graham *et al.*, 2002; Rice *et al.*, 2008) (Fig. 6.5b). Similar to *Epibulus*, *Oxycheilinus digrammus* usually feeds on highly evasive prey, but even when

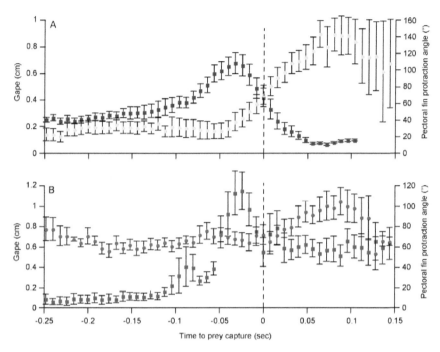

Fig. 6.5 Kinematic plots of jaw movement (squares) and pectoral fin braking movement (circles) in two species of wrasse with very different feeding strategies: (A) *Scarus quoyi*, an herbivore and (B) *Oxycheilinus digrammus*, a piscivore. The moment of prey capture is indicated by the dashed vertical line. For *S. quoyi*, pectoral fin braking is initiated before it makes contact with the food item and the jaws close, while for *O. digrammus*, pectoral fin braking is initiated after it makes contact with the prey item and the jaws close. *S. quoyi* data are from Rice and Westneat (2005) and *O. digrammus* data are from Rice *et al.* (2008).

feeding on non-moving prey items it still initiates the strike from a farther distance than *Cheilinus fasciatus* or *Epibulus insidiator* and attacks with a very high body velocity (Rice *et al.*, 2008).

The sister taxon relationship between cheilines and parrotfishes reveal further locomotor variation during the approach to a food item (Rice and Westneat, 2005). Both *Scarus quoyi* and *Sparisoma radians* approach their respective food items at an overall greater velocity than the cheilines, with more rapid, higher displacement of pectoral fin propulsive strokes (Rice and Westneat, 2005; Rice *et al.*, 2008). The increased approach velocity of the parrotfishes may be a mechanism to minimize the duration of the feeding event so as to reduce their own vulnerability to predation, since evasion of the prey item is not a concern. In contrast to parrotfishes approaching sessile prey, the cheilines may be limited in the speed at which they approach prey items so as to not startle potential prey items, even if some of them can only move slowly.

POST-PREY CAPTURE MANEUVERING AND LOCOMOTION

A change in the predator's position, direction of movement, and velocity are often required immediately after the strike. Canfield and Rose (1993) suggest that goldfish feeding at the surface use a Mauthner-cell elicited bend directly after capturing the prey and postulate that this moves them rapidly from a potentially vulnerable position at the surface of the water and decreases the chance that the prey will escape. It seems likely that other fish that are particularly vulnerable to predation during the feeding event may use similar strategies.

Though feeding on non-evasive prey generally requires a lower body velocity during the strike, a large braking maneuver is needed for benthic prey to prevent collision with the substrate. In *Cheilinus fasciatus* and *Scarus quoyi* a large, sweeping braking maneuver with the pectoral fins is a pronounced component of the food capture sequence (Rice and Westneat, 2005; Rice *et al.*, 2008). These fin movements are initiated well before contact with the prey item is made, and achieve a displacement as high as 85° in *Scarus quoyi* (Rice and Westneat, 2005). Additionally, the extended fin sweep of *Scarus* may create a backwards directed force which assists the jaws in excavation of the encrusting algae (Rice and Westneat, 2005). In contrast to *Scarus*, *Sparisoma radians* has a much smaller braking motion, likely because this species feeds on seagrass blades, where collision does not present a risk to the fish (Rice and Westneat, 2005). However, similarly to *Scarus*, *Sparisoma* may be using the motion of the pectoral fins to help exert increased force during body rotation after the bite to assist in tearing away edible pieces away from the blades of seagrass (Rice and Westneat, 2005).

Oxycheilinus digrammus does not display a large braking maneuver during the strike, but instead may use the pectoral fins to change direction, steering up and to the side of the prey location to avoid collision. Here, the proportionately small area of the fins of *O. digrammus* (Wainwright *et al.*, 2002) may be insufficient to completely stop the fish after such a high-velocity strike. *Epibulus insidiator* displays a dorsoventral braking motion rather than a anteroposterior pattern seen by the other species (Rice *et al.*, 2008). In this instance, the forward-directed braking stroke may not be necessary since the overall body velocity is relatively low, and instead, the ventrally-directed fin motion may be needed to maintain the proper body orientation due to a shift in the center of gravity when the jaws are propelled forward during the strike.

CONCLUSIONS

The importance of locomotion in feeding is evident in all stages of prey capture from foraging to body positioning prior to capture, to the actual prey capture event, to recovery and post-strike re-positioning. It is clear that to understand the evolution of locomotor morphology, behavior and its neural control and coordination, swimming movements need to be considered in the many behavioral contexts in which they are used. Some of the most common and important roles of locomotion are in the contexts of foraging and feeding. With a rich foundation of work (e.g., Keast and Webb, 1966; Rand and Lauder, 1981; Webb, 1982, 1984a, c; Webb and de Buffrénil, 1990; Canfield and Rose, 1993), interest in the relationships between feeding and locomotion has been growing (e.g., Collar *et al.*, 2007; Higham, 2007a, b; Higham *et al.*, 2007; Rice *et al.*, 2008; Rice, 2008); though many fundamental questions remain. Beyond basic questions of how the fins and axis function before and during feeding, broad issues remain to be addressed. These include: 1. how the ecology of predators and prey have shaped locomotor morphology, 2. issues of coevolution of the locomotor apparatus and the jaws (Rice, 2006; Collar *et al.*, 2007; Higham, 2007a), 3. mechanisms of neural coordination of jaws and the locomotor system and 4. how sensory input shapes feeding behaviors. By integrating studies on feeding and locomotion, work on predation and foraging in fishes promises to be an important system for understanding how distinct morphologies and kinematic systems can be coordinated to generate appropriate behavior.

ACKNOWLEDGEMENTS

ANR was supported by NIMH training grant 5-T32-MH15793. This material is also based upon work supported by the National Science Foundation under Grant IBN0238464 to MEH. We thank P. Domenici, B. Langerhans and M. Westneat for their input to the manuscript.

REFERENCES

Asaeda, T., T. Priyadarshana and J. Manatunge. 2001. Effects of satiation on feeding and swimming behaviour of planktivores. *Hydrobiologia* 443: 147–157.
Asaeda, T., T.K. Vu and J. Manatunge. 2005. Effects of flow velocity on feeding behavior and microhabitat selection of the stone moroko *Pseudorasbora parva*: a trade-off between feeding and swimming costs. *Transactions of the American Fisheries Society* 134: 537–547.
Bardach, J.E. and J. Case. 1965. Sensory capabilities of the modified fins of squirrel hake (*Urophycis chuss*) and searobins (*Prionotus carolinus* and *P. evolans*). *Copeia* 1965: 194–206.
Bellwood, D.R. and P.C. Wainwright. 2001. Locomotion in labrid fishes: implications for habitat use and cross-shelf biogeography on the Great Barrier Reef. *Coral Reefs* 20: 139–150.
Blake, R.W. 1983. Median and paired fin propulsion. In: *Fish Biomechanics*, P.W. Webb and D. Weihs (Eds.). Praeger Publishers, New York, pp. 214–247.

Block, B.A. and F.G. Carey. 1985. Warm brain and eye temperatures in sharks. *Journal of Comparative Physiology B* 156: 229–236.

Block, B.A. and J.R. Finnerty. 1994. Endothermy in fishes: a phylogenetic analysis of constraints, predispositions, and selection pressures. *Environmental Biology of Fishes* 40: 283–302.

Boisclair, D. and M. Tang. 1993. Empirical analysis of the influence of swimming pattern on the net energetic cost of swimming in fishes. *Journal of Fish Biology* 42: 169–183.

Borla, M.A., B. Palecek, S. Budick and D.M. O'Malley. 2002. Prey capture by larval zebrafish: evidence for fine axial motor control. *Brain, Behavior and Evolution* 60: 207–229.

Breder, C.M. 1926. The locomotion of fishes. *Zoologica* 4: 159–297.

Bruggemann, J.H., M.W.M. Kuyper and A.M. Breeman. 1994. Comparative analysis of foraging and habitat use by the sympatric Caribbean parrotfish *Scarus vetula* and *Sparisoma viride* (Scaridae). *Marine Ecology Progress Series* 112: 51–66.

Budick, S.A. and D.M. O'Malley. 2000. Locomotor repertoire of the larval zebrafish: swimming, turning and prey capture. *Journal of Experimental Biology* 203: 2565–2579.

Canfield, J.G. and G.J. Rose. 1993. Activation of Mauthner neurons during prey capture. *Journal of Comparative Physiology A* 172: 611–618.

Carey, F.G. 1982. A brain heater in the swordfish. *Science* 216: 1327–1329.

Carey, F.G. 1992. Through the thermocline and back again: heat regulation in big fish. *Oceanus* 35: 79–85.

Carey, F.G., J.W. Kanwisher, O. Brazier, G. Gabrielson, J.G. Casey and H.L. Pratt. 1982. Temperature and activities of a white shark, *Carcharodon carcharias. Copeia* 1982: 254–260.

Carroll, A.M., P.C. Wainwright, S.H. Huskey, D.C. Collar and R.G. Turingan. 2004. Morphology predicts suction feeding performance in centrarchid fishes. *Journal of Experimental Biology* 207: 3873–3881.

Choat, J.H. 1991. The biology of herbivorous fishes on coral reefs. In: *The Ecology of Fishes on Coral Reefs*, P.F. Sale (Ed.). Academic Press, San Diego, pp. 120–155.

Choi, S.H. and K. Gushima. 2002. Spot-fixed fin digging behavior in foraging of the benthophagous maiden goby, *Pterogobius virgo* (Perciformes: Gobiidae). *Ichthyological Research* 49: 286–290.

Clarke, R.D., E.J. Buskey and K.C. Marsden. 2005. Effects of water motion and prey behavior on zooplankton capture by two coral reef fishes. *Marine Biology* 146: 1145–1155.

Collar, D.C., P.C. Wainwright and M.E. Alfaro. 2007. Integrated diversification of locomotion and feeding in labrid fishes. *Biology Letters* 4: 84–86.

Compagno, L.J.V. 1984. *Sharks of the World: An Annotated and Illustrated Catalogue of Shark Species Known to Date. FAO Fisheries Synopsis* No. 125, Vol. 4. United Nations Development Programme, Rome.

Coughlin, D.J. and J.R. Strickler. 1990. Zooplankton capture by a coral reef fish: an adaptive response to evasive prey. *Environmental Biology of Fishes* 29: 35–42.

Crowder, L.B. and W.E. Cooper. 1982. Habitat structural complexity and the interaction between bluegills and their prey. *Ecology* 63: 1802–1813.

Cyrus, D.P. and S.J.M. Blaber. 1982. Mouthpart structure and function and the feeding mechanisms of *Gerres* (Teleostei). *South African Journal of Zoology* 17: 117–121.

Daniel, T.L. 1984. Unsteady aspects of aquatic locomotion. *American Zoologist* 24: 121–134.

Davis, W.P. and R.S. Birdsong. 1973. Coral reef fishes which forage in the water column: a review of their morphology, behavior, ecology and evolutionary implications *Helgoländer Wissenschaftliche Meeresuntersuchungen* 24: 292–306.

Diamond, J. 1971. The Mauthner cell. In: *Fish Physiology*, W.S. Hoar and D.J. Randall (Eds.). Academic Press, New York, Vol. 5: pp. 265–346.

Diehl, S. 1988. Foraging efficiency of three freshwater fishes: effects of structural complexity and light. *Oikos* 53: 207–214.

Domenici, P. 2001. The scaling of locomotor performance in predator-prey encounters: from fish to killer whales. *Comparative Biochemistry and Physiology A* 131: 169–182.

Domenici, P. 2002. The visually mediated escape response in fish: predicting prey responsiveness and the locomotor behaviour of predators and prey. *Marine and Freshwater Behaviour and Physiology* 35: 87–110.

Domenici, P. 2003. Habitat type, design and the swimming performance of fish. In: *Vertebrate Biomechanics and Evolution*, V.L. Bels, J.P. Gasc and A. Casinos (Eds.). BIOS Scientific Publishers Ltd, Oxford, pp. 137–160.

Domenici, P., H. Turesson, J. Brodersen and C. Brönmark. 2008. Predator-induced morphology enhances escape locomotion in crucian carp. *Proceedings of the Royal Society of London B* 275: 195–201.

Durbin, E.G. and A.G. Durbin. 1983. Energy and nitrogen budgets for the Atlantic menhaden, *Brevoortia tyrannus* (Pisces, Clupeidae), a filter-feeding planktivore. *Fishery Bulletin* 81: 177–199.

Drucker, E.G. and G.V. Lauder. 2003. Function of pectoral fins in rainbow trout: behavioral repertoire and hydrodynamic forces. *Journal of Experimental Biology* 206: 813–826.

Elshoud, G.C.A. and P. Koomen. 1985. A biomechanical analysis of spitting in archer fishes (Pisces, Perciformes, Toxidae). *Zoomorphology* 105: 240–252.

Ferry-Graham, L.A. and G.V. Lauder. 2001. Aquatic prey capture in ray-finned fishes: a century of progress and new directions. *Journal of Morphology* 248: 99–119.

Ferry-Graham, L.A., P.C. Wainwright, M.W. Westneat and D.R. Bellwood. 2001. Modulation of prey capture kinematics in the cheeklined wrasse *Oxycheilinus digrammus* (Teleostei: Labridae). *Journal of Experimental Zoology* 290: 88–100.

Ferry-Graham, L.A., P.C. Wainwright, M.W. Westneat and D.R. Bellwood. 2002. Mechanisms of benthic prey capture in wrasses (Labridae). *Marine Biology* 141: 819–830.

Flore, L. and H. Keckeis. 1998. The effect of water current on foraging behaviour of the rheophilic cyprinid *Chondrostoma nasus* (L.) during ontogeny: evidence of a trade-off between energetic gain and swimming costs. *Regulated Rivers: Research & Management* 14: 141–154.

Foreman, M.B. and R.C. Eaton. 1993. The direction change concept for reticulospinal control of goldfish escape. *Journal of Neuroscience* 13: 4101–4113.

Fulton, C.J. and D.R. Bellwood. 2002. Patterns of foraging in labrid fishes. *Marine Ecology Progress Series* 226: 135–142.

Geerlink, P.J. 1983. Pectoral fin kinematics of *Coris formosa* (Teleostei, Labridae). *Netherlands Journal of Zoology* 33: 515–531.

Gerstner, C.L. 1999. Maneuverability of four species of coral-reef fish that differ in body and pectoral-fin morphology. *Canadian Journal of Zoology* 77: 1102–1110.

Grobecker, D.B. 1983. The lie-in-wait feeding mode of a cryptic teleost, *Synanceia verrucosa*. *Environmental Biology of Fishes* 8: 191–202.

Grobecker, D.B. and T.W. Pietsch. 1979. High-speed cinematographic evidence for ultrafast feeding in antennariid anglerfishes. *Science* 205: 1161–1162.

Grubich, J.R. 2001. Prey capture in actinopterygian fishes: a review of suction feeding motor patterns with new evidence from an elopomorph fish, *Megalops atlanticus*. *American Zoologist* 41: 1258–1265.

Guinet, C., P. Domenici, R. de Stephanis, L. Barrett-Lennard, J.K.B. Ford and P. Verborgh. 2007. Killer whale predation on bluefin tuna: exploring the hypothesis of the endurance-exhaustion technique. *Marine Ecology Progress Series* 347: 111–119.

Hale, M.E., J.H. Long, M.J. McHenry and M.W. Westneat. 2002. Evolution of behavior and neural control of the fast-start escape response. *Evolution* 56: 993–1007.

Harden Jones, F.R. 1973. Tail beat frequency, amplitude, and swimming speed of a shark tracked by sector scanning sonar. *Journal du Conseil International pour l'Exploration de la Mer* 35: 95–97.

Harris, J.E. 1937. The mechanical significance of the position and movements of the paired fins in the Teleostei. *Papers from the Tortugas Laboratory of the Carnegie Institution of Washington* 31: 173–189.

Hart, P.J.B. 1997. Foraging tactics. In: *Behavioural Ecology of Teleost Fishes,* J.-G.J. Godin (Ed.). Oxford University Press, Oxford, pp. 104–133.

Heyman, W.D., R.T. Graham, B. Kjerfve and R.E. Johannes. 2001. Whale sharks, *Rhincodon typus,* aggregate to feed on fish spawn in Belize. *Marine Ecology Progress Series* 215: 275–282.

Higham, T.E. 2007a. The integration of locomotion and prey capture in vertebrates: Morphology, behavior, and performance. *Integrative and Comparative Biology* 47: 82–95.

Higham, T.E. 2007b. Feeding, fins and braking maneuvers: locomotion during prey capture in centrarchid fishes. *Journal of Experimental Biology* 210: 107–117.

Higham, T.E., C.D. Hulsey, O. Rican and A.M. Carroll. 2007. Feeding with speed: prey capture evolution in cichlids. *Journal of Evolutionary Biology* 20: 70–78.

Hobson, E.S. 1991. Trophic relationships of fishes specialized to feed on zooplankters above coral reefs. In: *The Ecology of Fishes on Coral Reefs,* P.F. Sale (Ed.). Academic Press, San Diego, pp. 69–95.

Hobson, E.S. and J.R. Chess. 1978. Trophic relationships among fishes and plankton in the lagoon at Enewetak Atoll, Marshall Islands. *Fishery Bulletin* 76: 133–153.

Hughes, N.F. and L.H. Kelly. 1996. A hydrodynamic model for estimating the energetic cost of swimming maneuvers from a description of their geometry and dynamics. *Canadian Journal of Fisheries and Aquatic Sciences* 53: 2484–2493.

James, A.G. and T. Probyn. 1989. The relationship between respiration rate, swimming speed and feeding behavior in the cape anchovy *Engraulis capensis* Gilchrist. *Journal of Experimental Marine Biology and Ecology* 131: 81–100.

James, P.L. and K.L. Heck, Jr. 1994. The effects of habitat complexity and light intensity on ambush predation within a simulated seagrass habitat. *Journal of Experimental Marine Biology and Ecology* 176: 187–200.

Jensen, G.C. 2005. A unique feeding method by a teleost fish, the fourhorn poacher *Hypsagonus quadricornis* (Agonidae). *Biological Bulletin* 209: 165–167.

Keast, A. and D. Webb. 1966. Mouth and body form relative to feeding ecology in fish fauna of a small lake, Lake Opinicon, Ontario. *Journal of the Fisheries Research Board of Canada* 23: 1845–1874.

Kendall, J.L., K.S. Lucey, E.A. Jones, J. Wang and D.J. Ellerby. 2007. Mechanical and energetic factors underlying gait transitions in bluegill sunfish (*Lepomis macrochirus*). *Journal of Experimental Biology* 210: 4265–4271.

Kotrschal, K. 1988. Evolutionary patterns in tropical marine reef fish feeding. *Zeitschrift für Zoologische Systematik und Evolutionforschung* 26: 51–64.

Kotrschal, K. and D.A. Thomson. 1989. From suckers towards pickers and biters: evolutionary patterns in trophic ecomorphology of tropical marine fishes. *Forschritte der Zoologie* 35: 564–568.

Langerhans, R.B. 2008. Predictability of phenotypic differentiation across flow regimes in fishes. *Integrative and Comparative Biology* 48: 750–768.

Lauder, G.V. 1982. Patterns of evolution in the feeding mechanism of actinopterygian fishes. *American Zoologist* 22: 275–285.

Lauder, G.V. 1983. Food capture. In: *Fish Biomechanics,* P.W. Webb and D. Weihs (Eds.). Praeger Publishers, New York, pp. 280–311.

Lauder, G.V. 1985. Functional morphology of the feeding mechanism in lower vertebrates. *Fortschritte der Zoologie* 30: 179–188.

Lauder, G.V. and E.G. Drucker. 2002. Forces, fishes, and fluids: hydrodynamic mechanisms of aquatic locomotion. *News in Physiological Sciences* 17: 235–240.

Lauder, G.V., J.C. Nauen and E.G. Drucker. 2002. Experimental hydrodynamics and evolution: function of median fins in ray-finned fishes. *Integrative and Comparative Biology* 42: 1009–1017.

Liem, K.F. 1979. Modulatory multiplicity in the feeding mechanism in cichlid fishes, as exemplified by the invertebrate pickers of Lake Tanganyika. *Journal of Zoology* 189: 93–125.

Matthews, L.H. and H.W. Parker. 1950. The anatomy and biology of the basking shark (*Cetorhinus maximus* (Gunner)). *Proceedings of the Zoological Society of London* 120: 535–576.

McElligott, M.B. and D.M. O'Malley. 2005. Prey tracking by larval zebrafish: axial kinematics and visual control. *Brain, Behavior and Evolution* 66: 177–196.

Nemeth, D.H. 1997. Modulation of attack behavior and its effect on feeding performance in a trophic generalist fish, *Hexagrammos decagrammus*. *Journal of Experimental Biology* 200: 2155–2164.

New, J., L. Alborg Fewkes and A. Khan. 2001. Strike feeding behavior in the muskellunge, *Esox masquinongy*: contributions of the lateral line and visual sensory systems. *Journal of Experimental Biology* 204: 1207–1221.

Norton, S.F. 1991. Capture success and diet of cottid fishes: the role of predator morphology and attack kinematics. *Ecology* 72: 1807–1819.

Norton, S.F. and E.B. Brainerd. 1993. Convergence in the feeding mechanics of ecomorphologically similar species in the Centrarchidae and Cichlidae. *Journal of Experimental Biology* 176: 11–29.

Nyberg, D.W. 1971. Prey capture in the largemouth bass. *American Midland Naturalist* 86: 128–144.

O'Brien, W.J., B.I. Evans and G.L. Howick. 1986. A new view of the predation cycle of a planktivorous fish, white crappie (*Pomoxis annularis*). *Canadian Journal of Fisheries and Aquatic Sciences* 43: 1894–1899.

O'Brien, W.J., B.I. Evans and H.I. Browman. 1989. Flexible search tactics and efficient foraging in saltatory searching animals. *Oecologia* 80: 100–110.

O'Brien, W.J., H.I. Browman and B.I. Evans. 1990. Search strategies of foraging animals. *American Scientist* 78: 152–160.

Pepin, P., J.A. Koslow and S. Pearre. 1988. Laboratory study of foraging by Atlantic mackerel, *Scomber scombrus*, on natural zooplankton assemblages. *Canadian Journal of Fisheries and Aquatic Sciences* 45: 879–887.

Priede, I.G. 1984. A basking shark (*Cetorhinus maximus*) tracked by satellite together with simultaneous remote sensing. *Fisheries Research* 2: 201–216.

Priyadarshana, T. and T. Asaeda. 2007. Swimming restricted foraging behavior of two zooplanktivorous fishes *Pseudorasbora parva* and *Rasbora daniconius* (Cyprinidae) in a simulated structured environment. *Environmental Biology of Fishes* 80: 473–486.

Priyadarshana, T., T. Asaeda and J. Manatunge. 2001. Foraging behaviour of planktivorous fish in artificial vegetation: the effects on swimming and feeding. *Hydrobiologia* 442: 231–239.

Rand, D.M. and G.V. Lauder. 1981. Prey capture in the chain pickerel, *Esox niger*: correlations between feeding and locomotor behavior. *Canadian Journal of Zoology* 59: 1072–1078.

Rice, A.N. 2005. Evolution of coordination in fish feeding: the intersection of functional morphology and ecology. *Integrative and Comparative Biology* 45: 1062.

Rice, A.N. 2006. *Sensorimotor Integration and Coordination of Feeding and Locomotion in Coral Reef Fishes*. Ph.D. Dissertation. University of Chicago, Chicago.

Rice, A.N. 2008. Coordinated mechanics of feeding, swimming, and eye movements in *Tautoga onitis*, and implications for the evolution of trophic strategies in fishes. *Marine Biology* 154: 255–267.

Rice, A.N. and M.W. Westneat. 2005. Coordination of feeding, locomotor, and visual systems in parrotfishes (Teleostei: Labridae). *Journal of Experimental Biology* 208: 3503–3518.

Rice, A.N., W.J. Cooper and M.W. Westneat. 2008. Diversification of coordination patterns during feeding behaviour in cheiline wrasses. *Biological Journal of the Linnean Society* 93: 289–308.

Rincón, P., M. Bastir and G.D. Grossman. 2007. Form and performance: body shape and prey-capture success in four drift-feeding minnows. *Oecologia* 152: 345–355.

Rossel, S., J. Corlija and S. Schuster. 2002. Predicting three-dimensional target motion: how archer fish determine where to catch their dislodged prey. *Journal of Experimental Biology* 205: 3321–3326.

Sanderson, S.L. and R. Wassersug. 1993. Convergent and alternative designs for vertebrate suspension feeding. In: *The Skull: Functional and Evolutionary Mechanisms*, J. Hanken and B.K. Hall (Eds.). University of Chicago Press, Chicago, Vol. 3: pp. 37–112.

Sanderson, S.L., J.J. Cech and A.Y. Cheer. 1994. Paddlefish buccal flow velocity during ram suspension feeding and ram ventilation. *Journal of Experimental Biology* 186: 145–156.

Savino, J.F. and R.A. Stein. 1982. Predator-prey interaction between largemouth bass and bluegills as influenced by simulated, submersed vegetation. *Transactions of the American Fisheries Society* 111: 255–266.

Savino, J.F. and R.A. Stein. 1989a. Behavioural interactions between fish predators and their prey: effects of plant density. *Animal Behaviour* 37: 311–321.

Savino, J.F. and R.A. Stein. 1989b. Behavior of fish predators and their prey: habitat choice between open water and dense vegetation. *Environmental Biology of Fishes* 24: 287–293.

Schaefer, J.F., W.I. Lutterschmidt and L.G. Hill. 1999. Physiological performance and stream microhabitat use by the centrarchids *Lepomis megalotis* and *Lepomis macrochirus*. *Environmental Biology of Fishes* 54: 303–312.

Schriefer, J.E. and M.E. Hale. 2004. Strikes and startles of northern pike (*Esox lucius*): a comparison of muscle activity and kinematics between S-start behaviors. *Journal of Experimental Biology* 207: 535–544.

Sims, D.W. 1999. Threshold foraging behaviour of basking sharks on zooplankton: life on an energetic knife-edge? *Proceedings of the Royal Society of London B* 266: 1437–1443.

Stoner, A.W. 1982. The influence of benthic macrophytes on the foraging behavior of pinfish, *Lagodon rhomboides* (Linnaeus). *Journal of Experimental Marine Biology and Ecology* 58: 271–284.

Taylor, L.R., L.J.V. Compagno and P.J. Struhsaker. 1983. Megamouth a new species, genus and family of laminid shark (*Megachasma pelagios*, family Magachasmidae) from the Hawaiian Islands. *Proceedings of the California Academy of Sciences* 43: 87–110.

Thorsen, D.H. and M.E. Hale. 2005. Development of zebrafish (*Danio rerio*) pectoral fin musculature. *Journal of Morphology* 266: 241–255.

Thys, T. 1997. Spatial variation in epaxial muscle activity during prey strike in largemouth bass (*Micropterus salmoides*). *Journal of Experimental Biology* 200: 3021–3031.

Timmermans, P.J.A. 2000. Catching in the archer fish: marksmanship, endurance of squirting at an aerial target. *Netherlands Journal of Zoology* 50: 411–423.

Tullis, A., B.A. Block and B.D. Sidell. 1991. Activities of key metabolic enzymes in the heater organs of scombroid fishes. *Journal of Experimental Biology* 161: 383–403.

Tytell, E.D. and G.V. Lauder. 2002. The C-start escape response of *Polypterus senegalus*: bilateral muscle activity and variation during stage 1 and 2. *Journal of Experimental Biology* 205: 2591–2603.

Videler, J.J. 1993. *Fish Swimming*. Chapman and Hall, London.

Vinyard, G.L. 1982. Variable kinematics of Sacramento perch (*Archoplites interruptus*) capturing evasive and non-evasive prey. *Canadian Journal of Fisheries and Aquatic Sciences* 208–211.

Wainwright, P.C. 1996. Ecological explanation through functional morphology: The feeding biology of sunfishes. *Ecology* 77: 1336–1343.

Wainwright, P.C. 2002. The evolution of feeding motor patterns in vertebrates. *Current Opinion in Neurobiology* 12: 691–695.

Wainwright, P.C. and G.V. Lauder. 1986. Feeding biology of sunfishes: patterns of variation in prey capture. *Zoological Journal of the Linnean Society* 88: 217–228.

Wainwright, P.C. and B.A. Richard. 1995. Predicting patterns of prey use from morphology of fishes. *Environmental Biology of Fishes* 44: 97–113.

Wainwright, P.C. and S.S. Shaw. 1999. Morphological basis of kinematic diversity in feeding sunfishes. *Journal of Experimental Biology* 202: 3101–3110.

Wainwright, P.C., M.W. Westneat and D.R. Bellwood. 2000. Linking feeding behaviour and jaw mechanics in fishes. In: *Biomechanics in Animal Behaviour,* P. Domenici and R.W. Blake (Eds.). BIOS Scientific Publishers Ltd., Oxford, pp. 207–221.

Wainwright, P.C., D.R. Bellwood and M.W. Westneat. 2002. Ecomorphology of locomotion in labrid fishes. *Environmental Biology of Fishes* 65: 47–62.

Wainwright, P.C., C.P. Sanford, S.M. Reilly and G.V. Lauder. 1989. Evolution of motor patterns—aquatic feeding in salamanders and ray-finned fishes. *Brain, Behavior and Evolution* 34: 329–341.

Webb, P.W. 1982. Locomotor patterns in the evolution of actinopterygian fishes. *American Zoologist* 22: 329–342.

Webb, P.W. 1984a. Body and fin form and strike tactics of four teleost predators attacking fathead minnow (*Pimephales promelas*) prey. *Canadian Journal of Fisheries and Aquatic Sciences* 41: 157–165.

Webb, P.W. 1984b. Chase response latencies of some teleostean piscivores. *Comparative Biochemistry and Physiology A* 79: 45–48.

Webb, P.W. 1984c. Body form, locomotion and foraging in aquatic vertebrates. *American Zoologist* 24: 107–120.

Webb, P.W. 1986. Locomotion and predator-prey relationships. In: *Predator-Prey Relationships: Perspectives and Approaches from the Study of Lower Vertebrates,* G.V. Lauder and M.F. Feder (Eds.). University of Chicago Press, Chicago, pp. 24–41.

Webb, P.W. 1988. Simple physical principles and vertebrate aquatic locomotion. *American Zoologist* 28: 709-725.

Webb, P.W. 1998. Swimming. In: *The Physiology of Fishes,* 2nd Edition, D.H. Evans (Ed.). CRC Press, Boca Raton, pp. 3–24.

Webb, P.W. and V. de Buffrénil. 1990. Locomotion in the biology of large aquatic vertebrates. *Transactions of the American Fisheries Society* 119: 629–641.

Webb, P.W. and J.M. Skadsen. 1980. Strike tactics of *Esox. Canadian Journal of Zoology* 58: 1462–1469.

Weihs, D. and P.W. Webb. 1983. Optimization of locomotion. In: *Fish Biomechanics,* P.W. Webb and D. Weihs (Eds.). Praeger Publishers, New York, pp. 339–371.

Werner, E.E. 1977. Species packing and niche complementarity in three sunfishes. *American Naturalist* 111: 553–578.

Westneat, M.W. 1991. Linkage biomechanics and evolution of the unique feeding mechanism of *Epibulus insidiator* (Labridae, Teleostei). *Journal of Experimental Biology* 159: 165–184.

Westneat, M.W. 2004. Evolution of levers and linkages in the feeding mechanisms of fishes. *Integrative and Comparative Biology* 44: 378–389.

Westneat, M.W. 2006. Skull biomechanics and suction feeding in fishes. In: *Fish Biomechanics,* R.E. Shadwick and G.V. Lauder (Eds.). Elsevier Academic Press, San Diego, pp. 29–25.

Westneat, M.W. and P.C. Wainwright. 1989. Feeding mechanism of *Epibulus insidiator* (Labridae, Teleostei): evolution of a novel functional system. *Journal of Morphology* 202: 129–150.

Westneat, M.W., M.E. Hale, M.J. McHenry and J.H. Long. 1998. Mechanics of the fast-start: muscle function and the role of intramuscular pressure in the escape behavior of *Amia calva* and *Polypterus palmas. Journal of Experimental Biology* 201: 3041–3055.

Wisenden, B.D., T.L. Lanfranconi-Izawa and M.H.A. Keenleyside. 1995. Fin digging and leaf lifting by the convict cichlid, *Cichlasoma nigrofasciatum*: examples of parental food provisioning. *Animal Behaviour* 49: 623-631.

Wöhl, S. and S. Schuster. 2006. Hunting archer fish match their take-off speed to distance from the future point of catch. *Journal of Experimental Biology* 209: 141–151.

Wöhl, S. and S. Schuster. 2007. The predictive start of hunting archer fish: a flexible and precise motor pattern performed with the kinematics of an escape C-start. *Journal of Experimental Biology* 210: 311–324.

Wolf, N.G., P.R. Swift and F.G. Carey. 1988. Swimming muscle helps warm the brain of lamnid sharks. *Journal of Comparative Physiology B* 157: 709–715.

7

Ecology and Evolution of Swimming Performance in Fishes: Predicting Evolution with Biomechanics

R. Brian Langerhans[1,*] and David N. Reznick[2]

INTRODUCTION

Residing within the immense diversity of fishes on earth is an equally impressive array of locomotor abilities. Some fish continuously swim virtually their entire lives; some move primarily in brief bursts of rapid acceleration; some gracefully maneuver through spatially complex habitats; and some even walk on land. What are the evolutionary root causes of such diversity in swimming abilities? Has swimming performance largely been shaped by random factors, evolving at the whim of genetic drift? If not, then what ecological mechanisms might be responsible for the evolution of swimming performance? Here we investigate some of the major ecological factors that might have shaped the evolution of locomotor performance in fishes. Using an integrative approach, we

Authors' addresses: [1]Museum of Comparative Zoology and Department of Organismic and Evolutionary Biology, Harvard University, Cambridge, MA 02138, USA. E-mail: langerhans@ou.edu
[2]Department of Biology, University of California, Riverside, CA 92521, USA. E-mail: david.reznick@ucr.edu
*Present Address: Biological Station and Department of Zoology, University of Oklahoma, Norman, OK 73019, USA.

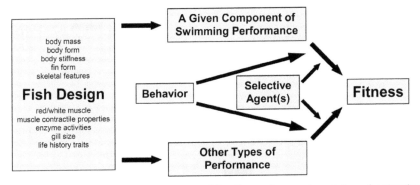

Fig. 7.1 Conceptual path diagram summarizing the ecology and evolution of swimming performance in fishes. The diagram is based on Arnold's (1983) morphology → performance → fitness paradigm, and largely represents an extension of Fig. 3 in Domenici (2003).

acknowledge the complexity of swimming performance, addressing how its evolution depends on numerous underlying organismal features, and is both influenced by and influences multiple fitness components.

At the outset, we should first describe what we mean by "swimming performance." In this chapter, swimming performance simply refers to any quantifiable component of locomotor ability that selection might act upon (e.g., endurance, metabolic rate, maximum acceleration, turning radius). As such, the ecology and evolution of swimming performance is complex, reflecting the net effects of numerous potential targets and agents of selection. Figure 7.1 illustrates the complexity of the evolution of swimming performance in fishes, and several important points can be deduced from this figure. First, as a whole-organism performance measure, swimming performance is influenced by numerous underlying traits (i.e., "Fish Design" in Fig. 7.1). Second, the same traits affecting swimming performance also influence other types of performance. Thus, selection on one type of performance can indirectly affect other types of performance. Third, because performance mediates fitness, selection acts directly on performance and not fish design itself (*sensu* Arnold, 1983). Fourth, behavior can modify the relationships between performance and fitness, most notably by influencing the manifestation of performance values in the wild (e.g., exhibiting sub-maximal performance) and by modifying the nature of the relationship between performance and fitness (e.g., behaviorally increasing or decreasing the frequency with which certain types of performances are employed). As a corollary, selection on one type of performance can indirectly cause behavioral shifts that alter the form of selection on other types of performance. Finally, the form of selection on swimming performance depends on various selective agents (e.g., predators, foraging mode, abiotic factors), and thus will vary across time and space as

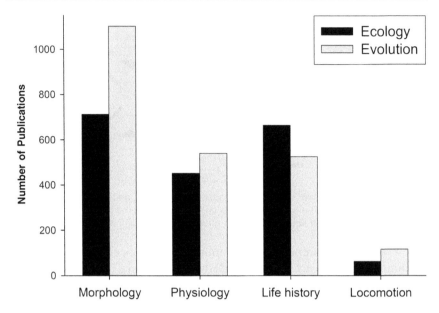

Fig. 7.2 The relative paucity of studies examining the ecology and evolution of fish locomotion compared to fish morphology, physiology, and life history. Data are from a search of the Institute for Scientific Information Science Citation Index, covering articles published between 1900 and 2007. Each search comprised the term "fish", a term for either "ecolog*" or "evolution*", and a term for a subdiscipline (either "morpholog*", "physiolog*", "life histor*", or "locomot*").

the relative importance of selective agents change. Because of all this complexity, acquiring a detailed understanding of the evolutionary ecology of swimming performance in fishes requires an appreciation of a multitude of interactions involving multiple traits and multiple environmental factors— while complex, we aim to demonstrate that such an understanding is possible.

To date, most research on the ecology and evolution of fish has centered on features of fish design, rather than aspects of locomotor performance (Fig. 7.2). This approach is interesting, as it is performance, and not design, that is directly linked to individual fitness. This has probably occurred, at least partially, because the underlying traits are typically much easier to measure than swimming performance for large sample sizes of multiple populations/species. However, the investigation of the evolutionary ecology of swimming performance is not so difficult that it is out of reach. Indeed, owing to a strong foundation that has grown in the realm of biomechanics of fish locomotion, the time is ripe for an explosion of ecological and evolutionary work on fish swimming performance.

Swimming performance lies in the middle of the general morphology → performance → fitness paradigm (Arnold, 1983). As part of this general interaction system, the generation of swimming abilities is understood by inspection of lower-level features (i.e., fish design), while the importance of these abilities is appreciated by understanding their influence on higher levels (i.e., fitness). A great deal of research has examined the links between fish design and swimming performance (e.g., reviewed in Webb and Weihs, 1983; Videler, 1993; Sfakiotakis *et al.*, 1999; Triantafyllou *et al.*, 2000; Lauder and Drucker, 2002; Blake, 2004; Colgate and Lynch, 2004; Müller and van Leeuwen, 2006; Shadwick and Lauder, 2006). Much of this work has had a bio-engineering focus, attempting to understand how specific structures work, largely in the absence of an explicit ecological or evolutionary context (typically lacking a comparative approach, which is the hallmark of ecological and evolutionary approaches; see Lauder *et al.*, 2003). Although considerably less work has explicitly examined the relationships between swimming performance and fitness (but see Katzir and Camhi, 1993; Walker *et al.*, 2005), copious working hypotheses and common assumptions abound (e.g., high burst speeds increase escape ability). Thus, there is no shortage of hypotheses regarding the form of selection on various types of swimming performance via various selective agents. While not originally developed for this purpose, the purely biomechanical approach can be combined with ecological knowledge regarding the links between swimming performance and fitness to generate detailed hypotheses regarding the form of natural selection and the course of evolution. Thus, we suggest evolutionary ecologists tap the wealth of biomechanical theory and empirical work (a field that is still rapidly growing) to formulate hypotheses testable within a comparative framework. Here, we review some key areas where strong biomechanical and ecological knowledge exists, and make some accurate and sometimes remarkably general evolutionary predictions (as well as generate other hypotheses yet to be tested). In the years to come, we encourage researchers in this arena to employ more pluralistic approaches as hypotheses become more refined to further uncover the multitude of mechanisms responsible for the evolution of swimming performance in fishes.

Formulating Evolutionary Predictions

Locomotor performance in fishes is shaped by so many factors that uncovering the ecological mechanisms actually responsible for its evolution might seem unattainable. For example, we know body shape is important in determining swimming performance; however a number of other aspects of organismal function are also inextricably linked to morphology (e.g., diet, mode of reproduction), and hence might also interact with the evolution of locomotor abilities (see Fig. 7.1). Not only can a multitude of design features influence various components of swimming performance, but selection likely operates on numerous types of performance simultaneously, and varies across time and space (e.g., Lowell, 1987; Dudley and Gans, 1991; Wainwright and Reilly, 1994; Lauder, 1996; Reznick and Travis, 1996; Vamosi, 2002; DeWitt and Langerhans, 2003; Ghalambor et al., 2003). One approach to understanding the ecology and evolution of swimming performance is to begin with a few design features and a few selective agents where we can generate well-supported hypotheses for the course of phenotypic evolution. That is, we can build generalized mechanistic models describing how a system operates based on a specified set of assumptions—i.e., how design features determine performance, how performance determines fitness, and consequently what the evolutionary response(s) should be. A similar approach has been taken with fish foraging and labriform locomotion (e.g., Wainwright and Richard, 1995; Westneat, 1995; Wainwright, 1996; Huckins, 1997; Shoup and Hill, 1997; Clifton and Motta, 1998; Fulton et al., 2001, 2005; Ferry-Graham et al., 2002; Grubich and Westneat, 2006; Fulton Chapter 12 in this book), and is proving highly successful. Here, our assumptions are that we correctly understand specific relationships linking fish design, swimming performance, and fitness (see below), and also that adequate genetic variation exists to allow significant evolutionary responses to selection (including the evolution of phenotypic plasticity). We can then test our models' predictions using comparative analyses. The degree of correspondence between predictions and observations should provide insight into the accuracy of our understanding of the biomechanics and evolutionary ecology of fish locomotion (e.g., Endler, 1986; Wainwright, 1988, 1996; Losos, 1990; Williams, 1992; Lauder, 1996; Walker, 1997; Koehl, 1999; Domenici, 2003). Because we cannot hope to cover in detail all factors relevant to the ecology and evolution of swimming performance in one chapter, we will narrow our focus and make evolutionary predictions for a specific set of attributes in a specific set of circumstances.

First, we will center on what is probably the most obvious component of fish design, morphology. Body and fin form is strongly linked to locomotor performance in fishes, and a long history of research has delved into understanding these relationships (e.g., Webb, 1982, 1984, 1986a; Weihs and Webb, 1983; Sfakiotakis et al., 1999; Triantafyllou et al., 2000; Blake, 2004; Lauder, 2005; Lauder and Tytell, 2006; Domenici Chapter 5 in this book). Because any attribute that affects morphology can also affect locomotor performance, we will further consider some components of fish design that can indirectly influence swimming performance via their effects on morphology (e.g., number and size of eggs/embryos, gut length). Second, we will center on only a few common selective agents for fishes: environmental structural complexity, water flow, and predation. Thus, our predictions will focus on divergent evolution of body and fin morphology based on hypotheses regarding the form of natural selection on swimming performance across these three major ecological gradients.

Our predictions derive from a hypothesized trade-off between the two primary swimming modes, steady and unsteady swimming. This trade-off is presumed to exist because of conflicts involving features of fish design, whereby traits that increase performance in one swimming mode necessarily decrease performance in the other (e.g., Blake, 1983, 2004; Webb, 1984; Videler, 1993; Reidy *et al.*, 2000; Domenici, 2003; Langerhans, 2006; Domenici chapter 5 in this book). With specific regard to fish morphology, this general trade-off between steady and unsteady swimming is not expected to apply to all fishes. Rather, the tradeoff should be stronger in fish employing more coupled locomotor systems—that is, cases in which the same morphological structures are used for propulsion during both steady and unsteady swimming. While this scenario applies to the great majority of fish because it is virtually impossible to completely decouple all propulsors, some species have evolved locomotor systems with varying degrees of independence among swimming modes. Specifically, some fish employ different body parts during different swimming activities—e.g., boxfish, *Ostracion meleagris*, use median-and-paired fin propulsion for low-speed cruising and body-and-caudal-fin propulsion for burst-and-coast swimming (Hove *et al.*, 2001). The generalized model we describe below is assumed to apply to fish using a variety of locomotor systems, as the predictions we focus on should apply equally to most fish—i.e., we center on body shape and caudal fin form, and fish have only one body and one tail. When formulating our predictions, we assume that existing theory (typically based on rigid bodies) and empirical work linking morphology and locomotor performance applies equally to fish using various sources of propulsion, excluding flatfish. Because the use of median and paired fins in locomotion varies considerably among fishes, we avoid making generalized predictions for these fins; although some clear predictions do exist for pectoral fin shape in labriform swimmers (e.g., see Wainwright *et al.*, 2002; Walker and Westneat, 2002; Fulton *et al.*, 2005). While other design features also influence swimming performance (e.g., muscle mass/type), and could complicate our morphological predictions, we will simply assume *ceteris paribus*, or at least that these potentially confounding factors do not overwhelm and conceal the predicted patterns. Because our predictions rely on a tradeoff between steady and unsteady locomotion, let us first assess exactly what is meant by these terms.

Steady swimming is constant-speed locomotion in a straight line. While this form of swimming specifically refers to laboratory measures of swimming performance, steady swimming is commonly employed in nature during routine cruising, as well as long-distance movements such as migrations. For instance, fish often utilize steady swimming when searching for food, chasing and obtaining mates, seeking favorable abiotic conditions, and holding station amidst water current (e.g., Plaut, 2001; Domenici, 2003; Blake, 2004; Rice and Hale Chapter 6 in this book). As steady swimming activities are often of critical importance, natural selection is believed to often favor various means of reducing the energetic cost of movement (see Fish Chapter 4 in this book). Steady swimming performance is typically assessed in the laboratory using various estimates of endurance or energy expenditure (e.g., Reidy *et al.*, 2000; Plaut, 2001; Nelson *et al.*, 2002). A wealth of theoretical and empirical work has developed some clear links between morphology and steady swimming (e.g., Wu, 1971; Lighthill, 1975; Webb, 1975, 1984; Blake, 1983; Weihs, 1989; Hobson, 1991; Videler, 1993; Vogel, 1994; Fisher and Hogan, 2007). Steady swimming is generally enhanced with a streamlined body shape (deep anterior body depth, tapering to a narrow caudal peduncle; see further details in Testing Evolutionary Predictions) and a high aspect ratio lunate caudal fin (long span with a short chord; $height^2$/surface area). These features act to maximize thrust while minimizing drag and recoil energy losses. While much of this work has been either theoretical/biomechanical in focus (e.g., using mathematics, physical models, manipulated organisms) or empirically compared distantly related and divergently shaped species, some recent research supports this work using only natural intraspecific variation. For instance, recent studies employing three-dimensional estimations of streamlining have revealed that more streamlined zebrafish (*Danio rerio*) exhibit lower drag coefficients (McHenry and Lauder, 2006), and more streamlined western mosquitofish (*Gambusia affinis*) exhibit higher endurance (R.B. Langerhans unpubl. data).

Unsteady swimming refers to more complicated locomotor patterns in which changes in velocity or direction occur, such as fast-starts, rapid turns, braking, and burst-and-coast swimming. In the wild, such activities are common during social interactions (e.g., courtship, antagonistic interactions), predator evasion, the capturing of evasive prey, and navigating structurally complex environments. For instance, the most commonly studied form of unsteady swimming is the Mauthner-cell initiated escape response present in most fish, called a "C-start" (e.g., Weihs, 1973; Eaton *et al.*, 1977; Domenici and Blake, 1997; Hale *et al.*, 2002; Blake, 2004; Domenici Chapter 5 in this book). During this fast-start, the fish body bends into a "C" shape and then produces a propulsive stroke of the caudal region in the opposite direction, resulting in a sudden, high-energy swimming burst. High unsteady performance is typically produced by a deep body (particularly in the caudal region; this might also be accomplished by median fins rather than the body) and a large caudal fin with a low aspect ratio (e.g., Blake, 1983, 2004; Webb, 1983, 1984, 1986b; Walker, 1997; Langerhans *et al.*, 2004; Domenici *et al.*, 2008). These features maximize thrust and stability during rapid bouts of swimming activity. Numerous theoretical and empirical studies have provided support for these relationships, and recent work at the intraspecific scale has also found supportive evidence. For instance, *G. affinis* individuals with larger caudal regions produce higher burst speeds (Langerhans *et al.*, 2004), and *G. hubbsi* individuals with larger caudal regions produce greater acceleration and angular velocity during fast-starts (R.B. Langerhans unpubl. data).

Because opposite suites of morphological traits optimize steady and unsteady swimming, this creates a scenario where no body form exists which can simultaneously optimize both swimming modes. Thus, fish bodies must reflect some form of compromise between competing swimming demands. Based on this knowledge, we can make general predictions for the course of evolution for fish inhabiting particular environments hypothesized to favor alternative swimming modes (Fig. 7.3). Specifically, we hypothesize that

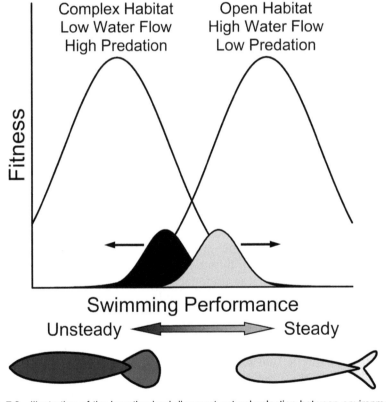

Fig. 7.3 Illustration of the hypothesized divergent natural selection between environments favoring either unsteady or steady swimming performance. Curves represent hypothesized fitness functions, with trait distributions for fishes inhabiting the alternative environments represented by the shaded areas. Arrows illustrate the direction that natural selection is predicted to pull trait means within each environment. The text above each curve describes the environmental gradients examined in this chapter where divergent natural selection is predicted to exist. Figures beneath the x-axis illustrate the general morphological predictions: more streamlined bodies and higher caudal fin aspect ratio in environments favoring steady swimming.

fish will evolve body forms which enhance steady swimming in environments where the importance of steady swimming outweighs unsteady swimming, such as open-water habitats, high-flow environments, and other circumstances where prolonged swimming behaviors are greatly important (e.g., areas with relatively high competition for patchily-distributed resources). In contrast, we predict fish will evolve body forms which enhance unsteady swimming performance in environments where the importance of unsteady swimming outweighs steady swimming, such as structurally complex habitats, low-flow environments, and areas with a high level of predation intensity. While it is certainly true that relationships between morphology and swimming ability can be quite complex—e.g., form-function relationships are often more complicated than predicted by theory, and multiple body designs can produce similar swimming performance (e.g., Wainwright and Reilly, 1994; Koehl, 1996; Lauder, 1996; Domenici and Blake, 1997; Schultz and Webb, 2002)— we focus here on relationships with particularly strong theoretical and empirical support in an effort to elucidate general and predictable trends.

Testing Evolutionary Predictions

The ideal scenario for testing evolutionary predictions would entail an experiment where similar organisms (preferably, genetic clones) are placed into alternative, *a priori* defined environments in a replicated fashion. Then, we would return to these environments many generations later to assess whether the predicted phenotypes evolved. Because this scenario is obviously infeasible for testing general trends across a group as ancient and diverse as fishes, we perform the next best approach for historical data, the comparative method. With the comparative method, we are essentially comparing the outcomes of nature's experiments—however, we must be careful to incorporate the evolutionary relationships among species into our analyses, as species are not independent of one another, but rather reflect a shared evolutionary history (e.g., Felsenstein, 1985; Harvey and Pagel, 1991; Rose and Lauder, 1996; Martins, 2000). The approach is most effective at establishing cause-and-effect (e.g., invasion of Environment A caused evolution of Trait 2) when comparisons are made between closely related organisms, as there are fewer potentially confounding variables to contend with (e.g., less time for development of phylogenetic structure, less divergence in non-focal traits). Historically, most work relevant to our predictions has lacked a strong comparative approach (e.g., commonly comparing few highly divergent taxa, ignoring phylogeny), and thus can only provide cautionary hints regarding relationships among morphology, swimming performance, and habitat. To provide broad and rigorous tests of our predictions, we will use the comparative method at various taxonomic scales (i.e., from distantly related species to inter-population variation within species), incorporating phylogenetic information where appropriate. Our tests will draw from both previously published work and new analyses conducted here for the first time.

Let us begin by explicitly stating our two general predictions: (1) fish should evolve more streamlined bodies and higher aspect-ratio caudal fins in environments where selection favors enhanced steady swimming performance, and (2) fish should evolve less streamlined bodies with a greater posterior body allocation (i.e., deeper/larger caudal peduncles, shallower/smaller heads) and lower aspect-ratio caudal fins in environments where selection favors enhanced unsteady swimming performance (see Fig. 7.3). We will test these predictions by comparing morphology among organisms inhabiting environments hypothesized to favor either steady or unsteady swimming abilities. For all analyses, we use one-tailed tests because we have directional hypotheses.

Before we test these predictions, we should first describe what is meant by the term streamlining. To be precise, a streamlined body is not simply elongate, but rather exhibits a fusiform shape that minimizes drag while maximizing volume (i.e., approximating an airfoil shape; Fig. 7.4a). Streamlining in fish has been measured in many ways, but the most common and straightforward method is the fineness ratio (FR) which is simply body length divided by maximum diameter (typically estimated as maximum body depth). While this is obviously a crude estimate—largely because maximum diameter might be located at various locations along the body— and primarily summarizes the general slenderness of a body, it has a long history in ichthyology, has proven quite useful for comparisons among taxa, and has a direct connection to streamlining theory (i.e., aerodynamics, hydrodynamics) as a symmetric airfoil which exhibits the minimum drag for the maximum volume has a FR around 4.5 (e.g., von Mises, 1945; Hoerner, 1965; Hertel, 1966; Alexander, 1968; Webb, 1975; Blake, 1983; Weihs and Webb, 1983). Thus, an optimum FR exists at the value of 4.5, providing an obvious target for evolutionary predictions of drag minimization. Moreover, the drag coefficient for a given volume can be estimated from FR (von Mises, 1945; Hoerner, 1965)—thus relating body shape directly to swimming performance (Fig. 7.4b). In an attempt to provide a more precise estimate of three-dimensional streamlining, McHenry and Lauder (2006) recently described a streamlining ratio comparing the volume distribution of a fish body to a body having the profile of a streamlined shape (i.e., an airfoil drawn from the U.S. National Advisory Committee for Aeronautics, NACA; see Fig. 7.4). Briefly, the procedure reconstructs the 3D shape of a fish body from dorsal/ventral and lateral views of the fish by approximating its shape as an ellipsoid, and then compares its volume distribution to that of a NACA-streamlined body of the same length, maximum depth, and maximum width, using second moments of area, $I_{fish}/$

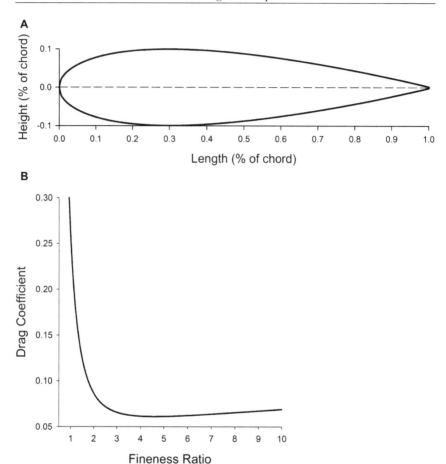

Fig. 7.4 Streamlining and its importance in drag reduction. **(A)** A streamlined shape illustrated using a foil without camber described by the 0000-series of foils from the U.S. National Advisory Committee for Aeronautics (NACA). Figure adapted from Jacobs *et al.* (1933). **(B)** Expected relationship between fineness ratio (length / maximum diameter) and the drag coefficient for a given volume, indicating that a streamlined body with least resistance has a fineness ratio around 4.5 (following Hoerner, 1965).

I_{NACA} (see McHenry and Lauder, 2006 for details). As with FR, an optimum streamlining ratio (SR) exists, where a value of 1.0 represents a perfect match in volume distribution between a given body and the airfoil. Thus, there are theoretical optima for both FR and SR (4.5 and 1.0, respectively), meaning that relationships between these ratios and steady swimming performance should be hump-shaped (see Fig. 7.5a for an empirical relationship with SR). This is because bodies can deviate from each optima in two different ways: either more elongate or rotund than a FR of 4.5, and either too much volume toward the leading or trailing edge compared to a

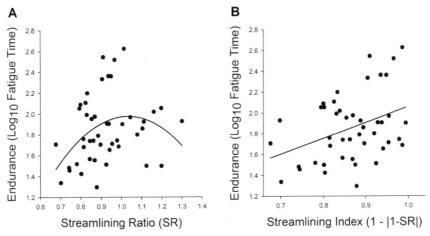

Fig. 7.5 Positive association between body streamlining and endurance in *Gambusia affinis* (R.B. Langerhans unpubl. data). Individuals were first and second generation lab-born descendents of wild-caught fish. **(A)** Matching *a priori* predictions, the raw streamlining ratio (SR) exhibits a hump-shaped relationship with endurance (quadratic regression; $F_{2,42} = 3.25$, one-tailed $P = 0.024$), where the maximum endurance is observed for fish with body forms most closely resembling an airfoil (predicted maximum = 1.0; modeled maximum = 1.03). SR values can be interpreted as follows: values less than 1 represent body shapes with more anterior volume relative to an airfoil, values greater than 1 have more posterior volume relative to an airfoil, and the value 1.0 represents a perfect match in volume distribution between a fish's estimated body form and an airfoil. **(B)** As predicted, the streamlining index (SI) exhibits a positively linear relationship with endurance (linear regression; $F_{1,43} = 7.13$, one-tailed $P = 0.005$). SI values can be interpreted as proportional similarity to the volume distribution of an airfoil (i.e., 0.9 reflects a 90% similarity to an airfoil, while 1.0 perfectly matches an airfoil).

SR of 1.0. Because we often prefer linear relationships in statistical analyses, we can perform simple transformations to these ratios to produce indices ranging from -" to 1, ignoring the specific nature of how a body deviates from the optima and linearizing their predicted relationship with steady swimming performance (see Fig. 7.5b): Fineness Index (FI) = $1 - | 1 - (FR/4.5) |$; Streamlining Index (SI) = $1 - | 1 - SR |$. Values for most fish fall between 0 and 1 for both indices. Note that this linearization will not hold across all of parameter space, but rather is most useful when comparing relatively similar shapes. That is, the relationships between these indices and drag are nearly linear across small ranges of shapes, but nonlinear across a broad range of shapes—this is because the increased drag experienced by bodies relative to the optimum is not symmetric about the optimum (von Mises 1945, Hoerner 1965; see Fig. 4b). Because these ratios and indices have clear and direct connections to streamlining theory, we will use these metrics when possible for investigations of streamlining. However, if our questions extend beyond streamlining *per se*, and greater detail is desired for lateral body shape (e.g., larger lateral surface area in the caudal region generates more thrust during fast starts), then we will employ other, more appropriate techniques, such as landmark-based geometric morphometrics.

Structural complexity

We will begin testing our predictions by comparing fish from environments of differing structural complexities. Fish inhabit environments greatly varying in structural complexity, ranging from pelagic environments, such as the open ocean, to the spatially complex networks of coral reefs and rocky outcrops. Our *a priori* hypothesis is that divergent natural selection between structure regimes arising via alternative locomotor demands has played an important role in driving divergent evolution in fishes. Domenici (2003) recently illustrated this particular hypothesis, where selection is predicted to favor steady swimming performance in structurally-open environments—where resources are widely distributed and continuous swimming is commonplace—but instead favor unsteady swimming performance in structurally-complex environments—where maneuvering, braking, and accelerating are of great importance for both foraging and escaping predation during brief encounters. We specifically predict greater streamlining and a higher caudal fin aspect ratio for fish inhabiting open habitats relative to those in more complex ones (e.g., Webb, 1983; Hobson, 1991; Videler, 1993).

Perhaps the most common—and most studied—axis of divergence in fishes falls under this rubric: phenotypic differentiation between benthic and limnetic environments. Benthic habitats refer to shallow, littoral areas where structural complexity often abounds from sources such as aquatic vegetation, woody debris, and rocks; limnetic habitats simply refer to the open-water environment, such as off-shore lake habitat. While studies of benthic-limnetic divergence have typically centered on foraging adaptations (e.g., number of gill rakers)—because benthic fish feed more commonly on benthic invertebrates, while limnetic fish feed heavily on plankton—body form divergence presumably is largely related to locomotor demands for diverse activities, including foraging, avoiding predation, and pursuing and attracting mates. This form of structure-driven divergence has been described under various names (e.g., benthic-limnetic; littoral-pelagic; stream-lake, where structurally complex, slow flowing inlet streams are compared to open-water areas of the lake), and has been implicated in numerous cases of intraspecific differentiation and recent or incipient speciation in fishes (e.g., Echelle and Kornfield, 1984; Ehlinger and Wilson, 1988; Robinson and Wilson, 1994; Schluter, 1996; Smith and Skúlason, 1996; Taylor, 1999; Robinson *et al.*, 2000; Jonsson and Jonsson, 2001; Hendry *et al.*, 2002; McKinnon and Rundle, 2002; Hendry and Taylor, 2004). Some of the classic examples include intra-lake differentiation in northern latitudes, such as stickleback (*Gasterosteus aculeatus*), whitefish (*Coregonus* spp.), arctic char (*Salvelinus alpinus*), pumpkinseed sunfish (*Lepomis gibbosus*), and bluegill sunfish (*Lepomis macrochirus*); although tropical examples exist as well (e.g., Meyer, 1990; Schliewen *et al.*, 2001; Barluenga *et al.*, 2006). In some cases, the predicted swimming performance tradeoff has been investigated— sometimes being confirmed. For instance, Taylor and McPhail (1986) found that anadromous threespine stickleback (open habitat) exhibit higher endurance, but lower fast-start performance than stream stickleback (complex habitat). Further, limnetic threespine stickleback (open habitat) exhibit lower drag coefficients and higher endurance than benthic stickleback (complex habitat) which evolved in the same lake (Blake *et al.*, 2005); although both limnetic and benthic forms produce similar fast-start performance (Law and Blake, 1996). In both of these examples, and with the vast majority of documented cases, the limnetic (open habitat) forms exhibit more slender bodies than the benthic (complex habitat) ones, matching *a priori* predictions (caudal fin form is very rarely examined). Thus, there is strong empirical support for this general hypothesis, where numerous distantly related lineages have repeatedly undergone parallel cases of predictable evolution. Note that phenotypic plasticity's role in these instances of phenotypic differentiation should not be underestimated, as inducible changes seem to represent an important component of observed patterns in many cases, and may facilitate subsequent genetic divergence (e.g., Day *et al.*, 1994; Robinson and Wilson, 1996; Robinson and Parsons, 2002; West-Eberhard, 2003; 2005; Svanbäck and Schluter unpubl. data). As an extension to this existing evidence, here we use comparative data to test the generality of this hypothesis across highly diverse fish taxa. This analysis is meant as a broad-brush approach to the question, seeking evidence for general trends across a diverse subset of fishes, spanning 11 orders.

A new test across diverse fishes

Domenici (2003) assembled a dataset of 32 fish species for which swimming performance data was available, assigned each species to a habitat category (complex, intermediate, open), and tested for the predicted performance tradeoff (see methodological details in Domenici, 2003). Results matched predictions for both steady and unsteady swimming performance, where fish inhabiting more open environments exhibit higher steady swimming abilities (endurance) but lower unsteady swimming abilities (turning radius and acceleration) than fish in more complex environments. If our predictions are accurate, then these performance differences derived, at least partially, from morphological differences. Here we use this dataset to test for the predicted morphological differences between habitats (see Fig. A1 in *Appendix* for species included in the dataset). We obtained morphological data (fineness ratio [FR] and caudal fin aspect ratio) from FishBase (Froese and Pauly, 2007). These variables were extracted from the "Morphometrics" section of the web pages for each species, and comprise measurements conducted on digital photographs of specimens. For three species (*Xenomystus nigri, Odontesthes regia,* and *Chromis punctipinnis*), no morphological data were available on FishBase. For these species, we used the same methods employed by FishBase to complete the dataset. We averaged cases where data from multiple specimens for a given species were available (eight species). Because FR ranged from less than to greater than the optimum of 4.5, we calculated fineness index (FI; see description above). We also estimated the drag coefficient based on volume (C_D) for each fish from its FR (following Hoerner, 1965; equation 36, pg. 6–19; assuming a common Re of 10^5). To ensure the methods for obtaining morphological variables were adequate for the examination of interspecific differences in morphology, we performed a repeatability analysis using the intraclass correlation coefficient from a model II ANOVA (Lessells and Boag, 1987; Sokal and Rohlf, 1995). We found significant repeatability for FI (intraclass correlation coefficient, $r = 0.94$, $P < 0.0001$), C_D ($r = 0.82$, $P = 0.002$), and caudal fin aspect ratio ($r = 0.56$, $P = 0.036$), suggesting the variables likely represent a fair approximation of morphological variation among these distantly related species.

We conducted both standard and phylogenetic ANOVAs, testing for differences in FI, C_D, and caudal fin aspect ratio between species inhabiting environments of varying structural complexity. We used the PDSIMUL and PDANOVA programs (Garland *et al.*, 1993) to perform phylogenetic ANOVAs. Using these programs, we simulated trait evolution as Brownian motion with the means and variances of the simulations set to the means and variances of the original data. We performed 1000 simulations, producing a null distribution of *F*-statistics against which the *F*-value from the actual data could be compared to assess statistical significance (i.e., determine how different the observed patterns were from that expected via genetic drift alone). We constructed a best-estimate phylogenetic hypothesis for this group of species based on previous morphological and molecular studies (Fig. A1 in *Appendix*; Johnson and Patterson, 1993; Nelson, 1994; Bernardi and Bucciarelli, 1999; Inoue *et al.*, 2001; O'Toole, 2002; Miya *et al.*, 2003; Crespi and Fulton, 2004; Smith and Wheeler, 2004; Shinohara and Imamura, 2007; Smith and Craig, 2007). All branch lengths were set equal to one. Some species were excluded from statistical analysis based on their highly divergent phenotypic values. For FI, *Anguilla anguilla* (European eel) was excluded, as it was an extreme outlier (4.3 standard deviations from the nearest species). For C_D, *Pterophyllum scalare* (freshwater angelfish) was an extreme outlier (5.3 standard deviations from the nearest species), and excluded from analysis. For caudal fin aspect ratio, *A. anguilla* and *X. nigri* (African knifefish) were excluded as these species essentially lack a caudal fin. In all cases, these procedures were conservative, as results would have more strongly supported predictions had these species been included.

Results matched *a priori* predictions for all three traits: fish in more open habitats tended to exhibit a higher FI ($F_{2,28}$ = 5.88, one-tailed P_{raw} = 0.0037, one-tailed P_{phy} = 0.025), a lower C_D ($F_{2,28}$ = 3.12, one-tailed P_{raw} = 0.030, one-tailed P_{phy} = 0.078), and a higher caudal fin aspect ratio ($F_{2,27}$ = 10.30, one-tailed P_{raw} = 0.0002, one-tailed, P_{phy} = 0.008) (Fig. 7.6). These results suggest that fish do indeed generally evolve predictable morphologies and swimming abilities in these alternative habitat types. Thus, despite the complicated nature of morphology → performance relationships, we can apparently still predict broad trends across a diverse group of fishes. Taken alone, these results provide cautionary evidence supporting our hypotheses of divergent natural selection between structure regimes and subsequent divergent evolution. However, when combined with the abundant supporting evidence from closely related groups of fishes (see above), we have strong confidence that structural complexity plays an important role in the evolution of body form and swimming performance—the nature of which is quite predictable.

Water flow

Fish inhabit environments greatly varying in the intensity of water movement, such as ponds, lakes, backwaters, calm tidal pools, stream riffles, rapid rivers, and wave-swept near-surface ocean waters. While fish virtually always contend with competing demands for steady and unsteady swimming performance, we hypothesize that this balance will swing toward favoring steady swimming in high-flow environments—where fish must constantly swim to maintain position and perform routine tasks under arduous conditions—but unsteady swimming in low-flow environments—where fish are largely freed from the severe demands on endurance and can instead exploit strategies requiring high acceleration or maneuverability. Thus, we predict greater streamlining and a higher caudal fin aspect ratio in fish found in high-flow habitats compared to fish inhabiting less-flowing waters. This general prediction has been outlined numerous times in the past (e.g., Blake, 1983, 2004; Webb, 1984; Videler, 1993; Vogel, 1994; Boily and Magnan, 2002), and empirical support for water-flow's role in fish evolution is non-trivial (e.g., Hubbs, 1941; Swain and Holtby, 1989; Claytor *et al.*, 1991; McLaughlin and Grant, 1994; Hendry *et al.*, 2000, 2006; Pakkasmaa and Piironen, 2000; Brinsmead and Fox, 2002; Imre *et al.*, 2002; Langerhans *et al.*, 2003, 2007a; McGuigan *et al.*, 2003; Neat *et al.*, 2003; Collyer *et al.*, 2005; Sidlauskas *et al.*, 2006; Zúñiga-Vega *et al.*, 2007). Let us briefly review the existing evidence to assess the degree to which evolution across flow regimes is predictable.

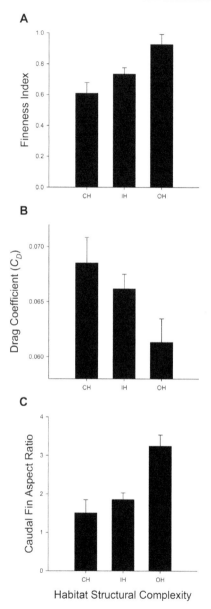

Fig. 7.6 Association between morphology and habitat structural complexity for 32 fish species spanning 11 orders. **(A)** Species inhabiting more open environments exhibit fineness index values closer to 1 (i.e., fineness ratios closer to the optimum of 4.5). **(B)** Drag coefficient per volume (estimated from fineness ratios, following Hoerner, 1965) tends to be lower for species inhabiting more open habitats. **(C)** Species in more open environments exhibit higher caudal fin aspect ratios. Least-squares means ± 1 S.E. are presented. CH = complex habitat, IH = intermediate habitat, OH = open habitat.

Examples of flow-driven phenotypic differentiation come from diverse groups of fishes—including characids, cichlids, cyprinids, salmonids, and poeciliids—and from numerous biogeographic regions—including all continents but Antarctica. For instance, in South American tropical systems, recent work has revealed intraspecific morphological variation within several species (Family Characidae: *Bryconops caudomaculatus*, *Bryconops* sp. cf. *melanurus*; Family Cichlidae: *Biotodoma wavrini*, *Cichla temensis*); in each case, a more streamlined body is evident in populations experiencing greater water flow (Langerhans *et al.*, 2003; Sidlauskas *et al.*, 2006; R.B. Langerhans unpubl. data). In rainforest streams of East Africa, *Barbus neumayeri* (Family Cyprinidae) populations experiencing higher water velocity exhibit both more fusiform bodies and higher aspect-ratio caudal fins (Langerhans *et al.*, 2007a). In central North America, recent work has revealed that *Cyprinella lutrensis* (red shiner, Family Cyprinidae) populations inhabiting fast-flowing inlet rivers exhibit more streamlined bodies than their dammed, non-flowing reservoir counterparts (R.A. Blaine and K.B. Gido, unpubl. data). In Canadian watersheds, two sunfish species (Family Centrarchidae), *L. gibbosus* (pumpkinseed) and *Ambloplites rupestris* (rock bass), exhibit more slender bodies in streams compared to conspecific populations inhabiting adjacent lakes (Brinsmead and Fox, 2002).

As these examples illustrate, a general pattern of increased streamlining for fish experiencing higher flow regimes is evident from the literature, however there are a number of cases that do not directly correspond to predictions. While some of these cases can be explained by species' natural histories (e.g., fish may not actually spend much time in the current, or other behaviors may diminish the predicted selective effects of water flow), others simply point to our lack of a complete understanding in this regard (e.g., McGuigan *et al.*, 2003; Neat *et al.*, 2003; Hendry *et al.*, 2006). Further, much fewer studies have used comparative data to test for greater caudal fin aspect ratio in higher-flow environments, although a number of studies have demonstrated taller caudal fins in such habitats (likely reflecting higher aspect ratios; e.g., Beacham *et al.*, 1989; Imre *et al.*, 2002; Peres-Neto and Magnan, 2004). Overall, velocity's effects on morphology appear to often, but not always, match our predictions based on divergent natural selection on swimming abilities.

Much of this empirical work has centered on intraspecific differentiation, at least some of which likely represents phenotypic plasticity. For instance, McLaughlin and Grant (1994) demonstrated a pattern in the field where brook charr (*Salvelinus fontinalis*; Family Salmonidae) exhibit more streamlined bodies and higher caudal fin heights (presumably resulting in higher caudal fin aspect ratio) in populations experiencing higher water velocity. Later, Imre *et al.* (2002) demonstrated that the observed differences in caudal fin height, but not body depth, could be due to phenotypic plasticity. Granted that the evolution of inducible adaptive phenotypes in alternative flow environments is impressive in its own right, we would also like to know whether evolutionary divergence between flow regimes is common and predictable. Thus, we will provide a new test of this hypothesis using comparative data across multiple congeneric species for which morphological differences have been shown to have a genetic basis (e.g., Hubbs and Springer, 1957; Greenfield *et al.*, 1982; Greenfield, 1983; Greenfield and Wildrick, 1984; Greenfield, 1985; Langerhans *et al.*, 2004, 2005; R.B. Langerhans unpubl. data).

A new test in Gambusia fishes

Of the approximately 50 species in the livebearing fish genus *Gambusia* (mosquitofishes; Family Poeciliidae), 15 can be readily classified as either inhabiting primarily low- or high-flow environments. We measured FR, C_D (calculated from FR; as described above), and lateral body shape for five adult males from each of 12 of these *Gambusia* species, ranging from Texas, Central America, the Caribbean, and southern Florida (see *Appendix 1* for species included and source of specimens). Species were classified to flow regime according to published accounts and museum records for all species (e.g., Hubbs, 1929; Hubbs and Springer, 1957; Rosen and Bailey, 1963; Hubbs and Peden, 1969; Miller and Minckley, 1970; Fink, 1971; Peden, 1973; Miller, 1975; Brune, 1981; Greenfield *et al.*, 1982; Rauchenberger, 1989; de León *et al.*, 2005; Tobler *et al.*, 2006), personal communications for three species (pers. comm. O. Dominguez, C. Hubbs), and personal observations (R.B.L.) for five species. The low-flow species inhabit lakes, ponds, tidal shores and pools, and slow-flowing springheads. The high-flow species inhabit rivers, streams, and fast-flowing spring runs. Although *Gambusia* species tend to seek slower-flowing microhabitats even within generally fast-flowing environments (with a few exceptions), the species included in the "high-flow" category must still regularly negotiate considerably stronger water velocities than the species included in the "low-flow" category which only rarely experience measurable flow.

To quantify lateral body shape, we digitized 10 landmarks (see Fig. 7.7b) on digital images of each specimen using tpsDig (Rohlf, 2006), and used geometric morphometric methods to examine morphological variation (Rohlf and Marcus, 1993; Marcus *et al.*, 1996; Zelditch *et al.*, 2004). For further description of landmarks and methods, see text and figures in Langerhans *et al.* (2004), Langerhans and DeWitt (2004), and Langerhans *et al.* (2007b). We used tpsRelw software (Rohlf, 2005) to obtain shape variables (relative warps) for analysis. Relative warps are principal components of geometric shape information. We visualized variation in landmark positions using the thin-plate spline approach, which maps deformations in shape from one object to another (Bookstein, 1991).

We calculated species averages for fineness ratio, drag coefficient, and relative warps using least-squares means from an ANCOVA with log standard length as the covariate (larger individuals tended to have deeper bodies) and species as the main effect (slopes were homogeneous)—i.e., we used size-adjusted trait values. We then conducted both standard and phylogenetic ANOVAs following methods described above, testing for differences in traits between species in low- and high-flow environments. Phylogenetic information was based on molecular and morphological data (Rivas, 1963; Rosen and Bailey, 1963; Rauchenberger, 1989; Lydeard *et al.*, 1995a; b; R.B. Langerhans, M.E. Gifford, O. Dominguez, I. Doadrio unpubl. data), and represents the best current phylogenetic hypothesis for these taxa (Fig. A2 in *Appendix*). All branch lengths were set equal to one.

We found that *Gambusia* species inhabiting higher velocity environments tended to exhibit FR values closer to the optimum of 4.5 ($F_{1,10}$ = 2.82, one-tailed P_{raw} = 0.062, one-tailed P_{phy} = 0.039; Fig. 7.7a), and thus lower drag coefficients ($F_{1,10}$ = 5.48, one-tailed P_{raw} = 0.021, one-tailed P_{phy} = 0.010)—matching *a priori* predictions. When examining overall body shape, we found significant differences between flow regimes for the first relative warp (RW 1; $F_{1,10}$ = 3.48, one-tailed P_{raw} = 0.046, one-tailed P_{phy} = 0.028), but not for other relative warps. RW 1 (explaining 49.95% of shape variation) was highly correlated with FR among species (r = 0.78, P = 0.003), however it also provided some additional morphological information (Fig. 7.7b). RW 1 results revealed that not only do high-flow species tend to be more slender, but they also differ in their location of maximum depth, which is more anteriorly positioned in higher-velocity environments. Further, much of the shape differences involve the caudal peduncle, where species in high-flow environments have longer and shallower caudal peduncles than those in low-flow habitats. All of these differences suggest that *Gambusia* species inhabiting high-flow environments exhibit increased streamlining compared to low-flow species. Combined with previous work discussed above, it seems that water velocity generally plays an important role in the phenotypic evolution of fishes.

Predation

For our final prediction, we will evaluate the role of predation in driving morphological and locomotor evolution in fishes. The strong role of predation in shaping phenotypic evolution in many, diverse taxa is undisputed (reviewed in Edmunds, 1974; Greene, 1988; Caro, 2005; Vamosi, 2005; Langerhans, 2006). Like most organisms, fishes inhabit environments that vary in predation intensity across time and space. Multiple types of selection might differ between predator regimes, but here we focus on one specific prediction: natural selection should favor steady swimming performance in environments with a low level of predation, but instead favor

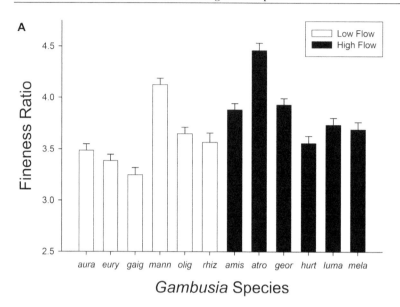

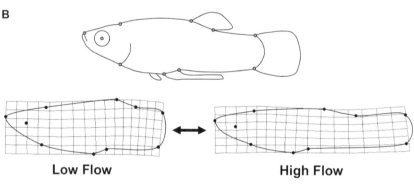

Low Flow   **High Flow**

Fig. 7.7 *Gambusia* species inhabiting environments with a greater magnitude of water flow tend to exhibit more streamlined bodies. **(A)** Species in high-flow environments have fineness ratios closer to the optimum of 4.5. Values represent least-squares means ± 1 S.E. from an analysis of covariance with log standard length as the covariate and species as the main effect (slopes were homogeneous). Species labels are from left to right: *G. aurata, G. eurystoma, G. gaigei, G. manni, G. oligosticta, G. rhizophorae, G. amistadensis, G. atrora, G. georgei, G. hurtadoi, G. luma,* and *G. melapleura.* **(B)** Landmarks used in the examination of lateral body shape (upper panel) and thin-plate spline transformation grids illustrating the nature of morphological differences between flow regimes as described by the first relative warp (lower panel; magnified 2×). Solid lines connecting outer landmarks are drawn to aid interpretation.

unsteady swimming abilities in high-predation environments. This prediction stems from ecological knowledge of prey populations inhabiting divergent predator regimes.

First, fitness should largely depend on competitive interactions in low-predation environments (e.g., low extrinsic mortality, high population density). Because steady swimming is used during most competitive activities (see *Formulating Evolutionary Predictions*), selection should favor enhanced steady swimming performance in low-predation environments (i.e., obtain and consume food more quickly, acquire mates more effectively, contain greater energy supplies for reproduction) (e.g., Vogel, 1994; Plaut, 2001; Roff, 2002; Domenici, 2003; Blake, 2004). Second, escape ability should be of paramount importance in high-predation environments. Because unsteady swimming activities, such as fast-starts and rapid turns, are highly important in escaping predation, fish with greater unsteady swimming performance should possess higher fitness in high-predation environments (e.g., Howland, 1974; Webb, 1986b; Domenici and Blake, 1997; Walker, 1997; Langerhans *et al.*, 2004; Walker *et al.*, 2005; Domenici Chapter 5 in this book). Third, predator regime can influence prey habitat use by altering the structural complexity of the environment experienced by prey—a factor already shown to affect fish evolution (see *Structural complexity*). This fact can reinforce divergent selection on steady and unsteady swimming performance between predator regimes. This is because many fishes utilize more open-water habitats in low-predation environments compared to high-predation environments, where fish are often restricted to complex habitats, such as the littoral zone, as a behavioral mechanism reducing predatory encounters (e.g., Horwood and Cushing, 1977; Stein, 1979; Mittelbach, 1981; Werner *et al.*, 1983; Tonn *et al.*, 1992; Winkelman and Aho, 1993; Eklov and Persson, 1996). Thus, as a combination of differences in competition, necessity of escape ability, and structural complexity, selection is predicted to favor alternative swimming modes in low- and high-predation environments.

Based on our understanding of how morphology influences swimming performance, we can make general predictions for morphological evolution in divergent predator regimes. We expect fish to evolve (or evolve the ability to developmentally induce) a less streamlined body with greater posterior allocation (i.e., shallower/smaller heads, deeper/larger caudal region, including median fins) and a larger, lower aspect-ratio caudal fin in high-predation environments compared to those inhabiting low-predation environments. These predictions derive from biomechanical predictions described above for enhancement of steady swimming performance in low-predation environments and enhancement of fast-start performance in high-predation environments via accentuation of thrust-producing regions and minimization of drag-producing regions. Here we review the existing evidence for this hypothesis. Because this prediction was only recently explicitly proposed (Walker, 1997; Langerhans *et al.*, 2004, 2007b), there are still relatively few studies which specifically test this hypothesis. However, as we reveal, there is a fair quantity of relevant work amassing. We will primarily focus our review on studies using comparisons among low- and high-predation environments defined by densities of piscivorous fish (often, presence vs. absence), as predictions are most clear for these interactions— and because most studies to date have involved predatory fish. However, note that other types of predators (e.g., invertebrates, snakes, birds) may be important in some systems, and may or may not generate similar selective pressures.

During the past decade, a number of examples of predator-associated morphological divergence have accumulated. All cases involve intraspecific differentiation between populations, and thus provide a set of comparisons that should be capable of establishing mechanistic relationships (e.g., presence of predators caused phenotypic changes in prey). Threespine stickleback (*G. aculeatus*) in the Cook Inlet region of Alaska, USA inhabit postglacial lakes differing in the presence of predatory fish. In the presence of predators, stickleback populations exhibit a relatively smaller head and larger median fin lengths compared with conspecific populations in lakes without predatory fish (Fig. 8A; Walker, 1997; Walker and Bell, 2000). In East African lakes, introduction of Nile perch (*Lates niloticus*)—a highly piscivorous fish—has led to subsequent morphological changes in multiple native fishes. For both a cichlid (*Pseudocrenilabrus multicolor*) and a cyprinid (*Rastrineobola argentea*), a posterior shift in body depth is apparent in the presence of the predator (R.B. Langerhans, L.J. Chapman, and T.J. DeWitt unpubl. data). These results were obtained using either temporal comparisons within lakes (before and after introductions) or contemporary comparisons among lakes with and without Nile perch. Perhaps the best case studies to date come from four species of poeciliid fishes (*Brachyrhaphis rhabdophora, G. affinis, G. hubbsi, Poecilia reticulata*), as these systems have the most thorough set of evidence relevant to the hypothesis at hand (Table 7.1). Here we will briefly review this evidence.

For all four of these species, conspecific populations inhabit environments differing in the presence of predatory fish. Table 7.1 summarizes the relevant results for these four species. In sum, there is strong evidence for predator-driven evolution in these fishes. In all cases, high-predation fish exhibit larger caudal regions than low-predation fish (see Fig. 7.8b for results in *G. affinis*). Moreover, differences in swimming performance match predictions in all tests to date, fitness advantages have been documented in two species, a genetic basis to phenotypic differences have been confirmed for three species, and molecular evidence suggests multiple events of predictable phenotypic evolution within three species

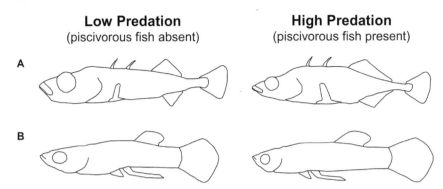

Fig. 7.8 Illustration of morphological differences between populations inhabiting low- and high-predation environments within two distantly related fish species. Note the smaller anterior/ head region and larger caudal region in populations coexisting with predatory fish. Figures represent results of thin-plate spline deformations from the grand mean in each species, and are magnified 2× to aid visualization. **(A)** Threespine stickleback, *Gasterosteus aculeatus*, from 40 lakes in Alaska, USA (data from Walker and Bell, 2000). **(B)** Western mosquitofish, *Gambusia affinis*, from six ponds in Texas, USA (data from Langerhans *et al.*, 2004).

(Table 7.1). Further, in both *G. affinis* and *G. hubbsi*, fish with larger caudal regions—irrespective of population of origin—are faster (Langerhans *et al.*, 2004; R.B. Langerhans unpubl. data), and in both *G. hubbsi* and *P. reticulata*, faster fish—irrespective of population of origin—exhibit higher survival with predators (Walker *et al.*, 2005; R.B. Langerhans unpubl. data). Additionally, more streamlined fish (like those found in low-predation populations) exhibit higher steady swimming abilities in *G. affinis* (R.B. Langerhans unpubl. data; see Fig. 5). Thus, our assumptions concerning the links between morphology, performance, and fitness have at least been partially confirmed for these livebearing fishes. While confounding variables do exist in some cases here (e.g., ponds and streams with predators tend to be larger and have higher productivity), at least for one species (*G. hubbsi*) these variables have been kept to a minimum. This is because *G. hubbsi* inhabit isolated blue holes (vertical solution caves) which are highly similar in most abiotic parameters other than predator regime (Langerhans *et al.*, 2007b), yielding a very strong test of the hypothesis.

While in some cases these differences are known to have a genetic basis (see Table 7.1), phenotypic plasticity has also been demonstrated in several cases (e.g., Brönmark and Miner, 1992; Andersson *et al.*, 2006; Eklov and Jonsson, 2007; Januszkiewicz and Robinson, 2007). So, divergent selection across predator regimes has apparently resulted in both the evolution of adaptive plasticity and local adaptation. For most cases, it is perhaps most likely that morphological differences reflect both plastic changes as well as genetic differentiation. Whether or not such intraspecific divergence translates into macroevolutionary patterns has not yet been

Table 7.1 Summary of predator-driven adaptive divergence within four livebearing fish species (Family Poeciliidae). Temporal persistence of differences are evident when results are consistent across multiple years of examination. Swimming performance refers to confirmation of predicted differences in either steady (S) or unsteady (U) swimming. Fitness advantages refer to elevated fitness of native fish over fish transplanted from the alternative predator regime, and can be tested within either low-predation (LP) or high-predation (HP) environments. Genetic basis refers to persistent differences in morphology, performance, fitness or some combination thereof in lab-born fish raised in a common environment for one (F1) or two generations (F2). Multiple origins refers to molecular evidence suggesting replicated trait evolution.

Species	Geographic Region	Populations Examined	Larger Caudal Region	Temporal Persistence	Swimming Performance	Fitness Advantage	Genetic Basis	Multiple Origins	References
Brachyrhaphis rhabdophora	Costa Rica	6	Y	NA	NA	NA	NA	Y	1,2
Gambusia affinis	Texas, USA	6	Y	Y	S,U	NA	F1, F2	NA	1,3,4
Gambusia hubbsi	Bahamas	12	Y	Y	U	HP	F1	Y	5,6
Poecilia reticulata	Trinidad	23	Y	Y	U	HP	F2	Y	1,7-12

Note: All *P. reticulata* populations from the Paria drainage were treated as a single unit of replication here because there are no high-predation localities within the entire drainage. Y = yes, N = no, NA = no data available. References: 1: Langerhans and DeWitt (2004), 2: Johnson (2001), 3: Langerhans *et al.* (2004), 4: R.B. Langerhans *et al.* (2007b), 6: R.B. Langerhans unpubl. data; 5: Langerhans *et al.* (2007b), 6: R.B. Langerhans unpubl. data; 7: Hendry *et al.* (2006), 8: Ghalambor *et al.* (2004), 9: Walker *et al.* (2005), 10: Carvalho *et al.* (1991), 11: Fajen and Breden (1992), 12: O'Steen *et al.* (2002).

revealed. However, evidence for ecological speciation as a by-product of divergent natural selection between predator regimes has recently been described for *Gambusia* fishes (Langerhans *et al.*, 2007b). Further, ongoing work is revealing that interspecific morphological divergence mirrors intraspecific divergence in the genus *Gambusia* (R.B. Langerhans unpubl. data), suggesting that predator-driven divergence can produce broader macroevolutionary patterns.

Unfortunately, no studies to date have examined predation's influence on caudal fin aspect ratio. However, our *a priori* predictions for body shape have been upheld in numerous empirical tests. Considering the predictions have been observed in distantly related species inhabiting different habitat types (e.g., ponds, lakes, streams) in widely distant geographic regions (e.g., Alaska, Bahamas, East Africa), we have fairly strong evidence that predation can drive predictable evolution (see Fig. 7.8). Yet for some species, researchers have made different predictions, such as increased defensive plates or spines and maximum body depth in environments with higher predation intensity (rather than a posterior shift in body depth as we predict here). This is because some species might be expected to enhance post-capture survival via increased handling time or even complete gape limitation. This prediction has been confirmed in multiple systems (e.g., Brönmark and Miner, 1992; Reimchen, 1994, 1995, 2000; Walker, 1997; Reimchen and Nosil, 2004; Eklov and Jonsson, 2007; Januszkiewicz and Robinson, 2007). Whether these morphological differences also entail a posterior shift in depth or increased fast-start performance is unknown in most cases. Moreover, when we might expect enhanced capture deterrence (i.e., fast-start escapes) versus consumption deterrence (i.e., post-capture survival) is often not clear in nature, and how these two selective forces might interact is also less than obvious. For instance, deeper bodies—which can gape limit some predators—might sometimes enhance maneuverability or fast-start performance. Thus, it may be possible for fish to evolve morphologies that simultaneously increase both handling time and unsteady swimming performance. Recent work in crucian carp (*Carassius carassius*) supports this proposition. Crucian carp are known to induce deeper bodies in the presence of pike (*Esox lucius*) predators; a phenotypic shift which reduces predation risk by increasing handling time of gape-limited piscivores (Brönmark and Miner, 1992; Brönmark and Pettersson, 1994; Nilsson *et al.*, 1995). Domenici *et al.* (2008) demonstrated that these deeper-bodied carp also produce higher velocity, acceleration, and turning rates during fast-start escapes than shallow-bodied individuals, suggesting that predator-induced crucian carp exhibit reduced vulnerability to predation for multiple reasons. Yet sometimes, defensive traits might incur costs associated with escape locomotor performance. For instance, a negative effect of defensive armor on fast-start performance has been observed in some fishes (Taylor and McPhail, 1986;

Andraso and Barron, 1995; Andraso, 1997; Bergstrom, 2002), suggesting a possible conflict between the evolution of armor and swimming performance in high-predation environments. More work is needed to better understand the integrated roles of these various defensive strategies.

Reproduction and diet

Reznick and Travis (1996) argued that adaptation is not simply optimizing some feature of an organism with regard to some feature of the environment. Natural selection acts on whole-organism performance. As a consequence, individual adaptations often represent a compromise among multiple types of selection and multiple competing internal features of an organism (see Fig. 7.1). Thus, the way any organism adapts to its environment should represent an integration of all of these external and internal factors. Up until now, we have focused on the role of body shape in determining swimming performance, and then on a few major environmental gradients where we might be capable of predicting evolution of performance and hence shape: structural complexity, water flow, and predation. Because of the strong relationship between shape and performance, we argue that any other feature of a fish's biology that influences shape will potentially be integrated with selection on swimming performance. While a number of factors might influence body shape in addition to direct selection on locomotor performance (e.g., sexual selection, selection for gape limitation, minimizing detection by prey via reduction of body profile), we focus here on two internal factors that might often alter shape and hence affect performance: reproduction and diet. There is not, at the present time, any experimental system that considers how the evolution of swimming performance represents a balance among all of these potentially competing functions. There is, however, sufficient characterization of enough pieces of this puzzle for us to argue that such an integrated study of the evolution of performance is justified.

The organ systems supporting both reproduction and digestion occupy the body cavity, and the size of each system can influence the shape and flexibility of a fish. Because swimming performance is influenced by shape and flexibility, the joint volume of the gut and gonads can influence performance. So an adaptation that causes an increase in the volume of either organ system can influence the evolution of the other organ system and/or the evolution of locomotor ability. Consequently, we can make the general prediction that increased volume in these organ systems will impair locomotion (see below). Thus, selection favoring larger gut or reproductive systems might constrain the evolution of swimming performance. We will begin assessing this prediction by examining the links between reproduction, body shape, and swimming performance. Then we will discuss similar relationships for diet—however, because few studies to date have directly assessed associations between diet, body shape, and locomotor ability, we can only briefly review and speculate about diet's potential significance in the evolution of swimming performance.

One obvious scenario where the reproductive system might have large effects on body shape in fishes is in the case of livebearers. This is because female livebearers not only carry eggs, but actually carry developing embryos which can sometimes be of considerable size or number. During pregnancy, a female's body shape can be significantly altered. In most livebearers in the family Poeciliidae, females carrying late term embryos are visibly robust and easily distinguished from non-pregnant females or females with young in early stages of development (Fig. 7.9a). Here we quantify this effect in pregnant females by presenting new data from our ongoing work in the livebearing fish genus *Gambusia*, establishing a relationship between reproduction and body shape.

We measured average embryo weight and fineness ratio (FR) of pregnant females for four *Gambusia* species (see *Appendix 2* for species included and source of specimens). To test whether each species exhibited a similar relationship between FR and embryo weight, we conducted an analysis of covariance with FR as the dependent variable, species as the main effect, and natural-log transformed mean embryo weight as the covariate. We found a significantly negative relationship between FR and mean embryo weight ($F_{3,66}$ = 16.46, P = 0.0001; Fig. 7.9b)—a trend which did not significantly differ among species (interaction term: $F_{3,66}$ = 1.22, P = 0.309). Thus, female *Gambusia* carrying larger embryos tend to exhibit deeper (less streamlined) bodies. Because we have already shown that body shape is well correlated with performance and have now shown that intraspecific variation in life histories can influence body shape, it is logical to propose that the evolution of life histories can have a collateral impact on the evolution of swimming performance.

To date, the most thorough examination of the relationship between life history and performance has been conducted using natural populations of guppies (Ghalambor *et al.*, 2004). Guppies (*P. reticulata*) reproduce continuously after attaining maturity, so mature females are almost always found with a brood of developing young. The quantity of tissue devoted to developing young can be quite large relative to their body sizes, with the dry mass of developing young sometimes exceeding 20% of the total dry mass of the female. The wet mass and volume of the young increases between 3.5 and 4 fold between the fertilization of the egg and birth because of a substantial increase in the water content of the embryos (Ghalambor *et al.*, 2004). This means that the reproductive burden increases throughout development (e.g., see Fig. 7.9b for effects on body depth). Such dramatic changes associated with pregnancy might impact swimming performance for

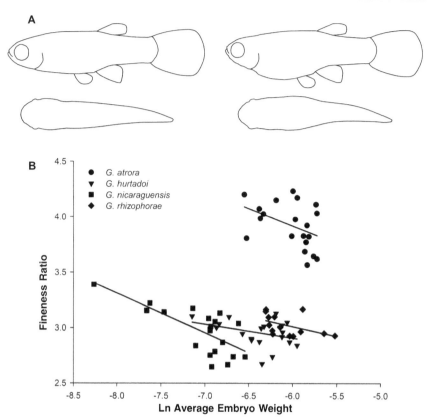

Fig. 7.9 Effects of pregnancy on body shape in poeciliid fish. **(A)** Lateral (top) and dorsal (bottom) illustrations of morphological differences between non-pregnant (left) and pregnant (right) female guppies (*P. reticulata*). Outlines are based on photographs of the same individual from the Oropuche drainage in Trinidad taken 1-day before parturition (right) and 1-day after parturition (left) (i.e., photographs were taken two days apart; photographs taken by C. Ghalambor). **(B)** *Gambusia* females carrying larger embryos exhibit lower fineness ratios (i.e., deeper bodies relative to their length). Slopes do not significantly differ among species (*P* = 0.31).

a number of reasons (e.g., Plaut, 2002, see below). In guppies, selection on both reproductive allocation and swimming performance is predicted to differ between predator regimes. Evolving adaptations in response to either selective pressure might conflict with one another. We will assess the integrated role of reproductive traits in the evolution of swimming performance within this context.

As discussed above (see *Predation*), guppies on the island of Trinidad are found in communities that differ in the risk of predation. High-predation communities are found in the lower portions of streams, where guppies co-occur with predatory fishes. Low-predation communities are found in the same drainages but above barrier waterfalls that exclude most piscivorous fish (Endler, 1978). Mark-recapture studies have shown that guppies in high-predation environments experience consistently higher mortality rates (Reznick *et al.*, 1996a; Reznick and Bryant, 2007). A guppy from a low-predation environment has a 20 to 30 fold higher probability of surviving for seven months than its counterpart from a high-predation environment (Reznick and Bryant, 2007). Life history theory predicts that the higher mortality rates that guppies experience in high predation environments will favor the evolution of earlier ages at maturity and an increase in the rate at which resources are allocated to reproduction (Charlesworth, 1994). Both of these changes will increase the rate of production of offspring early in life. These predictions are upheld in nature; high-predation guppies are significantly younger and smaller at sexual maturity, produce more offspring per litter, and have larger reproductive allocations (percent of total dry mass that consists of developing offspring) than low-predation guppies. These differences have been observed using comparisons of guppies from natural high- and low-predation environments (Reznick, 1982; Reznick and Endler, 1982; Reznick and Bryga, 1996; Reznick *et al.*, 1996b), as well as using experiments in which guppies were transplanted from high- to low-predation environments (Reznick and Bryga, 1987; Reznick *et al.*, 1990, 1997). This study system thus represents one of the rare occasions where it has been possible to experimentally test predictions derived from evolutionary theory in a natural setting.

Because selection is predicted to favor increased fast-start swimming performance in high-predation environments (see *Predation*), there could be a conflict between the way guppy life histories evolve and the way their swimming performance evolves in response to predation. For instance, the higher reproductive allocations seen in high-predation guppies means that at any stage of development, a guppy from a high-predation environment is carrying a reproductive burden that is 40% larger on average than a guppy from a low-predation environment (Ghalambor *et al.*, 2004). While predators favor the evolution of higher reproductive allocation, increasing the volume of reproductive tissues should have a concomitant impact on swimming performance for four reasons: (1) any increase in non-muscle mass associated with a larger ovary is expected to cause increased resistance to acceleration, (2) increased volume associated with a larger ovary can increase flexural stiffness and hence limit axial bending during acceleration (Beamish, 1978; James and Johnston, 1998), (3) any increase in surface and cross-sectional area caused by an enlarged ovary could increase drag (Beamish, 1978; Plaut, 2002), and (4) energy allocated to eggs and developing embryos could reduce the contractile properties of muscles and reduce power output (James and Johnston, 1998). For all of these reasons, selection by predators for increased allocation to reproduction could indirectly alter locomotor abilities, perhaps reducing performance at critical tasks, such as fast-start escapes. Or on the flip side, selection by predators for increased fast-start performance might constrain the evolution of increased reproductive allocation. The question is then whether or not we see evidence for an interaction between the evolution of life histories and the evolution of swimming performance.

Three alternative hypotheses exist (Fig. 7.10). First, high-predation guppies may be faster than low-predation guppies, with no effect of reproduction on performance (Fig. 7.10a). Second, reproduction might impair locomotion, but the impairment is the same for guppies from high- and low-predation communities. Because the mass of developing young increases throughout pregnancy, this impairment is expected to increase throughout pregnancy, and thus performance should progressively decline. The consequence is that high-predation guppies retain their superior fast-start performance regardless of where they are in the reproductive cycle (Fig. 7.10b). Third, guppies from high-predation environments might sustain a higher cost of reproduction than those from low-predation environments. In this case, high-predation guppies should experience a more rapid decline in fast-start performance as their litter progresses through development (Fig. 7.10c). These hypotheses were tested by evaluating performance in female guppies that were the second generation lab-born descendents of wild-caught females from four localities (two paired sets of high- and low-predation populations) (Ghalambor *et al.*, 2004). Rearing fish in a controlled environment for two generations means that any differences that are seen between populations can be interpreted as genetic differences, rather than a product of the environment in which they were reared.

Guppies from high-predation environments exhibit higher fast-start performance than those from low-predation environments: higher maximum acceleration, higher maximum velocity, and travel a greater distance during the 22 ms assessment period (Fig. 7.11). There is also a cost of reproduction, manifested in the maximum velocity and distance traveled. Females that were more advanced in their reproductive cycle attained a lower maximum velocity and shorter distance traveled; however, there was no effect of reproductive status on maximum acceleration (Fig. 7.11). There was a significant interaction between predation and the rate of decline in performance for maximum velocity and distance traveled; high-predation guppies declined in performance with advancing pregnancy more rapidly than low-predation guppies. Performance differences of this magnitude are known to have a significant affect on the ability of a guppy to survive a real encounter with a predator (Walker *et al.*, 2005).

We conclude that there is indeed evidence for a conflict between life history evolution and the evolution of performance. Guppies from high-predation environments are significantly faster than those from low-predation

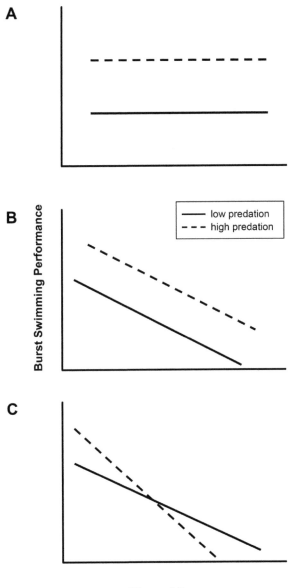

Stage of Pregnancy

Fig. 7.10 Three alternative outcomes for the effects of predation risk and pregnancy on swimming performance of guppies (*Poecilia reticulata*). **(A)** Fish in high-predation localities exhibit higher burst-swimming performance, with no cost of pregnancy. **(B)** High-predation guppies posses greater burst-swimming performance, and similar costs of pregnancy are experienced by all populations. **(C)** Although predators select for increased performance, a higher cost of pregnancy in high-predation populations (which have larger reproductive allocations) constrains its evolution. Adapted from Ghalambor *et al.* (2004).

environments, but they also suffer a higher cost of reproduction. Adaptation to the higher mortality rate that they experience in high-predation environments causes the evolution of a higher rate of offspring production and a larger reproductive allocation, but is accompanied by a more rapid decline in maximum velocity and distance traveled relative to guppies derived from low-predation environments. The high-predation guppies are faster when early in their reproductive cycle, but they lose this advantage and may even be slower when they are in an advanced stage of pregnancy. This statistical interaction thus shows that the evolution of reproductive performance and the evolution of escape performance have constrained one another. Guppies from high-predation environments are not as fast as they could be if selection did not also favor the evolution of higher reproductive allocation and/or they do not allocate as much as energy to reproduction as they might if selection did not also place a premium on being able to escape predators. The end product that we see reflects a compromise between these two components of fitness.

Although effects of reproductive traits on morphology and locomotion might generally be greater for females than males, there is evidence suggesting significant effects exist in males as well. For instance, in some livebearing fish, males possess modified fins utilized for the transfer of sperm (gonopodia). External, non-retractable copulatory organs represent obvious alterations of morphology, and can affect swimming performance. In poeciliids, the gonopodium is an elongate modification of the anal fin, and attains a large size in some taxa (e.g., as long as 70% of the body length; Rosen and Gordon, 1953; Rivas, 1963; Chambers, 1987). In *G. affinis*, males with larger gonopodia experience reduced burst-swimming speeds, presumably caused by drag incurred by the large organ (Langerhans *et al.*, 2005). Consequently, two *Gambusia* species are known to evolve smaller gonopodia in high-predation environments where selection favors increased escape speeds (Langerhans *et al.*, 2005). Thus, reproductive impacts on locomotion should not be neglected in males—it may indeed be commonplace at least for poeciliid fishes.

Diet has a similar potential to influence shape and performance, but there does not seem to have been the same formal analysis of the potential interaction between diet, morphology, and performance as there has been for reproduction. We present here some of what is known about the effects of diet on gut development to build an argument for the potential of such an interaction. First, selection arising from diet and foraging activities often targets jaw features which enhance consumption of particular prey items, but can also influence head size and shape (e.g., Smits *et al.*, 1996; Bouton *et al.*, 2002). Such morphological effects of diet can indirectly affect swimming performance, and thus foraging adaptations might often be intimately tied to locomotor adaptations. Second, efficient digestion of

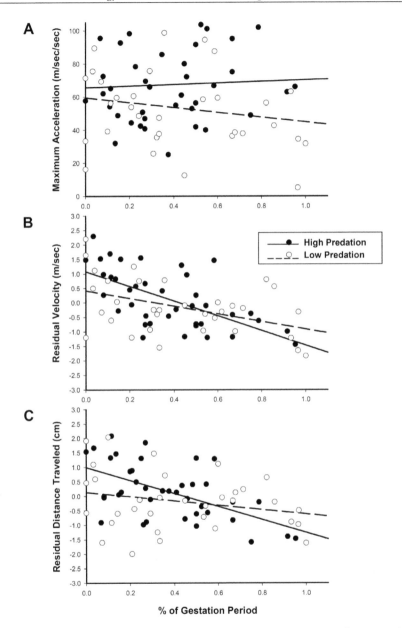

Fig. 7.11 Relationship between pregnancy (expressed as a percentage of the gestation period) and three components of fast-start swimming performance. **(A)** No cost of pregnancy for maximum acceleration. **(B)** Cost of pregnancy for residual maximum velocity (statistically controlling for effects of body mass and number of embryos). **(C)** Cost of pregnancy for residual distance traveled during the fast-start response (controlling for body mass). Adapted from Ghalambor *et al.* (2004).

various prey types often requires differences in gut length, which might influence swimming performance in a manner analogous to reproductive allocation. Kramer and Bryant (1995a; 1995b) quantified the relationship between gut length and diet, including correcting for the effects of body size. They found that there are non-overlapping values for the relative length of guts from carnivores, omnivores and herbivores. Carnivores have the shortest guts, omnivores are intermediate, and herbivores have the longest guts. Differences among diet type can exceed an order of magnitude. We could not find similar figures for the mass or volume of the gut relative to the body, but will assume for the sake of argument that the rank-ordering would be the same. If such differences in diet can have large impacts on gut volume, then it is reasonable to propose that they could impact shape and performance, just as differences in life history apparently do.

The herbivores and carnivores examined in Kramer and Bryant (1995a, b) were often distantly related organisms that differed in so many ways that it would be impossible to perform an unconfounded assessment of the interaction among diet, reproduction, shape and performance. However, similar diversity can be found among closely related species. For example, Kramer and Bryant (1995b) report on data for two poeciliid species: *Poecilia sphenops*, which is an herbivore, and *Brachyrhaphis cascajalensis*, which is an omnivore. These fish have similar body shapes and can be found in similar habitats. *Poecilia sphenops* has a standardized gut length that is more than ten times greater than *B. cascajalensis*. This difference is in part attributable to the larger average body size of *P. sphenops* but it exceeds that expected based on body size by a large margin. Other species in their study that had individuals in both size categories differed in relative gut length by only two to four fold. Such variation among closely related species, or within a species if it occurs, would exceed the volume differences associated with the ovaries of guppies from high- and low-predation localities and hence would be of a magnitude that could clearly cause an interaction among diet, reproduction, and swimming performance as adaptations to the local environment.

In conclusion, existing evidence for the interaction between the evolution of life histories, diet, body shape and swimming performance suggest that the evolution of locomotor ability does not happen independently of other adaptations. All of the factors that have been shown to influence the evolution of shape and performance, including environmental structure, water flow, predation, life histories, and diet, can interact with one another in shaping the response to selection. Using the approach illustrated above with the study on reproduction and performance in guppies, we can begin layering hypotheses on one another to better understand how multiple factors might interact with each another.

CONCLUSIONS

There are undoubtedly a multitude of factors involved in the ecology and evolution of swimming performance in fishes. While uncovering the various mechanisms responsible for the evolution of swimming performance seems quite complicated (and it truly is), it is not a futile endeavor. Here we constructed *a priori* predictions based on our current understanding of the relationships between morphology and swimming performance (built by biomechanical theory and experimental work) and between swimming performance and fitness in alternative environments (based on ecological theory and experimental work). It seems our predictions are fairly robust, as we found strong support in numerous, disparate systems and at several scales of analysis (within species, between closely related species, between distantly related species). Our results emphasize the utility of pinpointing predicted tradeoffs with strong theoretical and empirical support (e.g., between steady and unsteady locomotion), and using those tradeoffs as a starting point to formulate predictions testable with comparative data. We highlighted some selective agents that appear to be of widespread importance in the ecology and evolution of swimming performance in fishes, and pointed to some internal factors (e.g., reproductive allocation, gut length) which deserve greater attention in this regard. A major question remaining is how all these factors might interact with one another to influence locomotor abilities. We argue that increased employment of biomechanics-oriented research could provide key insights into ecological and evolutionary investigations of swimming performance in fishes. Building from the approach taken here, future studies might integrate predictions across multiple selective agents and multiple design features, strengthening our understanding of fish locomotion in its natural, albeit complex, context.

ACKNOWLEDGEMENTS

Firstly, we thank P. Domenici and B.G. Kapoor for inviting us to contribute to this book. We benefited from discussions with A. Hendry, G. Lauder, and J. Walker. P. Domenici and A. Rice provided comments on an earlier draft, and improved the chapter. M. Torres photographed and measured the life history traits for the female *Gambusia* specimens discussed in *Reproduction and diet*. We thank P. Domenici for providing information regarding the dataset of 32 fish species used in *Structural complexity*, and C. Ghalambor for providing photographs used in Fig. 7.9. R.B.L. was supported by Harvard University, an Environmental Protection Agency Science to Achieve Results fellowship, and a National Science Foundation Doctoral Dissertation Improvement grant. D.N.R. was supported by the University of California, Riverside and grants from the National Science Foundation (DEB-0416085 and DEB-0623632EF).

APPENDIX

1. Sampling for morphology of *Gambusia* males across velocity regimes

Abbreviations—RBL: R. Brian Langerhans personal collection; TNHC: Texas Natural History Collection, University of Texas, Austin; UMMZ: University of Michigan Museum of Zoology.

Specimens—*G. amistadensis*: TNHC 7247; *G. atrora*: TNHC 4570; *G. aurata*: RBL; *G. eurystoma*: UMMZ 184717; *G. gaigei*: TNHC 4213; *G. georgei*: TNHC 7203; *G. hurtadoi*: UMMZ 211112; *G. luma*: UMMZ 190612, UMMZ 197235; *G. manni*: RBL; *G. melapleura*: RBL; *G. oligosticta*: RBL; *G. rhizophorae*: UMMZ 213650.

2. Sampling for morphology and life history of *Gambusia* females

Abbreviations follow Appendix 1.

Specimens—*G. atrora*: UMMZ 169499 (12 pregnant females), 210724 (8 pregnant females); *G. hurtadoi*: UMMZ 211112 (20 pregnant females); *G. nicaraguensis*: UMMZ 199657 (10 pregnant females), UMMZ 199689 (10 pregnant females); *G. rhizophorae*: UMMZ 213650 (10 pregnant females), RBL (4 pregnant females).

Fig. A1 Phylogenetic hypothesis for 32 fish species inhabiting environments of varying structural complexities and belonging to 11 orders (Anguilliformes, Atheriniformes, Clupeiformes, Cypriniformes, Esociformes, Gadiformes, Gasterosteiformes, Osteoglossiformes, Perciformes, Salmoniformes, Scorpaeniformes). Species included in this dataset are identical to the dataset examined in Domenici (2003)—although because of an error in Table 1 of that paper, *Cymatogaster aggregata* was not listed. Open squares = open habitat, dotted squares = intermediate habitat, filled squares = complex habitat.

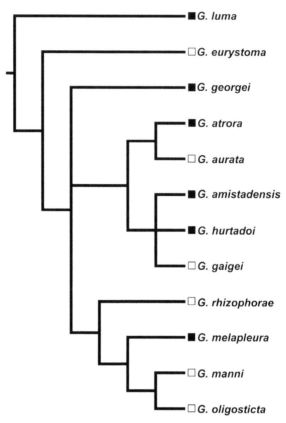

Fig. A2 Phylogenetic hypothesis for 12 *Gambusia* species inhabiting divergent velocity regimes. Open squares = low flow, filled squares = high flow.

REFERENCES

Alexander, R.M. 1968. *Animal Mechanics*, University of Washington Press, Seattle.

Andersson, J., F. Johansson and T. Soderlund. 2006. Interactions between predator- and diet-induced phenotypic changes in body shape of crucian carp. *Proceedings of the Royal Society of London* B273: 431–437.

Andraso, G.M. 1997. A comparison of startle response in two morphs of the brook stickleback (*Culaea inconstans*): further evidence for a trade-off between defensive morphology and swimming ability. *Evolutionary Ecology* 11: 83–90.

Andraso, G.M. and J.N. Barron. 1995. Evidence for a trade-off between defensive morphology and startle-response performance in the brook stickleback (*Culaea inconstans*). *Canadian Journal of Zoology* 73: 1147–1153.

Arnold, S.J. 1983. Morphology, performance and fitness. *American Zoologist* 23: 347–361.

Barluenga, M., K.N. Stolting, W. Salzburger, M. Muschick and A. Meyer. 2006. Sympatric speciation in Nicaraguan crater lake cichlid fish. *Nature* (London) 439: 719–723.

Beacham, T.D., C.B. Murray and R.E. Withler. 1989. Age, morphology, and biochemical genetic variation of Yukon River chinook salmon. *Transactions of the American Fisheries Society* 118: 46–63.

Beamish, F.W.H. 1978. Swimming capacity. In: *Fish Physiology*, W.S. Hoar and D.J. Randall (Eds.). Academic Press , New York, Vol. 7. pp. 101–187.

Bergstrom, C.A. 2002. Fast-start swimming performance and reduction in lateral plate number in threespine stickleback. *Canadian Journal of Zoology* 80: 207–213.

Bernardi, G. and G. Bucciarelli. 1999. Molecular phylogeny and speciation of the surfperches (Embiotocidae, Perciformes). *Molecular Phylogenetics and Evolution* 13: 77–81.

Blake, R.W. 1983. *Fish Locomotion*, Cambridge University Press, Cambridge.

Blake, R.W. 2004. Fish functional design and swimming performance. *Journal of Fish Biology* 65: 1193–1222.

Blake, R.W., T.C. Law, K.H.S. Chan and J.F.Z. Li. 2005. Comparison of the prolonged swimming performances of closely related, morphologically distinct three-spined sticklebacks *Gasterosteus* spp. *Journal of Fish Biology* 67: 834–848.

Boily, P. and P. Magnan. 2002. Relationship between individual variation in morphological characters and swimming costs in brook charr (*Salvelinus fontinalis*) and yellow perch (*Perca flavescens*). *Journal of Experimental Biology* 205: 1031–1036.

Bookstein, F.L. 1991. *Morphometric Tools for Landmark Data*, Cambridge University Press, New York.

Bouton, N., J. De Visser and C.D.N. Barel. 2002. Correlating head shape with ecological variables in rock-dwelling haplochromines (Teleostei: Cichlidae) from Lake Victoria. *Biological Journal of the Linnean Society* 76: 39–48.

Brinsmead, J. and M.G. Fox. 2002. Morphological variation between lake- and stream-dwelling rock bass and pumpkinseed populations. *Journal of Fish Biology* 61: 1619–1638.

Brönmark, C. and J.G. Miner. 1992. Predator-induced phenotypical change in body morphology in crucian carp. *Science* 258: 1348–1350.

Brönmark, C. and L.B. Pettersson. 1994. Chemical cues from piscivores induce a change in morphology in crucian carp. *Oikos* 70: 396–402.

Brune, G. 1981. *Springs of Texas, 1*, Branch-Smith, Inc., Fort Worth, Texas.

Caro, T.M. 2005. *Antipredator Defenses in Birds and Mammals*. University of Chicago Press, Chicago.

Carvalho, G.R., P.W. Shaw, A.E. Magurran and B.H. Seghers. 1991. Marked genetic divergence revealed by allozymes among populations of the guppy *Poecilia reticulata* (Poeciliidae), in Trinidad. *Biological Journal of the Linnean Society* 42: 389–405.

Chambers, J. 1987. The cyprinodontiform gonopodium, with an atlas of the gonopodia of the fishes of the genus *Limia*. *Journal of Fish Biology* 30: 389–418.

Charlesworth, B. 1994. *Evolution in Age-Structured Populations*, Cambridge University Press, New York. 2nd Edition.

Claytor, R.R., H.R. Maccrimmon and B.L. Gots. 1991. Continental and ecological variance components of European and North American Atlantic salmon (*Salmo salar*) phenotypes. *Biological Journal of the Linnean Society* 44: 203–229.

Clifton, K.B. and P.J. Motta. 1998. Feeding morphology, diet, and ecomorphological relationships among five Caribbean labrids (Teleostei, Labridae). *Copeia* 1998: 953–966.

Colgate, J.E. and K.M. Lynch. 2004. Mechanics and control of swimming: a review. *IEEE Journal of Oceanic Engineering* 29: 660–673.

Collyer, M.L., J.M. Novak and C.A. Stockwell. 2005. Morphological divergence of native and recently established populations of White Sands Pupfish (*Cyprinodon tularosa*). *Copeia* 2005: 1–11.

Crespi, B.J. and M.J. Fulton. 2004. Molecular systematics of Salmonidae: combined nuclear data yields a robust phylogeny. *Molecular Phylogenetics and Evolution* 31: 658–679.

Day, T., J. Pritchard and D. Schluter. 1994. A comparison of two sticklebacks. *Evolution* 48: 1723–1734.

de León, F.J.G., D. Gutiérrez-Tirado, D.A. Hendrickson and H. Espinosa-Pérez. 2005. Fishes of the continental waters of Tamaulipas: diversity and conservation status. In: *Biodiversity, Ecosystems, and Conservation in Northern Mexico*, J.-L.E. Cartron, G. Ceballos and R.S. Felger (Eds.). Oxford University Press, New York, pp. 138–166.

DeWitt, T.J. and R.B. Langerhans. 2003. Multiple prey traits, multiple predators: keys to understanding complex community dynamics. *Journal of Sea Research* 49: 143–155.

Domenici, P. 2003. Habitat, body design and the swimming performance of fish. In: *Vertebrate Biomechanics and Evolution*, V.L. Bels, J.-P. Gasc and A. Casinos (Eds.). BIOS Scientific Publishers Ltd, Oxford, pp. 137–160.

Domenici, P. and R.W. Blake. 1997. The kinematics and performance of fish fast-start swimming. *Journal of Experimental Biology* 200: 1165–1178.

Domenici, P., H. Turesson, J. Brodersen and C. Brönmark. 2008. Predator-induced morphology enhances escape locomotion in crucian carp. *Proceedings of the Royal Society of London* B275: 195–201.

Dudley, R. and C. Gans. 1991. A critique of symmorphosis and optimality models in physiology. *Physiological Zoology* 64: 627–637.

Eaton, R.C., R.A. Bombardieri and D.L. Meyer. 1977. The Mauthner-initiated startle response in teleost fish. *Journal of Experimental Biology* 66: 65–81.

Echelle, A.A. and I. Kornfield. 1984. *Evolution of Fish Species Flocks*. University of Maine Press, Orono.

Edmunds, M. 1974. *Defence in Animals: A Survey of Anti-predator Defences*, Longman, New York.

Ehlinger, T.J. and D.S. Wilson. 1988. Complex foraging polymorphism in bluegill sunfish. *Proceedings of the National Academy of Sciences, United States of America* 85: 1878–1882.

Eklov, P. and P. Jonsson. 2007. Pike predators induce morphological changes in young perch and roach. *Journal of Fish Biology* 70: 155–164.

Eklov, P. and L. Persson. 1996. The response of prey to the risk of predation: proximate cues for refuging juvenile fish. *Animal Behaviour* 51: 105–115.

Endler, J.A. 1978. A predator's view of animal color patterns. *Evolutionary Biology* 11: 319–364.

Endler, J.A. 1986. *Natural Selection in the Wild*. Princeton University Press, Princeton.

Fajen, A. and F. Breden. 1992. Mitochondrial DNA sequence variation among natural populations of the Trinidadian guppy, *Poecilia reticulata*. *Evolution* 46: 1457–1465.

Felsenstein, J. 1985. Phylogenies and the comparative method. *American Naturalist* 125: 1–15.

Ferry-Graham, L.A., D.I. Bolnick and P.C. Wainwright. 2002. Using functional morphology to examine the ecology and evolution of specialization. *Integrative and Comparative Biology* 42: 265–277.

Fink, W.L. 1971. A revision of the *Gambusia puncticulata* complex (Pisces: Poeciliidae). *Publications of the Gulf Coast Research Laboratory Museum* 2: 11–46.

Fisher, R. and J.D. Hogan. 2007. Morphological predictors of swimming speed: a case study of pre-settlement juvenile coral reef fishes. *Journal of Experimental Biology* 210: 2436–2443.

Froese, R. and D. Pauly, (Eds.). 2007. FishBase. World Wide Web electronic publication. www.fishbase.org, version (11/2007).

Fulton, C.J., D.R. Bellwood and P.C. Wainwright. 2001. The relationship between swimming ability and habitat use in wrasses (Labridae). *Marine Biology* 139: 25–33.

Fulton, C.J., D.R. Bellwood and P.C. Wainwright. 2005. Wave energy and swimming performance shape coral reef fish assemblages. *Proceedings of the Royal Society of London* B272: 827–832.

Garland, T., Jr., A.W. Dickerman, C.M. Janis and J.A. Jones. 1993. Phylogenetic analysis of covariance by computer simulation. *Systematic Biology* 42: 265–292.

Ghalambor, C.K., D.N. Reznick and J.A. Walker. 2004. Constraints on adaptive evolution: the functional trade-off between reproduction and fast-start swimming performance in the Trinidadian guppy (*Poecilia reticulata*). *American Naturalist* 164: 38–50.

Ghalambor, C.K., J.A. Walker and D.N. Reznick. 2003. Multi-trait selection, adaptation, and constraints on the evolution of burst swimming performance. *Integrative and Comparative Biology* 43: 431–438.

Greene, H.W. 1988. Antipredator mechanisms in reptiles. In: *Biology of the Reptilia*, C. Gans and R. B. Huey (Eds.). Alan R. Liss, Inc., New York, pp. 1–152.

Greenfield, D.W. 1983. *Gambusia xanthosoma*, a new species of poeciliid fish from Grand Cayman Island, BWI. *Copeia* 1983: 457–464.

Greenfield, D.W. 1985. Review of the *Gambusia yucatana* complex (Pisces, Poeciliidae) of Mexico and Central America. *Copeia* 1985: 368–378.

Greenfield, D.W., T.A. Greenfield and D.M. Wildrick. 1982. The taxonomy and distribution of the species of *Gambusia* (Pisces, Poeciliidae) in Belize, Central America. *Copeia* 1982: 128–147.

Greenfield, D.W. and D.M. Wildrick. 1984. Taxonomic distinction of the Antilles *Gambusia puncticulata* complex (Pisces, Poeciliidae) from the *Gambusia yucatana* complex of Mexico and Central America. *Copeia* 1984: 921–933.

Grubich, J.R. and M.W. Westneat. 2006. Four-bar linkage modelling in teleost pharyngeal jaws: computer simulations of bite kinetics. *Journal of Anatomy* 209: 79–92.

Hale, M.E., J.H. Long, M.J. McHenry and M.W. Westneat. 2002. Evolution of behavior and neural control of the fast-start escape response. *Evolution* 56: 993–1007.

Harvey, P.H. and M.D. Pagel. 1991. *The Comparative Method in Evolutionary Biology*, Oxford University Press, Oxford.

Hendry, A.P., M.L. Kelly, M.T. Kinnison and D.N. Reznick. 2006. Parallel evolution of the sexes? Effects of predation and habitat features on the size and shape of wild guppies. *Journal of Evolutionary Biology* 19: 741–754.

Hendry, A.P., and E.B. Taylor. 2004. How much of the variation in adaptive divergence can be explained by gene flow? An evaluation using lake-stream stickleback pairs. *Evolution* 58: 2319–2331.

Hendry, A.P., E.B. Taylor and J.D. McPhail. 2002. Adaptive divergence and the balance between selection and gene flow: lake and stream stickleback in the misty system. *Evolution* 56: 1199–1216.

Hendry, A.P., J.K. Wenburg, P. Bentzen, E.C. Volk and T.P. Quinn. 2000. Rapid evolution of reproductive isolation in the wild: evidence from introduced salmon. *Science* 290: 516–518.

Hertel, H. 1966. *Structure, Form, Movement*, Reinhold, New York.

Hobson, E.S. 1991. Trophic relationships of fishes specialized to feed on zooplankters above coral reefs. In: *The Ecology of Fishes on Coral Reefs*, P. F. Sale (Ed.). Academic Press, San Diego, pp. 69–95.

Hoerner, S.F. 1965. *Fluid-dynamic Drag*, published by the author, Midland Park.

Horwood, J.W. and D.H. Cushing. 1977. Spatial distributions and ecology of pelagic fish. In: *Spatial Patterns in Plankton Communities*, J.H. Steele (Ed.). Plenum Press, New York, pp. 355–383.

Hove, J.R., L.M. O'Bryan, M.S. Gordon, P.W. Webb and D. Weihs. 2001. Boxfishes (Teleostei: Ostraciidae) as a model system for fishes swimming with many fins: kinematics. *Journal of Experimental Biology* 204: 1459–1471.

Howland, H.C. 1974. Optimal strategies for predator avoidance: the relative importance of speed and maneuverability. *Journal of Theoretical Biology* 47: 333–350.

Hubbs, C. and A.E. Peden. 1969. *Gambusia georgei* sp. nov. from San Marcos, Texas. *Copeia* 1969: 357–364.

Hubbs, C. and V.G. Springer. 1957. A revision of the *Gambusia nobilis* species group, with descriptions of three new species, and notes on their variation, ecology, and evolution. *Texas Journal of Science* 9: 279–327.

Hubbs, C.L. 1929. Studies of the fishes of the order Cyprinodontes: VIII. *Gambusia gaigei*, a new species from the Rio Grande. *Occasional Papers of the Museum of Zoology The University of Michigan* 198: 1–11.

Hubbs, C.L. 1941. The relation of hydrological conditions to speciation in fishes. In: *A Symposium on Hydrobiology*, J. G. Needham, P. B. Sears and A. Leopold (Eds.). University of Wisconsin Press, Madison, pp. 182–195.

Huckins, C.J.F. 1997. Functional linkages among morphology, feeding performance, diet, and competitive ability in molluscivorous sunfish. *Ecology* 78: 2401–2414.

Imre, I., R.L. McLaughlin and D.L.G. Noakes. 2002. Phenotypic plasticity in brook charr: changes in caudal fin induced by water flow. *Journal of Fish Biology* 61: 1171–1181.

Inoue, J.G., M. Miya, K. Tsukamoto and M. Nishida. 2001. A mitogenomic perspective on the basal teleostean phylogeny: resolving higher-level relationships with longer DNA sequences. *Molecular Phylogenetics and Evolution* 20: 275–285.

Jacobs, E.N., K.E. Ward and R.M. Pinkerton. 1933. The characteristics of 78 related airfoil sections from tests in the variable-density wind tunnel. NACA report no. 460.

James, R.S. and I.A. Johnston. 1998. Influence of spawning on swimming performance and muscle contractile properties in the short-horn sculpin. *Journal of Fish Biology* 53: 485–501.

Januszkiewicz, A.J. and B.W. Robinson. 2007. Divergent walleye (*Sander vitreus*)-mediated inducible defenses in the centrarchid pumpkinseed sunfish (*Lepomis gibbosus*). *Biological Journal of the Linnean Society* 90: 25–36.

Johnson, G.D. and C. Patterson. 1993. Percomorph phylogeny: a survey of acanthomorphs and a new proposal. *Bulletin of Marine Science* 52: 554–626.

Johnson, J.B. 2001. Hierarchical organization of genetic variation in the Costa Rican livebearing fish *Brachyrhaphis rhabdophora* (Poeciliidae). *Biological Journal of the Linnean Society* 72: 519–527.

Jonsson, B. and N. Jonsson. 2001. Polymorphism and speciation in Arctic charr. *Journal of Fish Biology* 58: 605–638.

Katzir, G. and J.M. Camhi. 1993. Escape response of black mollies (*Poecilia shenops*) to predatory dives of a pied kingfisher (*Ceryle rudis*). *Copeia* 1993: 549–553.

Koehl, M.A.R. 1996. When does morphology matter? *Annual Review of Ecology and Systematics* 27: 501–542.

Koehl, M.A.R. 1999. Ecological biomechanics of benthic organisms: life history mechanical design and temporal patterns of mechanical stress. *Journal of Experimental Biology* 202: 3469–3476.

Kramer, D.L., and M.J. Bryant. 1995a. Intestine length in the fishes of a tropical stream: 1. Ontogenic allometry. *Environmental Biology of Fishes* 42: 115–127.

Kramer, D.L., and M.J. Bryant. 1995b. Intestine length in the fishes of a tropical stream: 2. Relationships to diet—the long and short of a convoluted issue. *Environmental Biology of Fishes* 42: 129–141.

Langerhans, R.B. 2006. Evolutionary consequences of predation: avoidance, escape, reproduction, and diversification. In: *Predation in Organisms: A Distinct Phenomenon*, A.M.T. Elewa (Ed.). Springer-Verlag, Heidelberg, pp. 177–220.

Langerhans, R.B., L.J. Chapman and T.J. Dewitt. 2007a. Complex phenotype-environment associations revealed in an East African cyprinid. *Journal of Evolutionary Biology* 20: 1171–1181.

Langerhans, R.B., and T.J. DeWitt. 2004. Shared and unique features of evolutionary diversification. *American Naturalist* 164: 335–349.

Langerhans, R.B., M.E. Gifford and E.O. Joseph. 2007b. Ecological speciation in *Gambusia* fishes. *Evolution* 61: 2056–2074.

Langerhans, R.B., C.A. Layman and T.J. DeWitt. 2005. Male genital size reflects a tradeoff between attracting mates and avoiding predators in two live-bearing fish species. *Proceedings of the National Academy of Sciences of the United States of America* 102: 7618–7623.

Langerhans, R.B., C.A. Layman, A.K. Langerhans and T.J. Dewitt. 2003. Habitat-associated morphological divergence in two Neotropical fish species. *Biological Journal of the Linnean Society* 80: 689–698.

Langerhans, R.B., C.A. Layman, A.M. Shokrollahi and T.J. DeWitt. 2004. Predator-driven phenotypic diversification in *Gambusia affinis*. *Evolution* 58: 2305–2318.

Lauder, G.V. 1996. The argument from design. In: *Adaptation*, G.V. Lauder and M.R. Rose (Eds.). Academic Press, San Diego, pp. 55–92.

Lauder, G.V. 2005. Locomotion. In: *The Physiology of Fishes*, D.H. Evans and J.B. Claiborne (Eds.). CRC Press, Boca Raton, pp. 3–46.

Lauder, G.V. and E.G. Drucker. 2002. Forces, fishes, and fluids: hydrodynamic mechanisms of aquatic locomotion. *News in Physiological Sciences* 17: 235–240.

Lauder, G.V. and E.D. Tytell. 2006. Hydrodynamics of undulatory propulsion. In: *Fish Biomechanics*, R. E. Shadwick and G. V. Lauder (Eds.). Academic Press, San Diego, pp. 425–468.

Lauder, G.V., E.G. Drucker, J.C. Nauen and C.D. Wilga. 2003. Experimental hydrodynamics and evolution: caudal fin locomotion in fishes. In: *Vertebrate Biomechanics and Evolution*, V.L. Bels, J.-P. Gasc and A. Casinos (Eds.). BIOS Scientific Publishers Ltd, Oxford, pp. 117–135.

Law, T.C. and R.W. Blake. 1996. Comparison of the fast-start performances of closely related, morphologically distinct threespine sticklebacks (*Gasterosteus* spp.). *Journal of Experimental Biology* 199: 2595–2604.

Lessells, C.M. and P.T. Boag. 1987. Unrepeatable repeatabilities: a common mistake. *Auk* 104: 116–121.

Lighthill, M.J. 1975. *Mathematical Biofluiddynamics*, Society for Applied and Industrial Mathematics, Philadelphia.

Losos, J.B. 1990. The evolution of form and function: morphology and locomotor performance in West Indian *Anolis* lizards. *Evolution* 44: 1189–1203.

Lowell, R.B. 1987. Safety factors of tropical versus temperate limpet shells: multiple selection pressures on a single structure. *Evolution* 41: 638–650.

Lydeard, C., M.C. Wooten and A. Meyer. 1995a. Cytochrome-B sequence variation and a molecular phylogeny of the live-bearing fish genus *Gambusia* (Cyprinodontiformes, Poeciliidae). *Canadian Journal of Zoology* 73: 213–227.

Lydeard, C., M.C. Wooten and A. Meyer. 1995b. Molecules, morphology, and area cladograms: a cladistic and biogeographic analysis of *Gambusia* (Teleostei, Poeciliidae). *Systematic Biology* 44: 221–236.

Marcus, L.F., M. Corti, A. Loy, G.J.P. Naylor and D.E. Slice. 1996. *Advances in Morphometrics*. Plenum Press, New York.

Martins, E.P. 2000. Adaptation and the comparative method. *Trends in Ecology & Evolution* 15: 296–299.

McGuigan, K., C.E. Franklin, C. Moritz and M.W. Blows. 2003. Adaptation of rainbow fish to lake and stream habitats. *Evolution* 57: 104–118.

McHenry, M.J. and G.V. Lauder. 2006. Ontogeny of form and function: locomotor morphology and drag in zebrafish (*Danio rerio*). *Journal of Morphology* 267: 1099–1109.

McKinnon, J.S. and H.D. Rundle. 2002. Speciation in nature: the threespine stickleback model systems. *Trends in Ecology & Evolution* 17: 480–488.

McLaughlin, R.L. and J.W.A. Grant. 1994. Morphological and behavioral differences among recently-emerged brook charr, *Salvelinus fontinalis*, foraging in slow-running vs. fast-running water. *Environmental Biology of Fishes* 39: 289–300.

Meyer, A. 1990. Ecological and evolutionary consequences of the trophic polymorphism in *Cichlasoma citrinellum* (Pisces, Cichlidae). *Biological Journal of the Linnean Society* 39: 279–299.

Miller, R.R. 1975. Five new species of Mexican poeciliid fishes of the genera *Poecilia, Gambusia*, and *Poeciliopsis*. *Occasional Papers of the Museum of Zoology The University of Michigan* 672: 1–44.

Miller, R.R. and W.L. Minckley. 1970. *Gambusia aurata*, a new species of poeciliid fish from northeastern Mexico. *Southwestern Naturalist* 15: 249–259.

Mittelbach, G.G. 1981. Foraging efficiency and body size: a study of optimal diet and habitat use by bluegills. *Ecology* 62: 1370–1386.

Miya, M., H. Takeshima, H. Endo, N.B. Ishiguro, J.G. Inoue, T. Mukai, T.P. Satoh, M. Yamaguchi, A. Kawaguchi, K. Mabuchi, S.M. Shirai and M. Nishida. 2003. Major patterns of higher teleostean phylogenies: a new perspective based on 100 complete mitochondrial DNA sequences. *Molecular Phylogenetics and Evolution* 26: 121–138.

Müller, U.K. and J.L. van Leeuwen. 2006. Undulatory fish swimming: from muscles to flow. *Fish and Fisheries* 7: 84–103.

Neat, F.C., W. Lengkeek, E.P. Westerbeek, B. Laarhoven and J.J. Videler. 2003. Behavioural and morphological differences between lake and river populations of *Salaria fluviatilis*. *Journal of Fish Biology* 63: 374–387.

Nelson, J.A., P.S. Gotwalt, S.P. Reidy and D.M. Webber. 2002. Beyond U-crit: matching swimming performance tests to the physiological ecology of the animal, including a new fish 'drag strip'. *Comparative Biochemistry and Physiology* A 133: 289–302.

Nelson, J.S. 1994. *Fishes of the World*. John Wiley & sons, New York. Third Edition.

Nilsson, P.A., C. Brönmark and L.B. Pettersson. 1995. Benefits of a predator-induced morphology in crucian carp. *Oecologia* 104: 291–296.

O'Steen, S., A.J. Cullum and A.F. Bennett. 2002. Rapid evolution of escape ability in Trinidadian guppies (*Poecilia reticulata*). *Evolution* 56: 776–784.

O'Toole, B. 2002. Phylogeny of the species of the superfamily Echeneoidea (Perciformes: Carangoidei: Echeneidae, Rachycentridae, and Coryphaenidae), with an interpretation of echeneid hitchhiking behaviour. *Canadian Journal of Zoology* 80: 596–623.

Pakkasmaa, S. and J. Piironen. 2000. Water velocity shapes juvenile salmonids. *Evolutionary Ecology* 14: 721–730.

Peden, A.E. 1973. Virtual extinction of *Gambusia amistadensis* n. sp, a poeciliid fish from Texas. *Copeia* 1973: 210–221.

Peres-Neto, P.R. and P. Magnan. 2004. The influence of swimming demand on phenotypic plasticity and morphological integration: a comparison of two polymorphic charr species. *Oecologia* 140: 36–45.

Plaut, I. 2001. Critical swimming speed: its ecological relevance. *Comparative Biochemistry and Physiology* A131: 41–50.

Plaut, I. 2002. Does pregnancy affect swimming performance of female Mosquitofish, *Gambusia affinis*? *Functional Ecology* 16: 290–295.

Rauchenberger, M. 1989. Systematics and biogeography of the genus *Gambusia* (Cyprinodontiformes: Poeciliidae). *American Museum Novitates* 2951: 1–74.

Reidy, S., S. Kerr and J. Nelson. 2000. Aerobic and anaerobic swimming performance of individual Atlantic cod. *Journal of Experimental Biology* 203: 347–357.

Reimchen, T.E. 1994. Predators and morphological evolution in threespine stickleback. In: *The Evolutionary Biology of the Threespine Stickleback*, M.A. Bell and S.A. Foster (Eds.). Oxford University Press, Oxford, pp. 240–276.

Reimchen, T.E. 1995. Predator-induced cyclical changes in lateral plate frequencies of *Gasterosteus. Behaviour* 132: 1079–1094.

Reimchen, T.E. 2000. Predator handling failures of lateral plate morphs in *Gasterosteus aculeatus*: functional implications for the ancestral plate condition. *Behaviour* 137: 1081–1096.

Reimchen, T.E. and P. Nosil. 2004. Variable predation regimes predict the evolution of sexual dimorphism in a population of threespine stickleback. *Evolution* 58: 1274–1281.

Reznick, D. and M. Bryant. 2007. Comparative long-term mark-recapture studies of guppies (*Poecilia reticulata*): differences among high and low predation localities in growth and survival. *Annales Zoologici Fennici* 44: 152–160.

Reznick, D. and J. Travis. 1996. The empirical study of adaptation in natural populations. In: *Adaptation*, M.R. Rose and G.V. Lauder (Eds.). Academic Press, San Diego, pp. 243–289.

Reznick, D.A., H. Bryga and J.A. Endler. 1990. Experimentally induced life-history evolution in a natural population. *Nature* (London) 346: 357–359.

Reznick, D.N. 1982. The impact of predation on life-history evolution in Trinidadian guppies: genetic basis of observed life history patterns. *Evolution* 36: 1236–1250.

Reznick, D.N. and H.A. Bryga. 1987. Life-history evolution in guppies (*Poecilia reticulata*): 1. Phenotypic and genetic changes in an introduction experiment. *Evolution* 41: 1370–1385.

Reznick, D.N. and H.A. Bryga. 1996. Life-history evolution in guppies (*Poecilia reticulata*: Poeciliidae). V. Genetic basis of parallelism in life histories. *American Naturalist* 147: 339–359.

Reznick, D.N. and J.A. Endler. 1982. The impact of predation on life history evolution in Trinidadian guppies (*Poecilia reticulata*). *Evolution* 36: 160–177.

Reznick, D.N., M.J. Butler, F.H. Rodd and P. Ross. 1996a. Life-history evolution in guppies (*Poecilia reticulata*). 6. Differential mortality as a mechanism for natural selection. *Evolution* 50: 1651–1660.

Reznick, D.N., F.H. Rodd and M. Cardenas. 1996b. Life-history evolution in guppies (*Poecilia reticulata*: Poeciliidae). IV. Parallelism in life-history phenotypes. *American Naturalist* 147: 319–338.

Reznick, D.N., F.H. Shaw, F.H. Rodd and R.G. Shaw. 1997. Evaluation of the rate of evolution in natural populations of guppies (*Poecilia reticulata*). *Science* 275: 1934–1937.

Rivas, L.R. 1963. Subgenera and species groups in the poeciliid fish genus *Gambusia* Poey. *Copeia* 1963: 331–347.

Robinson, B.W. and K.J. Parsons. 2002. Changing times, spaces, and faces: tests and implications of adaptive morphological plasticity in the fishes of northern postglacial lakes. *Canadian Journal of Fisheries and Aquatic Sciences* 59: 1819-1833.

Robinson, B.W. and D.S. Wilson. 1994. Character release and displacement in fishes: a neglected literature. *American Naturalist* 144: 596–627.

Robinson, B.W. and D.S. Wilson. 1996. Genetic variation and phenotypic plasticity in a trophically polymorphic population of pumpkinseed sunfish (*Lepomis gibbosus*). *Evolutionary Ecology* 10: 631–652.

Robinson, B.W., D.S. Wilson and A.S. Margosian. 2000. A pluralistic analysis of character release in pumpkinseed sunfish (*Lepomis gibbosus*). *Ecology* 81: 2799–2812.

Roff, D.A. 2002. *Life History Evolution*, Sinauer Associates Inc., Sunderland, MA.

Rohlf, F.J. 2005. TpsRelw. Department of Ecology and Evolution, State Univ. New York, Stony Brook.

Rohlf, F.J. 2006. TpsDig. Department of Ecology and Evolution, State Univ. New York, Stony Brook.

Rohlf, F.J. and L.F. Marcus. 1993. A revolution in morphometrics. *Trends in Ecology & Evolution* 8: 129–132.

Rose, M.R. and G.V. Lauder. 1996. *Adaptation*, Academic Press, San Diego.

Rosen, D.E. and R.M. Bailey. 1963. The poeciliid fishes (Cyprinodontiformes), their structure, zoogeography, and systematics. *Bulletin of the American Museum of Natural History* 126: 1–176.

Rosen, D.E. and M. Gordon. 1953. Functional anatomy and evolution of male genitalia in poeciliid fishes. *Zoologica* 38: 1–47.

Schliewen, U., K. Rassmann, M. Markmann, J. Markert, T. Kocher and D. Tautz. 2001. Genetic and ecological divergence of a monophyletic cichlid species pair under fully sympatric conditions in Lake Ejagham, Cameroon. *Molecular Ecology* 10: 1471–1488.

Schluter, D. 1996. Ecological speciation in postglacial fishes. *Philosophical Transactions of the Royal Society of London* B351: 807–814.

Schultz, W.W. and P.W. Webb. 2002. Power requirements of swimming: do new methods resolve old questions? *Integrative and Comparative Biology* 42: 1018–1025.

Sfakiotakis, M., D.M. Lane and J.B.C. Davies. 1999. Review of fish swimming modes for aquatic locomotion. *IEEE Journal of Oceanic Engineering* 24: 237–252.

Shadwick, R.E. and G.V. Lauder. 2006. *Fish Biomechanics*. Elsevier Academic Press, San Diego.

Shinohara, G. and H. Imamura. 2007. Revisiting recent phylogenetic studies of "Scorpaeniformes". *Ichthyological Research* 54: 92–99.

Shoup, D.E. and L.G. Hill. 1997. Ecomorphological diet predictions: an assessment using inland silverside (*Menidia beryllina*) and longear sunfish (*Lepomis megalotis*) from Lake Texoma. *Hydrobiologia* 350: 87–98.

Sidlauskas, B., B. Chernoff and A. Machado-Allison. 2006. Geographic and environmental variation in *Bryconops* sp cf. *melanurus* (Ostariophysi: Characidae) from the Brazilian Pantanal. *Ichthyological Research* 53: 24–33.

Smith, T.B. and S. Skúlason. 1996. Evolutionary significance of resource polymorphisms in fishes, amphibians, and birds. *Annual Review of Ecology and Systematics* 27: 111–133.

Smith, W.L. and M.T. Craig. 2007. Casting the percomorph net widely: the importance of broad taxonomic sampling in the search for the placement of serranid and percid fishes. *Copeia* 2007: 35–55.

Smith, W.L. and W.C. Wheeler. 2004. Polyphyly of the mail-cheeked fishes (Teleostei: Scorpaeniformes): evidence from mitochondrial and nuclear sequence data. *Molecular Phylogenetics and Evolution* 32: 627–646.

Smits, J.D., F. Witte and F.G. VanVeen. 1996. Functional changes in the anatomy of the pharyngeal jaw apparatus of *Astatoreochromis alluaudi* (Pisces, Cichlidae), and their effects on adjacent structures. *Biological Journal of the Linnean Society* 59: 389–409.

Sokal, R.R., and F.J. Rohlf. 1995. *Biometry*. Freeman, NY. 3rd Edition.

Stein, R.A. 1979. Behavioral response of prey to fish predators. In: *Predator-prey Systems in Fisheries Management*, R.H. Stroud and H. Clepper (Eds.). Sport Fishing Institute, Washington, D.C.

Swain, D.P. and L.B. Holtby. 1989. Differences in morphology and behavior between juvenile coho salmon (*Oncorhynchus kisutch*) rearing in a lake and in its tributary stream. *Canadian Journal of Fisheries and Aquatic Sciences* 46: 1406–1414.

Taylor, E.B. 1999. Species pairs of north temperate freshwater fishes: evolution, taxonomy, and conservation. *Reviews in Fish Biology and Fisheries* 9: 299–324.

Taylor, E.B. and J.D. McPhail. 1986. Prolonged and burst swimming in anadromous and freshwater threespine stickleback, *Gasterosteus aculeatus*. *Canadian Journal of Zoology* 64: 416–420.

Tobler, M., I. Schlupp, K.U. Heubel, R. Riesch, F.J.G. de Leon, O. Giere and M. Plath. 2006. Life on the edge: hydrogen sulfide and the fish communities of a Mexican cave and surrounding waters. *Extremophiles* 10: 577–585.

Tonn, W.M., C.A. Paszkowski and I.J. Holopainen. 1992. Piscivory and recruitment: mechanisms structuring prey populations in small lakes. *Ecology* 73: 951–958.

Triantafyllou, M.S., G.S. Triantafyllou and D.K.P. Yue. 2000. Hydrodynamics of fishlike swimming. *Annual Review of Fluid Mechanics* 32: 33–53.

Vamosi, S.M. 2002. Predation sharpens the adaptive peaks: survival trade-offs in sympatric sticklebacks. *Annales Zoologici Fennici* 39: 237–248.

Vamosi, S.M. 2005. On the role of enemies in divergence and diversification of prey: a review and synthesis. *Canadian Journal of Zoology* 83: 894–910.

Videler, J.J. 1993. *Fish Swimming*, Chapman and Hall, London.

Vogel, S. 1994. *Life in Moving Fluids*. Princeton University Press, Princeton. 2nd Edition.

von Mises, R. 1945. *Theory of Flight*. Dover Books, New York.

Wainwright, P.C. 1988. Morphology and ecology: functional basis of feeding constraints in Caribbean labrid fishes. *Ecology* 69: 635–645.

Wainwright, P.C. 1996. Ecological explanation through functional morphology: the feeding biology of sunfishes. *Ecology* 77: 1336–1343.

Wainwright, P.C. and S.M. Reilly. 1994. *Ecological Morphology: Integrative Organismal Biology*. University of Chicago Press, Chicago.

Wainwright, P.C. and B.A. Richard. 1995. Predicting patterns of prey use from morphology of fishes. *Environmental Biology of Fishes* 44: 97–113.

Wainwright, P.C., D.R. Bellwood and M.W. Westneat. 2002. Ecomorphology of locomotion in labrid fishes. *Environmental Biology of Fishes* 65: 47–62.

Walker, J.A. 1997. Ecological morphology of lacustrine threespine stickleback *Gasterosteus aculeatus* L. (Gasterosteidae) body shape. *Biological Journal of the Linnean Society* 61: 3–50.

Walker, J.A. and M.A. Bell. 2000. Net evolutionary trajectories of body shape evolution within a microgeographic radiation of threespine sticklebacks (*Gasterosteus aculeatus*). *Journal of Zoology* 252: 293–302.

Walker, J.A. and M.W. Westneat. 2002. Performance limits of labriform propulsion and correlates with fin shape and motion. *Journal of Experimental Biology* 205: 177–187.

Walker, J.A., C.K. Ghalambor, O.L. Griset, D. McKenney and D.N. Reznick. 2005. Do faster starts increase the probability of evading predators? *Functional Ecology* 19: 808–815.

Webb, P.W. 1975. Hydrodynamics and energetics of fish propulsion. *Bulletin of the Fisheries Research Board of Canada* 190: 1–159.

Webb, P.W. 1982. Locomotor patterns in the evolution of actinopterygian fishes. *American Zoologist* 22: 329–342.

Webb, P.W. 1983. Speed, acceleration and manoeuvrability of two teleost fishes. *Journal of Experimental Biology* 102: 115–122.

Webb, P.W. 1984. Body form, locomotion, and foraging in aquatic vertebrates. *American Zoologist* 24: 107–120.

Webb, P.W. 1986a. Effect of body form and response threshold on the vulnerability of four species of teleost prey attacked by largemouth bass (*Micropterus salmoides*). *Canadian Journal of Fisheries and Aquatic Sciences* 43: 763–771.

Webb, P.W. 1986b. Locomotion and predator-prey relationships. In: *Predator-Prey Relationships*, G.V. Lauder and M. E. Feder (Eds.). University of Chicago Press, Chicago, pp. 24–41.

Webb, P.W. and D. Weihs. 1983. *Fish Biomechanics*, Praeger Publishers, New York.

Weihs, D. 1973. The mechanism of rapid starting of slender fish. *Biorheology* 10: 343–350.

Weihs, D. 1989. Design features and mechanics of axial locomotion in fish. *American Zoologist* 29: 151–160.

Weihs, D. and P.W. Webb. 1983. Optimization of locomotion. In: *Fish Biomechanics*, P.W. Webb and D. Weihs (Eds.). Praeger Publishers, New York, pp. 339–371.

Werner, E.E., J.F. Gilliam, D.J. Hall and G.G. Mittelbach. 1983. An experimental test of the effects of predation risk on habitat use in fish. *Ecology* 64: 1540–1548.

West-Eberhard, M.J. 2003. *Developmental Plasticity and Evolution*. Oxford University Press, Oxford.

West-Eberhard, M.J. 2005. Developmental plasticity and the origin of species differences. *Proceedings of the National Academy of Sciences of the United States of America* 102: 6543–6549.

Westneat, M.W. 1995. Feeding, function, and phylogeny: analysis of historical biomechanics in labrid fishes using comparative methods. *Systematic Biology* 44: 361–383.

Williams, G.C. 1992.*Natural Selection: Domains, Levels and Challenges*, Oxford University Press, Oxford.

Winkelman, D.L. and J.M. Aho. 1993. Direct and indirect effects of predation on mosquitofish behavior and survival. *Oecologia* 96: 300–303.

Wu, T.Y. 1971. Hydromechanics of swimming propulsion. Part 1. Swimming of a two-dimensional flexible plate at variable forward speeds in an inviscid fluid. *Journal of Fluid Mechanics* 46: 337–355.

Zelditch, M.L., D.L. Swiderski, H.D. Sheets and W.L. Fink. 2004. *Geometric Morphometrics for Biologists: A Primer*. Elsevier Academic Press, London.

Zúñiga-Vega, J.J., D.N. Reznick and J.B. Johnson. 2007. Habitat predicts reproductive superfetation and body shape in the livebearing fish *Poeciliopsis turrubarensis*. *Oikos* 116: 995–1005.

8

Sexual Selection, Male Quality and Swimming Performance

Astrid Kodric-Brown

INTRODUCTION

Multiple selective pressures affect swimming performance of fish. Among these are abiotic conditions of the physical environment, such as temperature, salinity, and flow velocity, and biotic interactions, such as competition, predation, and parasitism. Both abiotic and biotic selective pressures have shaped the morphology and physiology of species and/or populations. In addition, sexual selection acts on male swimming performance to modify size and shape of the body and fins. Here, I focus on: (1) how sexual selection, acting within the framework of ecological, morphological and physiological constraints, influences swimming performance (2) how secondary sexual traits of males, such as enlarged fins, affect different aspects of swimming performance, such as endurance, agility, and burst speed (3) how environmental factors constrain the evolution of energetically expensive courtship displays and (4) how scaling issues, such as interactions between body size, sexual size dimorphism, secondary sexual traits, courtship behavior, and metabolism affect swimming performance.

Here, I review the importance of sexual selection in the elaboration of fins and courtship behavior in male fishes. Male secondary sexual traits,

Author's address: University of New Mexico, Albuquerque, NM, U.S.A.
E-mail: Kodric@unm.edu

such as elaborate fins, courtship, and aggressive behavior are metabolically costly and affect all aspects of swimming performance. Costs of secondary sexual traits are difficult to measure, and the various measures of swimming performance most likely underestimate the true cost of male ornaments. Carotenoids are often incorporated into secondary sexual traits and are positively associated with swimming performance, since they reflect overall health and physical condition. Predators exert a strong selection pressure on the elaboration of male ornaments; in environments with high densities of predators, male ornaments are reduced in size and conspicuousness, and females are less choosy, suggesting that both ecological and physiological constraints limit the elaboration of male ornaments. Sexually-selected traits such as elaborate fins and courtship are more likely to evolve in species with low steady and high unsteady (e.g., maneuverability) swimming performance living in complex environments than in pelagic species that have high steady and low unsteady swimming performance and inhabit less complex environments (Domenici, 2003). This pattern suggests that the aquatic medium imposes stringent limits to the operation of sexual selection.

Background

Swimming activities are typically defined based on the time a fish maintains a given swimming speed (Videler, 1993; Blake, 2004). Sustained swimming speed is defined as the speed that a fish can maintain over long periods, and includes various activities such as territorial patrolling, schooling, and migration. Another measure of performance is 'critical swimming' speed (U_{crit}), the maximum speed calculated by using a step protocol of increasing flow velocity in a flow chamber (Brett, 1964; Kolok, 1999). Steady swimming performance (U_{crit}) is correlated with active metabolism, physical condition, and health (Martinez *et al.*, 2003). Another activity is burst swimming, which is characterized by short periods of fast starts and quick turns. This type of swimming, often associated with predator avoidance (Domenici, 2001; O'Steen *et al.*, 2002), also is a frequent component of both courtship and mating, which may include displays performed to attract females as well as male-male contests such as chasing, and fighting. Another measure of performance is maneuverability, the ability to execute turns (Weihs, 1972; Blake 2004). Maneuverability is also an important aspect of both male courtship displays and agonistic interactions (Knapp and Kovach, 1991; Rosenthal *et al.*, 1996; Houde, 1997; Ptacek and Travis, 1997).

There is a trade-off between unsteady swimming (e.g., maneuverability) and sustained swimming, so specialization in one of these functions decreases performance in the other (Blake, 2004). Maneuverability depends on body flexibility (Breinerd and Patek,1998). It is often measured as the turning radius divided by body length (Webb, 1983), or the minimum turning radius observed in some large number of turns (Webb and de Buffrenil, 1990), or the minimum area required to turn (Walker, 2000). Species with high sustained swimming speeds have relatively stiff, fusiform bodies and fins designed to reduce drag. They have low maneuverability. For example, the turning radius of yellow tuna is 47% of its body length (Blake *et al.*, 1995). This compares with 7% for the highly maneuverable angelfish (Domenici and Blake, 1991).

Species with high maneuverability and low sustained swimming speeds vary in body morphology as well as fin size, shape and placement. Sexual selection can shape the morphology and physiology of these fish (Anderson, 1994; Oufiero and Garland, 2007). Maneuverability is accomplished via lateral flexing of the body and expanded median and dorsal fins. Maneuverability has typically been examined in the context of predator-prey interactions (Domenici and Blake, 1993). It has yet to be applied to courtship behavior and agonistic displays. Species that swim at low speeds in environments with low flow, such as coral reefs or still water of rivers, lakes or ponds often have elaborate courtship displays involving caudal, pectoral and dorsal fins. In these fishes, courtship consists of a series of abrupt turns or tight circular movements by males around, or in front of, females (Houde, 1997; Rosenthal and Evans, 1998; Suk and Choe, 2002). Turning increases swimming costs (Webb and Gerstner, 2000). However, these species are not subject to the constraints imposed by high sustained swimming speed, characteristic of pelagic, open-ocean fish such as tunas and mackerels (Scombridae), and sharks, or freshwater fish such as salmonids and minnows, that live in fast-flowing rivers and streams and swim continuously to maintain position.

A prediction resulting from adaptations to high swimming speed and low maneuverability is that courtship behavior in pelagic and riverine fishes would be less elaborate than in benthic and other species that typically swim at low speeds and are highly maneuverable. To date, there has not been a systematic attempt to examine this pattern, but available literature on species with elaborate courtship behavior, such as Siamese fighting fish (Anabantidae), cichlids (Cichlidae), blennies (Chaenopsidae), pupfishes (Cyprinodontidae), sticklebacks (Gasterosteidae), gobies (Gobiidae), wrasses (Labridae), blue-eyes (Melanotaeniidae), darters (Percidae), mollies, swordtails and guppies (Poeciliidae), damselfishes (Pomacentridae), pipefishes and seahorses (Syngnathidae) and tripplefins (Tripterygiidae)—all fish with high maneuverability and low swimming speeds (Blake, 2004)—suggest that this prediction is supported (Table 8.1). For the most part, these species are small (3–30 cm standard length), tend to defend territories, and live close to the substrate or in spatially complex environments.

Table 8.1 Examples of species in which males engage in elaborate courtship.

Family	Genus	Species	Courtship	Reference
Anabantidae	*Betta*	*splendens*	Tail beating, erect gill cover	Simpson, (1968)
Cichlidae	*Cichlasoma*	*nigrofasciatum*	Quivering, tail beating, lateral displays	Santangelo, (2005)
Cichlidae	*Neolamprologus*	*pulcher*	Tail beating, s-shaping, erect gill cover	Grantner and Taborsky, (1988)
Chaenopsidae	*Chaenopsis*	*schmitti*	S-shaping	Allen and Robertson, (1994)
Cyprinodontidae	*Jordanella*	*floridae*	Circles female with erect fins	Breder and Rosen, (1966); Mertz and Barlow, (1966)
Cyprinodontidae	*Cyprinodon*	*macularius*	s-shaping, arching, tail beating, circling female	Barlow, (1961)
Gasterosteidae	*Gasterosteus*	*aculeatus*	Zig-zag movement, sideways jumps	Van Iersel, (1953)
Gasterosteidae	*Culaea*	*inconstans*	vibrate pectoral and dorsal fins; erect dorsal fin, s-shaping, circling female	McLennan, (1993)
Gobiidae	*Bathygobius*	*fuscus*	Lateral display, vibrate body and caudal fin	Taru *et al.* (2002)
Labridae	*Pseudolabrus*	*fucicola*	Erect dorsal and anal fins	Ayling, (1980)
Labridae	*Semicossyphus*	*pulcher*	Circling female, head movements	Andreani *et al.* (2004)
Labridae	*Thalassoma*	*bifasciatum*	Circling female, vibrate pectoral and caudal fins	Dawkins and Guilford, (1994)
Melanotaeniidae	*Pseudomugil*	*gertrudae*	elaborate dorsal and anal fins	Merrick and Schmida, (1984)
Percidae	*Stizostedion*	*vitreum*	Circling female, head bob, spread pectoral fins	Breder and Rosen, (1966); Page, (1985)
Percidae	*Etheostoma*	*radiosum*	Erect dorsal, anal fins, fans pectoral fins, vibrates body	Scalet, (1973)
Poeciliidae	*Poecilia*	*reticulata*	Sigmoid display, erect fins, rapid erect fins, rapid turns	Liley, (1966); Houde, (1997)
Poeciliidae	*Poecilia*	*latipinna*	Erect large dorsal fin, quivering	Ptacek and Travis, (1997)
Poeciliidae	*Xiphophorus*	*nigrensis*	elongated caudal fin, rapid turns	Zimmerer and Kallman, (1989)

(contd. Table 8.1)

(*contd. Table 8.1*)

Family	Genus	Species	Courtship	Reference
Pomacentridae Haley,	*Stegastes*	*leucostictus*	Dips—swims forward and down, followed by jumps	Itzkowitz and (1999)
Syngnathidae	*Hippocampus*	*zosterae*	Erecting, vibrating pectoral fins	Masonjones and Lewis, (1996)
Tripterygiidae	*Ruanoho*	*decemdigitatus*	Erecting all fins, flaring of operculum, lateral displays	Wellenreuther *et al.* (2007)

Sexual Selection and the Evolution of Secondary Sexual Traits

Sexual selection is an important driving force in the evolution of male secondary sexual traits (Anderson, 1994). In fishes, male traits, such as elaborate fins (Basolo, 1990), color patterns (Houde, 1997), features of body size or shape (Barlow, 2000), and courtship behavior (Rosenthal *et al.*, 1996) enhance reproductive success either by conferring a competitive advantage against males or a mating advantage with females (Anderson, 1994). Fishes are excellent subjects to study the underlying mechanisms of sexual selection, and have been used to test predictions of several models of female choice. For example, we know how parasites and predators affect courtship displays and the expression of male carotenoid pigments in guppies and sticklebacks (Milinski and Bakker, 1990; Houde, 1997). Phylogenetic reconstruction provides insights into the evolution of elaborate male secondary sexual traits, such as swords in swordtails (Basolo, 1995).

Secondary sexual traits, such as elaborate fins, often have a dual function as signals in male-male competitive interactions and in female choice of males (Bisazza, 1993; Berglund *et al.*, 1996; Benson and Basolo, 2006). Elaborate fins are usually associated with courtship displays. In species with several male morphs, only males with ornaments typically also engage in courtship and aggressive interactions (e.g., Poeciliidae). The metabolic costs of such traits are paid by fitness benefits resulting from female choice for vigorous courtship displays (Nicoletto, 1993; Rosenthal *et al.*, 1996). There is mounting empirical evidence to suggest that sexually selected traits are accurate indicators of overall male genetic quality and physical condition, because they are costly to produce and maintain (Anderson, 1994; Knapp, 1995; Kotiaho, 2001; Vanhooydonck *et al.*, 2007). Numerous studies document the tendency of females to respond more strongly to males with high display rates (Rosenthal *et al.* 1996; Forsgren, 1997; Houde, 1997; Itzkowitz and Haley, 1999; Oliveira *et al.* 2000; Takahashi and Kohda 2001; Santangelo, 2005). Display rates are honest signals of physical condition, since only males in good physical condition can engage in high rates of courtship (Knapp and Kovach, 1991; Forsgren, 1997). Knapp (1995) experimentally manipulated energy reserves of male bicolor damselfish (*Stegastes partitus*) in the field through food augmentation, and demonstrated that fed males increased their courtship rate. Courtship displays also indicate health and condition, since they are affected by parasites (Brønseth and Folstad, 1997). In guppies, males that are parasitized with an external parasite (*Gyrodactylus*) had lower display rates, less intense carotenoid pigmentation, and are less attractive to females, and consequently had a lower mating success than unparasitized males (Houde, 1997).

It is surprising, therefore, that so little is known regarding the role of sexual selection in the design of fins that function in courtship displays. An unexplored aspect of the cost of secondary sexual traits in fishes is the effect of elaborate fins on increasing the drag associated by movement through water, and consequently the effects of sexually selected traits on other activities, such as foraging and predator avoidance, and on evolution of swimming performance. Surprisingly, there are few studies of how elaborate fins affect swimming performance, and how the contrasting demands of efficient swimming and elaborate courtship displays have influenced body shape and fin design.

Size and shape of fins should reflect their function in locomotion. Several experimental studies have examined the effects of fin shape and size on performance (e.g., Plaut, 2000; Billman and Pyron, 2005; Blake *et al.*, 2005; reviewed by Oufiero and Garland, 2007), but only a few have examined the performance costs of elaborate fins in the context of sexual selection (e.g., Nicoletto, 1991; Rosenthal and Evans, 1998; Basolo and Alcaraz, 2003; Karino *et al.* 2006). Plaut (2000) studied the swimming performance of strains of zebrafish (*Danio rerio*) differing in caudal fin length. He found a 22% reduction in the steady swimming performance (U_{crit}) in the long-finned strain compared to that of the smaller-finned wild type. He attributed such differences to a lower aspect ratio (square of fin span/fin area) of the caudal fin in the long-finned type. Longer caudal fins were more flexible than the caudal fins of the short-finned form, so they were probably more affected by drag forces. There also seems to be a trade-off between swimming speed and maneuverability. Zebrafish with longer caudal fins were more maneuverable, but had a lower steady swimming performance (U_{crit}). So there may be a trade-off between maneuverability selected for in courtship, and aspects of swimming performance related to other activities. In the zebrafish it is not known whether long-finned males have a reproductive advantage over short-finned ones.

Elaborate fins are likely to reflect a compromise between sexual selection for showy courtship displays and natural selection for other aspects of swimming performance, such as metabolic cost and predator avoidance.

Extensions of the caudal fin in swordtails (*Xiphophorus*; Basolo,1990), a freshwater goby (*Rhinogobius brunneus*; Suk and Choe, 2002) large dorsal fins in sailfin mollies (Jordan *et al.*, 2006), large dorsal and caudal fins in Siamese fighting fish (*Betta splendens*; Simpson, 1968), and large caudal fins in guppies (*Poecilia reticulata*; Karino *et al.*, 2006) feature prominently in male courtship displays. Males with longer swords or larger fins are more attractive to females and have a higher mating success, demonstrating that fin size is subject to sexual selection (Bischoff *et al.*, 1985; Ptacek and Travis, 1997; Suk and Choe, 2002; Jordan *et al.*, 2006).

Honest indicators of a male's genetic and phenotypic quality, such as elaborate fins, are expected to increase metabolic costs and reduce steady swimming. The few studies where metabolic costs or steady swimming performance have been measured support this prediction. Basolo and Alcaraz (2003) measured the metabolic cost of routine swimming and courtship in the Montezuma swordtail (*Xiphophorus montezumae*). In order to determine the metabolic cost of the ornament (sword), they compared metabolic rates of males with intact and artificially shortened swords. The metabolic rate of intact males was higher during both routine swimming (19.4%) and courtship swimming (30%) compared to males with shortened swords, showing that the sword imposes a substantial metabolic cost (Fig. 8.1). Another cost of long fins should be reduced steady swimming performance. Kruesi and Alcaraz (2007) measured swimming performance

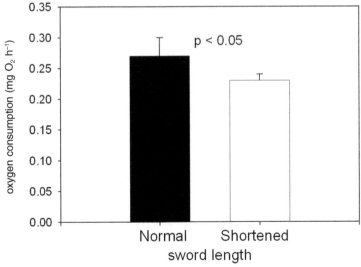

Fig. 8.1 Metabolic cost of routine swimming and courtship in the Montezuma swordtail (*Xiphophorus montezumae*). Comparisons are between males with normal swords and males with shortened swords. Values shown are mean± SE (from Basolo and Alcaraz, 2003). Males with intact swords have a higher oxygen consumption, suggesting that swords are metabolically costly ornaments.

in the Montezuma swordtail (*Xiphophorus montezumae*) and noted a 24% increase in U_{crit} for males after their swords were experimentally removed. Both these studies on swordtails indicate that the sword is an expensive ornament, because it increases metabolic costs and reduces steady swimming performance.

Studies measuring the metabolic costs of aggressive displays also show that they are metabolically costly. For example, the oxygen consumption of male Siamese fighting fish (*Betta splendens*) during aggressive displays, which involved the spread of the gill cover and dorsal fin, increased five-fold over the basal metabolic rate (0.6 mg O_2/h to 1.8 mg O_2/h; Castro *et al.*, 2006), Similarly, Grantner and Taborsky (1998) reported a four-fold increase in metabolic rate during agonistic displays, which included gill cover spread, tail beating and S-shaping of the body, in a cichlid fish (*Neolamprologus pulcher*). The problem with these studies on the metabolic costs of aggressive behavior is that it is unknown how they relate to various measures of swimming performance.

Trade-offs between steady swimming performance (U_{crit}) and fin size have been documented for guppies (Nicoletto,1991; Karino *et al.*, 2006). Males with long caudal fins had a lower U_{crit} than those with short caudal fins (Nicoletto,1991). Interestingly, the shape of the caudal fin does not seem to affect steady swimming performance in the guppy, suggesting that fin size, more than fin shape, is constrained by natural selection on this measure of swimming performance (Nicoletto, 1991). It is unclear whether females discriminate among males based on the shape of their caudal fin, or the extent to which caudal fin shape affects other measures of swimming performance, such as burst speed or maneuverability.

Other traits, such as gonopodial structures in poeciliids, may also represent a compromise between natural selection for performance and sexual selection for gonopodium length. In mosquito fish (*Gambusia sp.*) which lack courtship, long gonopodia increase mating success of males through sneak copulations. Males with longer gonopodia have a lower burst swimming speed, suggesting a potential cost of such structures in avoidance of predators (Langerhans *et al.*, 2005). It has yet to be determined whether such trade-offs exist between gonopodial length and various aspects of swimming performance in other poeciliid fishes such as mollies, swordtails, and guppies.

Trade-offs between fin elaboration and various measures of swimming performance may not always occur. In the filefish (*Paramonacanthus japonicus*) the elongated dorsal and anal fins of males function as hydrodynamic devices to enhance steady swimming performance (U_{crit}) (Kokita and Mizota, 2001). Costs of ornaments are best examined by comparing aspects of swimming performance between males in a population. Royle *et al.* (2006) noted that in the green swordtail (*Xiphophorus helleri*), males with the longest swords had higher burst swimming speeds than similar-sized individuals with smaller swords. Since burst swimming speed is correlated with the ability to escape predation, males with longer swords, which are also more attractive to females, may be less vulnerable to predators (Walker *et al.*, 2005). In the Pecos pupfish (*Cyprinodon pecosensis*), steady swimming (U_{crit}) is correlated with physical condition. Territorial males perform better in swimming trials and have a higher (U_{crit}) than similar-sized males that are not territorial (Fig. 8.2; Kodric-Brown and Nicoletto, 1993). Territorial males are preferred by females and have higher reproductive success (Kodric-Brown, 1983). Thus, individual variation in swimming performance reflected in the vigor and duration of courtship displays may be indicative of health and subject to sexual selection.

Steady swimming (U_{crit}) may not be the best indicator of metabolic scope or overall swimming performance since it only measures swimming endurance in a current, so it may underestimate the metabolic cost of routine swimming (reviewed in Blake, 2004). Nicoletto (1991) failed to detect an effect of caudal fin shape on swimming performance in guppies (*Poecilia reticulata*). Similarly, Ryan (1988) found no effect of sword length on swimming performance of male swordtails *X. nigrensis*. Both studies used steady swimming (U_{crit}) as a measure of swimming performance. Courtship, however, involves complex maneuvers, such as turning and backward swimming, and the sword, elaborate caudal fins, and enlarged dorsal fins tend to increase drag (Webb and Gerstner, 2000). Consequently, the cost of male ornaments should be measured using more than one performance criterion.

More integrative measures of swimming performance should incorporate measures of the energetic cost of routine swimming as well as the costs of sexually selected traits, such as courtship and agonistic displays, which are an integral part of the reproductive behavior of many fishes (Table 8.1). To date, the few studies on the effects of fin size and shape on male courtship behavior and mating success have been on species that live in relatively calm waters or occupy microhabitats with low flow velocity (e.g., guppies, swordtails, mollies, Siamese fighting fish, pupfish). The few studies where the hydrodynamic costs of sexually-selected traits, such as fins used in courtship displays, have been measured suggest that they increase the energetic costs of swimming by reducing thrust and/or increasing drag (Videler, 1993). One way fish may reduce the energetic costs of swimming, is by alternating bouts of swimming and gliding to maintain their position in currents with relatively low flow velocity (Kramer and McLaughlin, 2001). The few studies on swordtails and guppies suggest that complex maneuvering, such as turning, stopping, forward and backward swimming that is frequently seen during courtship and agonistic displays of many species, increases swimming costs (Basolo and Alcaraz, 2003; Karino *et al.*, 2006; Kruesi and Alcaraz, 2007).

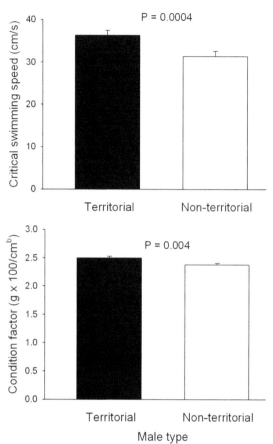

Fig. 8.2 Differences between territorial and non-territorial males of Pecos pupfish (*Cyprinodon pecosensis*): (A) Steady swimming performance (U_{crit}), and (B) Condition factor, a measure of condition. Values shown are mean± SE (from Kodric-Brown and Nicoletto, 1993). Territorial males show a higher swimming performance and are in better physical condition than non-territorial males.

Carotenoids and Swimming Performance

A general pattern emerging from studies of sexually selected traits in fishes is the prominent role played by carotenoids (Kodric-Brown, 1998). Many species of freshwater fishes incorporate carotenoids into secondary sexual traits, such as body skin color (stickleback: Bakker and Milinski, 1993; guppy: Houde, 1997) or fins (Kodric-Brown, 1998). Since carotenoids are acquired from the diet and cannot be synthesized, they provide a direct link between the expression of a male's ornament and his physical condition. Carotenoids have important roles as antioxidants (Bertrand *et al.*, 2006), in boosting the immune system

(Blas *et al.*, 2006), and have been linked with such indirect indicators of genetic quality and physical condition as increased sperm production (guppies: Pitcher and Evans, 1991). Several studies have found a positive relationship between steady swimming performance (U_{crit}) and diet, especially the expression of carotenoids. In a laboratory study of swimming performance in guppies, Nicoletto (1996) found that males with more intense carotenoid spots showed higher endurance during steady swimming than those will paler spots. A similar pattern was found for a population of guppies in Trinidad that experiences high predation (Nicoletto and Kodric-Brown, 1990).

Male Body Size, Ornament Allometry, and Swimming Performance

Both intra- and intersexual selection should favor large male size in fishes. Larger males have a competitive advantage in mating systems where males control resources or females (Anderson, 1994; Barlow, 2000). Females often prefer larger males because they are generally in better physical condition, have higher rates of courtship displays, are able to secure better nest sites, and provide better parental care (Knapp and Kovach, 1991; MacLaren *et al.*, 2004). Even in Poeciliids, where male investment in offspring is limited to providing sperm, females often prefer to mate with large males (guppies, mollies, swordtails). In guppies, females prefer larger males (Magellan *et al.*, 2005) and offspring of large males have higher growth rates (Reynolds and Gross, 1992). This is especially important in streams with predators that prey on young, because higher growth rates reduce the time window when young guppies are especially vulnerable.

Since ornament size scales with positive allometry with body size in fish as well as other animals (Kodric-Brown *et al.*, 2006), larger males have larger ornaments relative to their body size, and invest proportionally more resources in ornaments than smaller males. Poeciliids provide good examples. The allometric exponent indexes the way that a trait varies with body size. Exponents greater than one indicate positive allometry: differential investment in ornament with increasing size. Elaborate fins have allometric exponents between 1.25 and 2.37 for sailfin mollies and swordtails (Hankinson *et al.*, 2006; Kodric-Brown *et al.*, 2006).

Many male ornaments such as swords in swordtails, large pectoral fins in Siamese fighting fish, the sailfin in mollies, and long tails in guppies increase the apparent size of males. Experimental studies using models and video animations, which allow body size and ornament size to be manipulated separately, provide interesting insights into the kinds of traits that are attractive to females. In both swordtails (Rosenthal and Evans, 1998) and mollies (MacLaren *et al.*, 2004) females prefer large males. In swordtails, females discriminate between males with different sword lengths, but surprisingly not between images of males of identical standard length with and without swords. Similarly, female mollies show equal preferences when presented with a small male dummy with a large sailfin and a large male dummy with a small sailfin, if both had the same total lateral projection area (MacLaren *et al.*, 2004). Investment in an ornament is costly, because it reduces swimming performance and increases the metabolic cost of swimming, the cost is still less, however, compared to what it would be if males invested the resources into increasing body mass to achieve the same apparent body size.

Swimming Performance, Flow Conditions, and Mate Choice

Steady swimming (U_{crit}) is very sensitive to local physical conditions, especially flow velocity. Nicoletto (1996) subjected guppies to different velocities in the laboratory, and showed that individuals raised under high flow conditions had a higher U_{crit} than those raised under low flow conditions. Similarly, a comparison of steady swimming performance of male guppies from two streams in Trinidad reflected local flow conditions (Kodric-Brown and Nicoletto, 2005). Fish can compensate behaviorally by seeking microhabitats that reduce the metabolic costs of maintaining position in fast-flowing water. For example, guppies inhabiting streams in Trinidad frequent microhabitats with low flows, such as shallows and eddies (Kodric-Brown and Nicoletto, 2005; Magellan and Magurran, 2006).

Females also exert subtle effects on male swimming performance through their choice of microhabitats. Guppy females tend to select microhabitats with higher flow rates (Kodric-Brown and Nicoletto, 2005; Magellan and Magurran, 2006). Such habitats require males to swim faster to maintain position, thereby impose a higher metabolic cost of courting on the smaller males than the cost of maintaining position for the larger females, since U_{crit} increases with body size (Videler, 1993). Similarly, Takahashi and Kohda (2001) noted that males of a stream goby (*Rhinogobius sp.*) that court in microhabitats with faster flows are preferred as mates by females. Again, these studies suggest that these behavioral traits are due to sexual selection, because only males in good physical condition can pay the cost of courting in microhabitats with higher velocities (Karino *et al.*, 2006).

Costs of swimming also affect male mate choice. Whenever given a choice, males preferentially court large females, since these have a higher fecundity than smaller ones (Houde, 1997). In a laboratory study of mate choice in the Pacific blue-eye (*Pseudomugil signifer*) males preferentially courted the larger of the two females under low flow conditions. They switched to courting the smaller female when the flow to the compartment where the larger female was housed was increased (Wong and Jennions, 2003).

Predators, Sexual Selection, and Ornaments

Predation is a strong selective pressure influencing the evolution and expression of secondary sexual traits. Natural selection to avoid predation often conflicts with sexual selection, because many sexually selected traits increase swimming costs and reduce survival (Magurran and Seghers, 1990; Winemiller et al., 1990; Trexler et al., 1994; Rosenthal et al., 2001). The general evolutionary response to increased predation is reduced expression of energetically costly secondary sexual traits and courtship behavior (Endler and Houde, 1995; Langerhans et al., 2005). Secondary sexual traits may affect all aspects of swimming performance, such as steady swimming performance (U_{crit}), swimming endurance, and burst speed, by increasing metabolic costs as well as the ability to escape predators. Male ornaments, such as swords of swordtails, large dorsal fins of sailfin mollies, and elaborate caudal fins of guppies are not only attractive to females, but also to predators (Winemiller et al., 1990; Rosenthal et al., 2001). A common pattern seen in poeciliids is a reduction in male ornaments in environments with greater predation pressure (Endler and Houde, 1995; Ptacek and Travis, 1996; Langerhans et al., 2005). In both *Xyphophorus nigrensis* (Rosenthal; unpublished data) and *X. helleri* (Basolo and Wagner, 2004) males from populations with predators have shorter swords (Fig. 8.3); they invest relatively less in ornaments and more in body size (muscle mass), and consequently have lower allometric exponents (b = 1.44 and b = 1.51) compared to those from populations experiencing low predation (b = 2.20 and b = 2.06). Similar patterns have been reported for mollies (Trexler et al., 1994; Hankinson et al., 2006).These patterns suggest that male ornaments not only increase the metabolic costs of courtship, but also increase the risk of predation.

Studies on guppies, swordtails, mollies, and three-spine sticklebacks have shown that predation not only reduces male ornamentation, but may limit mate choice, especially if females incur additional costs by being choosy (Forsgren, 1992; Milinski and Bakker, 1992). In guppies, males in high-predation environments are less conspicuous, have reduced courtship, and females are less responsive to male courtship and less choosy (Magurran and Seghers, 1990; Houde, 1997).

CONCLUSIONS AND FUTURE DIRECTIONS

Male ornaments in fishes are costly to maintain. The few studies to date show that large fins and other structures favored by sexual selection increase metabolic costs of swimming and reduce steady swimming performance (Nicoletto, 1991; Basolo and Alcaraz, 2003; Karino et al., 2006; Kruesi and Alcaraz, 2007). Routine swimming under natural conditions is metabolically more expensive than laboratory measurements of swimming performance

A. *Xiphophorus nigrensis*

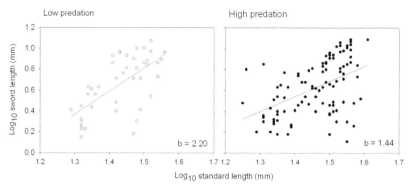

B. *Xiphophorus helleri*

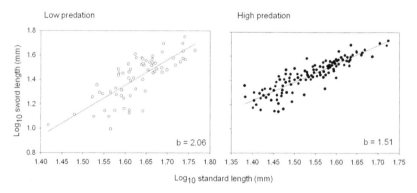

Fig. 8.3 Regressions of sword length against standard length for (A) two populations of the pygmy swordtail *Xiphophorus nigrensis* (data from G.G. Rosenthal, personal comm.) and (B) two populations of the green swordtail *Xiphophorus helleri* (data from Basolo and Wagner, 2004). Note that in both species the allometric exponents are higher for the low predation populations, suggesting that males in these populations invest more resources in ornaments than males from the high predation populations.

(Bisclair and Tancy, 1993; reviewed in Blake, 2004). On the other hand, large ornaments provide a selective advantage in both male-male competition and female choice. Males with large ornaments are often dominant and consequently have access to both females and resources, such as territorial sites (Kodric-Brown and Nicoletto, 1993). Females tend to preferentially mate with males with large ornaments and vigorous courtship displays (reviewed in Anderson, 1994). Since courtship displays involve both routine swimming and numerous turns, starts, changes in direction as well as body flexibility (s-shaping and body curvature), metabolic costs of such complex behavior are difficult to measure with any degree of accuracy, and are likely to underestimate actual costs. Perhaps it is not surprising that I know of only one study that measures the metabolic cost of courtship (Basolo and Alcaraz, 2003), and only a few that measure the effects of ornaments on steady swimming performance (e.g., Nicoletto, 1991, Karino *et al.*, 2006; Kruesi and Alcaraz, 2007). All these studies are limited to poeciliid species with males that are less than 50 mm in standard length. Additional studies are required on other groups before the generality of increased metabolic costs of ornaments, and the multiple and sometimes conflicting selective pressures operating on them are understood.

One question that remains unanswered is why large fish (greater than 500 cm in standard length) almost universally lack elaborate secondary sexual traits such as large fins and generally do not engage in metabolically costly courtship. The metabolic cost of swimming (ml O_2 g^{-1} km^{-1}) as a function of body mass, is obtained by dividing the mass-specific rate of oxygen consumption by swimming speed. The slope is approximately -0.3 (Schmidt-Nielsen, 1984) and this scaling relation holds for fish with different body shapes and swimming modes. This means that the cost of swimming per gram of body mass decreases with increasing body size, even though the cost per individual increases, but less than linearly, with increasing mass (with a slope of 0.7 on a log-log plot). Therefore, based on swimming energetics alone, it is unclear why large fish do not engage in elaborate courtship behavior. Many of them are capable of swimming at high speeds for sustained periods, and continuously, in the case of pelagic species such as tuna, mackerel, and swordfish. Large fish show lower maneuverability than small fish (Domenici, 2001), and this may be a reason why they do not engage in elaborate courtship.

Clearly, studies on other species are needed to confirm the patterns suggested here, namely that the energetic costs of courtship, which often involves rapid turns, backward swimming, or circling in front of the female, are too high for pelagic fish for sexual selection to favor either the elaboration of secondary sexual fins or courtship behavior. Future studies on aspects of courtship and mating behavior and swimming performance of pelagic species are needed to provide insights into the hydrodynamic and other factors that have shaped the evolution of reproductive behavior in these species. Large fish pose methodological challenges for measuring behavior associated with courtship and mating. It is difficult to create controlled laboratory conditions where these behaviors can be studied. Future studies could use indirect measurements of swimming performance, such as gill ventilation rates (Van Rooij and Videler, 1996).

Conclusions from this review are that sexually selected traits affect and are affected by swimming performance. Sexually selected traits, such as large fins and elaborate courtship displays, affect not only the metabolic cost of swimming, but also other components of performance, such as predator-avoidance. Under predation threat, sexually-selected traits will be reduced, and males will divert resources into muscle mass to increase both steady swimming performance and burst speed.

Most of the information about ways that elaborate sexually-selected traits affect swimming performance comes from studies of a very limited and biased sample, mostly poeciliids. These fish are of small size, readily reared and experimentally studied in the laboratory. Additional studies on other groups of fishes, especially pelagic ones, are needed to understand the tradeoffs between sexual selection for elaborate ornaments and natural selection for other activities that affect swimming performance.

ACKNOWLEDGEMENT

I thank J.H. Brown, P. Domenici, and three anonymous reviewers for commenting on an earlier draft of this chapter. This work was supported in part by National Science Foundation Grant IBN-0417338.

REFERENCES

Allen, G.R. and D.R. Robertson. 1994. *Fishes of the Tropical eastern Pacific*. University of Hawaii Press, Honolulu.

Anderson, M. 1994. *Sexual Selection*. Princeton University Press, Princeton.

Andreani, M.S., B.E. Erisman and R.R. Warner. 2004. Courtship and spawning behavior in the California sheephead, *Semicossyphus pulcher* (Pisces: Labridae). *Environmental Biology of Fishes*. 71: 13–19.

Ayling, A.M. 1980. Hybridization in the genus *Pseudolabrus* (Labridae). *Copeia* 1980: 176–180.

Bakker, T.C.M. and M. Milinski. 1993. The advantage of being red: sexual selection in the stickleback. *Marine Behavioral Physiology* 23: 287–300.

Barlow, G. 1961. Social behavior of the desert pupfish, *Cyprinodon macularius*, in the field and in the aquarium. *American Midland Naturalist* 65: 339–359.

Barlow, G. 2000. *The cichlid fishes: Nature's grand experiment in evolution*. Perseus Publishing, Cambridge, Mass.

Basolo, A.L. 1990. Female prefernce for male sword length in the green swordtail, *Xiphophorus helleri* (Pisces, Poeciliidae). *Animal Behaviour* 40: 332–338.

Basolo, A.L. 1995. Phylogenetic evidence for the role of a pre-existing bias in sexual selection. *Proceedings of the Royal Society of London* B259: 307–311.

Basolo, A.L. and G. Alcaraz. 2003. The turn of the sword: length increases male swimming costs in swordtails. *Proceedings of the Royal Society of London* B 270: 1631–1636.

Basolo, A.L. and W.E. Wagner. 2004. Covariation between predation risk, body size and fin elaboration in the green swordtail, *Xiphophorus helleri*. *Biological Journal of the Linnean Society* 83: 87–100.

Benson, K.E. and A.L. Basolo. 2006. Male-male competition and the sword in male swordtails, *Xiphophorus helleri*. *Animal Behaviour* 71: 129–134.

Berglund, A., A. Bisazza and A. Pilastro. 1996. Armaments and ornaments: an evolutionary explanation of traits of dual utility. *Biological Journal of the Linnean Society* 58: 385–399.

Bertrand, S., B. Faivre and G. Sorci. 2006. Do carotenoid-based sexual traits signal the availability of non-pigmentary antioxidants? *Journal of Experimental Biology* 209: 4414–4419.

Billman, E.J. and M. Pyron. 2005. Evolution of form and function: morphology and swimming performance in North American minnows. *Journal of Freshwater Ecology* 20: 221–232.

Bischoff, R.J., J.L. Gould and D.I. Rubenstein. 1985. Tail size and female choice in the guppy (*Poecilia reticulata*). *Behavioral Ecology and Sociobiology* 17: 253–255.

Bisazza, A. 1993. Male competition, female mate choice and sexual size dimorphism in poeciliid fishes. *Marine Behavior and Physiology* 23: 257–286.

Blake, R.W. 2004. Fish functional design and swimming performance. *Journal of Fish Biology* 65: 1193–1222.

Blake, R.W., T.C. Law, K.H.S. Chan and J.F.Z. Li. 2005. Comparison of the prolonged swimming performances of closely related, morphologically distinct three-spined sticklebacks *Gasterosteus* spp. *Journal of Fish Biology* 67: 834–848.

Blake, R.W., L.M. Chatters and P. Domenici. 1995. Turning radius of yellowfin tuna (*Thunnus albacares*) in unsteady swimming maneuvers. *Journal of Fish Biology* 46: 536–538.

Blas, J. Perez-Rodriguez, G.R.L. Bortolotti, J. Vinuela and T.A. Marchant. 2006. Testosterone increases bioavailability of carotenoids: insights into the honesty of sexual signaling. *Proceedings of the National Academy of Sciences U.S.A.* 103: 18633–18637.

Breder, C.M. and D.E. Rosen. 1966. *Modes of reproduction in fishes*. The Natural History Press, Garden City.

Breinerd, E.L. and S.N. Patek. 1998. Vertebral column morphology, C-start curvature, and the evolution of mechanical defenses in tetradontiform fishes. *Copeia* 1998: 971–984.

Brett, J.R. 1964. The respiratory metabolism and swimming performance of young sockeye salmon. *Journal of the Fisheries Research Board of Canada* 21: 1183–1226.

Brønseth, T and I. Folstad. 1997. The effects of parasites on courtship dance in three-spined sticklebacks: more than meets the eye? *Canadian Journal of Zoology* 75: 589–894.

Castro, N., A.F.H. Ros, K. Becker and R.F. Oliveira. 2006. Metabolic costs of aggressive behaviour in the Siamese fighting fish, *Betta splendens*. *Aggressive Behavior* 32: 474–480.

Dawkins M.S. and T. Guilford. 1994. Design of an intention signal in the bluehead wrasse (*Thalassoma bifasciatum*). *Proceedings of the Royal Society London* B257: 123–128.

Domenici, P. 2001. The scaling of locomotor performance in predator-prey encounters: from fish to killer whales. *Comparative Biochemistry and Physiology* A131: 169–182.

Domenici, P. and R.W. Blake. 1991. The kinematics and performance of the escape response in the angelfish (*Pterophyllum eimekei*). *Journal of Experimental Biology* 156: 187–205.

Domenici, P. and R.W. Blake. 1993. Escape trajectories in the angelfish (*Pterophyllum eimekei*). *Journal of Experimental Biology* 177: 253–272.

Domenici, P. 2003. Habitat, body design, and the swimming performance of fish. In: *Vertebrate Biomechanics and Evolution*. V.L. Bels, J.P. Gasc and A. Casions (Eds.). BIOS Scientific Publishers, Ltd. Oxford.

Endler, J.A. and A.E. Houde. 1995. Geographic variation in female preferences for male traits in *Poecilia reticulata*. *Evolution* 49: 456–468.

Forsgren, E. 1997. Mate sampling in a population of sand gobies. *Animal Behaviour* 56: 267–276.

Forsgren, E. 1992. Predation risk affects mate choice in a gobiid fish. *American Naturalist* 140: 1041–1049.

Grantner, A. and M. Taborsky. 1998. The metabolic rates associated with resting, and with the performance of agonistic, submissive and digging behaviours in the cichlid fish *Neolamprologus pulcher* (Pisces: Cichlidae) *Journal of Comparative Physiology* B168: 427–433.

Hankinson, S.J. and M.B. Ptacek. 2007. Within and between species variation in male mating behaviors in the Mexican sailfin mollies *Poecilia velifera* and *P. petensis*. *Ethology* 113: 802–812.

Hankinson, S.J., M.J. Cildress, J.J. Schmitter-Soto and M.B. Ptacek. 2006. Morphological divergence within and between populations of the Mexican sailfin mollies *Poecilia velifera* and *P. petensis*. *Journal of Fish Biology* 68: 1610–1630.

Houde, A.E. 1997. *Sex, Color, and Mate Choice in Guppies*. Princeton University Press, Princeton.

Itzkowitz, M. and M. Haley. 1999. Are males with more attractive resources more selective in their mate preferences? A test in a polygynous species. *Behavioral Ecology* 10: 366–371.

Jordan, R., D. Howe, T. Knight and J. Gould. 2006. Female choice linked to male dorsal fin height in a shortfin molly. *Journal of Ethology* 24: 301–304.

Karino, K., K. Orita and A. Sato. 2006. Long tails affect swimming performance and habitat choice in the male guppy. *Zoological Science* 23: 255–260.

Knapp, R.A. 1995. Influence of energy reserves on the expression of a secondary sexual trait in male bicolor damselfish *Stegastes partitus*. *Bulletin of Marine Science* 57: 672–681.

Knapp, R.A. and J.T. Kovach. 1991. Courtship as an honest indicator of male parental quality in the bicolor damselfish, *Stegastes partitus*. *Behavioral Ecology* 2: 295–300.

Kodric-Brown, A. 1983. Determinants of male reproductive success in pupfish (*Cyprinodon pecosensis*). *Animal Behaviour* 31: 128–137.

Kodric-Brown, A. 1998. Sexual dichromatism and temporary color changes in the reproduction of fishes. *American Zoologist* 38: 70–81.

Kodric-Brown, A. and P.F. Nicoletto. 1993. The relationship between physical condition and social status in pupfish (*Cyprinodon pecosensis*). *Animal Behaviour* 46: 1234–1236.

Kodric-Brown, A. and P.F. Nicoletto. 2005. Courtship behavior, swimming performance, and microhabitat use of Trinidadian guppies. *Environmental Biology of Fishes* 73: 299–307.

Kodric-Brown, A., R. Sibly and J.H. Brown. 2006. The allometry of ornaments and weapons. *Proceedings of the National Academy of Sciences of the United States of America* 103: 8733–8738.

Kokita, T. and T. Mizota. 2001. Male secondary sexual traits are hydrodynamic devices for enhancing swimming performance in a monogamous filefish (*Paramonacanthus japonicus*). *Journal of Ethology* 20: 35–42.

Kolok, A.S. 1999. Interindividual variation in the prolonged locomotor performance of ectothermic vertebrates: a comparison of fish and herpetofaunal methodologies and a brief review of the recent fish literature. *Canadian Journal of Fisheries and Aquatic Sciences* 56: 700–710.

Kotiaho, J.S. 2001. Costs of sexual traits: a mismatch between theoretical considerations and empirical evidence. *Biological Reviews* 76: 365–376.

Kramer, D.L. and R.L. McLaughlin. 2001. The behavioral ecology of intermittent locomotion. *American Zoologist* 41: 137–153.

Kruesi, K. and G. Alcaraz. 2007. Does a sexually selected trait represent a burden in locomotion? *Journal of Fish Biology* 70: 1161–1170.

Langerhans, R.B., C.A. Layman and T.J. DeWitt. 2005. Male genital size reflects a trade-off between attracting mates and avoiding predators in two live-bearing fish species. *Proceedings of the National Academy of Sciences, of the United States of America* 102: 7618–7623.

Liley, N.R. 1966. Ethological isolating mechanisms in four sympatric species of poeciliid fishes. *Behaviour* (Supplement) 13: 1–197.

MacLaren, D.R., W.J. Rowland and N. Morgan. 2004. Female preferences for sailfin and body size in the sailfin molly, *Poecilia latipinna*. *Ethology* 110: 363–379.

Magellan, K., L.B. Pettersson and A.E. Magurran. 2005. Quantifying male attractiveness and mating behaviour through phenotypic size manipulation in the Trinidadian guppy, *Poecilia reticulata*. *Behavioural Ecology and Sociobiology* 58: 366–374.

Magellan, K. and A.E. Magurran. 2006. Habitat use mediates the conflict of interest between the sexes. *Animal Behaviour* 72: 75–81.

Magurran, A.E. and B.H. Seghers. 1990. Risk sensitive courtship in the guppy (*Poecilia reticulata*). *Behaviour* 112: 194–201.

Martinez, M., H. Guderley, J.D. Dutil, P.D. Winger, P. He and S.J. Walsh. 2003. Condition, prolonged swimming performance and muscle metabolic capacities of cod *Gadus morhua*. *Journal of Experimental Biology* 206: 503–511.

Masonjones, H.D. and S.M. Lewis. 1996. Courtship behavior in the dwarf seahorse, *Hippocampus zosterae*. Copeia. 1996: 634–640.

McLennan, D.A. 1993. Phylogenetic relationships in the Gasterosteidae: An updated tree based on behavioral characteristics with a discussion of homoplasy. *Copeia*. 1993: 318–326.

Merrick, J.R. and G.E. Schmida, 1984. *Australian Freshwater Fishes*. Griffin Press Limited. Netley, Australia.

Mertz, J. C. and G.W. Barlow. 1966. On the reproductive behavior of *Jordanella floridae* (Pisces: Cyprinodontidae) with special reference to a quantitative analysis of parental fanning. *Zeitschrift für Tierpsychologie*. 23: 537–554.

Milinski, M. and T.C.M. Bakker. 1990. Female sticklebacks use male coloration in mate choice and hence avoid parasitized males. *Nature* (London) 344: 330–333.

Milinski, M. and T.C.M. Bakker. 1992. Costs influence sequential mate choice in sticklebacks, *Gasterosteus aculeatus*. *Proceedings of the Royal Society London* B250: 229–233.

Nicoletto, P.F. 1991. The relationship between male ornmanetation and swimming performance in the guppy *Poecilia reticulata*. *Behavioral Ecology and Sociobiology* 28: 365–370.

Nicoletto, P.F. 1996. The influence of water velocity on the display behavior of male guppies, *Poecilia reticualta*. *Behavioral Ecology* 7: 272–278.

Nicoletto, P.F. and A. Kodric-Brown. 1990. The relationship between swimming performance, courtship behavior, and carotenoid pigmentation of guppies in four rivers of Trinidad. *Environmental Biology of Fishes* 55: 277–235.

Oufiero, C.E. and Jr. T. Garland. 2007. Evaluating performance costs of sexually selected traits. *Functional Ecology*, 21: 676–689.

Oliveira, R.F., J.A. Miranda, N. Carvalho, E.J. Goncalves, M.S. Grober and R.S. Santos. 2000. Male mating success in the Azorean rock-pool blenny: the effects of body size, male behaviour and nest characteristics. *Journal of Fish Biology*. 57: 1416–1428.

O'Steen, S., A.J. Cullum and A.F. Bennett. 2002. Rapid evolution of escape ability in Trinidadian guppies (*Poecilia reticulata*). *Evolution* 56: 776–784.

Page, L.M. 1985. Evolution of reproductive behaviors in Percid fishes. Ill. *Natural History Survey Bulletin*. 33: 275–295.

Pitcher, T.E. and J.P. Evans. 2001. Male phenotype and sperm number in the guppy (*Poecilia reticulata*). *Canadian Journal of Zoology* 79: 1891–1896.

Plaut, I. 2000. Effects of fin size on swimming performance, swimming behavior and routine activity of zebrafish *Danio rerio*. *Journal of Experimental Biology* 203: 8130–820.

Ptacek, M.B. and J. Travis. 1996. Inter-population variation in male mating behaviours in the sailfin molly, *Poecilia latipinna*. *Animal Behaviour* 52: 59–71.

Ptacek, M.B. and J. Travis. 1997. Mate choice in the sailfin molly, *Poecilia latipinna*. *Evolution* 51: 1217–1231.

Reynolds, J.D. and M.R. Gross. 1992. Female mate preference enhances offspring growth and reproduction in a fish, *Poecilia reticulata*. *Proceedings of the Royal Society of London* B250: 57–62

Rosenthal, G.G., C.S. Evans and W.L. Miller. 1996. Female preference for dynamic traits in the green swordtail, *Xiphophorus helleri*. *Animal Behaviour* 51: 811–820.

Rosenthal, G.G. and C.S. Evans. 1998. Female preference for swords in *Xiphophorus helleri* reflects a bias for large apparent size. *Proceedings of the National Academy of Sciences of the United States of America* 95: 4431–4436.

Rosenthal, G.G., T.Y. Flores Martínez, F.J. García de León and M.J. Ryan. 2001. Shared preferences by predators and females for male ornameants in swordtails. *The American Naturalist* 158: 146–154.

Royle, N.J., N.B. Metcalfe and J. Lindström. 2006. Sexual selection, growth compensation and fast-start swimming performance in Green Swordtails, *Xiphophorus helleri*. *Functional Ecology* 20: 662–669.

Ryan, M.J. 1988. Phenotype, genotype, swimming endurance and sexual selection in a swordtail *X. nigrensis. Copeia* 1988: 484–487.

Santangelo, N. 2005. Courtship in the monogamous convict cichlid; what are individuals saying to rejected and accepted mates? *Animal Behaviour.* 69: 143–149.

Scalet, C.G. 1973. Reproduction of the orangebelly darter, *Etheostoma radiosum cyanorum* (Osteichthyes: Percidae). *American Midland Naturalist.* 89: 156–165.

Schmidt-Nielsen, K. 1984. *Scaling: Why is animal size so important?* Cambridge University Press, Cambridge.

Simpson, M.J.A. 1968. The display of the Siamese fighting fish B*etta splendens. Animal Behaviour Monographs* 1: 1–73.

Suk, H.Y. and J.C. Choe. 2002. Females prefer males with larger first dorsal fins in the common freshwater goby. *Journal of Fish Biology* 61: 899–914.

Videler, J.J. 1993. *Fish Swimming.* Chapman and Hall, London.

Takahashi, D. and M. Kohda. 2001. Females of a stream goby choose mates that court in fast water currents. *Behaviour* 138: 937–946.

Taru, M., T. Kanda and T. Sunobe. 2002. Alternative mating tactics of the gobiid fish *Bathygobius fuscus. Journal of Ethology* 20: 9–12.

Trexler, J.C., R.C. Tempe and J. Travis. 1994. Size-selective predation of sailfin mollies by two species of heron. *Oikos* 69: 250–258.

Van Iersel, J.J.A. 1953. An analysis of the paternal behaviour of the male three-spined stickleback (*Gasterosteus aculeatus* L.). *Behaviour* (Supplement 3): 1–159.

Van Rooij, J.M. and J.J. Videler. 1996. Estimating oxygen uptake rate from ventilation frequency in the reef fish *Sparisoma viridae.* Marine Ecology Progress Series. 132: 31–41.

Vanhooydonck, B., R. Van Damme, A. Herrel and D.J. Irschick. 2007. A performance based approach to distinguish indices from handicaps in sexual selection studies. Functional Ecology 21: 645–652.

Walker, J.A. 2000. Does a rigid body limit maneuverability? Journal of Experimental Biology 203: 3391–3396.

Walker, J.A., C.K. Ghalambor, O.L. Grisset, D. McKenney and D.N. Resnick. 2005. Do faster starts increase the probability of evading predators? *Functional Ecology* 19: 808–815.

Webb, P.W. 1983. Speed, acceleration and maneuverability of two teleost fishes *Journal of Experimental Biology* 102: 115–122.

Webb, P.W. and V. de Buffrenil, 1990. Locomotion in the biology of large aquatic vertebrates. *Transactions of the American Fisheries Society* 119: 629–641.

Webb, P.W. and C.L. Gerstner. 2000. Fish swimming behavior: predictions from physical principles. In: Biomechanics in Animal Behavior, P. Domenici and R.W. Blake (Eds.). BIOS Scientific Publishers. Oxford, pp. 59–77

Webb, P.W. and R.S. Keyes. 1981. Division of labour between median fins in swimming dolphin (Pisces: Coryphaeidae) *Copeia* 1981: 901–904.

Weihs, D. 1972. A hydrdodynamical analysis of fish turning manoeuvres. *Proceedings of the Royals Society London* B182: 50–72.

Wellenreuther, M., C. Syms and K.D. Klements. 2008. Body size and ecological diversification in a sister species pair of triplefin fishes. *Evolutionary Ecology.* 22: 578–592.

Winemiller, K.O, M.A. Leslie and R. Roche. 1990. Phenotypic variation in male guppies from natural inland populations: corroboration of Haskin's sexual selection/predation hypothesis. *Environmental Biology of Fishes* 29: 179–191.

Wong, B.B.M. and M.D. Jennions. 2003. Costs influence male mate choice in a freshwater fish. *Proceedings of the Royal Society of London,* B (Supplement) 270: S36–S38.

Zimmerer, E.J. and K.D. Kallman. 1989. Genetic basis for alternative reproductive tactics in the pygmy swordtail, *Xiphophorus nigrensis. Evolution* 43: 1298–1307.

9

Environmental Influences on Unsteady Swimming Behaviour: Consequences for Predator-prey and Mating Encounters in Teleosts

R.S. Wilson,[1] C. Lefrançois,[2] P. Domenici[3] and I.A. Johnston[4]

INTRODUCTION

The aquatic environment of fish can vary over developmental and evolutionary time scales, eliciting responses at levels of organization ranging from the gene to the whole-animal (Johnston and Wilson, 2006). Temperature and environmental variables such as oxygen availability, pH, salinity, pollution and turbidity can directly and/or indirectly influence physiological processes and behaviour in fish (Bennett, 1990; Prosser, 1991). Many of these factors,

Authors' addresses: [1]School of Biological Sciences, The University of Queensland, St Lucia Qld 4072, Australia.
[2]LIENSs-UMR 6250 (University of La Rochelle, CNRS). 2 rue Olympe de Gouges 17000 La Rochelle, France.
[3]IAMC-CNR, c/o localita Sa Mardini 09072, Torregrance (OR), Italy.
[4]School of Biology, Scottish Oceans Institute, University of St Andrews, Fife KY16 8LB, Scotland, UK.
Corresponding author: E-mail: r.wilson@uq.edu.au

such incidences of hypoxia and pollution, are also becoming more commonplace in coastal areas, and can have profound effects on fish swimming behavior (Domenici *et al.*, 2007a). Although numerous studies have successfully elucidated the functional consequences of environmental variation on locomotion, and their underlying mechanisms, the link between such variation and fitness is not often established (Garland and Carter, 1994; Irschick and Garland, 2001; Wilson, 2005). In this context, studies of prey capture/escape success and reproductive performance are also likely to be of direct relevance.

Unsteady swimming is considered to be a major determinant of predator-prey interactions, although in open habitats these often result in chases that also require high levels of steady swimming performance (Guinet *et al.*, 2007). Unsteady swimming performance is also central to the attraction of potential mates and during male territorial behavior. While unsteady and steady swimming are undoubtedly crucial to reproductive success, the importance of each mode of swimming is most likely related to a species' mating system and habitat preferences. As such, territorial defensive behaviour and mate attraction may depend on the speed and maneuverability of the fish (unsteady swimming) and their ability to sustain these activities over extended periods (steady swimming).

Although swimming performance has a major role in determining the outcome of predator-prey and mating encounters, other aspects of fish behaviour can also be important. For example, the timing of an escape response can be crucial for avoiding predation, and aggressiveness can affect mating performance. In this chapter, we advocate an integrative approach that considers both swimming performance and behaviour within an ecological context, in order to evaluate the effect of environmental variation on mating and predator-prey interactions.

THE EFFECT OF ENVIRONMENTAL FACTORS ON STRIKES AND ESCAPES

Strikes and escapes are unsteady swimming behaviours that are relatively well characterized, particularly from a biomechanical standpoint. Both strikes and escape responses are fast-starts, which can be considered as a form of burst swimming starting either from rest or following steady swimming (Domenici and Blake, 1997). While this chapter includes the specific effects of environmental factors on swimming behaviour, further details on the various modes of swimming behaviour are also covered in chapters 5 (Domenici) and 6 (Rice and Hale) in this book. Briefly, strikes are used by predators to intercept mobile prey. They consist of a sudden acceleration towards the prey, usually accomplished by bending the body into an S-shape caused by bilateral contraction of the axial musculature

(S-starts). There is variability in S-starts, such that predators may begin the strike with a straight body bending rapidly into an S-shape and then accelerate towards the prey, or they may hold the S-shape in the body for a few seconds prior to initiating the strike (Chapter 6 by Rice and Hale). Escape responses usually exhibit a unilateral contraction of the body (i.e., a C-shape) opposite to the side of the stimulation. This contraction corresponds to stage 1, which is commonly followed by a second, contralateral contraction (stage 2). The onset of an escape response is usually triggered by one of a pair of giant hindbrain neurons (the Mauthner cells). Like strikes, escape responses can show high variability, e.g., stage 2 may be absent (in single bend responses, as opposed to double bend responses), S-start escapes may occur in addition to C-starts in certain species, and non-Mauthner cell responses have also been observed (Chapter 5 by Domenici).

Temperature

Temperature is perhaps the most studied environmental factor affecting unsteady swimming performance and has direct, and indirect, implications for the escape and prey capture of fish. Fast-start performance associated with escape behaviour generally increases with rising temperature with the maximum velocity and acceleration attained showing a Q_{10} ranging from 1.1 to 2.8 (Beddow *et al.*, 1995; Johnson and Bennett, 1995; Temple and Johnston, 1998; Fernández *et al.*, 2002). Variable responses of temperature on cumulative turning angle and angular velocity during the fast-start have been reported with both variables showing a much reduced thermal dependence compared to maximum velocity, except at very low temperatures in some species (e.g., long-spined sea scorpion, *Taurus bubalis*, at 0°C) (Temple and Johnston, 1998). Acute reductions in swimming temperature result in an increase in spine curvature during the initial tail-beat of the fast start and an increase in the strain of fast myotomal muscle fibres across a range of distantly related species (Wakeling and Johnston, 1998; Wakeling *et al.*, 2000).

Quantitative estimates of the effects of temperature on escape velocity among species are complicated by interspecific variation in body size, lack of phylogenetic control in studies, as well as the effects of past thermal history (Temple and Johnston, 1998; Johnston and Temple, 2002). In addition, other factors, such as condition index and/or reproductive status (James and Johnston, 1998; Martinez *et al.*, 2004) and growth rate (Alvarez and Metcalfe, 2007), may interact with temperature to independently affect performance. The power output of the fast myotomal muscle is thought to limit fast-start performance of fish (Wakeling and Johnston, 1998). Muscle power output can be measured *in vitro* using the strain and electrical stimulation patterns recorded *in vivo* during fast-starts (Johnston *et al.*, 1995;

Franklin and Johnston, 1997). Such studies do not provide evidence for major evolutionary adjustments in maximum muscle power output and hence swimming speed at any given temperature (Wakeling and Johnston, 1998), although minor adjustments cannot be excluded for the reasons discussed above in relation to kinematic analyses. However, what is more certain is that the range of temperatures at which fast-starts can be successfully elicited is strongly related to an organism's environmental temperature, and performance often declines precipitously towards the thermal limits of a species. For small larvae (~< 1 cm), some of the effects of temperature on performance can be attributed to the physical properties of the water with increased viscosity and drag at low temperatures, though such effects are not significant for juvenile and adult fish (Fuiman and Batty, 1997).

Thermal history has a major impact on all aspects of fast-start behaviour during prey capture (Beddow *et al.*, 1995) and escape responses (Temple and Johnston, 1998). The ability to respond to long-term changes in environmental temperature by modifying underlying physiological state is a process known as thermal acclimation. In some, mostly eurythermal species, receiving stable seasonal temperature signals, several weeks of cold acclimation can result in major changes in muscle biochemistry and contractile properties involving hundreds of changes in gene expression (Gracey *et al.*, 2004), with concomitant effects on fast-start performance (reviewed in Johnston and Temple, 2002). There is evidence that continuous changes in protein expression, e.g., myosin heavy chains (Imai *et al.*, 1997), result in threshold effects on contractile properties which serve to increase the thermal range and hence fast-start performance close to the temperature limits for eliciting the behaviour (Wakeling *et al.*, 2000). For this reason the magnitude of temperature acclimation responses in fast-start behaviour are typically greater in species with a large (e.g., goldfish, *Carassius auratus*) than a more restricted thermal range (e.g., rainbow trout, *Oncorhynchus mykiss*) (Johnson and Bennett, 1995). For example, 4 weeks after transferring rainbow trout from 5°C to 20°C, the maximum linear swimming velocity during fast-starts at 20°C had not increased significantly whereas there were modest improvements in the distance moved (24%) and in angular velocity (15%) (Johnson and Bennett, 1995). In many species (e.g., goldfish, Johnson and Bennett, 1995), an improvement in escape swimming speed at low temperatures with cold-acclimation is associated with a decline in performance at high temperatures, however, such trade-offs are not observed in all species (e.g., short-horn sculpin *Myoxocephalus scorpio*, Temple and Johnston, 1998). The ability to show an acclimation response with respect to fast-start performance is often acquired during ontogeny and is associated with seasonal warming during development (e.g., common carp *Cyprinus carpio*; Wakeling *et al.*, 2000; Cole and Johnston, 2001) or a change in habitat (e.g., short-horn sculpin, Temple and Johnston, 1998).

A characteristic of temperature acclimation responses in juvenile and adult stages is that they are completely reversible. In contrast, changes in temperature during the early stages of development can result in either transient or persistent effects on the phenotype (Johnston and Temple, 2002). In the Atlantic herring, *Clupea harengus*, exposure to different temperatures at the embryo stage resulted in heterochronies in dorsal and anal fin development later in larval life (Johnston *et al.*, 2001). In herring, the adult form of the anal and dorsal fins, and associated fin ray musculature, gradually develops between 12 and 22 mm total length as the trunk becomes more laterally compressed. This coincides with a switch from an anguilliform to a more carangiform mode of steady swimming in which head movements decrease and the amplitude of the trunk increases markedly towards the caudal fin (Batty, 1984). Metamorphosis from a transparent larva to the silver juvenile stage is usually completed at 35–42 mm TL and marks the completion of the adult body plan. Prior to this stage the development of the dorsal and anal fins, and associated muscles that function to stiffen and alter the position of the fin rays, significantly increases the maximum velocity during fast-starts (Morley and Batty, 1996). Dorsal and anal fin ray development was found to occur at shorter body lengths in herring reared at 12 than 5°C until hatching and then transferred to a common temperature, resulting in significantly greater fast-start performance at certain body lengths (Fig. 9.1). Once the unpaired fins are fully developed the difference between temperature groups disappeared (Johnston *et al.*, 2001).

Heterochronies in morphological characters supporting fast-start performance may be of considerable ecological importance during the early larval stages when mortality is high due to starvation and predation pressure. There are several reports indicating that development temperature influences the maximum escape velocity of amphibian tadpoles (Watkins, 2000; Arendt and Hoang, 2005; Niehaus *et al.*, 2006). However, the maximum speed and tail-beat frequency of yolk-sac herring larvae performing C-starts in response to tactile stimulation were found to be dependent on test temperature (e.g., tail beat frequency increased from 19 Hz at 5°C to 37 Hz at 17°C), but were independent of development temperature over the natural range of 5 to 12°C (Batty *et al.*, 1993). In contrast, for feeding strikes (S-starts) on slow moving prey such as artemia, herring larvae reared at low temperature were generally using higher maximum velocities for a given attack distance and had more curved trunks at the start of the attack than those reared at slightly higher temperatures (Morley and Batty, 1996). However, there was no significant relationship between the maximum speed of the S-strike and test temperature, suggesting the larvae were able to alter the characteristics of the strike and compensate for temperature changes, perhaps thereby minimizing energy use whilst maintaining an adequate capture success (Morley and Batty, 1996). The results of this interesting study support the

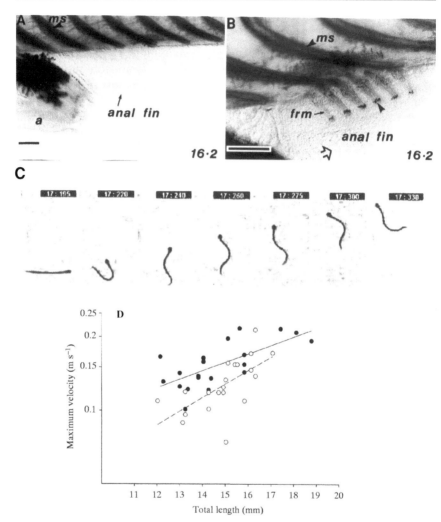

Fig. 9.1 Influence of rearing temperature on dorsal and anal fin ray muscle development and fast-start performance in Atlantic herring (*Clupea harengus*). Embryos were incubated at 5°C (open circles) or 12°C (filled circles) until first feeding and then reared at ambient temperature. (A, B) Whole-mount larvae of 16.2 mm total length stained for acetylcholinesterase activity following the 5°C (A) and 12°C (B) rearing regimes. Abbreviations: *a*, anus; *frm*, fin ray muscles; *ms*, myosepta. (C) Escape response of Clyde herring larva (*Clupea harengus*) 18 mm total length, filmed at 200 frames s⁻¹. (D) Maximum velocity attained during escape responses in Clyde herring larvae reared at 5°C (open circles, dashed line) and 12°C (filled circles, solid line) until first feeding, when they were transferred to ambient seawater temperatures. Each point represents an escape response from one larva. First-order linear regressions were fitted to the data. 5°C, replace with: $U_{max}=-3.12TL^{1.92}$, $r^2=0.42$, $P=0.002$; 12°C, $U_{max}=2.20TL^{1.21}$, $r^2=0.48$, $P=0.001$), where U_{max} is maximum swimming velocity (m s⁻¹) and TL is total length (mm). From Johnston *et al.* (2001) with permission from Marine Ecology Progress Series.

hypothesis that herring reared at cooler temperatures are less capable to compensate strike characteristics and, therefore, have a reduced behavioural flexibility than larvae reared at warmer temperature. Thus, it is likely that many types of unsteady swimming behaviour are influenced by thermal history during early development. This is an important area for future research.

Temperature can also have important effects on other behaviours of importance for predator-prey outcomes. Among these is schooling, which is used by many species as an anti-predator behaviour (Pitcher and Parrish, 1993). In the sand smelt (*Atherina mochon*), transient heat shocks at the embryo and larval stage produce a loss of normal schooling behaviour causing schools to take longer to form, persist for a shorter time and with fish spaced further apart (Williams and Coutant, 2003). Temperature also affects the shoaling decisions of guppies, *Poecilia reticulata*, under predation threat from a confined cichlid predator (Weetman *et al.*, 1999). Guppies under threat joined the larger of two schools in warm water (26°C), but showed no preference for larger school size in colder water (22°C). Other studies have shown that temperature may affect the reactivity of prey to predator attack. Webb and Zhang (1994) investigated the effect of heat shock on the escape response of goldfish (*Carassius auratus*) attacked by rainbow trout (*Oncorhyncus mykiss*). Goldfish acclimated at 15°C were exposed to a temperature of 34–39°C for 2 minutes before the experiment. This treatment had no effect on escape speed but decreased the reaction distance of the prey to the attacking trout (Webb and Zhang, 1994), thereby increasing vulnerability to predators.

Temperature is also known to have an effect on escape latencies, which is the interval between stimulus onset and the first detectable movement leading to the escape of the animal. Latency generally increases at lower temperatures (Webb, 1978; Preuss and Faber, 2003) mainly due to the effect of temperature on neural conduction speed. A late response may increase prey vulnerability, especially when facing endothermic predators whose reaction speeds are largely independent of temperature. Preuss and Faber (2003) examined the effect of acute temperature change on escape responses triggered by a clicking sound. They found that cooling had contrasting effects on escape performance, with negative effects on locomotor performance, latency and directionality (i.e., by increasing the proportion of response towards the stimulus instead of away from it), but an increased responsiveness (i.e., the proportion of fish responding to stimulation over the total). Preuss and Faber (2003) attributed the high responsiveness and the low directionality caused by cooling to changes in the intrinsic properties of the Mauthner cell. At the ecological level, increased responsiveness at colder temperatures might in part compensate for the reduced or more sluggish execution of the escape response (Preuss and Faber, 2003). Overall, these studies emphasize the importance of several aspects of behaviour in addition to swimming performance for the outcome of predator-prey interactions (see also Yocom and Edsall, 1974; Webb and Zhang, 1994).

Oxygen

Many aquatic environments exhibit extreme temporal and/or spatial variation in the partial pressure of dissolved oxygen. Oxygen level in aquatic environments is critical to the survival of all fishes and can influence their physiology, behaviour, distribution and predator-prey relationships (Fig. 9.2). Since muscle contraction in fast-starts is fuelled anaerobically (Domenici and Blake, 1997), we may expect no effect of hypoxia on the swimming performance during strikes or escapes. However, recent studies showed that cumulative distance and maximum swimming speed during escape responses were reduced in golden grey mullet (*Liza aurata*) at 10% of air saturation. This alteration in performance was associated with a change in kinematics, particularly the number of body bends per fast start (single bend and double bend responses, Lefrancois *et al.*, 2005). Lefrançois *et al.* (2005) suggested that the hypoxia-related increase in occurrence of low performance responses (i.e., single bend) may be due to changes in the balance between physiological exhaustion and the need to escape from a predator attack. Interestingly, such effects were mitigated if the mullet was permitted to engage in aquatic surface respiration (Kramer *et al.*, 1983; ASR, Kramer, 1987). In contrast, in European sea bass (*Dicentrarchus labrax*) cumulative distance, maximum swimming speed and acceleration were not affected by hypoxia (Lefrançois and Domenici, 2006).

In addition to locomotor performance, escape success depends on the sensory performance of the prey (Domenici *et al.*, 2007a). Responsiveness was reduced in hypoxia, in both *L. aurata* and *D. labrax*, (Lefrançois *et al.*, 2005; Lefrançois and Domenici, 2006; Fig. 9.3), suggesting a reduction of acoustic/visual sensitivity and/or motivation to escape (Webb, 1984). In both species, responsiveness was affected at 10% of air saturation, but to different extents (37% of responders in seabass and 69% in grey mullets). The hypoxia-induced alteration of responsiveness may have important ecological consequences for predation success, since unless predators make an error, the absence of an escape attempt leads unavoidably to prey capture. In contrast, escape latency was independent of oxygen in both species, suggesting that the time course to initiate the response was not affected by hypoxia.

The partial pressure of oxygen can also influence the directionality of the escape response (Lefrançois *et al.*, 2005; Lefrançois and Domenici, 2006). Directionality is determined at the onset of the escape response and represents the total proportion of "away responses" (i.e., responses with the C-bend oriented away from the stimulus). In *L. aurata*, directionality was impaired at

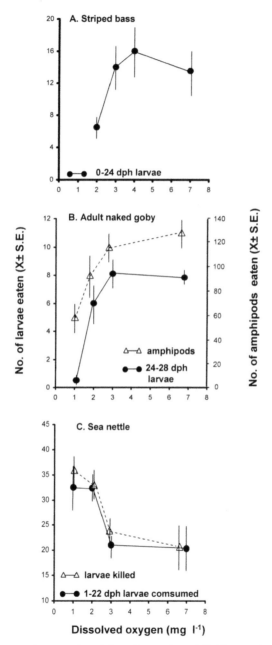

Fig. 9.2 Influence of oxygen (mg. l⁻¹) on the number of *Gobiosoma bosc* larvae eaten by three different types of predator. A–*Morone saxatilis*, B–*Gobiosoma bosc* adult, C–*Chrysaora quinquecirrha*. (dph means days posthatch). From Breitburg *et al.* (1994) with permission from Marine Ecology Progress Series.

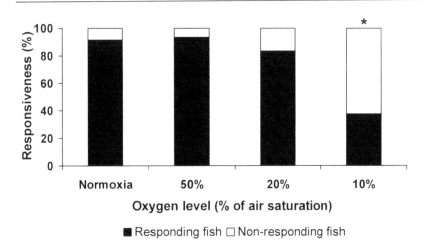

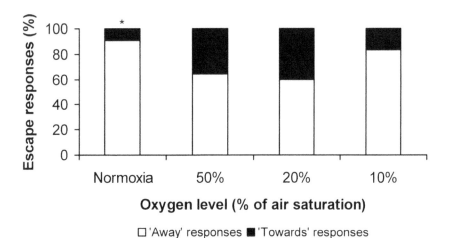

Fig. 9.3 Effect of oxygen on sea bass anti-predator behaviour. **A**–Sea bass responsiveness, i.e. the proportion (%) of individuals that responded with a sudden acceleration after being startled by a mechano-acoustic stimulus. Filled rectangle: responding fish, Open rectangle: non-responding fish * indicates significant difference from normoxia (p<0.05). **B**–Sea bass directionality, i.e. the proportion (%) of individuals that showed a the first detectable movement of the head directed away or towards the mechano-acoustic stimulus. Filled rectangle: 'towards' responses, Open rectangle: 'away' responses. *Indicates non-random responses. From Lefrançois and Domenici (2006) With kind permission of Springer Science+Business Media. © Springer-Verlag 2006.

relatively low hypoxia, i.e., at ≤ 20% of air saturation (Lefrançois *et al.*, 2005), while changes were observed in *D. labrax* at ≤ 50% of air saturation (Lefrançois and Domenici, 2006; Fig. 9.3). At these oxygen levels, fish showed a random directionality, i.e., the proportion of 'away: towards' escape response was not different from '50: 50', suggesting a significant impairment of the left-right discrimination in the individuals tested. Even if the final escape trajectory in sea bass and golden grey mullet tends to be away from the stimulus, an early mistake may have important consequences for predator escape. Making a tactical error by bending towards the predator at the initiation of the fast start may induce a significant delay in the effort of the prey to escape away from the predator. Since the first milliseconds of the escape response may be crucial for surviving a predator attack, a high proportion of 'towards' response may reduce the probability of escape success.

Oxygen content can also affect schooling behaviour. This effect has been studied in a number of gregarious species, such as grey mullet (*Mugil cephalus*), anchovies (*Engraulis mordax*), and herring (*Clupea harengus*) (McFarland and Moss, 1967; Moss and McFarland, 1970; Domenici *et al.*, 2000, 2002), for which schooling can be an effective anti-predator behaviour. Field work on wild fish showed that oxygen level decreases from the front to the back of large schools (McFarland and Moss, 1967; Domenici *et al.*, 2007a). As a consequence, when encountering moderately hypoxic waters in the field, individuals at the back of the school experience limiting conditions and tend to disperse (McFarland and Moss, 1967). Laboratory observations showed that moderate hypoxia can induce an increase in the volume occupied by a school of herring, and can change the shuffling behaviour of individuals, while severe hypoxia may lead to the disruption of the school (Domenici *et al.*, 2000, 2002). Thus, as suggested by Domenici *et al.* (2007a), the effects of hypoxia on school structure and dynamics may affect the synchronization and execution of antipredator manoeuvres.

When the predators are insensitive to hypoxia (e.g., birds, marine mammals), variation in the ability of the prey to escape is the main component that regulates the effects of hypoxia on predator-prey interactions. Situations in which both the predator and prey are hypoxia-sensitive species also need to be considered. In these cases, oxygen appears to affect predator-prey interactions mainly through its effect on hunger level, by reducing the occurrence of strikes [e.g., Breitburg *et al.* (1994, 1999), Sandberg and Bonsdorff (1996), Chabot and Dutil (1999), Robb and Abrahams (2002), Shimps *et al.* (2005), Shoji *et al.* (2005)]. For instance, in cod *Gadus morhua*, the ingestion rate decreased progressively with mild hypoxia and was reduced by half in individuals exposed to ~ 50% of air saturation (Chabot and Dutil, 1999). Similar hypoxia-related reductions in appetite have been observed in other piscivorous species, such as sea bass, *Dicentrarchus labrax* (~ 40% of air saturation, Thetmeyer *et al.*, 1999), and turbot, *Scophthalmus maximus* (60% of air saturation, Pichavant *et al.*, 2000). Furthermore, Breitburg *et al.* (1994) showed that the attack rates of two species of fish predators, the juvenile striped bass (*Morone saxatilis*) and the adult naked goby (*Gobiosoma bosc*), decreased with oxygen level (~ 28% and ~ 14% of air saturation, respectively,

Fig. 9.2). A similar oxygen-related change in predation rate was observed in the Spanish mackerel while feeding on red seabream larvae (~ 13% of air saturation, *Pagrus major*, Shoji *et al.*, 2005).

Salinity

In anadromous fish, such as Salmonids, juveniles experience huge changes in salinity during their seaward migration. The associated changes in salinity lead to an osmotic stress that juveniles mitigate via osmoregulation processes contributing to the re-establishment of homeostasis. A further stress that juveniles have to deal with is the high predation pressure characterizing the habitats they have to pass through (e.g., estuaries). By exposing freshwater acclimated *Salmo salar* smolts to seawater, Handeland *et al.* (1996) experimentally induced an osmotic stress and evaluated the consequences with respect to predation. They observed a reduction in the reactive distance in osmotically stressed smolts when attacked by an approaching predator and suggested that this less effective anti-predator behaviour may be related to the high energy costs of the escape, as well as to the accumulation of metabolic end-products such as lactate. Therefore, stressed fish may show a higher threshold (i.e., shorter reactive distance) for escaping, since too frequent reactions to predators may result in increased lactate accumulation in muscle, with negative consequences for later attacks and the osmoregulatory capacity (Handeland *et al.*, 1996).

Salinity is also known to affect schooling behaviour. Handeland *et al.* (1996) found that the degree of schooling (i.e., the proportion of fish forming a school) in freshwater acclimated salmon smolts decreased when exposed to seawater for 12–24 hours. Schooling behaviour was partially re-established when fish were allowed to acclimate to seawater for >48 hours.

Turbidity

Turbidity may occur because of suspended and dissolved particulates, as well as plankton blooms. These can be of natural or anthropogenic origin. Turbidity is expected to influence prey perception and therefore reactivity. Meager *et al.* (2006) investigated the escape responses of juvenile Atlantic cod (*Gadus morhua*) attacked by a model predator under various conditions of turbidity. Juvenile Atlantic cod showed a shorter reaction distance to an approaching model predator and a lower responsiveness when tested in turbid than clear water. Turbidity also influenced escape locomotion. Cod exhibited lower turning rates (characteristic of slow escape responses) in turbid than in clear water when confronted by an approaching model predator (Meager *et al.*, 2006; Fig. 9.4). Low locomotor performance at high turbidity

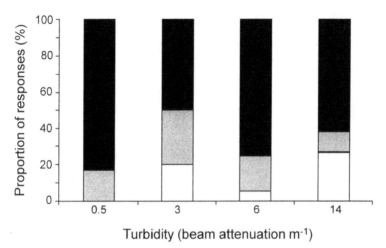

Fig. 9.4 Effect of turbidity on kinematics of the escape response in juvenile cod startled by an approaching predator model. White bar: responses with slow turning rate, grey bar: responses with intermediate turning rate, black bar: responses with fast turning rate. From Meager *et al.* (2006) Reproduced with permission of the Journal of Exprimental Biology.

levels seems counterintuitive since the predator model was closer to the fish at the time of the response. Meager *et al.* (2006) suggest that this result may be due to lower amount of visual information from the predator model (i.e., reduced contrast in turbidity) over a shorter duration of time. Hence, fish may show low performance because they have not received sufficient 'threat' information. It is therefore likely that in turbid water prey 'opt' for a less energy-costly escape response when risk was not correctly evaluated. As underlined by Meager *et al.* (2006), different neural circuits or muscle activation patterns may be involved in triggering slow compared to fast escape responses.

The ecological consequences of turbidity-related effects on fish escape performance have been considered by establishing putative escape success (defined as the likelihood of a prey to escape the attack of the predator model; Meager *et al.*, 2006). The probability of prey survival was mainly related to its responsiveness and reactive distance. As explained above, the outcome of the prey-predator interactions also depends on the predator tolerance to the environmental factor of concern. Turbidity may act on the searching effort of the predator (e.g., swimming speed, volume explored) and also influence its behaviour because of reduced visibility (Utne-Palm, 2002).

Light cycle

Visual processes are among the biological mechanisms affected by circadian rhythm. Li and Downling (1998) investigated in zebrafish (*Danio rerio*) how induced changes in visual sensitivity may influence escape response to a threatening object. They studied the threshold of light intensity required to evoke the escape response and found that it depended on, (i) the light-dark cycle the fish is acclimated to, and (ii) the time of the day. The threshold of light intensity was lower in late afternoon than in early morning hours in individuals acclimated to a natural light-dark cycle, i.e., zebrafish were more sensitive to visual stimuli in late afternoon. Li and Downling (1998) suggested that visual sensitivity in zebrafish was regulated by an endogenous circadian clock, thereby modulating the threshold of light intensity.

Pollutants

Work on the effects of pollutants on unsteady swimming has been limited to ammonia and some heavy metals. More work on other pollutants would be relevant since a number of chemicals has been found to affect fish swimming performance (see Mackenzie and Claireaux, Chapter 10)

Ammonia: Ammonia is perhaps the most pervasive pollutant of aquatic habitats worldwide (Randall and Tsui, 2002). It is toxic to all vertebrates and impairs nervous and muscle function in fish (Raabe and Lin, 1985; Ip *et al.*, 2001; Randall and Tsui, 2002; McKenzie *et al.*, 2003). McKenzie *et al.* (2008) demonstrated that sublethal levels of ammonia impair performance of the escape response in grey mullet (*Liza aurata*), with negative effects on both reactivity and locomotion. In particular, latency increased when exposed to high levels of ammonia, although responsiveness and directionality were not affected. Therefore, the effects of ammonia on reactivity were different from those observed in hypoxia (Lefrancois *et al.*, 2005; see also above), suggesting a different mechanism of interfering with the nervous system. Similar to the effects of hypoxia on grey mullet (Lefrancois *et al.*, 2005), ammonia had a negative effect on performance by decreasing the proportion of the double bend responses. As a consequence, all locomotor variables studied (turning rate, distance covered, maximum speed and acceleration) were significantly affected by exposure to ammonia (McKenzie *et al.*, 2008).

Heavy metals: Pollutants, such as metallic compounds (e.g., cadmium and mercury), originate through direct discharges from industry, as well as atmospheric deposition due to fuel combustion or incineration of industrial metal-containing wastes. Metallic compounds affect responsiveness, latency and speed in startled sea bass, *Dicentrarchus labrax* (Faucher *et al.*, 2006) and zebrafish, *Danio rerio* (Weber, 2006). Faucher *et al.* (2006) suggested that the low responsiveness observed in cadmium-treated sea bass was due to damages of the lateral line neuromasts involved in detection of mechano-acoustic stimuli. In mercury-exposed zebrafish, dysfunction of the lateral line, as well as of the several sensory neurons triggering the Mauthner cells (M-cells), were also suggested to be largely responsible for the low responsiveness (Weber, 2006). In zebrafish, a possible delayed M-cell information processing could explain the long latency observed, while the dose-dependent decline in speed was attributed to the alterations of the neuromuscular junction and associated trunk muscles (Weber, 2006).

THE EFFECT OF ENVIRONMENTAL FACTORS ON MATING AND COURTSHIP BEHAVIOUR

Temperature

Few studies have considered the implications of variation in locomotory performance traits on reproductive performance or mating behaviour (Garland and Carter, 1994; Irschick and Garland, 2001). The locomotor performance of ectotherms is highly sensitive to changes in temperature (see above), and many of the physiological mechanisms underlying these temperature effects have been well documented (Rome, 1983; Johnston *et al.*, 1985; Fleming *et al.*, 1990; Johnston and Temple, 2002). However, most studies have simply assumed that locomotor performance is an adequate proxy for reproductive success, rather than actually testing the consequences of temperature-mediated variation in performance. By examining the effects of acute temperature changes on the ability of males to either attract mates or directly achieve matings, the fitness consequences of temperature variation can be examined.

The mating behaviour of the eastern mosquito fish (*Gambusia holbrooki*) has been utilised extensively to study the consequences of variation in body temperature for reproduction (Wilson, 2005). The eastern mosquitofish is a live-bearing poeciliid fish native to the south-eastern United States, but introduced throughout many of the world's tropical and temperate freshwater waterways to control mosquito populations (Arthington and Lloyd, 1989; Cabral and Marks, 1999). Wilson (2005) found that mosquito fish possess one of the broadest reproductively-active temperature ranges for any ectotherm and obtain copulations across a range of at least 14° to 38°C (Fig. 9.5). The mating system of the eastern mosquito fish is dominated by male sexual coercion (Bisazza *et al.*, 1989; McPeek, 1992). Copulations only occur when males approach females with stealth and rapidly thrust their gonopodium (modified anal fin) into the female genital pore to release their spermatozoa (Farr, 1989; McPeek, 1992; Bisazza *et al.*, 1996; Pilastro *et al.*, 1997). Females are assumed never to cooperate, but to avoid mating by strategies including fleeing, lying against an object, adopting a position that thwarts the attempt, or attacking (Bisazza *et al.*, 1996; Pilastro *et al.*, 1997;

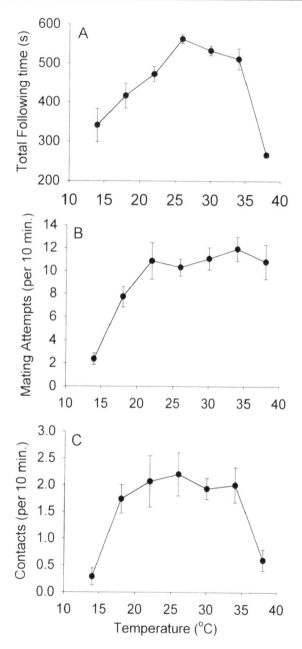

Fig. 9.5 Influence of temperature on **(A)** the total time spent following females, and **(B)** the total number of mating attempts, and **(C)** the number of copulations, by male eastern mosquito fish (*Gambusia holbrooki*) during a 10 min observation period. Data represent means ± S.E. Adapted from Wilson (2005) with permission from Elsevier.

Pilastro *et al.*, 2003). The males best able to compete against female avoidance achieve the greatest number of copulations and increase their chances of producing more offspring (Cunningham and Birkhead, 1998). Although the ability of males to obtain copulations over an extended period relies heavily on sustained aerobic performance, each copulation event itself requires speed and maneuverability. Thus, burst locomotor performance and maneouvrability are some of the primary determinants of coercive mating performance in male mosquitofish (Wilson, 2005). In one of the first studies to examine the effects of temperature on reproductive behaviour, Wilson (2005) reported that temperature markedly affected both the propensity of male *G. holbrooki* to follow females and their ability to obtain coercive matings (Fig. 9.5).

Hammill *et al.* (2004) reported that male *G. holbrooki* possessed the capacity to acclimate their maximum sustainable swimming performance over the range from 18 to 30°C. Wilson *et al.* (2007a) tested whether male mosquitofish could also respond to long-term changes in temperatures by improving their ability to obtain coercive copulations. They exposed the fish to either 18 or 30°C for six weeks and, then, tested their ability to obtain copulations in either the presence or absence of male-male competition (Fig. 9.6). They found copulation success was greater for acclimated than non-acclimated males at both temperatures when individual males were tested without competing males. Performance under these circumstances is likely to be mediated directly by burst swimming performance and maneuverability. In contrast, when males from the different acclimation treatments competed against each other for copulations with a single female, the 30°C-acclimated males were more aggressive and obtained a greater number of copulations at both test temperatures. Therefore, behavioural interactions among competing males effectively removed the benefits for mating success that were associated with cold acclimation.

As it is assumed that female *G. holbrooki* indiscriminately resist matings from all males, temperature should also influence the ability of females to avoid unwanted copulations. Like males, the ability of females to resist copulations will rely on both speed and agility for each individual attempt, but also sustained aerobic metabolism across extended periods. This could theoretically lead to an evolutionary arms race in the thermal acclimation of coercive mating performance of males versus the thermal acclimation of avoidance of matings by females. The influence of temperature on the ability of female mosquitofish to avoid unwanted copulations was examined for both virgin and non-virgin females (Condon and Wilson, 2006; Wilson *et al.*, 2007b). Initially, Condon and Wilson (2006) tested this prediction by examining the mating behaviour of virgin female *G. holbrooki* after acclimation to 16°C and 32°C for 5 weeks. In contrast with their prediction, females acclimated to warm temperatures experienced higher copulation rates across

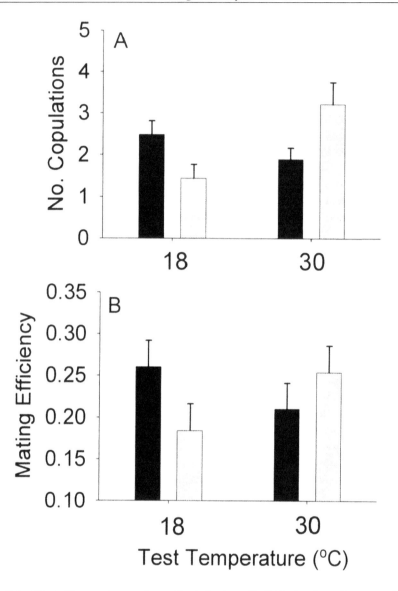

Fig. 9.6   Effect of temperature on the mating behaviour of individual male eastern mosquito fish (*Gambusia holbrooki*) following acclimation to 18º or 30ºC for 6 weeks. The total number of copulations obtained (A), and the number of copulations as a proportion of the total number of mating attempts (successful + unsuccessful), were recorded for each male at 18º and 30ºC during a 6 min. observation period with 2 females. Dark bars represent cool-acclimated males and grey-bars represent warm-acclimated males. Data represent means ± S.E. Adapted from Wilson *et al.* (2007a).

all test temperatures than females acclimated to cool temperatures. Although females did not appear to accept any copulations and every mating appeared to be coercive, it seems female *G. holbrooki* are able to play a greater role in determining the outcome of sexual coercive mating interactions than previously thought. When tested with non-virgin females, warm-acclimated females again experienced a higher copulation rate than cool-acclimated females. Thus, females of *G. holbrooki* appear to modify their reproductive behaviour following warm-acclimation and allow a greater rate of coercive copulations. It is apparent that in this context burst swimming performance was not the dominant factor determining copulation rate.

Oxygen

Despite the diversity of studies demonstrating the physiological mechanisms underlying improved uptake of oxygen under hypoxic conditions, few studies have addressed the reproductive consequences of short- or long-term changes in the partial pressure of oxygen in the aquatic environment (Kramer and Mehegan, 1981; Hale *et al.*, 2003). Carter and Wilson (2006) tested the ability of male eastern mosquito fish to modify their reproductive and locomotor performance to compensate for long-term changes in the partial pressure of oxygen. They exposed male mosquito fish to normoxic or hypoxic conditions for four weeks and tested their maximum sustained swimming performance, as well as their ability to obtain coercive matings under both normoxic and hypoxic conditions. They found the sustained swimming performance of male mosquito fish was greater in a hypoxic environment following long-term exposure to low partial pressures of oxygen. In a non-competitive environment, male mosquito fish acclimated to hypoxic conditions spent a greater amount of time following females and obtained more copulations than normoxic-acclimated males when tested in low partial pressures of oxygen (Fig. 9.7). When males competed against each other for copulations, there was no effect of long-term exposure to different partial pressures of oxygen on mating behaviour.

Turbidity

Due to the importance of visual signals for reproduction in many fish, water turbidity has the potential to markedly affect both the mating behaviour and reproductive success of a wide range of species. This may be particularly important for species that use extensive and complex courtship rituals and/or females that select males on the basis of their colour or contrast in the environment (Seehausen *et al.*, 1997; Wong *et al.*, 2007). Attention has been recently directed towards this possibility due to the widespread human-induced eutrophication of many environments. Extensive algal blooms can lead to

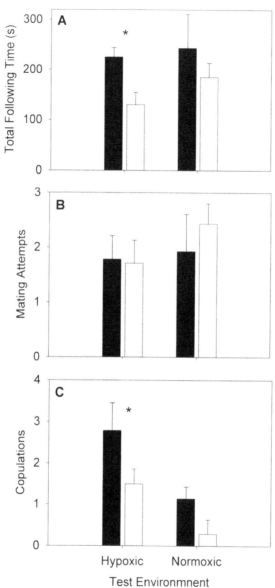

Fig. 9.7 Effect of partial pressure of oxygen on the mating behaviour (non-competitive environment) of male eastern mosquito fish (*Gambusia holbrooki*) acclimated to either hypoxic or normoxic conditions for at least 4 weeks. (A)The total time males spent following females, (B) total number of attempted copulations, and (C) the total number of successful copulations recorded during a 10 minute observation period in a non-competitive environment. Data represent means ± S.E. * denotes significant results at level of P < 0.05. Dark bars indicate hypoxic-treatment fish and grey bars are normoxic- treatment fish. Adapted from Carter and Wilson (2006) with permission of the Journal of Exprimental Biology.

increases in turbidity in many freshwater and marine habitats. Recent studies show that these increases in water turbidity can directly influence the courtship activity of male fish (Engstrom and Candolin, 2007), responsiveness of females to male displays (Engstrom and Candolin, 2007; Candolin *et al.*, 2008), extent of male-male aggression (Candolin *et al.*, 2008) and intensity of sexual selection (Seehausen *et al.*, 1997; Jarvenpaa and Lindstrom, 2004). These changes in turbidity can also have complex and unpredictable effects on population dynamics via their role in reproductive behaviour. For example, phytoplankton algal blooms in some stickleback habitats result in female choice for higher rates of courtship (Engstrom and Candolin, 2007), but a decrease in male aggression due to difficulties in detecting other males. This leads to overall increases in reproductive output in turbid waters. In addition, studies of the mating behaviour of the sand goby (*Pomatoschistus minutus*) found that under turbid conditions sexual selection was relaxed because mating success was less skewed towards larger males. For cichlids in lakes, dull coloration, fewer colour morphs and lower species diversity are found in areas that have increased levels of turbidity due to eutrophication (Seehausen *et al.*, 1997). Clearly, the role of turbidity on the intensity of sexual selection and mating behaviour in fish is likely to become more important with increased anthropogentic effects on aquatic environments.

Pollutants

Courtship behavior can be affected by pollution, as reviewed by Jones and Reynolds (1997). Effects on courtship include changes in the frequency of display, increased courtship duration, or performance of male-like behavior in masculinized females (Jones and Reynolds, 1997). These effects were caused by a number of pollutants such as acidification, herbicides, insecticides and pesticides across a variety of fish families (Cichlidae, Poecilidae, Gasterosteidae, Cyprinidae). Specific effects of pollution were also shown by more recent laboratory and field studies. Laboratory work by Nakayama *et al.* (2004) shows that the number of male medaka that performed "following " and "dancing" courtship behavior was decreased after treatment with TBT (tributyltin, a pollutant deriving from paper mill plants and marine paints), although PCBs (polychlorinated biphenyls) had no effect. Field work also suggests that the presence of chemicals with antiandrogenic effects when comparing different lakes was the main determinant of body testosterone concentration and the intensity of male courtship behavior (Toft *et al.*, 2003). Specific work that links the effect of pollutants on courtship in relation with swimming performance is lacking. However, since pollutants were found to affect swimming performance per se (section on the effect of environmental factors on strikes and escape, and Chapter 10 by McKenzie and Claireaux), it is possible that at least part of the detrimental effects of pollutants on courtship behavior may be due to an effect on swimming performance.

FUTURE DIRECTIONS

There is clearly a need for increasing the integration of various disciplines of biology to improve our understanding of the effect of environmental factors on predator-prey and mating encounters. This integration should include various fields such as behavioural ecology, sensory biology, evolution, genetics, and environmental physiology. Similarly, while work carried out in the laboratory allows precise control of environmental variables, this work needs to be supplemented by field studies in order to place results into a clear ecological context. To this aim, new techniques for studying behaviour and environmental physiology are now available for use in the field, such as field miniature cameras, data loggers and telemetry (Domenici *et al.*, 2007b). An integrated approach is necessary to predict impacts of both natural and anthropogenic global change on organisms. In order to increase our ability to predict the consequences of environmental change it will be necessary to understand how environmental responses are modulated according to ontogenetic state. Future studies should take into account population variation in physiological (swimming) traits, especially if these reflect underlying changes in gene frequencies across the geographic range of species (e.g., Schulte *et al.*, 2000).

REFERENCES

Alvarez, D. and N.B. Metcalfe. 2007. The trade-off between catch-up growth and escape speed: variation between habitats in the cost of compensation. *Oikos* 116: 1144–1151.

Arendt, J. and L. Hoang. 2005. Effect of food level and rearing temperature on burst speed and muscle composition of Western spadefoot toad (*Spea hammondii*). *Functional Ecology* 19: 982–987.

Arthington, A.H. and L.N. Lloyd. 1989. Introduced poeciliids in Australia and New Zealand. In: *Ecology and Evolution of Livebearing Fishes (Poeciliidae)*, By G.K. Meffe and F.F. Snelson Jnr. (Eds.). Prentice Hall, New Jersey, pp. 333–348.

Batty, R.S. 1984. Development of swimming movements and musculature of larval herring (*Clupea harengus*). *Journal of Experimental Biology* 110: 217–229.

Batty, R.S., J.H.S. Blaxter and K. Fretwell. 1993. Effect of temperature on the escape responses of larval herring, *Clupea harengus*. *Marine Biology* 115: 523–528.

Beddow, T. A., J.L. Van Leuween and I.A. Johnston. 1995. Swimming kinematics of fast starts are altered by temperature-acclimation in the marine fish *Myoxocephalus scorpius*. *Journal of Experimental Biology* 198: 203–208.

Bennett, A.F. 1990. Thermal dependence of locomotor capacity. *American Journal of Physiology* 259: R253–R258.

Bisazza, A., A. Isazza, A. Marconato and G. Marin. 1989. Male mate preferences in the mosquitofish *Gambusia holbrooki*. *Ethology*. 83: 335–343.

Bisazza, A., A. Pilastro, R. Palazzi and G. Marin. 1996. Sexual behaviour of immature male eastern mosquitofish: A way to measure intensity of intra-sexual selection? *Journal of Fish Biology* 48: 726–737.

Breitburg, D.L., N. Steinberg, S. DuBeau, C. Cooksey and E.D. Houde. 1994. Effects of low dissolved oxygen on predation on estuarine fish larvae. *Marine Ecology Progress Series* 104: 235–246.

Breitburg, D.L., K.A. Rose and J.H. Cowan. 1999. Linking water quality to larval survival: predation mortality of fish larvae in an oxygen-stratified water column. *Marine Ecology Progress Series* 178: 39–54.

Cabral, J.A. and J.C. Marks. 1999. Life history, population dynamics and production of eastern mosquitofish, *Gambusia holbrooki* (Pisces, Poecilidae), in rice fields of the lower Mondego River Valley, western Portugal. *Acta Oecologia* 20: 607–620.

Candolin, U., J. Engstroem-Oest and T. Salesto. 2008. Human-induced eutrophication enhances reproductive success through effects on parenting ability in sticklebacks. *Oikos* 117: 459–465.

Carter, A.J. and R.S. Wilson. 2006. Improving sneaky-sex in a low oxygen environment: reproductive and physiological responses of male mosquito fish to chronic hypoxia. *Journal of Experimental Biology* 209: 4878–4884.

Chabot, D. and J.D. Dutil. 1999. Reduced growth of Atlantic cod in non-lethal hypoxic conditions. *Journal of Fish Biology* 55: 472–491.

Cole, N.J. and I.A. Johnston. 2001. Plasticity of myosin heavy chain expression is gradually aquired during ontogeny in the common carp (*Cyprinus carpio* L.). *Journal of Comparative Physiology* 171: 321–326.

Condon, C.H.L. and R.S. Wilson. 2006. Effect of thermal acclimation on female resistance to forced matings in the eastern mosquitofish. *Animal Behaviour* 72: 585–593.

Cunningham, E.J.A. and T.R. Birkhead. 1998. Sex roles and sexual selection. *Animal Behaviour* 56: 1311–1321.

Domenici, P. Escape responses in Fish: Kinematic, performance and behaviour. In: *Fish Locomotion: An eco-ethological perspective*, P. Domenici and B.G. Kapoor (Eds.). Science Publishers, Enfield(NH), USA (In Press).

Domenici, P. and R. W. Blake. 1997. The kinematics and performance of fish fast start swimming. *Journal of Experimental Biology* 200: 1165–1178.

Domenici, P., J.F. Steffensen and R.S. Batty. 2000. The effect of progressive hypoxia on swimming activity and schooling in Atlantic herring. *Journal of Fish Biology* 57: 1526–1538.

Domenici, P., R.S. Ferrari, J.F. Steffensen and R.S. Batty. 2002. The effects of progressive hypoxia on school structure and dynamics in Atlantic herring *Clupea harengus*. *Proceedings of the Royal Society of London* B269: 2103–2111.

Domenici, P., C. Lefrançois and A. Shingles. 2007a. The effect of hypoxia on the antipredator behaviours of fishes. *Philosophical Transaction of the Royal Society* 362: 2105– 2121.

Domenici, P., G. Claireaux and D.J. McKenzie. 2007b. Environmental constraints upon locomotion and predator-prey interactions in aquatic organisms: an introduction. *Philosophical Transaction of the Royal Society*. 362: 1929–1936.

Engstrom, O.J. and U. Candolin. 2007. Human-induced water turbidity alters selection on sexual displays in sticklebacks. *Behavioural Ecology* 18: 393–398.

Farr, J.A. 1989. Sexual selection and secondary differentiation in poeciliids: determinants of male mating success and the evolution of female mate choice. In: *Ecology and Evolution of Livebearing Fishes (Poeciliidae)*, By G.K. Meffe and F.F. Snelson Jnr. (Eds.).Prentice Hall, New Jersey, pp. 91–124.

Faucher, K., D. Fichet, P. Miramand and J.P. Lagardère. 2006. Impact of acute cadmium exposure on the trunk lateral line neuromasts and consequences on the "C-start" response behaviour of the sea bass (*Dicentrarchus labrax* L.; Teleostei, Moronidae). *Aquatic Toxicology* 76: 278–294.

Fernandez, D.A., J. Calvo, J.M. Wakeling, F.A. Vanilla and I.A. Johnston. 2002. Escape performance in the sub-Antarctic notothenioid fish *Eleginops maclovinus*. *Polar Biology* 25: 914–920.

Fleming, J.R., T. Crockford, J.D. Altringham and I.A. Johnston. 1990. The effects of temperature acclimation on muscle relaxation in the carp: a mechanical, biochemical and ultrastructural study. *Journal of Experimental Biology* 255: 286–295.

Franklin, C.E. and I.A. Johnston. 1997. Muscle power output during escape responses in an Antarctic fish. *Journal of Experimental Biology* 200: 703–712.

Fuiman, L.A. and R.S. Batty. 1997. What a drag it is getting cold: portioning the physical and physiological effects of temperature on fish swimming. *Journal of Experimental Biology* 200: 1745–1755.

Garland, T. Jnr. and P.A. Carter. 1994. Evolutionary physiology. *Annual reviews of Ecology and Systematics* 56: 579–621.

Gracey, A.Y., E.J. Fraser, W.Z. Li, Y.X. Fang, R.R. Taylor, J. Rogers, A. Brass and A.R. Cossins. 2004. Coping with cold: an integrative, multitissue analysis of the transcriptome of a poikilothermic vertebrate. *Proceedings National Academy Sciences of the United States of America* 101: 16970–16975.

Guinet, C., P. Domenici, P. de Stephanis, R.L. Barrett-Lennard, J.K.B Ford and P. Verborgh. 2007. Killer whale predation on bluefin tuna: exploring the hypothesis of the endurance-exhaustion technique. *Marine Ecology Progress Series* 347: 111–119.

Hale, M. and A. Rice. Roles of Locomotion in feeding. In: *Fish locomotion: An Eco-Ethological Perspective*, P. Domenici and B.G. Kapoor (Eds.). Science Publishers, Enfield(NH),USA. (In Press).

Hale, R.E., C.M. St. Mary and K. Lindstrom. 2003. Parental responses to change in costs and benefits along an environmental gradient. *Environmental Biology of Fishes* 67: 107–116.

Hammill, E., R.S. Wilson and I.A. Johnston. 2004. Sustained swimming performance and muscle structure are altered by thermal acclimation in male mosquitofish. *Journal of Thermal Biology* 29: 251–257.

Handeland, S.O., T. Järvi, A. Fernö and S.O. Stefansson. 1996. Osmotic stress, antipredator behaviour, and mortality of Atlantic salmon (*Salmo salar*) smolts. *Canadian Journal of Fisheries and Aquatic Sciences* 53: 2673–2680.

Imai, J.-I., Y. Hirayama, K. Kikuchi, M. Kakinuma and S. Watabe. 1997. cDNA cloning of myosin chain heavy chains from carp skeletal muscle and their gene expression associated with temperature acclimation. *Journal of Experimental Biology* 200: 27–34.

Ip, Y.K., S.F. Chew and D.J. Randall. 2001. Ammonia toxicity, tolerance and excretion.In: *Fish Physiology*, P.A. Wright and P.M. Anderson, (Eds.)., *Ammonia*, Academic Press, San Diego, Vol. 20, pp. 109–148

Irschick, D.J. and Jr. T. Garland. 2001. Integrating function and ecology in studies of adaptation: Investigations of locomotor capacity as a model system. *Annual Reviews of Ecology and Systematics* 32: 367–396.

Jarvenpaa, M. and K. Lindstrom. 2004. Water turbidity by algal blooms causes mating system breakdown in a shallow-water fish, the sand goby *Pomatoschistus minutus*. *Proceedings of the Royal Society of London Series* B 271: 2361–2365.

James, R.S. and I.A. Johnston. 1998. Influence of spawning on swimming performance and muscle contractile properties in the short-horn sculpin. *Journal of Fish Biology* 53: 485–501.

Johnson, T.P. and A.F. Bennett. 1995. The thermal acclimation of burst escape performance in fish: an integrated study of molecular and cellular physiology and organismal performance. *Journal of Experimental Biology* 198: 2165–2175.

Johnston, I.A. and G.K. Temple. 2002. Thermal plasticity of skeletal muscle phenotype in ectothermic vertebrates and its significance for locomotory behaviour. *Journal of Experimental Biology* 205: 2305–2322.

Johnston, I.A. and R.S. Wilson. 2006. Temperature induced developmental plasticity. In: *Comparative Developmental Plasticity: Contributions, Tools and Trends*. S.J. Warburton, W.W. Burggren, B. Pelster, C.L. Reiber and J. Spicer (Eds.). Oxford University Press, Oxford, pp. 124–138.

Johnston, I.A., B.D. Sidell and W.R. Driedzic. 1985. Force-velocity characteristics and metabolism of carp muscle fibres following temperature acclimation. *Journal of Experimental Biology* 119: 239–249.

Johnston, I.A., J.L. van Leewen, M.L.F. Davies and T. Beddow. 1995. How fish power predation fast-starts. *Journal of Experimental Biology* 198: 1851–1861.

Johnston, I.A., V.L.A. Vieira and G.K. Temple. 2001. Functional consequences and population differences in the developmental plasticity of muscle to temperature in Atlantic herring *Clupea harengus*. *Marine Ecology Progress Series* 213: 2285–300.

Jones, J.C. and Reynolds J.D. 1997. Effects of pollution on reproductive behaviour of fishes. *Reviews in Fish Biology and Fisheries* 7: 463–491.

Kramer, D.L. 1987. Dissolved oxygen and fish behaviour. *Environmental Biology of Fishes* 18: 81–92.

Kramer, D.L. and J.P. Mehegan. 1981. Aquatic surface respiration, an adaptive response to hypoxia in the guppy, *Poecilia reticulata*. *Environmental Biology of Fishes* 6: 299–313.

Kramer, D.L., D. Manley and R. Bourgeois. 1983. The effect of respiratory mode and oxygen concentration on the risk of aerial predation in fishes. *Canadian Journal of Zoology*. 61: 653–665.

Lefrançois, C. and P. Domenici. 2006. Locomotor kinematics and responsiveness in the escape behaviour of European sea bass (*Dicentrarchus labrax*) exposed to hypoxia. *Marine Biology* 149: 969–977.

Lefrançois, C., A. Shingles and P. Domenici. 2005. The effect of hypoxia on locomotor performance and behaviour during escape in the golden grey mullet (*Liza aurata*). *Journal of Fish Biology* 67: 1711–1729.

Li L. and J.E. Downling. 1998. Zebrafish visual sensitivity is refulated by a circadian clock. *Visual Neuroscience* 15: 851–857.

McFarland W. N. and S.A. Moss. 1967. Internal behavior in fish schools. *Science* 156: 260–262.

McKenzie, D.J., A. Shingles and E.W. Taylor. 2003. Sub-lethal plasma ammonia accumulation and the swimming performance of salmonids. *Comparative Biochemistry and Physiology* A135: 515–526.

McKenzie, D.J., A. Shingles, G. Claireaux and P. Domenici. 2008. Sub-lethal concentrations of ammonia impair performance of the teleost fast-start escape response. *Physiological and Biochemical Zoology*. (In Press).

McPeek, M.A. 1992. Mechanisms of sexual selection operating on body size in the mosquito fish (*Gambusia holbrooki*). *Behavioural Ecology* 3: 1–12.

Martinez, M., M. Bedard, J.D. Dutil and H. Guderley. 2004. Does condition of Atlantic cod (*Gadus morhua*) have a greater impact upon swimming performance at U-crit or sprint speeds? *Journal of Experimental Biology* 207: 2979–2990.

Meager, J.J., P. Domenici, A. Shingles and A.C. Utne-Palm. 2006. Escape responses in juvenile Atlantic cod *Gadus morhua* L.: the effects of turbidity and predator speed. *Journal of Experimental Biology* 209: 4174–4184.

Morley, S. and R.S. Batty. 1996. The effects of temperature on "S-strike" feeding of larval herring, *Clupea harengus* L. *Marine Freshwater Behaviour Physiology* 28: 123–136.

Moss, S.A. and W.N. McFarland. 1970. The influence of dissolved oxygen and carbon dioxide on fish schooling behaviour. *Marine Biology* 5: 100–107.

Nakayama, K., Y. Oshima, T. Yamaguchi, Y. Tsuruda, I.J. Kang, M. Kobayashi, N. Imada and T. Honjo. 2004. Fertilization success and sexual behavior in male medaka, *Oryzias latipes*, exposed to tributyltin: *Chemosphere*. 555: 1331–1337.

Niehaus, A.C., R.S. Wilson and C.E. Franklin. 2006. Short- and long-term consequences of thermal variation in the larval environment of anurans. *Journal of Animal Ecology* 75: 686–692.

Pichavant, K., J. Person-Le-Ruyet, N. Le Bayon, A. Sévère, A. Le Roux, L. Quémerer, L, V. Maxime, G. Nonotte and G. Bœuf. 2000. Effects of hypoxia on growth and metabolism of juvenile turbot. *Aquaculture* 188: 103–104.

Pilastro, A., S. Benetton and A. Bisazza, A. 2003. Female aggregation and male competition reduce costs of sexual harassment in the mosquitofish *Gambusia holbrooki*. *Animal Behaviour* 65: 1161–1167.

Pilastro, A., E. Giacomello and A. Bisazza. 1997. Sexual selection for small size in male mosquitofish (*Gambusia holbrooki*). *Proceedings of the Royal Society of London* B 264: 1125–1129.

Pitcher, T.J. and J.K. Parrish. 1993. Functions of shoaling behaviour in teleosts. In: *Behaviour of Teleost Fishes*, T.J. Picher, (Ed.). Chapman and Hall London, pp. 363–439.

Preuss, T. and D.S. Faber. 2003. Central cellular mechanisms underlying temperature-dependent changes in the goldfish startle escape behaviour. *Journal of Neuroscience* 23: 5617–5626.

Prosser, C.L. 1991. *Environmental and Metabolic Animal Physiology: Comparative Animal Physiology*. Wiley-Liss, New York. 4th Edition.

Raabe, W. and S. Lin. 1985. Pathophysiology of ammonia intoxication. *Experimental Neurology* 87: 519–532.

Randall, D.J. and T.K.N. Tsui. 2002. Ammonia toxicity in fish. *Marine Pollution Bulletin* 45: 17–23.

Robb, T. and M.V. Abrahams. 2002. The influence of hypoxia on risk of predation and habitat choice by the fathead minnow, *Pimephales promelas*. *Behavioural Ecology and Sociobiology* 52: 25–30.

Rome, L.C. 1983. The effect of long-term exposure to different temperatures on the mechanical performance of frog muscle. *Physiological Zoology* 56: 33–40.

Sandberg, E. and E. Bonsdorff. 1996. Effects of predation and oxygen deficiency on different age classes of the amphipod *Monoporeia affinis*. *Journal of Sea Research* 35: 345–351.

Schulte, P. M., H.C. Glemet, A.A. Fiebig and D.A. Powers. 2000. Adaptive variation in lactate dehydrogenase-B gene expression: Role of a stress-responsive element. *Proceedings of the National Academy of Sciences of the United States of America* 97: 6597–6602.

Seehausen, O., J.J.M. vanAlphen and F. Witte. 1997. Cichlid fish diversity threatened by eutrophication that curbs sexual selection. *Science* 277: 1808–1811.

Shimps, E.L., J.A. Rice and J.A. Osborne. 2005. Hypoxia tolerance in two juvenile estuary-dependent fishes. *Journal of Experimental Marine Biology and Ecology* 325: 146–162.

Shoji, J., R. Masuda, Y. Yamashita and M. Tanaka. 2005. Effect of low dissolved oxygen concentrations on behavior and predation rates on red sea bream *Pagrus major* larvae by the jellyfish *Aurelia aurita* and by juvenile Spanish mackerel *Scomberomorus niphonius*. *Marine Biology* 147: 863–868.

Temple, G.K. and I.A. Johnston. 1998. Testing hypotheses concerning the phenotypic plasticity of escape performance in fish of the family Cottidae. *Journal of Experimental Biology* 201: 317–331.

Thetmeyer, H., U. Waller, K.D. Black, S. Inselmann and H. Rosenthal. 1999. Growth of European sea bass *Dicentrarchus labrax* L. under hypoxic and oscillating oxygen condition. *Aquaculture* 174: 355–367.

Toft, G., T.M. Edwards, E. Baatrup and L.J. Guillette. 2003. Disturbed sexual characteristics in male mosquitofish (Gambusia holbrooki) from a lake contaminated with endocrine disruptors. *Environmental Health Perspectives* 111: 695–701.

Utne-Palm, A.C. 2002. Visual feeding of fish in a turbid environment: physical and behavioural aspects. *Marine and Freshwater Behaviour and Physiology* 35: 111–128.

Wakeling, J.M. and I.A. Johnston. 1998. Muscle power output limits fast-start performance in fish. *Journal of Experimental Biology* 201: 1505–1526.

Wakeling, J.M., N.J. Cole, K.M. Kemp and I.A. Johnston. 2000. The biomechanics and evolutionary significance of thermal acclimation in the common carp, *Cyprinus carpio*. *American Journal Physiology—Regulatory Integrative and Comparative Physiology* 279: R657–R665.

Watkins, T.B. 2000. The effects of acute and developmental temperature on burst swimming speed and myofibrillar ATPase activity in tadpoles of the pacific tree frog, *Hyla regilla*. *Physiological and Biochemical Zoology* 73: 354–364.

Webb, P.W. 1978. Temperature effects on acceleration of rainbow trout *Salmo gairdneri*. *Journal of the Fisheries Research Board of Canada* 35: 1417–1422.

Webb, P.W. 1984 Body form, locomotion and foraging in aquatic vertebrates. *American Zoologist* 24: 107–120.

Webb, P.W. and H. Zhang. 1994. The relationship between responsiveness and elusiveness of heat-shocked goldfish (*Carassius auratus*) to attacks by rainbow trout (*Oncorhynchus mykiss*). *Canadian Journal of Zoology* 72: 423–426.

Weber, D.N. 2006. Dose-dependent effects of developmental mercury exposure on C-start escape responses of larval zebrafish *Danio rerio*. *Journal of Fish Biology* 69: 75–94.

Weetman, D., D. Atkinson and J.C. Chubb. 1999. Water temperature influences the shoaling decisions of guppies, *Poecilia reticulata*, under predation threat. *Animal Behaviour* 58: 735–741.

Williams, M.A. and C.C. Coutant. 2003. Modification of schooling behaviour in larval atherinid fish *Atherina mochon* by heat exposure of eggs and larvae. *Transactions of American Fisheries Society* 132: 638–645.

Wilson, R.S. 2005. Temperature influences the coercive mating and swimming performance of male eastern mosquitofish. *Animal Behaviour* 70: 1387–1394.

Wilson, R.S., E. Hammill and I.A. Johnston. 2007a. Competition moderates the benefits of thermal acclimation to reproductive performance in male eastern mosquito fish. *Proceedings of the Royal Society of London* Series B 274: 1199–1204.

Wilson, R.S., C.H.L. Condon and I.A. Johnston. 2007b. Consequences of thermal acclimation for the mating behaviour and swimming performance of female mosquito fish. *Philosophical Transactions of the Royal Society* 362: 2131–2139

Wong, B.B.M., U. Candolin and K. Lindstrom. 2007. Environmental deterioration compromises socially enforced signals of male quality in three-spined sticklebacks. *American Naturalist* 170: 184–189.

Yocom, T.G. and T.A. Edsall. 1974. Effect of acclimation temperature and heat shock on the vulnerability of fry of lake whitefish (*Coregonus clupeaformis*) to predation. *Journal of the Fisheries Research Board of Canada* 31: 1503–1506.

10

The Effects of Environmental Factors on the Physiology of Aerobic Exercise

D.J. McKenzie[1,*] and G. Claireaux[2]

INTRODUCTION

Fishes can be found in almost any body of water, be it an underground cave, beneath the arctic ice, in the deepest marine trench, in a fast moving mountain stream, or on a tropical reef. In each of these environments, fish have evolved combinations of physiological, morphological and biomechanical traits which adapt them to their particular habitat. In particular, almost all fish swim. Many functional traits are integrated to perform swimming activities, at many different levels of organisation ranging from cell energy production, to myocyte and muscle contraction, to cardiac and respiratory function, and to the systems for neural and endocrine regulation, integration and control. These traits have presumably all co-evolved, and contribute not only to locomotion but also to environmental adaptation in its broadest sense. It is generally held that the diversity of fish swimming modes and performance is a direct reflection of their ecological diversity. For instance, illuminated open

Authors' addresses: [1]Université de Montpellier 2, Institut des Sciences de l'Evolution, UMR 5554 CNRS-UM2, Station Méditerranéenne de l'Environnement Littoral, 1 quai de la Daurade, 34200 Sète, France.
[2]ORPHY, Université Européenne de Bretagne - Campus de Brest, UFR Science et Technologies, 6 avenue Le Gorgeu, 29285 Brest, France.
Corresponding author: E-mail: David.Mckenzie@univ-montp2.fr

oceans may select for high locomotory capacity, with a streamlined drag-minimising, torpedo-shaped morphology. In the abyss, on the other hand, the absence of light and the reduction in the distance over which predators and prey interact may have relaxed selection pressures on the body shapes and swimming abilities of the resident fish (Verity *et al.*, 2002; Seibel and Drazen, 2007). It seems intuitive that swimming ability should contribute directly to ecological performance and evolutionary fitness in fishes, although there is currently very little direct experimental evidence for this (Billerbeck *et al.*, 2001; Lankford *et al.*, 2001; Ghalambor *et al.*, 2003). Nonetheless, the assumption that swimming performance is of ecological significance has been the impetus for many studies to understand the anatomical and physiological mechanisms that contribute to swimming performance, and how performance is influenced by factors in the environment (Domenici *et al.*, 2007).

There are a number of reasons why environmental factors can be expected to influence swimming performance in fishes. The vast majority of fishes are poikilotherms whose biochemical and physiological rates are controlled by water temperature (the exceptions being some large oceanic species such as the tunas and lamnid sharks, see Bernal *et al.*, 2001; Block *et al.*, 2001; Graham and Dickson, 2004, for reviews). Water is not ideal as a respiratory medium, because of its high specific density and low capacitance for dissolved oxygen. This means that fishes ventilate high volumes of water across a very large surface area of delicate respiratory epithelium. As a result, variations in factors such as water salinity, dissolved carbon dioxide, and also pollution, can challenge physiological homeostasis. Finally, fishes interact chemically with their environment through their diet, with profound effects on their body composition and physiology. By investigating how all such environmental factors influence fish swimming performance, we should improve our understanding of their life history strategies and our ability to predict the impact of environmental change.

Muscle Types and Categories of Exercise

All fish groups possess two prominent myotomal muscle masses on either side of their body which are used for swimming with undulations of the body and caudal fin (Bone, 1978; Shadwick and Gemballa, 2006; McKenzie *et al.*, 2007a). Extensive reviews of the structure, innervation and function of these muscles in fishes are provided by Bone (1978), Sänger and Stoiber (2001), Shadwick and Gemballa (2006) and Syme (2006). Of significance in the context of the current review is the distinct anatomical and metabolic division of the muscle fibres. "Slow oxidative" fibres have, as the name implies, a slow twitch frequency that relies almost exclusively upon oxidative phosphorylation to provide ATP. They are well-vascularised to provide the O_2 and exogenous nutrients for oxidative work, and contain myoglobin, hence the common name "red muscle". Red muscle typically represents a minor proportion of the muscle mass and is used for steady aerobic swimming at relatively slow speeds. "Fast glycolytic" fibres have a fast twitch frequency and rely almost exclusively upon endogenous anaerobic fuels, initially phosphagen hydrolysis and then anaerobic glycolysis. These "white muscle" fibres are less well vascularised and lack myoglobin, they comprise the main part of the myotome and are used for short bursts of high-speed swimming (Bone, 1978; Webb, 1998; Sänger and Stoiber, 2001). Some fish species also possess intermediate fibres, "pink muscle", which is intermediate in terms of its distribution, energy metabolism, and role in steady versus burst swimming. Bone (1978) has argued that the large muscle mass reflects the density of the medium in which fish live, which exerts great friction drag, while the anatomically discrete muscle types reflect the conflicting demands of allowing both low speed cruising economy and also short bursts of high speed. Fish can also swim by movements of median and paired fins (Webb, 1998; Korsmeyer *et al.*, 2001; McKenzie *et al.*, 2007a). These swimming modes have been much less well studied than body/caudal fin swimming. In teleosts these types of locomotion appear to be predominantly aerobic (Webb, 1998; Korsmeyer *et al.*, 2001), and the muscles tend to be slow twitch and oxidative (Webb, 1998).

Rhythmic body/caudal fin swimming activities have been categorised according to the muscle types and energetic substrates that they rely upon, and to their duration (Beamish, 1978; Webb, 1993, 1998). Two extremes are recognised. Sustained aerobic swimming is dependent upon red muscle function, with nutrients (primarily lipids) and O_2 supplied in the bloodstream (Jones and Randall, 1978; Richards *et al.*, 2002). Thus, performance of red muscle also depends upon cardiac muscle function and pumping capacity of the heart (Jones and Randall, 1978; Claireaux *et al.*, 2005). This type of swimming comprises relatively low tailbeat frequencies and speeds during, for example, foraging, maintaining position against currents, and migrating against little or no opposing water currents. It can, in theory, be maintained indefinitely without any eventual muscular fatigue (Webb, 1998). The other extreme is burst swimming, which is an anaerobic activity dependent upon white muscle function that can only be sustained for short periods (seconds) (Webb, 1998). This comprises the much higher tailbeat frequencies and swimming speeds that are achieved by recruiting the large white muscle mass, for example during predator-prey chases or when negotiating velocity barriers during upriver migration. The duration of burst swimming is presumably limited, ultimately, by the availability of endogenous white muscle stores of high energy phosphagens and glycogen (Wood, 1991; Richards *et al.*, 2002). Intermediate between these extremes are those speeds at which fish can swim for prolonged periods, but not indefinitely. Prolonged swimming presumably can involve the recruitment of all muscle fibre types and therefore is a mixed aerobic and anaerobic activity. Such prolonged swimming speeds would be used for particular periods of up-river migration (Standen *et al.*, 2004) or perhaps elements of pursuit for predator prey interactions. The duration of prolonged swimming may be limited by cardiorespiratory capacity for O_2 and nutrient supply to the oxidative muscle and myocardium (Jones and Randall, 1978; Farrell, 2002, 2007) and/or when white muscle recruitment is too frequent and/or intense (Peake and Farrell, 2004). This chapter focuses upon the effects of environmental factors upon aerobic exercise, hence how these factors influence the ability of fish to perform sustained and prolonged swimming. The effects of environmental factors on anaerobic swimming performance are reviewed in Chapter 9, by Wilson *et al.* in this book.

Measuring the Performance of Aerobic Exercise: an Evaluation of the U_{crit} Protocol

The physiology and performance of aerobic swimming has been studied quite extensively in laboratory studies, using flumes and swimming respirometers, with the vast majority of work confined to salmonids (see reviews by Beamish, 1978; Jones and Randall, 1978; Webb, 1993, 1998; Videler, 1993). Various exercise protocols using flumes have been developed to measure traits such as maximum sustained swimming speed (see below) and maximum prolonged speed in fishes (Videler, 1993; Wilson and Egginton, 1994; Claireaux *et al.*, 2006), but the vast majority of studies have utilised the "critical speed" (U_{crit}) protocol described by Brett (1964). Brett (1964) developed the protocol to mimic upriver migration by anadromous salmonids, presuming that these would be swimming against currents of ever higher velocity as they approached their spawning grounds in headwaters. Thus, the protocol submits a fish to incremental stepwise increases in swimming speed in a swim flume until it fatigues, and U_{crit} is then interpolated as a single performance trait (Brett, 1964; Randall and Brauner, 1991; Farrell, 2007; Fig. 10.1A).

During the initial stages of the protocol the fish utilises red muscle, provided with O_2 and nutrients by the heart. This allows elements of aerobic metabolic performance and energetics to be measured. The red muscle generates constant rhythmic tailbeats, which show a linear increase in frequency with swimming speed. This aerobic muscular work is associated with an exponential increase in O_2 demand as swimming speed increases (Brett, 1964; Beamish, 1978; Webb, 1993). Measurement of rates of O_2 uptake (M_{O2}) reveal that the maximum aerobic metabolic rate during the U_{crit} protocol is typically at least three-fold higher than basal (standard) metabolism (reviewed by Jones and Randall, 1978 and Randall, 1982 for salmonids, see also Webber *et al.*, 1998; McKenzie *et al.*, 2001a, 2003a; Chatelier *et al.*, 2005, 2006 for examples of other species), with the highest maximum aerobic metabolic rate probably being in the tunas (see Korsmeyer and Dewar, 2001, for a review of the comparative respiratory metabolism and energetics of the tunas). This increased O_2 demand is met by an approximate doubling of water flow over the gills (whether by active or ram ventilation), and an up to five-fold increase in cardiac output in most teleosts (Jones and Randall, 1978; Thoraresen *et al.*, 1996; Gallaugher *et al.*, 2001; Chatelier *et al.*, 2005, 2006; Claireaux *et al.*, 2005), with, once again, the tunas probably having the highest cardiac outputs of all teleosts during exercise (see Brill and Bushnell, 2001, for a review) . The rise in cardiac output causes lamellar recruitment and so augments effective exchange surface at the gills, and blood O_2 carrying capacity may also be increased by release of erythrocytes from the spleen (Jones and Randall, 1978; Randall, 1982). There is, however, also a large increase in rates of O_2 extraction by the working muscle and, despite the concomitant rise in cardiac output, there is a profound decline in the O_2 tension of venous blood returning to the heart, and therefore a much larger arterio-venous O_2 partial pressure gradient across the gills (Stevens and Randall, 1967; Jones and Randall, 1978; Randall, 1982; Farrell and Clutterham, 2003).

At a certain speed, however, these aerobic adjustments are no longer sufficient and fish will be obliged to recruit the large white muscle mass in order to maintain position in the current (Wilson and Egginton, 1994). This is visible as a discrete change in swimming behaviour where short spurts of very high tailbeat frequencies are alternated with moments of passive coasting, such that the fish bursts forward and coasts back within the swim flume ("burst and coast" behaviour). This gait transition and recruitment of white muscle is typically associated with a plateau (asymptote) in both M_{O2} and rates of cardiac output, which can begin at speeds as early as 70% of the subsequent U_{crit} (reviewed by Farrell, 2002, 2007; see also Wilson and Egginton, 1994; Lee *et al.*, 2003; Chatelier *et al.*, 2005). Thus, white muscle recruitment can contribute up to 30% to the top end of U_{crit} performance, followed by fatigue, with the fish falling against the back screen of the flume. The actual U_{crit} that is derived can depend upon the magnitude and duration of the increments in swimming speed, which should ideally be increments of between 0.5 and 1 bodylengths per second over at least 30 minutes (Beamish, 1978). The protocol is, therefore, an index of prolonged swimming performance that involves first red and then white muscle use (Beamish, 1978).

There is some debate about whether fish reach their maximum aerobic cardiorespiratory and exercise capacity during a U_{crit} protocol, and about the factors which cause the transition from red to white muscle activity. By far the most information is available for salmonids of the genus *Oncorhynchus*, in particular the rainbow trout, *O. mykiss*. Farrell (2007) argues that, prior to the transition to burst-and-coast swimming in salmonids performing a U_{crit} test, arterial blood is being pumped at the maximum rate by the heart, this arterial blood is fully saturated with O_2, and the venous blood has been depleted of O_2 to the maximum possible extent. Thus, the rates of M_{O2} measured at this point can be considered a good estimate of maximum aerobic metabolic rate and aerobic exercise capacity for these species (Farrell, 2007). There is excellent agreement between maximum *in-vivo* measures of cardiac output during a U_{crit} test and maximum *in vitro* measures of output in heart maximally stimulated with adrenaline in these salmonids (Claireaux *et al.*, 2005; Farrell, 2007). This matching of *in vivo* and *in vitro* cardiac pumping capacities has recently also been demonstrated for a non-salmonid, the European sea bass, *Dicentrarchus labrax* (Chatelier *et al.*, 2005, 2006; Farrell *et al.*, 2007). In rainbow trout, blood remains fully saturated with O_2 throughout a U_{crit} protocol (reviewed by Jones and Randall, 1978; Farrell, 2007), and the capacity for O_2 uptake at the gills and transport in the blood seems to be maximised at gait transition. Indeed, aquatic hyperoxia does not increase maximum aerobic metabolic rate or improve U_{crit} performance (Kiceniuk and Jones, 1977; reviewed by Jones and Randall, 1978; Farrell, 2007), and blood doping (polycythemia) only leads to minor improvements to these performance traits (Gallaugher *et al.*, 1995; reviewed by Farrell, 2007). Thus, the O_2 demands of increased red muscle work during a U_{crit} protocol depend almost entirely upon the increase in cardiac output. This has led Farrell (2002, 2007) to propose that it is cardiac pumping capacity which defines maximum aerobic metabolic rate and aerobic swimming performance in salmonids. Perhaps the most direct evidence for this is that rainbow trout selected for high maximal cardiac output (*in vivo* and *in vitro*) can achieve higher maximum aerobic metabolic rate (and U_{crit}) than animals with poorer maximal cardiac performance (Claireaux *et al.*, 2005).

McKenzie *et al.* (2004) measured partial pressures of O_2 (PO_2) in the red muscle of rainbow trout during U_{crit} swimming and found no evidence that O_2 was depleted at the gait transition to white muscle use nor, indeed, at fatigue from U_{crit}. Thus, it would appear that it is not inadequate O_2 supply to working red muscle which elicits the transition to anaerobic swimming. Farrell and Clutterham (2003) measured venous O_2 tensions continuously during U_{crit} swimming and found that these never fell below a PO_2 of about 3 kPa (20 mm Hg) and that the trout transitioned to burst-and-coast swimming when this lower asymptotic threshold was reached. The heart of trout derives its O_2 supply in large part from venous return, and it will be the PO_2 gradient from plasma to myocardium that will assure this supply (reviewed by Farrell, 2002, 2007). Thus, it has been proposed that the transition to white muscle occurs because myocardial O_2 supply becomes limited (Farrell and Clutterham, 2003; Farrell, 2007). These authors suggest that the transition is implemented to protect the heart from becoming hypoxic (Farrell and Clutterham, 2003; Farrell, 2007). This would imply that it is a behavioural response based upon information from, for example, a venous O_2-sensitive chemo-receptor (Barratt and Taylor, 1984). Presumably, fishes with greater cardiac pumping capacity would maintain a higher venous PO_2 for any given degree of red muscle work, and hence the gait transition would be deferred to a higher speed. This remains to be demonstrated.

At fatigue from U_{crit} the red muscle of trout does not exhibit major metabolic imbalances (Richards $et\ al.$, 2002) or O_2 depletion (McKenzie $et\ al.$, 2004). This raises the question of the selective advantage of having a system where maximum aerobic swimming performance is not limited by O_2 delivery to red muscle but, rather, by venous O_2 delivery to the heart. One possibility is that this allows the fish to repeat bouts of aerobic exercise with little intervening recovery. That is, as soon as the red muscle stops working, O_2 delivery to the heart would be restored and a new bout of aerobic exercise can commence. Indeed many fish, not just salmonids, can repeat U_{crit} performance after only a very short period of intervening recovery (Jain $et\ al.$, 1998; McKenzie $et\ al.$, 2007b). It is not clear whether maximum aerobic swimming performance of other fish groups and species is also defined by cardiac pumping capacity, and venous O_2 supply to the heart, not least because of a lack of information. There is some circumstantial evidence that cardiac performance limits maximum aerobic metabolic rate and U_{crit} in the European sea bass (Chatelier $et\ al.$, 2005, 2006; Claireaux $et\ al.$, 2006; Farrell $et\ al.$, 2007).

Given that the protocol was developed to mimic salmonid spawning migrations (Brett, 1964), the ecological significance of U_{crit} as a performance trait for other fish groups and species has been questioned (Plaut, 2001; Nelson $et\ al.$, 2002). Furthermore, because the experimenter effectively imposes the gait transition upon the fish, it has been suggested that fatigue is behavioural rather than due to physiological limitations such as substrate depletion in white muscle (McFarlane and McDonald, 2002; Peake and Farrell, 2006). That is, the fish may not be able to recruit white muscle effectively within the confined space of a swim flume, and so may prefer to fall back against the screen (Peake and Farrell, 2006). Thus, the U_{crit} protocol may be a profoundly artificial situation for a fish, not least because when swimming spontaneously against natural currents fish often swim with positive ground speed, especially following white muscle recruitment (Standen $et\ al.$, 2002, 2004; Peake and Farrell, 2004), whereas in the flume they must maintain position at zero ground speed as current velocity is increased.

Thus, aerobic swimming performance may be maximised during a U_{crit} protocol, but the actual trait itself is not a direct measure of maximum aerobic swimming performance because white muscle provides up to 30%

to the top end of the speed range. Then, it is not clear what actually causes fatigue, whether it is behavioural or physiological or both. Recent studies on wild migrating salmonids, and on fish spontaneously negotiating velocity barriers in long raceways, indicate that animals switch between red and white muscle recruitment as required, and may be able to achieve very much higher swimming speeds than those measured in a U_{crit} test without the subsequent metabolic imbalances that characterise anaerobic exercise (Standen *et al.*, 2002, 2004; Peake and Farrell, 2004, 2006; Castro-Santos, 2005).

Nonetheless, if the U_{crit} protocol is associated with measures of O_2 uptake, it provides valid information about swimming metabolism and energetics, and therefore underlying aerobic performance. Individual U_{crit} performance is consistent and repeatable in a diversity of fish species, both in the short term (Jain *et al.*, 1998; McKenzie *et al.*, 2007b) and the long-term (Claireaux *et al.*, 2007), indicating that the protocol can be used to gauge and compare performance. Furthermore, U_{crit} performance is sensitive to a variety of environmental conditions. Beamish (1978) and Randall and Brauner (1991) reviewed the impact of water temperature, salinity, dissolved gases and pollutants, upon aerobic swimming, primarily measured as U_{crit}. In particular, Randall and Brauner (1991) argued that the U_{crit} performance of salmonid species is impaired when environmental factors deviate from particular optimal values (Fig. 10.1). The current chapter will provide an update of knowledge upon how major abiotic environmental factors influence aerobic and U_{crit} performance, and will also review evidence that diet quality is an environmental factor that can significantly influence these elements of physiological performance.

Temperature

Temperature is probably the most significant abiotic environmental factor for fish, with a controlling influence upon all aspects of their physiology and metabolism (Fry, 1947, 1971; Brett, 1971). It has a profound effect upon aerobic swimming performance that has been the subject of specific reviews, including with reference to potential effects of global warming (Taylor *et al.*, 1997; Farrell, 2007; Rome, 2007). The reader is advised to consult these and only a summary of the major effects of temperature is provided here.

Brett (1971) and Brett and Glass (1973) demonstrated a unimodal bell-shaped relationship between temperature and U_{crit} performance in sockeye salmon (*Oncorynchus nerka*), with low performance at the extremes of the species' temperature range, and a clear thermal optimum at which performance was maximal. There is good evidence to indicate that these temperature effects on U_{crit} reflect changes in aerobic performance, and the bell-shaped relationship is generally accepted as the typical effect of

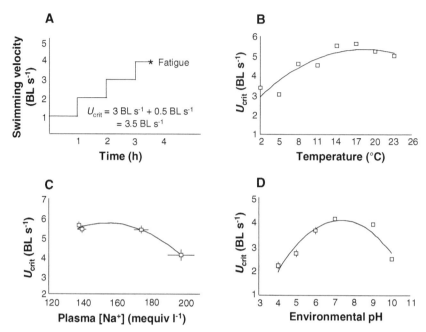

Fig. 10.1 Calculation used to determine critical swimming speed (U_{crit}) (A), and the effect of water temperature (B), elevated plasma Na+ concentration (C) and water pH (D) on U_{crit} performance in various salmonid species. BL, body length. Figure redrafted from Randall and Brauner 1991, with permission from the Company of Biologists.

temperature on maximum aerobic swimming performance in all fishes, with species-specific optima (see Beamish, 1978; Randall and Brauner, 1991, and Farrell, 2007, for reviews). Lee *et al.* (2003) demonstrated this bell-shaped relationship for adult sockeye and coho (*O. kisutch*) salmon migrating up the Fraser river (British Columbia, Canada), where different stocks exhibited temperature optima for maximum metabolic rate during swimming, aerobic metabolic scope, and U_{crit}, that were very similar to the average ambient water temperatures of their natal streams (Fig. 10.2). This may indicate that thermal adaptation has occurred over an evolutionary time scale, although temperature sensitivity of performance was not particularly marked and the fish would retain at least 90% of their maximum aerobic performance over a range of temperatures spanning as much as 5°C. This would allow the fish to respond to annual variations in water temperature (Lee *et al.*, 2003; Farrell, 2007).

Temperature influences aerobic swimming performance by direct effects on muscle function (including myocardial function) and on metabolic rate and consequent O_2 demand. The relatively poor performance at low temperatures may result from an overall depression of all physiological

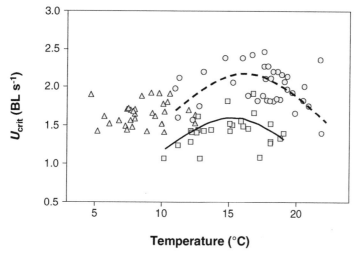

Fig. 10.2 Relationships between critical swimming speed (U_{crit}) and water temperature (t) for different species and stocks of adult Pacific salmon captured during their migration up the Fraser River, Canada. Circles are the Gates Creek stock of sockeye (*Oncorynchus nerka*) salmon, where the dotted line is $U_{crit} = 2.17/1+[(t-16.15)/9.59]^2$ ($P < 0.05$; $r^2 = 0.41$). Squares are Weaver Creek sockeye salmon, for which the solid line is $U_{crit} = 1.60/1+[(t-15.18)/8.52]^2$ ($P < 0.05$; $r^2 = 0.27$). The triangles are for Chehalis coho (*O. kisutch*) salmon, where there was no statistically significant relationship with temperature over the range examined. Data from Lee *et al.* 2003, with permission from the Company of Biologists.

processes, but in particular of muscle function. Rome *et al.* (1984) demonstrated that the reduced swimming performance of carp (*Cyprinus carpio*) acclimated to cold water occurs because there is a compression of the speed range over which red and then white muscles are recruited, with early recruitment of white fibres at much lower swimming speeds. Working on swimming carp and scup (*Stenotomus chrysops*), Rome (1990) and Rome *et al.* (1992) demonstrated that temperature has a direct effect on the maximum velocity of shortening of muscle fibres and upon their maximum power production. The power production of red muscle is depressed at low temperatures and this requires the early recruitment of white fibres.

As temperature rises towards the optimum, positive effects on muscle function and associated power generation contribute to an increase in swimming performance (Temple and Johnston, 2002; Rome, 2007, for reviews). Claireaux *et al.* (2006) demonstrated, however, that the net cost of swimming was relatively stable in sea bass seasonally acclimatised to temperatures beyond 11°C, increased swimming performance at higher temperatures was due to an increase in maximum aerobic metabolic rate and aerobic metabolic scope. That is, there was a tight relationship between these aerobic performance traits and maximum aerobic swimming speed (U_{ms}, defined as the maximum swimming speed that the sea bass could achieve before exhibiting burst and coast swimming behaviours, Claireaux *et al.*, 2006). The authors suggested, therefore, that improved swimming capacity with increasing temperature resulted largely from increased energy fluxes and support metabolism, and therefore in scope for activity, rather than in properties of muscle function per se (Rome, 1990; Taylor *et al.*, 1997). However, there are some examples in the literature where swimming costs increased with water temperature. For instance, Dickson *et al.* (2002) found that the energetic costs of swimming at a given speed increased with temperature in chub mackerel *Scomber japonicus*, and McKenzie *et al.* (2007b) found similar results for the cyprinid chub, *Leuciscus cephalus*.

The reasons for the relative decline in aerobic performance as temperature exceeds the optimum are many. Firstly, because temperature increases all metabolic processes, it elicits an exponential increase in metabolic rate in fishes (Fry, 1971), both of maintenance metabolism and all other routine metabolic activities. There is not, however, a coincident increase in maximum aerobic metabolic rate because, presumably, of intrinsic constraints to maximum cardiorespiratory performance. As a result, as temperature increases, the demands for maintenance and routine activities rise inexorably towards the maximum possible rate, with a coincident decline in aerobic metabolic scope, until a lethal temperature where they coincide (Brett, 1971; Fry, 1971; Taylor *et al.*, 1997). The intrinsic constraints to maximum aerobic performance at high temperature have been discussed in detail before (Taylor *et al.*, 1997; Farrell, 2007) and may include diffusion limitations at the gills and an inability of the heart to meet the very elevated oxygen demands of the tissues. These same limitations presumably also underlie the lethal effects of the highest temperatures (Taylor *et al.*, 1997).

What is clear, however, is that the form of the bell-shaped temperature curve is strongly influenced by rates of temperature acclimation, being most pronounced in fishes that are exposed to acute temperature change, and increasingly less pronounced with time of acclimation. At low temperatures, muscle performance may improve with acclimation because of adaptive changes to the isomeric composition of ion channels in the sarcolemma, with the insertion of channels that work better at low temperatures (Temple and Johnston, 2002). Similar processes of molecular acclimation presumably also occur in the myocardium. The relative improvements to U_{crit} at higher temperatures will also reflect changes at this level but presumably also in organismal metabolic temperature adaptation, so that costs for maintenance metabolism and routine activities drop progressively as fish acclimate to the increased water temperature, and so the metabolic scope available for aerobic swimming is extended (Fry, 1971; Taylor *et al.*, 1997).

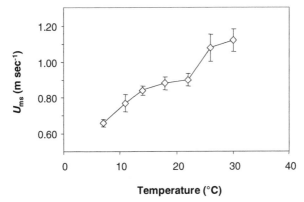

Fig. 10.3 Maximum sustainable aerobic swimming speed (U_{ms}) in European sea bass (*Dicentrarchus labrax*) seasonally acclimatised to seven water temperatures over an annual cycle encompassing their seasonal thermal range in the Bay of Biscay (Mid-Atlantic). Data are mean (± SEM) of at least 7 animals at each temperature, and are derived from Claireaux *et al.* (2006).

Relatively few studies have investigated the effect of seasonal thermal acclimatisation (rather than laboratory acclimation) on aerobic swimming performance in fish, and these have rarely measured the effects of more than three environmental temperatures (Butler *et al.*, 1992; Taylor *et al.*, 1996; Adams and Parson, 1998; Day and Butler, 2005; Claireaux *et al.*, 2006; McKenzie *et al.*, 2007b). Claireaux *et al.* (2006) measured U_{ms} in European sea bass, *Dicentrarchus labrax*, seasonally acclimatised to seven temperatures between 7°C and 30°C, these being the annual thermal range that they experience in coastal waters of the mid Atlantic (Bay of Biscay, France). As defined above, U_{ms} should be a true measure of maximum aerobic performance at the prevailing temperature. Claireaux *et al.* (2006) did not find a pronounced bell-shaped curve but, rather, swimming performance increased markedly with temperature from a minimum at 7°C up to about 22°C, beyond which performance levelled off, although the maximum was at 30°C (Fig. 10.3). This shows a remarkable ability of the sea bass to maintain performance over its natural thermal range, although it is to be presumed that temperatures higher than 30°C would eventually cause a decline in performance. Other studies have found little effect of temperature upon U_{crit} in fish that were seasonally acclimatised, indicating that some species may in fact be capable of almost perfect compensation of performance (Butler *et al.*, 1992; Taylor *et al.*, 1996; McKenzie *et al.*, 2007b). Interestingly, Day and Butler (2005) demonstrated that brown trout acclimated to reversed seasonal temperature displayed reduced U_{crit}, suggesting that factors other than temperature (the authors suggest photoperiod) contribute to seasonal compensation of performance.

A discussion of the effects of temperature on aerobic swimming in fishes would be incomplete without some mention of fish with regional endothermy, such as the tunas and lamnid sharks, which use vascular counter-current heat-exchangers (the *retia mirabilia* in tunas) to maintain their red muscle at temperatures above ambient. The comparative metabolism, energetics and performance of aerobic exercise in these fish groups have been the subject of in-depth reviews (Bernal *et al.*, 2001; Block *et al.*, 2001; Graham and Dickson, 2004), and shall only be given a brief mention here. Tunas perform continuous sustained aerobic exercise, to ram ventilate their gills and generate lift, and as part of a life history that involves constant foraging and wide-scale migrations. The red muscle is located in a different position to other fishes, being towards the centre of the body, and the heat produced by the continuous swimming activity is retained by the *rete* (see Graham and Dickson, 2001, for a review). Tunas are generally considered to be amongst the most powerful of all aerobic swimmers, and their exceptional sustained aerobic performance is at least in part due to increased power output and contraction rates in their warm red muscle (Graham and Dickson, 2001, 2004), coupled with an exceptional cardiac pumping capacity (Brill and Bushnell, 2001). Lamnid sharks show some interesting evolutionary convergences with the tunas, but have been relatively less well studied (see Bernal *et al.*, 2001, 2005; Donley *et al.*, 2004). Although these adaptations allow such animals to swim at relatively high sustained speeds throughout their lives, the few studies of tuna performance in a U_{crit} protocol have not revealed higher relative U_{crit}, in bodylengths per second, than other large fish athletes, such as Chinook salmon (Lee *et al.*, 2003; Graham and Dickson, 2004). Furthermore, how water temperature influences aerobic swimming performance in fishes with regional endothermy remains to be described (e.g., Graham and Dickson, 2004). Bernal *et al.* (2005) reported that isolated red muscle fibres of the salmon shark, *Lamna ditropis*, were designed for high power production at very high body temperatures (26°C or above) only, and did not produce useful work and power below 20°C, only 6°C below their operating temperature *in vivo* but still over 10°C warmer than their ambient water.

It is clear that the effects of temperature on aerobic swimming performance may be of great ecological significance for fishes. They could be very profound if fish encounter sudden changes in water temperature, which may impede their capacity for sustained exercise. This may limit foraging behaviours but, most significantly, may impede migration in the vicinity of, for example, thermal power stations which use surface waters to cool their turbines. Furthermore, although many species are clearly able to acclimatise to the seasonal changes in temperature in their environment, there is some risk that the increased maximum summer temperatures which have accompanied overall global warming may exceed the adaptive range of some and so, potentially, modify their migratory cycles and behaviours (Taylor *et al.*, 1997; Lee *et al.*, 2003). One ecological effect that has already been observed, which must be linked to the performance of sustained aerobic exercise, is the change in distribution patterns of various marine species in response to global warming. This may be a facultative behavioural response of habitat selection, which has been experimentally demonstrated in species such as Atlantic cod, *Gadus morhua* (Claireaux *et al.*, 1995a). Some species may expand their range to colonise areas that were once too cold (Perry *et al.*, 2005; Hiddinck and ter Hofstede, 2008) whereas other may do the opposite, with increasing temperature constraining their distribution (Perry *et al.*, 2005). There is evidence that reductions in species range may be mediated by limitations to their aerobic metabolic scope (Pörtner and Knust, 2007).

Dissolved O_2 and CO_2

Water is intrinsically rather poor in dissolved O_2, and this has long been recognised as the most important environmental factor limiting aerobic metabolic scope in fishes (Fry, 1947, 1971). Hypoxia occurs quite frequently in aquatic habitats, especially freshwater environments, when overall respiration rates exceed O_2 production by aurotrophs and O_2 diffusion from the atmosphere. Hypoxic events due to anthropogenic eutrophication are now growing in frequency in aquatic habitats worldwide (see Diaz, 2001; Wu, 2002; Diaz *et al.*, 2004, for reviews).

Figure 10.4 demonstrates the limiting effect that hypoxia has on maximum aerobic metabolic rate in two species of marine teleost, the European sea bass and the common sole (*Solea solea*), seasonally acclimatised to a range of different temperatures (Claireaux and Lagardère, 1999; Lefrançois and Claireaux, 2003). Hypoxia has been shown to limit the performance of various species in the U_{crit} test (Davis *et al.*, 1963; Dahlberg *et al.*, 1968; Jones, 1971), in the rainbow trout this is coupled with the expected limitation to maximum aerobic metabolic rate and metabolic scope (Bushnell *et al.*, 1984). Dutil *et al.* (2007) found that the transition from steady swimming to burst and coast movements occurred at lowered speeds in Atlantic cod (*Gadus morhua*) swimming under hypoxic conditions compared to individuals in normoxia.

Presumably, the fish are limited in their ability to provide oxygen to the red muscles and, in turn, to their heart via the venous blood, so they resort to anaerobic swimming mode earlier and quit at lower swimming speeds in the U_{crit} test. Dahlberg *et al.* (1968) demonstrated interspecific diversity in sensitivity to the limiting effect of hypoxia, with largemouth bass (*Macropterus salmoides*) being less sensitive than Chinook salmon (*O. tsawytscha*), a difference which presumably parallels their overall hypoxia tolerance.

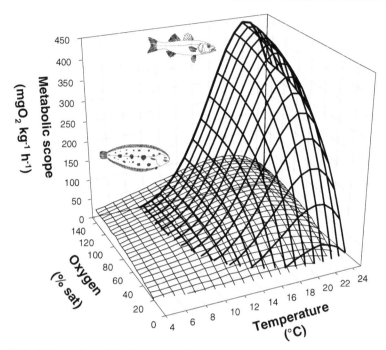

Fig. 10.4 Surface plots of the interactive effects of water oxygen saturation (%) and water temperature upon the aerobic metabolic scope of two species of marine teleost, the European sea bass, *Dicentrarchus labrax*, (steep upper surface curve) and the Dover sole, *Solea solea* (shallow lower surface curve). Models derive from data reported in Claireaux and Lagardère (1999) and Lefrançois and Claireaux (2003).

Bushnell *et al.* (1984) found that acclimation of rainbow trout for two weeks to various levels of hypoxia did not improve their ability to perform aerobic exercise, indicating that this species might not be able to compensate for chronic environmental hypoxia, except perhaps on an evolutionary time scale.

Hypoxia can also influence the intensity of spontaneous aerobic swimming activity in fishes. In some species such as crucian carp, *Carassius carassius* (Nilsson *et al.*, 1994), Atlantic cod (Schurmann and Steffensen, 1994) and Adriatic sturgeon, *Acipenser naccarii* (McKenzie *et al.*, 1995), spontaneous aerobic swimming activity is reduced in hypoxia, presumably as an energy-saving response. In other species such as Atlantic herring, *Clupea harengus*, hypoxia causes increased activity, which presumably reflects a desire to escape and may involve both aerobic and anaerobic components (Domenici *et al.*, 2000; Herbert and Steffensen, 2006).

A number of teleost fish species have evolved the ability to gulp air at the water surface and store this in a variety of air-breathing organs (ABO), to then extract the O_2 (Graham, 1997). These adaptations for bimodal respiration (i.e., to breathe both water with gills and air with an ABO) are believed to have evolved, at least in part, as a response to environmental hypoxia in tropical habitats (Randall *et al.*, 1981; Graham, 1997). Farmer and Jackson (1998) reported that the bowfin, *Amia calva*, a primitive bony fish which uses a modified swimbladder as an ABO, exhibits an exponential increase in air-breathing events when exposed to a step-wise U_{crit} protocol. There are also anecdotal reports that other fish with bimodal respiration use air-breathing during activity (reviewed in McKenzie *et al.*, 2007a) and thus it would be interesting to determine whether this allows them to avoid any negative effects of aquatic hypoxia on their aerobic exercise performance.

Aquatic hypoxia may have significant ecological implications for water-breathing fishes through its effects on aerobic swimming performance. Clearly, foraging and migratory performance may be limited if fishes encounter hypoxic areas. The behavioural responses to hypoxia may also have ecological implications, some species may reduce all activities that depend upon aerobic swimming until conditions improve, and other species may migrate away and change towards less hypoxic areas. Indeed, fish such as the Atlantic cod can show clear behavioural responses of habitat selection to avoid hypoxia (Claireaux *et al.*, 1995b).

The partial pressure of CO_2 (PCO_2) in surface waters is typically less than $0.015 \cdot kPa$ (~ $1 \cdot mm$ Hg), but this can increase as a consequence of microbial metabolism and inadequate surface gas exchange, causing hypercapnia. There is now a growing interest in the potential effects on fish of the accumulation of CO_2 in all aquatic habitats worldwide, which reflects the accumulation of this molecule in the atmosphere as a whole (Ishimatsu *et al.*, 2005). Fish, of course, produce CO_2 from aerobic metabolism, which then diffuses across the gills into the surrounding water. Aquatic hypercapnia inhibits excretion, and can even cause CO_2 to diffuse into the blood across the gills. This can cause the blood to become more acid and its O_2-carrying capacity to be reduced (reviewed by Heisler, 1984, 1993). Severe levels of hypercapnia may also cause acidosis of muscle and cardiac tissue, and impair their metabolism (Heisler, 1984, 1993; Driedzic and Gesser, 1994).

Dahlberg *et al.* (1968) demonstrated that acute exposure to hypercapnia, with no time given for acid-base compensation, limited U_{crit} performance in largemouth bass and Chinook salmon, with salmon being more sensitive than bass. McKenzie *et al.* (2003a) found, on the other hand, that chronic hypercapnia had no effect on the U_{crit} performance of European eels, *Anguilla anguilla*, despite a 50% reduction in O_2 carrying capacity. Thus, this very limited database does nonetheless indicate that there are wide inter-specific differences in the effects of hypercapnia on U_{crit} performance in teleosts. The impaired U_{crit} performance in hypercapnic salmon and bass (Dahlberg *et al.*, 1968) may have reflected limitations to tissue O_2 supply and/or direct negative effects of extracellular and intracellular acidosis on the function of skeletal and cardiac muscle (Driedzic and Gesser, 1994).

Teleosts can compensate for hypercapnic acidosis, and consequent effects on blood O_2 carrying capacity and tissue intracellular pH, by ion-exchange at the gills, accumulating bicarbonate (HCO_3^-) ions at the expense of an equimolar loss of chloride (Cl^-), although there are species-specific limits to the extent to which this compensation can occur (Heisler, 1984, 1993; McKenzie *et al.*, 2003a). Thus, CO_2 accumulation in global waters worldwide may have significant ecological implications through effects on aerobic swimming performance (Ishimatsu *et al.*, 2005). In particular, if this accumulation exceeds the capacity for compensation of a species, it will have no means of escape and will exhibit permanently impaired scope to forage and migrate.

Salinity

Many fish species swim through waters that differ markedly in salinity, some as an obligatory component of their spawning migrations (e.g., salmon, eel and sturgeon), and others as a facultative element of foraging strategies (e.g., sea bass, sea bream and mullet). A number of studies have shown a direct link between the ability to perform aerobic exercise and the homeostatic regulation of body fluid osmolality (Brauner *et al.*, 1992, 1994; McKenzie *et al.*, 2001a, b; Chatelier *et al.*, 2005). In coho salmon (*O. kisutch*) smolts, and Adriatic sturgeon juveniles, a "seawater challenge" comprising direct transfer from freshwater (at a salinity of 0‰) to seawater (28‰) caused a significant reduction in U_{crit} at 24 h later. This reduced performance was directly related to an accumulation of plasma ions and large increases in plasma osmolality (Brauner *et al.*, 1992, 1994; McKenzie *et al.*, 2001a,b; Fig. 10.5). It has been suggested that this was due to impaired cardiac and skeletal muscle function consequent to ionic imbalances and loss of tissue moisture, with additional strain upon the heart from haemoconcentration (Brauner *et al.*, 1992, 1994; McKenzie *et al.*, 2001a, b). These effects remain to be demonstrated. Acclimation of the Adriatic sturgeon to slightly hypertonic brackish water improved their ability to regulate iono-osmotic status following the salinity challenge and, consequently, improved their post-challenge U_{crit} (McKenzie *et al.*, 2001b). This indicates that a period spent in the brackish waters of estuaries would improve the ability of this sturgeon species to tolerate excursions into full strength seawater, although *A. naccarii* is not reported actually to colonise the marine environment for extended periods (Rochard *et al.*, 1992). No such beneficial effect of prior salinity acclimation has been demonstrated in salmon smolts of the genus *Oncorhynchus* (Morgan and Iwama, 1992), although these species inhabit the marine environment as they grow from smolts to maturity (Hoar, 1976).

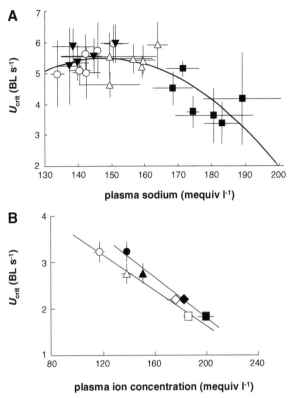

Fig. 10.5   A. Second-order regression of the relationship between mean (± SEM) plasma sodium at rest and subsequent mean (± SEM) critical swimming speed (U_{crit}) in coho salmon (*Oncorhynchus kisutch*) parr. Fish derive from four experimental treatments: freshwater controls (closed triangles); at 24 h following abrupt transfer to seawater at 28 gl⁻¹ (closed squares), at 5 days following transfer to seawater at 28 gl⁻¹ (open triangles), and at 24 h following transfer back from seawater to freshwater (open circles). N = at least 5 per data point, r^2 = 0.68. Figure reproduced from Brauner *et al.* (1992).
B. Least-squares linear regression analysis describing the relationship of mean (±SE) U_{crit} versus mean (±SE) Na⁺ (open symbols) and Cl⁻ (closed symbols) measured in the plasma of Adriatic sturgeon (*Acipenser naccarii*) following exercise to fatigue. When reading from left to right down the regression lines, the symbols circles represent a group raised in freshwater (circles), a group raised in brackish water at 11 g l⁻¹ (triangles), the brackish water group at 24 h following abrupt transfer to seawater at 28 gl⁻¹ (diamonds), and freshwater group following the same abrupt transfer (squares). N = at least 6 animals for each point. See McKenzie *et al.* (2001b) for further details. Figures reproduced with permission from the NRC Research Press.

Nelson *et al.* (1996) studied Atlantic cod populations from brackish water lakes (20 ‰) or the marine environment (38 ‰, and found that reciprocal acclimation to the two salinities had no effect on the U_{crit} of either population, although it increased variability in performance. The gulf killifish (*Fundulus grandis*) is an estuarine species and Kolok and Sharkey (1997) found that killifish acclimated to freshwater had a significantly lower U_{crit} than fish maintained in brackish water at a salinity of 10‰, which the authors attributed to the osmotic stresses the animals suffered in freshwater. Chatelier *et al.* (2005) investigated the effects of direct transfer from sea water (35‰) to freshwater on European sea bass U_{crit} and associated cardiac performance. They found that the sea bass was able to maintain both elements of performance unchanged at 24 h following this transfer, and the same was true of a reciprocal transfer from fresh to sea water. This was linked to an exceptional ability of sea bass to regulate plasma osmotic homeostasis and tissue water balance in the face of sudden changes in ionic and osmotic gradients between water and body fluids. The mechanism by which the sea bass do this is unknown. Varsamos *et al.* (2002) also found that sea bass had an exceptional capacity for osmotic regulation, and suggested that this might reflect extremely plastic morphofunctional adaptations of mitochondria rich ("chloride") cell populations in the gills. Thus, the sea bass exhibited a very different response to that observed in the salmon and sturgeon, but this ability would clearly allow them to protect their swimming ability while moving throughout estuaries, and between marine, brackish water and freshwater habitats (Pickett and Pawson, 1994). It would be interesting to investigate whether other species which perform facultative movements can also regulate iono-osmotic homeostasis and thereby protect exercise performance.

Toxicants

The use of aerobic swimming performance as an indicator of the sub-lethal toxic effects of aquatic pollution was first proposed many years ago (Cairns, 1966; Sprague, 1971). Since then, studies have demonstrated that U_{crit} is impaired by pollutants such as low pH (Ye and Randall, 1989; Butler *et al.*, 1992), dissolved metals (e.g., Waiwood and Beamish, 1978; Wilson *et al.*, 1994, Beaumont *et al.*, 1995a, b, 2003; Alsop *et al.*, 1999; Pane *et al.*, 2004), ammonia (e.g., Shingles *et al.*, 2001; Wicks *et al.*, 2002; McKenzie *et al.*, 2003b), and various non-natural toxic organic chemicals such as organo-phosphate pesticides (Peterson, 1974), organo-chlorine fungicides (MacKinnon and Farrell, 1992; Nikl and Farrell, 1993; Wood *et al.*, 1996) and bleached kraft pulp-mill effluents (Howard, 1975; MacLeay and Brown, 1979). The majority of these studies were performed upon the rainbow trout.

Many of the pollutants cause direct damage to the gills if they exceed certain concentrations, and this leads to a decline in aerobic swimming performance by interfering with gas transfer and so limiting maximum aerobic metabolic rate (Nikl and Farrell, 1993; Pane *et al.*, 1994; Wood *et al.*,

1996). Some heavy metals such as aluminium and nickel can cause an elevation of standard metabolic rate at concentrations that do not damage the gills, and this metabolic loading factor then reduces metabolic scope and hence U_{crit} (e.g., Wilson *et al.*, 1994; Pane *et al.*, 2004). In brown trout, *Salmo trutta*, exposure to sub-lethal concentrations of copper (in slightly acidic water, at pH 5) impairs U_{crit} performance by interfering with ammonia excretion, causing plasma ammonia accumulation (Beaumont *et al.*, 1995a, b). Zinc may have a similar mode of action in rainbow trout (Alsop *et al.*, 1999). The negative relationship between plasma ammonia concentration and U_{crit} in copper-exposed trout (Beaumont *et al.*, 1995a, b) is very similar to that observed in rainbow trout and brown trout exposed to ammonia alone (Shingles *et al.*, 2001; McKenzie *et al.*, 2003b). Beaumont *et al.* (2000a, b) demonstrated that the plasma ammonia accumulation caused a depolarization of white muscle, which then compromised its recruitment at the highest speeds (Beaumont *et al.*, 2003; McKenzie *et al.*, 2006). Thus, the toxic effects of ammonia accumulation on U_{crit} performance, whether from environmental ammonia pollution or effects of heavy metals on excretion of the metabolic waste product, are not actually an impairment of aerobic performance (Beaumont *et al.*, 2003; McKenzie *et al.*, 2006).

Clearly, the objective of ecotoxicological studies that investigate effects of pollutants on swimming performance is to establish ecologically relevant limits for pollutant concentrations in natural environments (e.g., MacKinnon and Farrell, 1992; Nikl and Farrell, 1993; Beaumont *et al.*, 1995a; Wicks *et al.*, 2002). The literature investigating the effects of pollutants upon fish exercise performance is, however, almost exclusively laboratory-based and, with few exceptions (Howard, 1975; MacLeay and Brown, 1979; Hopkins *et al.*, 2003; Taylor *et al.*, 2004), comprises the exposure of salmonid species to single toxicants. Very little is known about the potential physiological effects of exposure to polluted natural environments, which can contain a complex mixture of chemicals.

McKenzie *et al.* (2007b) investigated whether exposure to the mixtures of chemicals which prevail in polluted natural environments (urban rivers) had an impact upon the U_{crit} performance of two species of cyprinid fish, the chub and the carp. These are both lowland species characterising the "cyprinid reach", where water flows are relatively slow and temperatures and O_2 levels are often variable, even in the absence of pollution. Portable swimming respirometers were used to compare exercise performance and respiratory metabolism of fish exposed in cages for three weeks to either clean or polluted sites on three urban European river systems, in Italy, the UK and the Netherlands. Passive sampling devices were used to measure the bioavailable fraction of pollutants at the sites during the caging exposures. These analyses demonstrated that the polluted sites on the British and Italian rivers contained complex mixtures of bioavailable heavy metals and organics, whereas the Dutch river was severely polluted by mixtures of bioavailable organics. Swimming performance was assessed with a "repeat-exercise" protocol, which measures the ability of fish to perform two sequential U_{crit} tests with a brief intervening recovery interval (Jain *et al.*, 1998). These authors demonstrated that rainbow trout can repeat U_{crit} performance perfectly after only 45 min recovery, with a "repeat ratio" of unity between first and second U_{crit} tests. Impairments to the second U_{crit} test proved to be a much more sensitive indicator of fish health and water quality than a single exercise test alone (Farrell *et al.*, 1998; Jain *et al.*, 1998).

In both Italy and the UK, indigenous chub exposed to clean or polluted sites swam equally well in the first U_{crit} test, but the chub from polluted sites could not repeat performance in the second test after a 45 min recovery interval (Fig. 10.6). This was due to impairments to their aerobic exercise performance, they were unable to raise metabolic rate and allocate oxygen towards exercise in the second trial. As a result, they exhibited significantly lower maximum aerobic metabolic rate and metabolic scope, and an approximately 30% reduction in U_{crit}, with a repeat ratio of about 0.7. Similar results were obtained over two successive campaigns in Italy, and chub at the polluted sites in GB showed qualitatively similar responses (Fig. 10.6). Parallel studies on chub caged at the polluted sites demonstrated an accumulation of organics and metals that was directly related to the bioavailable fractions of these pollutants in the water (Winter *et al.*, 2004; Garofalo *et al.*, 2005). Thus, the impaired aerobic swimming performance could be linked to the increased bioavailable pollution at the sampling sites (McKenzie *et al.*, 2007b). The most likely explanation for this reduced performance in the second trial was impairments to the function of red and/or cardiac muscle (McKenzie *et al.*, 2007b), although extensive laboratory testing based upon the bioavailable pollutant mixtures prevailing at the sites would be necessary to verify this.

Exposure of cultured cloned carp to polluted sites on the Dutch river caused a marked accumulation of organic pollutants in their tissues (Verweij *et al.*, 2003) but did not, however, affect their repeat swimming performance (McKenzie *et al.*, 2007b). This may indicate either that the prevailing chemical mixture in this river was without effects on swimming performance, unlike the Italian and British rivers (which contained metals as well as organics), or that carp were more tolerant of the organic pollution. However, measurements of oxygen uptake during swimming revealed increased rates of routine aerobic metabolism in both chub and carp at polluted sites in all of the rivers studied, indicating a sub-lethal metabolic loading effect. It is not clear whether this metabolic loading was due to the presence of a particular pollutant or pollutant combination in the complex bioavailable mixtures in the rivers, because it can reflect an aspecific stress response in teleosts (Wendelaar Bonga, 1997). Seasonal studies on the chub in Italy

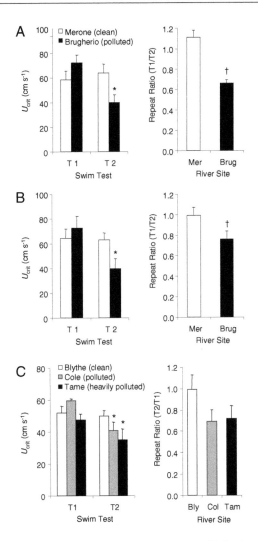

Fig. 10.6 Swimming performance of chub (*Leuciscus cephalus*) following exposure for three weeks in submerged cages to differently polluted sites on rivers in Italy or the United Kingdom. In each case, mean (± SEM) critical swimming speed (U_{crit}) was measured twice, with the second swim test (T2) measured 40 min following fatigue in the first swim test (T1), to calculate the corresponding repeat performance ratio (U_{crit} T2/U_{crit} T1). Panels A and B show the effects of exposure of chub to two sites on the river Lambro near Milan: a relatively unpolluted site (Merone) or a polluted site (Brugherio), in June 2001 (A) and then in September 2001 (B). Panel C shows the effects of exposure of chub to one of three confluent rivers in Birmingham: the Blythe (unpolluted), the Cole (polluted) and the Tame (heavily polluted), in June 2002. n = 6 in all cases, * denotes a significant difference between T1 and T2 for that river site; + indicates a significant difference in repeat performance ratio between sites on the river Lambro. Bly, Blythe; Col, Cole; Tam, Tame. Data taken from McKenzie *et al.* (2007b) with permission from the Royal Society.

and the UK also revealed that sub-lethal physiological effects of aquatic pollution were most obvious in fish acclimatised to warm spring/summer temperatures.

Therefore, the studies demonstrated that physiological traits of exercise performance and metabolic rate can indeed be used to reveal sub-lethal effects of aquatic pollution in natural environments. These physiological impairments may cause fish to be unable to colonise polluted areas successfully, in particular if seasonal spates and floods require them to perform repeated bouts of high-intensity aerobic exercise. Interestingly, many of the criteria required of biomarkers (Stegeman *et al.*, 1992; van der Oost *et al.*, 2003) are met by complex physiological traits of performance and metabolism such as the ability to repeat a U_{crit} swim test, and routine metabolic rate (McKenzie *et al.*, 2007b). Although they may be less specific than biochemical or molecular biomarkers, which can in theory reveal responses to specific pollutants (van der Oost *et al.*, 2003), and the exact mechanism by which exposure to polluted river sites caused the physiological impairments was not clear, these traits of aerobic metabolism are reliable integrated measures of the responses of many physiological systems, and can provide insight into why fish fail to colonise some polluted habitats. The physiological biomarkers may, therefore, be particularly useful within programs of ecological risk assessment which also comprise analyses of a suite of other biochemical and molecular markers.

Diet quality

All animals interact profoundly with their environment through their diet, which provides all of the structural materials and energy required for expression of the phenotype. Diet is both an abiotic and a biotic factor, hence in a category unto itself. The interaction between animals and their environment through their diet is extremely complex, as revealed by the large number of "essential" nutrients that animals require in their food. The syndromes associated with deficiency in these compounds have been the focus of much research. There is also, however, a growing body of evidence to indicate that the relative intake of specific nutrients can have a profound effect on the physiology of well-nourished animals (Crawford and Marsh, 1989; Arts *et al.*, 2001; McKenzie, 2001, 2005), including upon their aerobic exercise performance (Beamish, 1988; McKenzie *et al.*, 1998; Wagner *et al.*, 2004; Chatelier *et al.*, 2006).

Beamish *et al.* (1988) investigated the potential effects of variation in dietary composition of major nutrient groups. They demonstrated that rearing lake trout (*Salvelinus namaycush*) on a protein-rich diet improved their U_{crit} swimming performance by comparison with conspecifics fed a lipid-rich or carbohydrate-rich diet. They attributed this to a larger muscle mass in the lake trout fed the protein-rich diet.

There has also been some interest in the relative dietary intake of certain fatty acids, in particular following the reports that long-chain highly unsaturated fatty acids of the n-3 and n-6 series have a major influence on human cardiovascular health (Bang and Dyerberg, 1985; British Nutrition Task Force, 1992) and also influence the respiratory and cardiovascular physiology of fishes (Bell *et al.*, 1991, 1993; Randall *et al.*, 1992). The long-chain highly unsaturated arachidonic acid (ARA, C20: 4n-6), eicosapentaenoic acid (EPA, C20: 5n-3), and docosahexaenoic acid (DHA, C22: 6n-3) are physiologically essential for the normal structure and function of cell membranes and are precursors of the autocrine eicosanoids (Sargent *et al.*, 1999). In all animals, the composition of fatty acids in the tissues reflects long-term intake in the diet, because the fatty acids are stored as fuels and also inserted into membrane phospholipids (hence the old axiom "we are what we eat"). Thus, studies can be performed to change tissue fatty acid composition by providing dietary lipids as specific natural oils, chosen for their relative composition of particular fatty acids (McKenzie *et al.* 2005). Marine fish oils are well-known to be rich in the essential fatty acids, especially EPA and DHA, whereas terrestrial plant oils can provide a spectrum of others (Sargent *et al.* 1999; McKenzie 2005).

Early work on the Adriatic sturgeon indicated that accumulation of EPA and DHA from dietary fish oil influenced metabolic rate and improved tolerance of hypoxia, both by the whole animal and the *in vitro* working heart (Randall *et al.*, 1992; McKenzie *et al.*, 1995; Agnisola *et al.*, 1996, reviewed in McKenzie, 2005). This led to an exploration of the potential positive effects of these essential fatty acids upon U_{crit} performance in Atlantic salmon, *Salmon salar* (Dosanjh *et al.*, 1998; McKenzie *et al.*, 1998). These studies investigated the effects of substituting dietary fish oil with increasing amounts of a vegetable oil (Dosanjh *et al.* 1998). It did not in fact reveal any positive effect of these fatty acids but, unexpectedly, revealed a direct positive relationship between U_{crit} and muscle levels of two unrelated ones, namely oleic acid (18: 1n-9) and linoleic acid (18: 2n-6) from the vegetable oil (McKenzie *et al.* 1998, Fig. 10.7). Chatelier *et al.* (2006) subsequently found that tissue accumulation of these same 18-carbon unsaturates from specific dietary vegetable oils improved the U_{crit} performance of European sea bass, and extended the observations to demonstrate that this was linked to improved maximum cardiac performance, higher maximum aerobic metabolic rate and greater metabolic scope (Fig. 10.8). The mechanistic basis for this effect on exercise and cardiac performance is unknown. It is known that aerobic metabolism, including work by red and cardiac muscle, is fuelled primarily by β-oxidation of fats in fish (Hochachka and Somero, 1984; Richards *et al.*, 2002), and there is *in vitro* evidence to suggest that, amongst the many

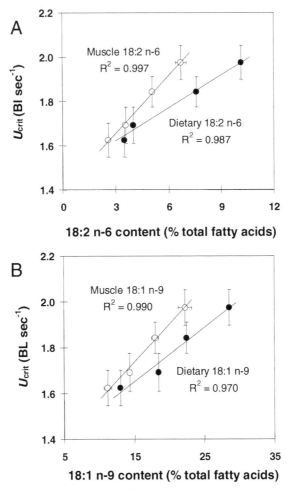

Fig. 10.7   Least squares linear regression analysis of the relationship between critical swimming speed (U_{crit}) and the content of two fatty acids, (A) linoleic acid, 18: 2n-6 and (B) oleic acid, 18: 1 n-9, in the lipids of the muscle and the diet of Atlantic salmon (*Salmo salar*). Four groups of salmon had been fed one of four diets where the dietary lipid was composed of differing proportions of fish (menhaden) oil and vegetable (canola) oil. Open symbols, relation of mean (± SE) muscle % fatty acid (FA) content versus mean (± SE) U_{crit}; closed symbols, mean dietary content versus mean (± SE) U_{crit}. N = 9 for muscle FA analyses, between 6 and seven for U_{crit}, 3 for dietary FA analyses. The group with the highest dietary and muscle content of these two FAs had a significantly higher U_{crit} than the group with the lowest levels. Modified from McKenzie *et al.* (1998) with permission from Elsevier.

fatty acids in the tissues, oleic and linoleic acids are preferred substrates for β-oxidation (Henderson and Sargent, 1985; Sidell and Driedzic, 1985; Egginton, 1996). It is conceivable, therefore, that higher levels of these preferred substrates in the tissues might allow the animals to achieve higher

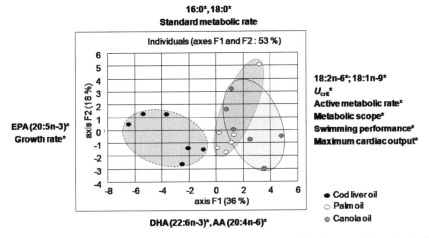

Fig. 10.8 Principal components analysis (PCA) of the relationships between the content of particular fatty acids in the heart of individual European sea bass (*Dicentrarchus labrax*) and their measured values for various physiological traits, including critical swimming speed (U_{crit}) and associated traits of cardiorespiratory performance. Three groups of sea bass had been fed one of three diets where the dietary lipid was provided as either fish (cod liver) oil, or one of two vegetable oils (palm or canola). The asterisks next to the variables on the PCA axes indicate where the mean values differed between the cod liver oil and vegetable oil groups by one-way analysis of variance. Data are taken from Chatelier *et al.* (2006).

rates of aerobic work (McKenzie, 2001, 2005). On the other hand, an interspecific study has demonstrated a positive relationship between skeletal muscle content of linoleic acid and running speed in terrestrial vertebrates, and has proposed that this is due to an effect of this fatty acid on the activity of the sarcoplasmatic $Ca^{2+}Mg^{2+}$-ATPase (Ruf *et al.*, 2006). This remains an interesting area for future research.

McKenzie *et al.* (1998) and Chatelier *et al.* (2006) found that accumulation of EPA and DHA was associated with relatively poor exercise performance. These results are, however, in direct contrast to those reported by Wagner *et al.* (2004), who found improved U_{crit} performance in Atlantic salmon fed fish oils rich in these n-3 fatty acids. The explanation for these opposing results presumably lies in the enormous complexity of factors within such diet studies. In particular, because many of the fatty acids have specific biological roles that interact with each other, and the dietary fatty acids are provided as a complex mixture in natural oils (Sargent *et al.*, 1999). Thus, diet studies are very complex to interpret and each study is effectively unique unless great care is taken to match the ingredients. There may also be minimum threshold of action for some fatty acids (Chatelier *et al.*, 2006).

There is greater variation in tissue fatty acid phenotype within wild fish populations, including for EPA, DHA, oleic and linoleic acids (e.g., Jangaard *et al.*, 1967; Klungsøyr *et al.*, 1989; Kuusipalo and Käkelä, 2000; Budge *et al.*,

2002; Iverson *et al.*, 2002), than those associated with the differences in exercise performance described by McKenzie *et al.* (1998), Wagner *et al.* (2004) and Chatelier *et al.* (2006). If such diversity in phenotype of wild fish is associated with diversity in physiological traits similar to that observed in the laboratory studies, then dietary fatty acid intake may have significant ecological implications. Many coastal marine species can exploit lagoons and estuaries during their life cycle, in which foodwebs can have larger inputs of fatty acids from terrestrial sources, hence richer in oleic and linoleic acids (Galois *et al.*, 1996). This diet may, therefore, contribute to their aerobic swimming performance. Diversities in dietary and tissue fatty acid composition within populations may contribute in some way to their overall physiological diversity, and so influence their responses to other environmental factors, such as temperature and hypoxia.

CONCLUSIONS AND PERSPECTIVES

This review demonstrates quite clearly that the aerobic swimming performance of fishes (in particular as measured within the context of the U_{crit} protocol) is profoundly influenced by all of the most important environmental factors, namely water temperature, dissolved O_2 and CO_2, salinity, toxicant load, and also by the chemistry of their diet. Clearly, the effects of these environmental factors on aerobic performance could influence such behaviours as searching for food, migrating and searching for mates.

The actual ecological significance of maximal aerobic exercise capacity remains, however, to be demonstrated, and very few studies have explored the relationship between aerobic swimming performance and fitness in natural populations (Billerbeck *et al.*, 2001; Lankford *et al.*, 2001). Nonetheless, Nelson and Claireaux (2005) and Claireaux *et al.* (2007) reported that U_{crit} was repeatable and stable over a significant portion of the lifespan of a European sea bass, such that the underlying traits of aerobic performance might indeed be potential targets for natural selection (Endler, 1986; Dohm, 2002). Thus, the major challenge for the future is to demonstrate the intuitive links between aerobic swimming and fitness.

Over the last 20–30 years, biologists have increasingly been solicited to provide legislators and conservation managers with information about the causal mechanisms underlying, for instance, a decline in biodiversity or the collapse of a population. This is particularly true for the world fishing industry, where there is an urgent need for reliable models for the sustainable management of aquatic resources. Mechanistic ecophysiology could play a critical role in integrated approaches to ecosystem management, by contributing to predictive models of how the environment influences the physiology of its inhabitants, and how this influences the fitness of their populations.

REFERENCES

Adams, S.R. and G.R. Parsons. 1998. Laboratory-based measurements of swimming performance and related metabolic rates of field-sampled smallmouth buffalo (*Ictiobus bubalus*): A study of seasonal changes. *Physiological Zoology* 71: 350–358.

Agnisola, C., D.J. McKenzie, E.W. Taylor, C.L. Bolis and B. Tota. 1996. Cardiac performance in relation to oxygen supply varies with dietary lipid composition in sturgeon. *American Journal of Physiology* 271: R417–425.

Alsop, D.H., J.C. McGeer, D.G. McDonald and C.M. Wood. 1999. Costs of chronic waterborne zinc exposure and the consequences of zinc acclimation on the gill/zinc interactions of rainbow trout in hard and soft water. *Environmental Toxicology and Chemistry* 18: 1014–1025.

Arts, M.T., R.G. Ackman and B.J. Holub. 2001. "Essential fatty acids" in aquatic ecosystems: a crucial link between diet and human health and evolution. *Canadian Journal of Fisheries and Aquatic Sciences* 58: 122–137.

Bang, H.O. and J. Dyerberg. 1985. Fish consumption and mortality from coronary heart disease. *New England Journal of Medicine* 313: 822–823.

Barrett, D.J. and E.W. Taylor. 1984. Changes in heart-rate during progressive hyperoxia in the dogfish *Scyliorhinus canicula* L—evidence for a venous oxygen receptor. *Comparative Biochemistry and Physiology* A78: 697–703.

Beamish, F.W.H. 1978. Swimming Capacity. In: *Fish Physiology*, W.S. Hoar and D.J. Randall (Eds.). Academic Press, New York, Vol. 7: Locomotion, pp. 101–187.

Beamish, F.W.H., J.C. Howlett and T.E. Medland. 1988. Impact of diet on metabolism and swimming performance in juvenile lake trout, *Salvelinus namaycush*. *Canadian Journal of Fisheries and Aquatic Sciences* 46: 384–388.

Beaumont, M.W., P.J. Butler and E.W. Taylor. 1995a. Exposure of brown trout, *Salmo trutta*, to sub-lethal copper concentrations in soft acidic water and its effects upon sustained swimming performance. *Aquatic Toxicology* 33: 45–63.

Beaumont, M.W., P.J. Butler and E.W. Taylor. 1995b. Plasma ammonia concentration in brown trout (*Salmo trutta*) exposed to acidic water and sublethal copper concentrations and its relationship to decreased swimming performance. *Journal of Experimental Biology* 198: 2213–2220.

Beaumont, M.W., P.J. Butler and E.W. Taylor. 2000a. Exposure of brown trout, *Salmo trutta*, to a sub-lethal concentration of copper in soft acidic water: effects upon muscle metabolism and membrane potential. *Aquatic Toxicology* 51: 259–272.

Beaumont, M.W., E.W. Taylor and P.J. Butler. 2000b. The resting membrane potential of white muscle from brown trout (*Salmo trutta*) exposed to copper in soft, acidic water. *Journal of Experimental Biology* 203: 2229–2236.

Beaumont, M.W., P.J. Butler and E.W. Taylor. 2003. Exposure of brown trout, *Salmo trutta*, to a sub-lethal concentration of copper in soft acidic water: effects upon gas exchange and ammonia accumulation. *Journal of Experimental Biology* 206: 153–162.

Bell, J.G., A.H. McVicar, M. Park and J.R. Sargent. 1991. High dietary linoleic acid affects fatty acid compositions of individual phospholipids from tissues of Atlantic salmon (*Salmo salar*): association with stress susceptibility and cardiac lesion. *Journal of Nutrition* 121: 1163–1172.

Bell, J.G., J.R. Dick, A.H. McVicar, J.R. Sargent and K.D. Thomson. 1993. Dietary sunflower, linseed and fish oils affect phospholipid fatty acid composition, development of cardiac lesions, phospholipase activity and eicosanoid production in Atlantic salmon (*Salmo salar*). *Prostaglandins, Leukotrienes and Essential Fatty Acids* 49: 665–673.

Bernal, D., K.A. Dickson, R.E. Shadwick and J.B. Graham. 2001. Analysis of the evolutionary convergence for high performance swimming in lamnid sharks and tunas. *Comparative Biochemistry and Physiology* A129: 695–726.

Bernal, D., J.M. Donley, R.E. Shadwick and D.A. Syme. 2005. Mammal-like muscles power swimming in a cold-water shark. *Nature* (London) 437: 1349–1352.

Billerbeck, J.M., T.E. Lankford and D.O. Conover. 2001. Evolution of intrinsic growth and energy acquisition rates. I. Trade-offs with swimming performance in *Menidia menidia. Evolution* 55: 1863–1872.

Block, B.A. and E.D. Stevens (Eds.). 2001. *Tuna: Physiology, Ecology and Evolution, Fish Physiology* Vol. 19. Academic Press, San Diego.

Bone, Q. 1978. Locomotor muscle. In: *Fish Physiology* , W.S. Hoar and D.J. Randall (Eds.). Academic Press, New York, Vol. 7: Locomotion, pp. 361–424.

Brauner, C.J., J.M. Shrimpton and D.J. Randall. 1992. The effect of short duration seawater exposure on plasma ion concentrations and swimming performance in coho salmon (*Oncorhynchus kisutch*) parr. *Canadian Journal of Fisheries and Aquatic Sciences* 49: 2399–2405.

Brauner, C.J., G.K. Iwama and D.J. Randall. 1994. The effect of short-duration seawater exposure on the swimming performance of wild and hatchery-reared juvenile coho salmon (*Oncorhynchus kisutch*) during smoltification. *Canadian Journal of Fisheries and Aquatic Sciences* 51: 2188–2194.

Brett, J.R. 1964. The respiratory metabolism and swimming performance of young sockeye salmon. *Journal of the Fisheries Research Board of Canada* 21: 1183–1226.

Brett, J.R. 1971. Energetic responces of salmon to temperature. A study of some thermal relations in the physiology and freshwater ecology of sockeye salmon (*Oncorhynchus nerka*). *American Zoologist* 11: 99–113.

Brett, J.R. and N.R. Glass. 1973. Metabolic rates and critical swimming speeds of sockeye salmon (*Oncorhynchus nerka*) in relation to size and temperature. *Journal of the Fisheries Research Board of Canada* 30: 379–387.

Brill, R.W. and P.G. Bushnell. 2001. The cardiovascular system of tunas. In: *Tuna: Physiology, Ecology and Evolution,* B.A. Block and S.D. Stevens (Eds.). Academic Press, San Diego, Fish Physiology, Vol. 19: pp. 79–120.

British Nutrition Task Force. 1992. *Unsaturated Fatty Acids: Nutritional and Physiological Significance.* Chapman and Hall, London.

Budge, S.M., S.J. Iverson, W.D. Bowen and R.G. Ackman. 2002. Among- and within-species variability in fatty acid signatures of marine fish and invertebrates on the Scotian Shelf, Georges Bank and southern Gulf of St. Lawrence. *Canadian Journal of Fisheries and Aquatic Sciences* 59: 886–898.

Bushnell, P.G., J.F. Steffensen and K. Johansen. 1984. Oxygen consumption and swimming performance in hypoxia-acclimated rainbow trout *Salmo gairdneri. Journal of Experimental Biology* 113: 225–235.

Butler, P.J., N. Day and K. Namba. 1992. Interactive effects of seasonal temperature and low pH on resting oxygen uptake and swimming performance of adult brown trout *Salmo trutta. Journal of Experimental Biology* 165: 195–212.

Cairns, J. 1966. Don't be half-safe—The current revolution in bioassay techniques. *Engineering Bulletin of Purdue University Proceedings* 21: 559–567.

Castro-Santos, T. 2005. Optimal swim speeds for traversing velocity barriers: an analysis of volitional high-speed swimming behavior of migratory fishes. *Journal of Experimental Biology* 208: 421–432.

Chatelier A., D.J. McKenzie and G. Claireaux. 2005. Effects of changes in salinity upon exercise and cardiac performance in the European sea bass *Dicentrarchus labrax. Marine Biology* 147: 855–862.

Chatelier, A., D.J. McKenzie, A. Prinet, R. Galois, J. Robin, J. Zambonino and G. Claireaux. 2006. Associations between tissue fatty acid composition and physiological traits of performance and metabolism in the sea bass (*Dicentrarchus labrax*). *Journal of Experimental Biology* 209: 3429–3439.

Claireaux, G. and J.-P. Lagardère. 1999. Influence of temperature, oxygen and salinity on the metabolism of European sea bass. *Journal of Sea Research* 42: 157–168.

Claireaux, G., D.M. Webber, S.R. Kerr and R.G. Boutilier 1995a. Physiology and behaviour of free-swimming Atlantic cod, *Gadus morhua*, facing fluctuating temperature conditions. *Journal of Experimental Biology* 198: 49–60.

Claireaux, G., D.M. Webber, S.R. Kerr and R.G. Boutilier 1995b. Physiology and behaviour of free-swimming Atlantic cod, *Gadus morhua*, facing fluctuating salinity and oxygenation conditions. *Journal of Experimental Biology* 198: 49–60.

Claireaux, G., D.J. McKenzie, G. Genge, A. Chatelier and A.P. Farrell. 2005. Linking swimming performance, cardiac performance and cardiac morphology in rainbow trout. *Journal of Experimental Biology* 208: 1775–1784.

Claireaux, G., C. Couturier and A.L. Groison. 2006. Effect of temperature on maximum swimming speed and cost of transport in juvenile European sea bass, *Dicentrarchus labrax*. *Journal of Experimental Biology* 209: 3420–3428.

Claireaux, G., C. Handelsman, E. Standen and J.A. Nelson. 2007. Thermal and temporal stability of swimming performance in the European sea bass. *Physiological and Biochemical Zoology* 80: 186–196.

Crawford, M. and D. Marsh. 1989. *The Driving Force: Food, Evolution and the Future*. Harper and Row, New York.

Dahlberg, M.L., D.L. Shumway and P. Doudoroff. 1968. Influence of dissolved oxygen and carbon dioxide on swimming performance of largemouth bass and coho salmon. *Journal of the Fisheries Research Board of Canada* 25: 49–70.

Davis, G.E., J. Foster, C.E. Warren and P. Doudoroff. 1963. The influence of oxygen concentration on the swimming performance of juvenile Pacific salmon at various temperatures. *Transaction of the American Fisheries Society* 92: 111–24.

Day, N. and P.J. Butler. 2005. The effects of acclimation to reverse seasonal temperatures on the swimming performance of adult brown trout *Salmo trutta*. *Journal of Experimental Biology* 208: 2683–2692.

Diaz, R.J. 2001. Overview of hypoxia around the world. *Journal of Environmental Quality* 30: 275–81.

Diaz, R.J., J. Nestlerode and M.L. Diaz. 2004. A global perspective on the effects of eutrophication and hypoxia on aquatic biota. In: *The 7th International Symposium on Fish Physiology, Toxicology, and Water Quality*, Tallinn, Estonia, G.L. Rupp and M.D. White (Eds.). Athens, GA: U.S. Environmental Protection Agency Research Division, EPA 600/R-04/049, pp. 1–33.

Dickson, K.A., J.M. Donley, C. Sepulveda and L. Bhoopat. 2002. Effects of temperature on sustained swimming performance and swimming kinematics of the chub mackerel (*Scomber japonicus*). *Journal of Experimental Biology* 203: 3103–3116.

Dohm, M.R. 2002. Repeatability estimates do not always set an upper limit to heritability. *Functional Ecology* 16: 273–280.

Domenici, P., J.F. Steffensen and R.S. Batty. 2000. The effect of progressive hypoxia on swimming activity and schooling in Atlantic herring. *Journal of Fish Biology* 57: 1526–1538.

Domenici, P., G. Claireaux and D.J. McKenzie. 2007. Environmental constraints upon locomotion and predator-prey interactions in aquatic organisms. *Philosophical Transactions of the Royal Society* B362: 1929–1936.

Donley, J.M., C.A. Sepulveda, P. Konstantinidis, S. Gemballa and R.E. Shadwick. 2004. Convergent evolution in mechanical design of lamnid sharks and tunas. *Nature* (London) 429: 61–5.

Dosanjh, B., D.A. Higgs, D.J. McKenzie, G. Deacon and D.J. Randall. 1998. Influence of blends of menhaden oil and canola oil on the performance, muscle lipid composition and thyroidal status of Atlantic salmon (*Salmo salar*) in seawater. *Fish Physiology and Biochemistry* 19: 123–134.

Driedzic, W.R. and H. Gesser. 1994. Energy metabolism and contractility in ectothermic vertebrate hearts: hypoxia, acidosis and low temperature. *Physiological Reviews* 74: 221–258.

Dutil, J.-D., E.-L. Sylvestre, L. Gamache, R. Laroche and H. Guderley. 2007. Burst and coast use, swimming performance and metabolism of Atlantic cod *Gadus morhua* in sub-lethal hypoxic conditions. *Journal of Fish Biology* 71: 363–375.

Egginton, S. 1996. Effect of temperature on optimal substrate for β-oxidation. *Journal of Fish Biology* 49: 753–758.

Endler, P. 1986. *Natural Selection in the Wild.* Princeton University Press, Princeton.

Farmer, C.G. and D.C. Jackson. 1998. Air-breathing during activity in the fishes *Amia calva* and *Lepisosteus oculatus. Journal of Experimental Biology* 201: 943–948.

Farrell, A.P. 2002. Cardiorespiratory performance in salmonids during exercise at high temperature: insights into cardiovascular design limitations in fishes. *Comparative Biochemistry and Physiology* A132: 797–810.

Farrell, A.P. 2007. Cardiorespiratory performance during prolonged swim tests with salmonids: a perspective on the temperature effects and potential analytical pitfalls. *Philosophical Transactions of the Royal Society* B362: 2017–2030.

Farrell, A.P. and S.M. Clutterham. 2003. On-line venous oxygen tensions in rainbow trout during graded exercise at two acclimation temperatures. *Journal of Experimental Biology* 206: 487–496.

Farrell, A.P., A.K. Gamperl and I.K. Birtwell. 1998. Prolonged swimming, recovery and repeat swimming performance of mature sockeye salmon *Oncorhynchus nerka* exposed to moderate hypoxia and pentachlorophenol. *Journal of Experimental Biology* 201: 2183–2193.

Farrell, A.P., M. Axelsson, J. Altimiras, E. Sandblom and G. Claireaux. 2007. Maximum cardiac performance and adrenergic sensitivity of the sea bass *Dicentrarchus labrax* at high temperatures. *Journal of Experimental Biology* 210: 1216–1224.

Fry, F.E.J. 1947. The effects of the environment on animal activity. *University of Toronto Studies, Biological Series* 55. Publication of the Ontario Fisheries Research Laboratory 68: 1–62.

Fry, F.E.J. 1971. The effect of environmental factors on the physiology of fish. In: *Fish Physiology*, W.S. Hoar and D.J. Randall (Eds.). Academic Press, New York, Vol. 6, pp. 1–98.

Gallaugher, P.E., H. Thorarensen and A.P. Farrell. 1995. Hematocrit in oxygen transport and swimming in rainbow trout (*Oncorhynchus mykiss*). *Respiration Physiology* 102: 279–292.

Gallaugher, P. E., H. Thorararensen, A. Kiessling and A.P. Farrell. 2001. Effects of high intensity exercise training on cardiovascular function, oxygen uptake, internal oxygen transfer and osmotic balance in chinook salmon (*Oncorhynchus tshawytscha*) during critical speed swimming. *Journal of Experimental Biology* 204: 2861–2872.

Galois, R., P. Richard and B. Fricourt. 1996. Seasonal variations in suspended particulate matter in the Marennes-Oléron Bay, France, using lipids as biomarkers. *Estuarine and Coastal Shelf Science* 43: 335–357.

Garofalo, E., S. Ceradini and M.J. Winter. 2004. The use of diffusive gradients in thin-film (DGT) passive samplers for the measurement of bioavailable metals in river water. *Annali di Chimica* 94: 515–20.

Ghalambor, C.K., J.A. Walker and D.N. Reznick. 2003. Multi-trait selection, adaptation, and constraints on the evolution of burst swimming performance. *Integrative and Comparative Biology* 43: 431–438.

Graham, J.B. 1997. *Air-breathing Fishes: Evolution, Diversity, and Adaptation.* Academic Press, San Diego.

Graham, J.B. and K.A. Dickson. 2001. Anatomical and physiological specializations for endothermy. In: *Tuna: Physiology, Ecology and Evolution*, B.A. Block and S.D. Stevens (Eds.). Academic Press, San Diego, Fish Physiology, Vol. 19, pp. 121–165.

Graham, J.B. and K.A. Dickson. 2004. Tuna comparative physiology. *Journal of Experimental Biology* 207: 4015-4024.

Heisler, N. 1984. Acid-base regulation in fishes. In: *Fish Physiology*, W.S. Hoar and D.J. Randall (Eds.). Academic Press, New York, Vol. 10A: Acid-Base Balance, pp. 315–401.

Heisler, N. 1993. Acid-base regulation. In: *The Physiology of Fishes*, D.H. Evans (Ed.). CRC Press, Boca Raton, pp. 343–378.

Henderson, R.J. and J.R. Sargent. 1985. Chain-length specificities of mitonchondrial and peroxisomal β-oxidation of fatty acids in livers of rainbow trout (*Salmo gairdneri*). *Comparative Biochemistry Physiology* B82: 79-85.

Hiddink, J.G. and R. ter Hofstede 2008. Climate induced increases in species richness of marine fishes. *Global Change Biology* 14: 453–460.

Hoar, W.S. 1976. Smolt transformation: Evolution, behaviour and physiology. *Journal of the Fisheries Research Board of Canada* 33: 1234–1252.

Hochachka, P.W. and G.N. Somero. 1984. *Biochemical Adaptation*. Princeton University Press, Princeton.

Hopkins, W.A., J.W. Snodgrass, B.P. Staub, B.P. Jackson, J.D. Congdon. 2003. Altered swimming performance of a benthic fish (*Erimyzon sucetta*) exposed to contaminated sediments. *Archives of Environmental Contamination and Toxicology* 44: 383–389.

Howard, T.E. 1975. Swimming performance of juvenile coho salmon (*Oncorhynchus kisutch*) exposed to bleached kraft pulpmill effluent. *Journal of the Fisheries Research Board of Canada* 32: 789–793.

Ishimatsu, A., M. Hayashi, K.S. Lee, T. Kikkawa, and J. Kita. 2005. Physiological effects on fishes in a high-CO$_2$ world. *Journal of Geophysical Research-Oceans* 110: 68–79.

Iverson, S.J., K.J. Frost and S.L.C. Lang. 2002. Fat content and fatty acid composition of forage fish and invertebrates in Prince William Sound, Alaska: factors contributing to among- and within species variability. *Marine Ecology Progress Series* 241: 161–181.

Jain, K.E., I.K. Birtwell and A.P. Farrell. 1998. Repeat swimming performance of mature sockeye salmon following a brief recovery period: a proposed measure of fish health and water quality. *Canadian Journal of Zoology* 76: 1488–1496.

Jangaard, P.M., R.G. Ackman and J.C. Sipos. 1967. Seasonal changes in fatty acid composition of cod liver, flesh, roe and milt lipids. *Journal of the Fisheries Research Board of Canada* 24: 613–627.

Johnston, I.A. and G.K. Temple. 2002. Thermal plasticity of skeletal muscle phenotype in ectothermic vertebrates and its significance for locomotory behaviour. *Journal of Experimental Biology* 205: 2305–2322.

Jones, D.R. and D.J. Randall. 1978. The respiratory and circulatory systems during exercise. In: *Fish Physiology*, W. S. Hoar and D.J. Randall (Eds.). Academic Press, New York, Vol. 7: Locomotion, pp. 425–501.

Jones, D.R. 1971. The effect of hypoxia and anaemia on the swimming performance of rainbow trout (*Salmo gairdneri*). *Journal of Experimental Biology* 55: 541–51.

Kiceniuk, J.W. and D.R. Jones. 1977. The oxygen transport system in trout (*Salmo gairdneri*) during sustained exercise. *Journal of Experimental Biology* 69: 247–260.

Klungsøyr, J., S. Tilseth, S. Wilhelmsen, S. Falk-Petersen and J.R. Sargent. 1989. Fatty acid composition as an indicator of food intake on cod larvae *Gadus morhua* from Lofoten, Northern Norway. *Marine Biology* 102: 183–188.

Kolok, A.S. and D. Sharkey. 1997. Effect of freshwater acclimation on the swimming performance and plasma osmolarity of the euryhaline gulf killifish. *Transaction of the American Fisheries Society* 126: 866–870.

Korsmeyer, K.E. and H. Dewar. 2001. Tuna metabolism and energetics. In: *Tuna: Physiology, Ecology and Evolution*, B.A. Block and S.D. Stevens (Eds.). Academic Press, San Diego, Fish Physiology, Vol. 19, pp. 35–78.

Korsmeyer, K.E., N.C. Lai, R.E. Shadwick and J.B. Graham. 1997. Oxygen transport and cardiovascular responses to exercise in yellowfin tuna *Thunnus albacores*. *Journal of Experimental Biology* 200: 1987–1997.

Korsmeyer, K.E., J.F. Steffensen, and J. Herskin. 2002. Energetics of median and paired fin swimming, body and caudal fin swimming, and gait transition in parrotfish (*Scarus schlegeli*) and triggerfish (*Rhinecanthus aculeatus*). *Journal of Experimental Biology* 205: 1253–1263.

Kuusipalo, L. and R. Käkelä. 2000. Muscle fatty acids as indicators of niche and habitat in Malawian cichlids. *Limnology and Oceanography* 45: 996–1000.

Lankford, T.E., J.M. Billerbeck and D.O. Conover. 2001. Evolution of intrinsic growth and energy acquisition rates. II. Trade-offs with vulnerability to predation in *Menidia menidia. Evolution* 55: 1873–1881.

Lee, C.G., A.P. Farrell, A. Lotto, M.J. MacNutt, S.G. Hich and M.C. Healey. 2003. The effect of temperature on swimming performance and oxygen consumption in adult sockeye (*Oncorhynchus nerka*) and coho (*O. kisutch*) salmon stocks. *Journal of Experimental Biology* 206: 3239–3251.

Lefrançois, C. and G. Claireaux. 2003. Influence of ambient oxygenation and temperature on metabolic scope and scope for heart rate in the common sole, *Solea solea. Marine Ecology Progress Series* 259: 273–284.

McFarlane, W.J. and D.G. McDonald. 2002. Relating intramuscular fuel use to endurance in juvenile rainbow trout. *Physiological and Biochemical Zoology* 75: 250–259.

MacKinnon, D.L. and A.P. Farrell. 1992. The effects of 2-(thiocyanomethylthio) benzothiazole on juvenile coho salmon (*Oncorhynchus kisutch*) - sublethal toxicity testing. *Environmental Toxicology and Chemistry* 11: 1541–1548.

McKenzie, D.J. 2001. Effects of dietary fatty acids on the respiratory and cardiovascular physiology of fish. *Comparative Biochemistry and Physiology* A128: 607–621.

McKenzie, D.J. 2005. Effects of dietary fatty acids on the physiology of environmental adaptation in fish. In: *Physiological and Ecological Adaptations to Feeding in Vertebrates*, J.M. Starcke and T. Wang (Eds.). Science Publishers, Enfield, (NH), pp. 363–388.

McKenzie D.J., G. Piraccini, J.F. Steffensen, C.L. Bolis, P. Bronzi and E.W. Taylor. 1995. Effects of diet on spontaneous locomotor activity and oxygen consumption in the Adriatic sturgeon (*Acipenser naccarii*). *Fish Physiology and Biochemistry* 14: 341–355.

McKenzie D.J., G. Piraccini, N. Papini, C. Galli, P. Bronzi, C.G. Bolis and E.W. Taylor. 1997. Oxygen consumption and ventilatory reflex responses are influenced by dietary lipids in sturgeon. *Fish Physiology and Biochemistry* 16: 365–379.

McKenzie, D.J., D.A. Higgs, B. Dosanjh, G. Deacon and D.J Randall. 1998. Dietary lipid composition influences swimming performance in Atlantic salmon (*Salmo salar*) in seawater. *Fish Physiology and Biochemistry* 19: 111–122.

McKenzie, D.J., E. Cataldi, S. Owen, E.W. Taylor and P. Bronzi. 2001a. Effects of acclimation to brackish water on the growth, respiratory metabolism and exercise performance of Adriatic sturgeon (*Acipenser naccarii*). *Canadian Journal Fisheries and Aquatic Sciences* 58: 1104–1112.

McKenzie D.J., E. Cataldi, E.W. Taylor, S. Cataudella and P. Bronzi. 2001b. Effects of acclimation to brackish water on tolerance of salinity challenge by Adriatic sturgeon (*Acipenser naccarii*). *Canadian Journal of Fisheries and Aquatic Sciences* 58: 1113–1120.

McKenzie D.J., A.Z. Dalla Valle, M. Piccolella, E.W. Taylor and J.F. Steffensen. 2003a. Tolerance of chronic hypercapnia by the European eel (*Anguilla anguilla*). *Journal of Experimental Biology* 206: 1717–1726.

McKenzie, D.J., A. Shingles and E.W. Taylor. 2003b. Sub-lethal plasma ammonia accumulation and the swimming performance of salmonids. *Comparative Biochemistry and Physiology* A135: 515–526.

McKenzie D.J., S. Wong, D.J. Randall, S. Egginton, E.W. Taylor and A.P. Farrell. 2004. The effects of sustained exercise and hypoxia upon oxygen tensions in the red muscle of rainbow trout. *Journal of Experimental Biology* 207: 3629–3637.

McKenzie D.J., A. Shingles, E.W. Taylor and P. Domenici 2006. Effects of ammonia on the locomotor performance of fishes. In: *Fish Physiology, Fish Toxicology and Fisheries Management*, C.J. Brauner, K. Suvajdzic, G. Nilsson, D.J. Randall (Eds.). Ecosystem Research Division publication EPA/600/R-07/010, United States Environmental Protection Agency, Athens, Georgia, pp. 49–64.

McKenzie D.J., M. Hale and P. Domenici. 2007a. Locomotion in primitive fishes. In: *Primitive Fishes*, D.J. McKenzie, C.J. Brauner and A.P Farrell (Eds.). Academic Press, San Diego, California, Fish Physiology, Vol. 26, pp. 162–224.

McKenzie, D.J., E. Garofalo, M.J. Winter, F. Verweij, S. Ceradini, N. Day, R. van der Oost P.J. Butler, J.K. Chipman and E.W. Taylor. 2007b. Complex physiological traits as biomarkers of the sub-lethal toxicological effects of pollutant exposure in fishes. *Philosophical Transactions of the Royal Society* B362: 2043–2059.

McLeay, D.J. and D.A. Brown. 1979. Stress and chronic effects of untreated and treated bleached kraft pulpmill effluent on the biochemistry and stamina of juvenile coho salmon (*Oncorhynchus kisutch*). *Journal of the Fisheries Research Board of Canada* 36: 1049–1059.

Morgan, J.D. and G.K. Iwama. 1991. Effects of salinity on growth, metabolism and ion regulation in juvenile rainbow and steelhead trout (*Oncorhynchus mykiss*) and fall chinook salmon (*O. tshawytscha*). *Canadian Journal of Fisheries and Aquatic Sciences* 48: 2083–2094.

Nelson, J.A. and G. Claireaux. 2005. Sprint swimming performance of juvenile European sea bass. *Transactions of the American Fisheries Society* 134: 1274–1284.

Nelson, J.A., P.S. Gotwalt, S.P. Reidy and D.M. Webber. 2002. Beyond U_{crit}: matching swimming performance tests to the physiological ecology of the animal, including a new fish 'drag strip'. *Comparative Biochemistry and Physiology* A133: 289–302.

Nelson, J.A., Y. Tang and R.G. Boutilier. 1996. The effects of salinity change on the exercise performance of two Atlantic cod (*Gadus morhua*) populations inhabiting different environments. *Journal of Experimental Biology* 199: 1295–1309.

Nikl, D.L. and A.P. Farrell. 1993. Reduced swimming performance and gill structural changes in juvenile salmonids exposed to 2-(thiocyanomethylthio)benzothiazole. *Aquatic Toxicology* 27: 245–263.

Nilsson, G.E., P. Rosen and D. Johansson. 1993. Anoxic depression of spontaneous locomotor activity in crucian carp quantified by a computerized imaging technique. *Journal of Experimental Biology* 180: 153–162.

Pane, E.F., A. Haque, G.G. Goss and C.M. Wood. 2004. The physiological consequences of exposure to chronic, sub-lethal waterborne nickel in rainbow trout (*Oncorhynchus mykiss*): exercise vs resting physiology. *Journal of Experimental Biology* 207: 1249–1261.

Peake, S.J. and A.P. Farrell. 2004. Locomotory behavior and post-exercise physiology in relation to swimming speed, gait transition and metabolism in free-swimming smallmouth bass (*Micropterus dolomieu*). *Journal of Experimental Biology* 207: 1563–1575.

Peake, S.J. and A.P. Farrell. 2006. Fatigue is a behavioural response in respirometer confined smallmouth bass. *Journal of Fish Biology* 68: 1742–1755.

Perry, A.L., P.J. Low, J.R. Ellis and J.D. Reynolds 2005. Climate change and distribution shifts in marine fishes. *Science* 308: 1912–1915.

Petersen, R.H. 1974. Influence of fenitrothion on swimming velocities of brook trout (*Salvelinus fontinalis*). *Journal of the Fisheries Research Board of Canada* 31: 1757–1762.

Pickett, G.D. and M.G. Pawson. 1994. *Sea bass: biology, exploitation and conservation. Fish and Fisheries Series 12.* Chapman and Hall, London.

Plaut, I. 2001. Critical swimming speed: its ecological relevance. *Comparative Biochemistry and Physiology* A131: 41–50.

Portner, HO and R. Knust. 2007. Climate change affects marine fishes through the oxygen limitation of thermal tolerance. *Science* 315: 95–97.

Randall, D.J. 1982. The control of respiration and circulation in fish during exercise and hypoxia. *Journal of Experimental Biology* 100: 175–188.

Randall, D.J. and C. Brauner. 1991. Effects of environmental factors on exercise in fish. *Journal of Experimental Biology* 160: 113–126.

Randall, D.J., W.W. Burggren, A.P. Farrell and M.S. Haswell. 1981. *The Evolution of Air-Breathing in Vertebrates.* Cambridge University Press, Cambridge.

Randall D.J., D.J. McKenzie, G. Abrami, G.P. Bondiolotti, F. Natiello, P. Bronzi, L. Bolis and E. Agradi. 1992. Effects of diet on responses to hypoxia in the sturgeon (*Acipenser naccarii*). *Journal of Experimental Biology* 170: 113–125.

Richards, J.G., G.J.F. Heigenhauser and C.M. Wood. 2002. Lipid oxidation fuels recovery from exhaustive exercise in white muscle of rainbow trout. *American Journal of Physiology* 282: R89–R99.

Rochard, E., P. Williot, G. Castelnaud and M. Lepage. 1991. Elements de systematique et de biologie des populations sauvage d'esturgeons. In *Acipenser, Actes du Premier Colloque International sur l'Esturgeon*, P. Williot, (Ed.). CEMAGREF, Bordeaux, France, pp. 475–507.

Rome, L.C. 1990. Influence of temperature on muscle recruitment and muscle function *in vivo*. *American Journal of Physiology* 28: R210–R222.

Rome, L.C. 2007. The effect of temperature and thermal acclimation on the sustainable performance of swimming scup. *Philosophical Transactions of the Royal Society* B362: 1995–2016.

Rome, L.C., P.T. Loughna and G. Goldspink. 1984. Muscle fibre activity in carp as a function of swimming speed and muscle temperature. *American Journal of Physiology* 247: 272–279.

Rome, L.C., A.A. Sosnicki and I. Choi. 1992. The influence of temperature on muscle function in the fast swimming scup. I. Shortening velocity and muscle recruitment during swimming. *Journal of Experimental Biology* 163: 259–279.

Ruf, T., T. Valencak, F. Tataruch and W. Arnold. 2006. Running speed in mammals increases with muscle n-6 polyunsaturated fatty acid content. *PLoS ONE* 1: e65. doi: 10.1371/journal.pone.0000065

Sänger, A.M. and W. Stoiber. 2001. Muscle fiber diversity and plasticity. In: *Muscle Development and Growth*, I.A. Johnston (Ed.). Academic Press, San Diego, Fish Physiology, Vol. 18, pp. 187–250.

Sargent, J., G. Bell, L. McEvoy, D. Tocher and A. Estevez. 1999. Recent developments in the essential fatty acid nutrition of fish. *Aquaculture* 177: 191–199.

Schurmann, H. and J.F. Steffensen. 1994. Spontaneous swimming activity of Atlantic cod *Gadus morhua* exposed to graded hypoxia at three temperatures. *Journal of Experimental Biology* 197: 29–142.

Seibel, B.A. and J.C. Drazen. 2007. Thermal substitution and aerobic efficiency: measuring and predicting effects of heat balance on endotherm diveing energetics. *Philosophical Transactions of the Royal Society* B362: 2079–2093.

Shadwick, R.E. and S. Gemballa. 2006. Structure, kinematics, and muscle dynamics in undulatory swimming. In: *Fish Biomechanics*, R.E. Shadwick and G.V. Lauder (Eds.). Academic Press/Elsevier, San Diego, Fish Physiology, Vol. 23, pp. 241–280.

Shingles, A., D.J. McKenzie, E.W. Taylor, A. Moretti, P.J. Butler and S. Ceradini. 2001. Effects of sub-lethal ammonia exposure on swimming performance in rainbow trout (*Oncorhynchus mykiss*). *Journal of Experimental Biology* 204: 2699–2707.

Sidell, B.D. and W.R. Driedzic. 1985. Relationship between cardiac energy metabolism and cardiac work demand in fishes. In: *Circulation, Respiration and Metabolism*, G. Gilles (Ed.). Springer-Verlag, Berlin, pp. 381–401.

Sprague, J.B. 1971. Measurement of pollutant toxicity to fish—III. Sublethal effects and "safe" concentrations. *Water Research* 5: 245–266.

Standen, E.M., S.G. Hinch and P.S. Rand. 2004. Influence of river speed on path selection by migrating adult sockeye salmon (*Oncorhynchus nerka*). *Canadian Journal of Fisheries and Aquatic Sciences* 61: 905–912.

Standen, E.M., S.G. Hinch, M.C. Healey and A.P. Farrell. 2002. Energetic costs of migration through the Fraser River Canyon, British Columbia, in adult pink (*Oncorhynchus gorbuschka*) and sockeye (*O. nerka*) salmon as assessed by EMG telemetry. *Canadian Journal of Fisheries and Aquatic Sciences* 59: 1809–1818.

Stegeman, J.J., K.W. Rento, B.R. Woodin, Y.-S. Zhang and R.F. Addison. 2002. Molecular responses to environmental contamination: enzyme and protein systems as indicators of chemical exposure and effect. In: *Biomarkers: Biochemical, Physiological and Histological*

Markers of Anthropogenic Stress, R.J. Hugget, R.A. Kimerly, M. Mehrle and H.L. Bergman (Eds.). Lewis Publishers, Chelsea MI, USA, pp. 235–335.

Stevens, E.D. and D.J. Randall. 1967. Changes in gas concentrations in blood and water during moderate swimming activity in rainbow trout. *Journal of Experimental Biology* 46: 329–337.

Syme, D.A. 2006. Functional properties of skeletal muscle. In: *Fish Biomechanics*, R.E. Shadwick and G.V. Lauder (Eds.). Academic Press/Elsevier, San Diego, Fish Physiology, Vol. 23, pp. 179–240.

Taylor, E.W., S. Egginton, S.E. Taylor and P.J. Butler. 1997. Factors which may limit swimming performance at different temperature. In: *Global warming: implications for freshwater and marine fish*, C.M. Wood and D.G. McDonald (Eds.). Society for Experimental Biology Seminar Series 61, Cambridge University Press, Cambridge, pp. 105–133.

Taylor, L.N., W.J. McFarlane, G.G. Pyle, P. Couture and D.G. McDonald. 2004. Use of performance indicators in evaluating chronic metal exposure in wild yellow perch (*Perca flavescens*). *Aquatic Toxicology.* 67: 371–385.

Taylor, S.E., S. Egginton and E.W. Taylor. 1996. Seasonal temperature acclimatisation of rainbow trout: cardiovascular and morphometric influences on maximum sustainable exercise level. *Journal of Experimental Biology* 199: 835–845.

Thorarensen, H., P.E. Gallaugher and A.P. Farrell. 1996. Cardiac output in swimming rainbow trout, *Oncorhynchus mykiss*, acclimated to seawater. *Physiology Zoology* 69: 139–153.

Van der Oost, R., J. Beyer and N.P.E. Vermeulen. 2003. Fish bioaccumulation and biomarkers in environmental risk assessment: a review. *Environmental Toxicology and Pharmacology* 13: 57–149.

Varsamos, S., J.P. Diaz, G. Charmantier, G. Flik, C. Blasco and R. Connes. 2002. Branchial chloride cells in seabass (*Dicentrarchus labrax*) adapted to freshwater, seawater and doubly concentrated seawater. *Journal of Experimental Zoology* 293: 12–26.

Verity, P.G., V. Smetacek and T.J. Smayda. 2002. Status, trends and the future of the marine pelagic ecosystem. *Environmental Conservation* 29: 207–237.

Verweij, F., K. Booij, K. Satumalay, N. van der Molen and R. van der Oost. 2004. Assessment of bioavailable PAH, PCB and OCP concentrations in water, using semipermeable membrane devices (SPMDs), sediments, and caged carp. *Chemosphere* 54: 1675–1689.

Videler, J.J. 1993. *Fish Swimming*. Chapman and Hall, London.

Wagner, G.N., S.K. Balfry, D.A. Higgs, S.P. Lall and A.P. Farrell. 2004. Dietary fatty acid composition affects the repeat swimming performance of Atlantic salmon in seawater. *Comparative Biochemistry and Physiology* A137: 567–576.

Waiwood, K.G. and F.W.H. Beamish. 1978. Effects of copper, pH and hardness on the critical swimming performance of rainbow trout (*Salmo gairdneri* Richardson). *Water Research* 12: 611–619.

Webb, P.W. 1993. Swimming. In: *The Physiology of Fishes*, D.H. Evans (Ed.). CRC Press, Boca Raton, pp. 47–73.

Webb, P.W. 1998. Swimming. In: *The Physiology of Fishes*, 2nd Edition, D.H. Evans (Ed.). CRC Press, Boca Raton, pp. 3–24.

Webber, D.M., R.G. Boutilier and S.R. Kerr. 1998. Cardiac output as a predictor of metabolic rate in cod *Gadus morhua*. *Journal of Experimental Biology* 201: 2779–2789.

Wendelaar Bonga, S.E. 1997. The stress response in fish. *Physiological Reviews* 77: 591–625.

Wicks, B.J., R. Joensen, Q. Tang and D.J. Randall. 2002. Swimming and ammonia toxicity in salmonids: the effect of sub lethal exposure on the swimming performance of coho salmon and the acute toxicity of ammonia in swimming and resting rainbow trout. *Aquatic Toxicology.* 59: 55–69.

Wilson, R.W. and S.E. Egginton. 1994. Assessment of maximum sustainable swimming performance in rainbow trout (*Oncorhynchus mykiss*). *Journal of Experimental Biology* 192: 299–305.

Wilson, R.W., H.L. Bergman and C.M. Wood. 1994. Metabolic costs and physiological consequences of acclimation to aluminium in juvenile rainbow trout (*Oncorhynchus mykiss*). 2. Gill morphology, swimming performance, and aerobic scope. *Canadian Journal of Fisheries and Aquatic Sciences* 51: 527–535.

Winter, M.J., F. Verweij, E. Garofalo, S. Ceradini, D.J. McKenzie, M.A. Williams, E.W. Taylor, P.J. Butler, R. van der Oost and J.K. Chipman. 2005. Tissue levels and biomarkers of organic contaminants in feral and caged chub (*Leuciscus cephalus*) from rivers in the West Midlands, UK. *Aquatic Toxicology* 73: 394–405.

Wood, A.W., B.D. Johnston, A.P. Farrell and C.J. Kennedy. 1996. Effects of didecyldimethylammonium chloride (DDAC) on the swimming performance, gill morphology, disease resistance, and biochemistry of rainbow trout (*Oncorhynchus mykiss*). *Canadian Journal of Fisheries and Aquatic Sciences* 53: 2424–2432.

Wood, C.M. 1991. Acid-base and ion balance, metabolism, and their interactions, after exhaustive exercise in fish. *Journal of Experimental Biology* 160: 285–308.

Wu, R.S.S. 2002. Hypoxia: from molecular responses to ecosystem responses. *Marine Pollution Bulletin* 45: 35–45.

Ye, X. and D.J. Randall. 1991. The effect of water pH on swimming performance in rainbow trout (*Salmo gairdneri* R.). *Fish Physiology and Biochemistry* 9: 15–21.

11

Swimming Speeds in Larval Fishes: from Escaping Predators to the Potential for Long Distance Migration

Rebecca Fisher[1] and Jeffrey M Leis[2]

INTRODUCTION

The vast majority of marine, teleost fishes undergo a life history phase involving some kind of larval stage, typically spending at least some time in the open ocean. For many reef associated fishes, this represents a distinct and ecologically different phase in their life cycle, potentially instilling a unique set of demands on their swimming capabilities. Survival of larval fishes is very low, and even small changes in survival rates can have a substantial influence on the success of a cohort and subsequent levels of recruitment into the adult population (Houde, 1989b). Small size at birth makes larval fishes highly vulnerable to predation and the rapid development of swimming capabilities is essential for escaping from predators. In addition, their small size, along with the necessity for rapid growth to enhance survival, means larval fishes will have high metabolic requirements and will be vulnerable to starvation (Nilsson *et al.*, 2007). Coupled

Authors' addresses: [1]Australian Institute of Marine Science, University of Western Australia, 35 Stirling Highway, Crawley, 6009, Australia. E-mail: r.fisher@aims.gov.au
[2]Ichthyology, Australian Museum, 6 College St, Sydney, NSW 2010, Australia.

with the apparent paucity and patchiness of available food in the ocean the development of cruising swimming capabilities and efficient search behaviour is critical to their foraging success, dramatically influencing search volumes and prey encounter rates.

In addition to these challenges for predator avoidance and foraging, a bipartite life cycle may also place demands on the swimming ability of reef fish larvae when they attempt to adopt a benthic lifestyle. Many species require suitable adult and/or juvenile habitat in which to settle. Although some species may be able to delay metamorphosis and avoid settlement for extended periods (McCormick, 1999), finding a settlement site is ultimately essential to their survival. Because currents in the ocean and around reefs can be quite strong (Hamner and Hauri, 1981; Andrews, 1983; Wolanski and Pickard, 1983; Frith *et al.*, 1986; Cowen and Castro, 1994; Pitts, 1994), larvae may be carried large distances in the several days to months before they are ready to leave the pelagic environment and transition to the adult/ juvenile habitat. Consequently, larvae may need to cover considerable distances in order to locate suitable settlement habitat, requiring substantial sustained swimming capabilities. If the settlement habitat is particularly rare or isolated, avoiding advection altogether may be the only viable way of achieving settlement and ensuring cohort survival (Cowen *et al.*, 2000). Clearly, an understanding of the ability of larvae to relocate spatially on relatively large scales, or, alternatively, to remain near their natal reef, will be important to understanding their potential to behaviourally influence their horizontal location in the ocean, and their overall dispersal patterns. Furthermore, their swimming abilities will define the degree to which settlement stage larvae are able to locate and choose specific habitats, and the extent to which the timing of settlement is an active versus passive process. With the increasing trend towards the implementation of marine protected areas as a management tool designed to enhance production and maximize the number of returning larvae, an understanding of connectivity of marine populations through larval dispersal and recruitment patterns is crucial to ensure sufficient gene flow among reserve networks as well as adequate self-sustainability, and to gauge the potential for larval spill over into non-reserve areas (Man *et al.*, 1995; Warner *et al.*, 2000; Botsford *et al.*, 2001; Gell and Roberts, 2003; Halpern and Warner, 2003).

Traditionally, fish larvae were perceived to be weak swimmers, moving in a high-drag viscous environment (Blaxter, 1986). Because of their small size and slow speeds relative to adults, young larvae typically swim in a low Reynolds number environment (< 1000) and swimming activity is energetically expensive (Muller *et al.*, 2000). Most researchers concluded that larvae would swim for only limited times and distances. As a consequence, the feeble efforts of larvae to swim were thought to have limited, if any, influence on their dispersal (Williams *et al.*, 1984), although their ability to use vertical migration to alter dispersal trajectories was well recognised (Norcross and Shaw, 1984). However, many species of fish span two or three orders of magnitude in length during the growth from larvae to adults, and this change dramatically influences their locomotor performance (McHenry and Lauder, 2005). The discovery, initially in the tropics, that settlement-stage larvae of many demersal fish species could swim faster than ambient currents and over 10's of km (Stobutzki and Bellwood, 1994; Leis *et al.*, 1996; Leis and Carson-Ewart, 1997; Stobutzki, 1997b, 1998; Stobutzki and Bellwood, 1997) stimulated an interest in dispersal, retention, and settlement biology of larval fish, and considerable debate about the extent to which coral reef fish larvae might influence dispersal patterns (Roberts, 1997; Bellwood *et al.*, 1998; Sale and Cowen, 1998). As a consequence, there has been increased attention on swimming of fish larvae, but for reasons largely unconnected with traditional interests in trophic ecology and behaviour. A greater understanding of larval swimming abilities and behaviour on a large spatial scale is crucial for developing accurate bio-physical models of dispersal and connectivity (Cowen *et al.*, 2000; James *et al.*, 2002; Paris and Cowen, 2004; Cowen *et al.*, 2006).

Although we can identify three types of swimming behaviour of ecological relevance to larval fishes (escaping, foraging and spatial location), obtaining empirical data to quantify their abilities within each ecological context can be quite challenging. Endurance of swimming activity decreases rapidly as swimming speed is increased, and the swimming capacity of fishes represents a continuum from slow speeds that can be maintained for hundreds of hours (indefinitely) up to fast speeds that can only be maintained in the order of seconds (Fig. 11.1). Swimming ability in fishes is generally classed into three types, including: (1) burst, which uses exclusively anaerobically powered muscle and lasts for less than 20 seconds (2) prolonged, which may include both aerobic and anaerobic muscle activity lasting from 20 seconds to 200 minutes or (3) sustained, consisting of aerobically powered muscle activity lasting for longer than 200 min (Videler, 1993; Webb, 1994; Kolok, 1999; Plaut, 2001). It is logistically impossible to measure behavioural traits of animals in a natural environment, completely without disturbances, and measuring swimming ability in particular can be fraught with difficulties. Although measurements of burst swimming abilities in the laboratory may be considered relatively straightforward, especially with the advent of high speed video techniques, a multitude of methods have developed for the measurement of prolonged and sustained swimming abilities of fishes, and several of these have been applied to larval stages. The most common measurements of swimming capability include routine speed, critical speed (U-crit), *in situ* measures of swimming speed, and endurance swimming trials. The variety of methods

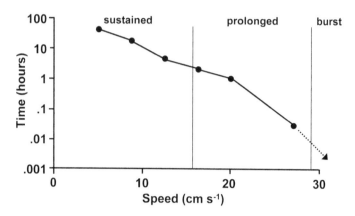

Fig. 11.1 The relationship between swimming speed and endurance for settlement stage (8 days) *Amphiprion melanopus* larvae reared in the laboratory. Data are from Fisher and Bellwood (2002). Swimming activity can be divided in to three broad categories based on the duration over which the activity can be sustained. No data are available on the burst swimming speeds of these larvae.

that exist make comparing values of prolonged and sustained swimming ability from different studies difficult. Frequently, methodological differences are overlooked or confused, leading to comparisons of swimming ability among taxa or locations that are misleading, or to applications of inappropriate measures of ability for a particular ecological context. Furthermore, the wide variety of techniques for which data are available confuses the ecological interpretation of research findings.

This chapter will examine and review our knowledge of the swimming capabilities of fish larvae and attempt to identify which methods are behaviourally and ecologically relevant to particular questions. Life history stages can be defined either morphologically, or ecologically. Here we use an ecological definition of the term larvae to refer generally to the younger life stages of fishes when they are subject to pelagic dispersal. We include data from taxa that show a distinct metamorphosis into a juvenile stage, along with others that merely increase in size, eventually becoming adults. We have concentrated on the pelagic larval stages of demersal fishes, with an emphasis on tropical species, largely because of our experience in this area. However, frequent reference is made to temperate and non-demersal fish literature.

A considerable amount of research has been done on the swimming capabilities of larval fishes over the last decade, utilising a range of techniques. Because of the difficulty in comparing swimming data of different types, we chose to examine each type of swimming measure in turn. Each method is part of the repertoire of swimming ability in fishes, and each is relevant in different ecological contexts and scales. Our purpose is to review the different methodologies used to measure swimming ability of fish larvae, describe how they work and their ecological relevance, and then discuss potential limitations of each. We also examine existing empirical data for larval fishes, emphasising the ecological importance of research findings. Finally, in order to gain a more holistic understanding of the ecological importance of different larval fish swimming measures, we finish with a synthesis of the ecological relevance of the different measures of swimming ability, and a discussion of their relationship to one another.

Swimming Ability Measures in Larval fishes, their Ecological Relevance, and our Existing Knowledge for Coral Reef Fishes

Spontaneous Swimming

Spontaneous swimming is the undisturbed day-to-day swimming speeds and behaviour of larvae. There are two different measures of spontaneous swimming activity in the literature—these are routine swimming measurements in the laboratory, and *in situ* swimming speed measurements in the field,

Laboratory measurements

In the laboraotory , this type of swimming has been measured in a variety of ways, but most often by placing larvae in a relatively small (50–400 cm² surface area) shallow (1–18 cm depth) dish and filming the larva(e) from below or above using either regular or high speed video (Hunter, 1972; Fuiman and Batty, 1997; Fuiman *et al.*, 1999; Plaut, 2000; Fuiman and Cowan, 2003; Smith and Fuiman, 2004). The experimental protocols adopted generally involve acclimation periods of varying lengths (5 min to 2 hours) to allow the larvae to become accustomed to their new environment. In some instances routine speeds have also been measured directly from larvae swimming undisturbed in their rearing tanks (Hunter, 1972; Hunter and Kimbrell, 1980; Fisher and Bellwood, 2003). Filming sequences are usually done over relatively short time intervals (3–10 min), although the assumption is that the behaviour being recorded is carried out continuously during that part of the diurnal cycle. Measures of routine speed usually include periods when the larvae do not move (although this depends on the feeding behaviour of the species being examined), thus, they include a behavioural component other than the larva's ability to reach and maintain a particular speed. Most studies report routine speeds that are integrated over the "pause-travel" movements of some larvae (but see (Hunt Von Herbing *et al.*, 2001)) and as such would represent an underestimate of actual swimming speed for species that utilize this swimming mode. However, the resultant measure of speed is a measure of the overall distances larvae travel during their "routine" activity. It is the speed which represents the distance they would cover during searching behaviour and/or horizontal movements.

The primary function of routine swimming is generally considered to be foraging, as well as migration and predator avoidance (Plaut, 2001). Routine swimming is important for calculating ecologically relevant parameters such as search volume and metabolic expenditure that are essential for developing individual based models of larval survival and growth (Letcher *et al.*, 1996). Most of the swimming speed values included in prominent reviews (Blaxter, 1986; Miller *et al.*, 1988) are routine speed, and the generally slow values initially reported for temperate species (1–2 body lengths per second) lead to the historical belief that horizontal swimming by larvae would have little influence on their position in the open ocean. Routine swimming is arguably the most relevant type of swimming ability in the context of the ability of larvae to influence their dispersal patterns, because it provides an estimate of the undisturbed, day-to-day speeds that larvae swim. However, because the measurements can only be made in an enclosed environment, these estimates of the behaviour and swimming speeds of larvae may differ from those under natural conditions.

Probably one of the biggest difficulties in comparing routine measures of swimming speed among different species and assessing the ecological relevance of current estimates of routine swimming is the variety of methodologies that have been used. Potential affects include filming tank size and volume, food concentration and stomach fullness, light conditions, turbulence, and the length of the acclimation period. We do know that fishes change their activity in response to food availability (Hecht *et al.*, 1996; Kasumyan *et al.*, 1998; Dou *et al.*, 2000; Killen *et al.*, 2007a), light (Batty, 1987), the density of conspecifics (Cooke *et al.*, 2000) and predators (Killen *et al.*, 2007c) and it will be essential to further explore the effect of these factors on larval behaviour, so that laboratory measures of routine speed can be better placed within an ecological context. Laboratory tanks typically contain much greater densities of both food and conspecifics than are found in the sea. Higher food densities may lead to less time spent swimming (Munk, 1995; Hecht *et al.*, 1996; Puvanendran *et al.*, 2002), although this is not always the case (Marchand *et al.*, 2002). Unfed larvae also appear to swim faster than those that have been recently fed (Fukuhara, 1985, 1987). Larvae are also known to increase their feeding activity in response to low levels of turbulence (MacKenzie and Kiorboe, 1995). The size of the containers in which routine speeds have been measured vary, but are frequently very small. Few studies have considered the influence container size and other laboratory conditions may have on routine speed measurements, but at least two have noted that swimming speed of larvae is lower in small containers than in large ones (von Westernhagen and Rosenthal, 1979; Theilacker and Dorsey, 1980), albeit with larvae of pelagic, not demersal, species (e.g., anchovy larvae swam 3–4 times faster in large than in small containers). The small container sizes, high food densities and lack of turbulence used in many laboratory measures of routine speed suggests that published estimates of routine speed may be conservative relative to natural undisturbed swimming speeds in the sea. This is likely to be further exacerbated by the lack of other stimuli and cues likely to be present in the sea, but absent under laboratory conditions, that may also stimulate swimming behaviour.

Published larval-fish routine speeds are typically quite low, and are in the vicinity of 1 to 5 body lengths per second (bl s^{-1}), seldom exceeding 5 cm s^{-1} except in very large larvae (Leis, 2006) (Fig. 11.2). Absolute routine swimming speeds (cm s^{-1}) appear to increase relatively linearly throughout larval life in relationship to size, whereas relative speed (bl s^{-1}) remains relatively constant (Fig. 11.2). Considerable variability exists in the development of routine swimming speed among families (Fig. 11.2). In general, it appears that routine speeds of larvae of perciform fishes are greater than those of clupeiform, gadiform or pleuronectiform fishes, especially when expressed as bl s^{-1} (Leis, 2006). Speeds recently reported for temperate rockfishes at parturition (*Sebastes* spp.) are in the order of 0.3 cm s^{-1} (0.5 bl s^{-1}), and varied among species (Fisher *et al.*, 2007).

Very few studies have examined the routine swimming speeds of coral reef fish larvae. One study of two families (Apogonidae and Pomacentridae) (Fisher and Bellwood, 2003) found that routine speeds and their development during ontogeny can vary considerably among taxa, with the Pomacentridae increasing their swimming speeds faster than the Apogonidae during ontogeny (Fig. 11.2). The only other known study of routine swimming in a coral reef species is from the family Lutjanidae (Doi *et al.*, 1998), the larvae of which appear to increase routine swimming speeds considerably slower than the Pomacentridae, showing a developmental rate similar to the apogonids (Fig. 11.2). But, this comparison may be confounded by differences in the methodology between these two studies. It appears that for at least one of the three coral reef fish families examined (the Pomacentridae) routine speeds are sufficient to facilitate an active behavioural mechanism for self-recruitment, even if they never swim faster than their measured routine daily swimming speeds throughout ontogeny (Fisher and Bellwood, 2003).

The undisturbed swimming speeds of younger stages of some of the coral reef fish species examined (Fisher and Bellwood, 2003) (Fig. 11.2 Pomacentridae) were considerably higher than in other subtropical (Fuiman

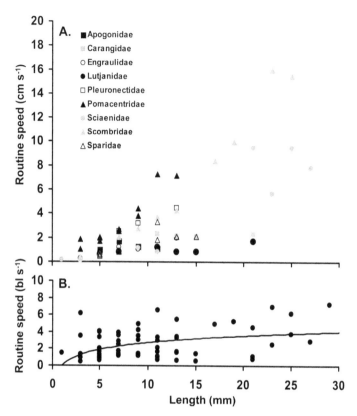

Fig. 11.2 The ontogeny of routine swimming speeds in cm s⁻¹ (A) and body lengths per second (bl s⁻¹, B) with respect to larval length, for several fish families. Associated water temperatures (where reported), along with the data sources are for each family as follows: Apogonidae–29°C (Fisher and Bellwood, 2003); Carangidae—(Sakakura and Tsukamoto, 1996); Engraulidae–17.5°C (Hunter, 1972); Lutjanidae–25–30°C (Doi *et al.*, 1998); Percichthyidae–23.5°C (Chick and Van Den Avyle, 2000); Pleuronectidae–7°C (Ryland, 1963); Pomacentridae–29°C (Fisher and Bellwood, 2003); Sciaenidae–26°C (Fuiman *et al.*, 1999); Scombridae–19°C (Hunter and Kimbrell, 1980), 17.5°C (Hunter, 1981), (Masuda *et al.*, 2002); Sparidae–13°C (Fukuhara, 1985, 1987). Note that length estimates are a mix of standard and total length and this will introduce a small error.

et al., 1999) and temperate studies (Hunter, 1972; Hunter and Kimbrell, 1980). This may be due to the fact that routine swimming speeds of coral reef fish larvae were measured in rearing aquaria, which provided a larger volume as well as a turbid environment due to the "green water" rearing techniques used. Turbidity is known to influence the activity levels of larvae in the laboratory: herring larvae swam a higher proportion of the time, and the duration of swimming was greater in turbid than in non-turbid water (Utne-Palm, 2004). However, the potential effects of environment are more likely to influence older, rather than younger larvae, because they have greater sensory abilities and because the rearing tank decreases in size relative to the size of the larvae during ontogeny. Given the potentially large effects of only small changes in temperature on swimming ability (Fuiman and Batty, 1997) and development (Green and Fisher, 2004), it could be that these coral reef fish larvae growing in warmer waters increase their swimming speeds per unit increase in size more quickly than their subtropical and temperate counterparts. However, speeds for the other two tropical reef fish families are much slower (Lutjanidae and Apogonidae), and are comparable to studies from colder waters (Fig. 11.2). It may simply be that members of the family Pomacentridae are exceptionally good swimmers (Leis *et al.*, 2007b) and show precocious development of swimming performance (Bellwood and Fisher, 2001) and generally higher levels of activity (Fisher and Bellwood, 2003). Similar taxonomic differences in swimming speeds and feeding strategies have been reported for temperate species (MacKenzie and Kiorboe, 1995; Hillgruber *et al.*, 1997; Fisher *et al.*, 2007). Slower routine swimming speeds, with larger periods spent inactive, imply a lower food encounter rate as well as decreased metabolic requirements, and suggest larvae may adopt different energetic strategies whilst in the pelagic environment (Hunter, 1981). Trade-offs between swimming capacity and other larval traits may encourage the evolution of different energetic strategies depending on a larva's specific environment. It becomes apparent that not all larval fish are equivalent in their behaviour and ecology, despite their occupation of similar habitats. Of even greater interest is the ability of some species to switch their feeding behaviour from an active search strategy when prey are scare to a 'sit-and-wait' strategy when prey are abundant(Killen *et al.*, 2007a).

Routine swimming data provides critically important information on the undisturbed swimming speeds of larval fishes, which may be key to understanding the minimum speeds at which larvae swim in nature, and provides insights into the energetic strategies adopted by larvae. Current data suggests routine speeds do vary, perhaps with temperature and certainly among taxa. Although routine or "cruising speeds" have been the dominant focus of swimming ability studies in the temperate literature, currently, there are few estimates of routine swimming speeds of larval reef fishes, especially in tropical waters. Given that all other available measures of swimming ability are either disturbed or forced, it is clear that more information on the routine swimming speeds of larvae are urgently needed for assessing the potential effects of behaviour on larval dispersal patterns, gaining a better understanding of the variability in the energetic strategies of larvae, and for parameterising individual based models of feeding, growth and survival. At the same time, efforts to understand the potential effects of laboratory environments on routine speed are also required.

Field measurements

Ideally, swimming speeds used in the context of dispersal should be based on measurements of larval fishes swimming in the sea. This is obviously challenging due to their dilute concentrations, small size and transparency. At least for late-stage larvae, these challenges have been met by an *in situ* observation methodology wherein larvae are released into the sea and then followed by SCUBA divers who observe their behaviour and measure the speed of the larvae through the water with a modified plankton-net flow meter (Leis *et al.*, 1996; Leis and Carson-Ewart, 1997). Called *in situ* speed, this is a measure of cruising speed. The released and observed larvae may be either laboratory-reared (Leis *et al.*, 2006a, b), or wild individuals that had been captured with light traps or nets (Leis and Carson-Ewart, 1997; Trnski, 2002).

The speeds at which larvae actually swim in the sea are a result of both their capacity and their motivation. In the sea, various sensory cues can provide the stimulus to swim faster or slower, or not at all. In the laboratory (where routine speed is measured), these cues may be absent or be replaced or over-ridden by cues provided, either wittingly or unwittingly, by the investigator. The greatest advantage of measuring swimming speed *in situ* is that it provides a measure of swimming speed in the sea, to some extent avoiding the complications and ambiguities of laboratory measurements. The larva itself chooses the swimming speed. However, the presence of the observer diver in close proximity to, and following the larva, has an unknown effect on its swimming speed. Unfortunately, it is difficult to directly measure the effect, if any, of the observer diver as observing the behaviour of the larva without the presence of the diver is not possible at present. Several features of the *in situ* speeds reported thus far provide circumstantial evidence that *in situ* speed is a reasonable measure of what larvae actually do in the sea (Leis and Carson-Ewart, 1998). Within a species, *in situ* speeds are slower than U-crit (by 33 to 50%) and much slower than burst speeds (Leis and Fisher, 2006), demonstrating that larvae being observed by divers are not trying to escape and are probably swimming at speeds that can be sustained aerobically. Furthermore, the fact that larvae swim at different speeds in different situations (locations or directions) while under observation by divers suggests that the observer presence does not completely override the behaviour of larvae. Larvae also feed while being observed, and appear to feel so unthreatened by the observers that observers are sometimes used as shelter when large fishes are in the vicinity (Leis and Carson-Ewart, 1998; Trnski, 2002; Hindell *et al.*, 2003).

A possibility also exists that the behaviour of the larva is influenced by its sudden release into the sea from a small container. Again, this is difficult to measure, although observations of speed over time since release might provide insight. Most studies of *in situ* swimming speed have reported speed over the full period of observation—typically 10 min. Little attention has been paid to how speed might vary over longer periods, and there are only two published estimates of how speed might vary within the 10-min observation period. For the 8–10 mm larvae of a reef pomacentrid (*Amblyglyphidodon curacao*), speed in the first 5 min averaged 2.4 cm s^{-1} faster (ie, 16% faster) than in the second 5 min (Leis *et al.*, 2007). In a second study (Leis *et al.*, 2009), swimming speeds of larvae of two non-reef, tropical marine species across a range of sizes (5–20 mm) were measured. Larvae of a leiognathid (*Leiognathus equulus*), swam 3 cm s^{-1} faster (i.e., 20% faster) in the first five minutes, whereas, larvae of a polynemid (*Eleutheronema tetradactylum*) did not differ significantly in speed between the first and second five-minute periods. Thus, there is apparently no consistency among species in changes in speed with time, and in the two instances where such variation has been found, speed was slower in the second half of the observation period by 2–3 cm s^{-1}.

Finally, as *in situ* speed is measured for only a relatively short time (normally, 10 minutes), without supporting measures of endurance at the speeds exhibited, it is not entirely clear that *in situ* speed is a measure of cruising speed. However, the speeds at which larvae chose to swim in the sea are similar to those used in measures of swimming endurance in the laboratory, in which larvae can swim for hours to days, covering 10s of kilometres (Stobutzki and Bellwood, 1997).

Most available measurements of *in situ* speed are of settlement-stage wild larvae, mostly of coral reef fishes (Leis and Carson-Ewart, 1997, 2001, 2003; Leis and Fisher, 2006), but some data on warm-temperate species are also available (Trnski, 2002; Hindell *et al.*, 2003; Leis *et al.*, 2006a), and a few opportunistic estimates of *in situ* speed have been made of cool temperate clupeiform species swimming near the surface based on observations from wharves (von Westernhagen and Rosenthal, 1979). The speeds of these settlement-stage larvae are high relative to ocean currents and, like other estimates of swimming speed, vary widely among species. This variation will reflect not only variation in the swimming capacity of these late stage larvae, but also variation in motivation, which may in turn also be related to variation in sensory cues. Species means for late stage larval coral reef fishes measured on the Great Barrier Reef and French Polynesia were 2–66 cm s^{-1} (2–34 bl s^{-1}), with one holocentrid swimming as fast as 66 cm s^{-1}, and several species with mean speeds in excess of 40 cm s^{-1} (Leis and Fisher, 2006).

Relative to the later settlement stages and early juveniles, it is much more difficult to work with smaller larvae *in situ*, and it may not be possible at all to work *in situ* with the smallest larvae. As a result, there is very limited information on the ontogeny of swimming based on *in situ* speed, and all of this is based on reared larvae. *In situ* swimming speeds of 3–10 bl s^{-1} were found in reared larvae of two families (three species) of warm-temperate (Leis *et al.*, 2006b) fishes and one family of tropical fish (Leis *et al.*, 2006a), over sizes of 5–20 mm, depending on species. Size vs speed relationships were approximately linear, and increased at a rate of 0.4–2.6 cm s^{-1} per 1 mm increase in length.

Speed measured *in situ* varies on a taxonomic and spatial basis. *In situ* speed differs both among families and among species (Leis and Fisher, 2006). Within a species, spatial differences in *in situ* speed have been documented: between lagoon and ocean (Leis and Carson-Ewart, 2001); between windward and leeward sides of an island (Leis and Carson-Ewart, 2003); between bay and ocean (Leis *et al.*, 2006b); and among coastal locations (Leis and Lockett, 2005). In addition, settlement-stage larvae of several species swim faster when swimming away from a reef than when swimming toward or over it (Leis and Carson-Ewart, 2002; Leis and McCormick , 2002). Swimming speeds may differ in the presence of sensory cues that differ in quality or intensity, such as underwater sounds (Leis *et al.*, 2002). Underwater reef noises are known to attract late stage larval fishes near both temperate and tropical reefs (Tolimieri *et al.*, 2000; Simpson *et al.*, 2004; Leis and Lockett, 2005; Simpson *et al.*, 2005b), and appear to be an important cue for settlement (Simpson *et al.*, 2004).

In situ measurements of swimming speed of fish larvae provide the only information presently available on swimming speeds of larvae in the sea. These observations allow insight into factors that would be extremely difficult to investigate in the laboratory, such as the influence that different spatial factors (e.g., near or away from settlement sites), environmental factors (e.g., time of day or sensory cues) or swimming directions have on swimming speed. The relatively few studies that exist on the *in situ* behaviour of late stage coral reef fish larvae, clearly show the importance of such *in situ* environmental cues on swimming speeds of these fishes, and highlight the need for a greater understanding of the factors influencing larval behaviour in the open ocean. Unfortunately, *in situ* observations can be conducted over only relatively limited time periods, only during the day, only in the upper portions of the water column, and not all species or developmental stages are amenable to this methodology.

Forced Swimming Performance Measures

While routine and in-situ measurements of speed deal with the spontaneous swimming abilities of larvae, there are a wide range of studies that have utilized swimming flumes for measuring swimming abilities. These provide information on the swimming performance of fishes and define the limits within which animals can chose what to do. Two flume-based methods of swimming capability for which there is the most information for larvae include critical speed (U-crit) and endurance swimming.

Critical Speed

U-crit (also known as "critical speed") is a widely used measure of swimming performance throughout the fish physiology literature (Jones *et al.*, 1974; Hartwell and Otto, 1991; Kolok, 1991; Hawkins and Quinn, 1996; Lowe, 1996; Myrick and Cech, 2000), and has also been used more recently to estimate the potential effects of swimming behaviour on dispersal patterns of coral (Stobutzki and Bellwood, 1994; Fisher *et al.*, 2000, 2005; Fisher, 2005; Hogan *et al.*, 2007; Leis *et al.*, 2007a), sub-tropical and temperate reef fishes (Clark *et al.*, 2005). This technique first appeared in the 1960's and was intended to be a way of measuring aerobic swimming capacity of fishes (Brett, 1964). The technique involves placing a fish in a swimming flume and increasing the speed incrementally over time until it can no longer maintain position. U-crit is then calculated as: U-crit = U + (t/t_i * U_i), where U is the penultimate speed, U_i is the velocity increment, t is the time swum in the final velocity increment and t_i is the set time interval for each velocity increment.

There is considerable debate about the type of swimming that is measured, and the ecological relevance of U-crit as a measure of swimming performance. Although originally establish to measure aerobic or sustained swimming capacity of fishes (Brett, 1964), it is now generally agreed that U-crit is a measure of prolonged swimming that may incorporate both aerobic and anaerobic muscle activity (Reidy *et al.*, 2000). Although sometimes referred to as "sustained swimming", to distinguish it from burst speeds, U-crit does not appear to be sustainable over long periods (Fisher and Wilson, 2004).

It has been suggested that this measure of swimming has limited utility for ecological or evolutionary inferences, because a final U-crit value is a complex product of multiple swimming modes and changing metabolic support (Nelson *et al.*, 2002). On its own, U-crit probably most directly relates to the ability of the larvae to cover short distances (100's of meters), at high speeds (up to 100 cm s^{-1}) (Fisher *et al.*, 2005) over short time scales (minutes). These abilities may be important in terms of the potential for larvae to move between locations on a small scale, such as: away from reefs at hatching, within slicks or between plankton patches in the open ocean, or between habitats at settlement (Fisher *et al.*, 2000). Swimming abilities on the scale measured by U-crit would also allow extensive rapid vertical migration to access different current regimes or prey (Fisher *et al.*, 2000). Although many species have U-crit values in excess of average ambient current speeds (Fisher, 2005; Fisher *et al.*, 2005; Hogan *et al.*, 2007), it is unlikely these speeds could be maintained for the length of time required to substantially affect overall dispersal patterns. Swimming constantly at U-crit speed would be disadvantageous since larvae would be devoting almost all of their limited aerobic scope to swimming activity (Post and Lee, 1996; Killen *et al.*, 2007b), and would therefore have no oxygen delivery capacity left for processes involved with growth and maintenance. As such, critical speed is not directly applicable to dispersal and the control of spatial location on large scales. However, there is evidence that U-crit is positively correlated with routine swimming speeds (Fisher and Bellwood, 2003), sustainable swimming speeds (maintainable for 24 hours) (Fisher and Wilson, 2004) and *in situ* speed (Fulton *et al.*, 2005; Leis and Fisher, 2006; Leis *et al.*, 2006a). It seems that U-crit may be useful for estimating other ecologically relevant measures of swimming ability, such as endurance or *in situ* speed, which can be far more time consuming, and logistically impossible to measure in some species or at some developmental stages.

U-crit has been severely criticized in several reviews because of the potential confounding effects of methodology, especially with respect to it's sensitivity to the time intervals used, with shorter time intervals leading to increased U-crit estimates (Kolok, 1999). However, a quick survey of the available studies for a range of species suggests that this effect is minimal, and in the order of 7 to 9 % (Farlinger and Beamish, 1977; Hartwell and Otto, 1978; Williams and Brett, 1987; Hartwell and Otto, 1991) Another study on salmon parr indicated that performance was unaffected by the time intervals (2, 5 and 10 min) used (Peake and McKinley, 1998). For larval coral reef fishes, two studies have found little effect of U-crit methodology on estimates of speed (Fisher *et al.*, 2005; Hogan *et al.*, 2007). In the first, swimming speeds estimated using time intervals of 2 minutes and speeds increments of 3 body lengths were found to be similar to those estimated using time intervals of 5 minutes and speed increments of 1.6 cm s^{-1} despite experiments using the later method lasting on average 6 times longer (Fisher *et al.*, 2005). Hogan and co-authors (Hogan *et al.*, 2007) also found no difference in estimates of U-crit measured using 2 and 15 minute time intervals in settlement stage larvae of six species of coral reef fishes in the Caribbean.

Gait transition speed (e.g., the speed at which pectoral fin locomotion is supplemented by body caudal fin locomotion) is an alternative measure of swimming speed that may be potentially more ecologically relevant. It is typically measured on labriform swimmers, where propulsion is achieved using pectoral fin locomotion (Drucker, 1996; Drucker and Jensen, 1996). Although labriform swimming is the primary mode of locomotion for many substratum-associated taxa, and is particularly common among the vast assemblage of reef-dwelling perciform fishes (Lindsey, 1978; Blake, 1983; Drucker, 1996, Chapter 12 of this book), this is only used by very late stage larvae, with well developed pectoral fins. In addition, the young larvae of some fishes swim using anguilliform motion, which simply increases in frequency at higher swim speeds without a change in gait. Gait transition speed could not be used to examine the development of swimming throughout ontogeny, or to compare swimming abilies of species with different gaits, so while it may have greater ecological relevance, its utility for examining the swimming abilities of larval fishes is limited.

Of all the measures of swimming speed discussed here, U-crit is probably the most widely studied in larval fishes. At settlement, U-crit speeds range widely among families (0.3 to 85 cm s^{-1}), but the majority of families have speeds in excess of ambient currents in the ocean (Table 11.1), suggesting considerable short term control of their position. In general, the fastest families appear to belong to the Perciformes, although the fastest observed swimming individual (100 cm s^{-1}; Table 11.1) is a holocentrid belonging to the order Beryciformes. Although such taxonomic trends do exist in swimming performance, much of these differences can be explained and predicted relatively well using external morphological measurements (Fisher and Hogan, 2007). There do not appear to be any consistent differences in the swimming abilities of settlement stage tropical marine fishes with respect to adult habitat (pelagic versus demersal) or adult swimming mode (caudal, pectoral or dorso-ventral) once morphological differences have been taken into account (Fisher and Hogan, 2007).

Compared to settlement stage larvae, U-crit data on younger individuals are relatively rare, although measurements do exist throughout development for a handful of tropical, warm temperate and temperate families (Fig. 11.3). Swimming abilities increase more or less linearly with larval length, although like burst and routine speed measures, relative U-crit speeds (bl s^{-1}) tend to rapidly reach a maximum (Fig. 11.3). Considerable variability exists among families, with the warm-water coral reef fishes generally showing the fastest rates of development (e.g., Pomacentridae, Lutjanidae and Serranidae). Data from families studied at the coldest temperatures (Pleuronectidae and Percidae) show the slowest developmental rate of swimming ability with length (Fig. 11.3). Comparisons of developmental rates of U-crit speed with temperature are, however, confounded by taxonomic differences in studies from different climates. Because of the large variability among groups in the development of U-crit swimming performance, the size that must be attained by larvae to overcome local currents varies substantially. Speeds great enough to influence position on a reasonable scale (> 5 cm s^{-1}) occur more or less after notochord flexion, and concomitant formation of the caudal fin, which takes place around 5 mm in the seven marine perciform species studied (Leis, 2006). By 10–15 mm in length, a size at which settlement takes place in many coral reef species, U-crit speeds of 10–30 cm s^{-1} are observed (Leis, 2006) (Fig.11.3).

Table 11.1 U-crit swimming speeds of 31 families of settlement stage reef fishes captured in light traps and crest nets on coral reefs in the Caribbean and Indo-Pacific. Also indicated are the adult swimming modes (C–caudal, DV–dorsoventoral, P–pectoral and PC–Pectoral/Caudal), dominant adult habitat (D–Demersal, P–pelagic) and the taxonomic order to which each family belongs to (B–Beryciformes, C–Clupeiformes, G–Gasterosteioformes, L–Lophiiformes, P–Perciformes and T–Tetraodontiformes). Data are from (Fisher, 2005; Fisher *et al.*, 2005; Hogan *et al.*, 2007). Water temperatures at the locations were these data were collected were 28–30°C. Swimming modes are derived from (Webb, 1984; Fulton, 2005), CJ Fulton personal communication, and from personal observations (R Fisher). Habitat and taxonomic information are derived from Fishbase (www.fishbase.org).

Family	Speed (cm s⁻¹)			Total Length (bl s⁻¹) (mm)			Mode	Habitat	Order
	Ave	SE	Max	Ave	Ave	SE			
Acanthuridae	52.2	3.7	61.5	15.8	33.1	1.7	P	D	P
Antennariidae	5.1	3.2	8.4	5.4	9.6	2.4	C	D	L
Apogonidae	20.5	1.2	30.0	14.0	14.6	0.8	PC	D	P
Balistidae	32.0	19.5	51.5	18.3	17.5	6.8	DV	D	T
Blennidae	15.4	9.6	34.7	8.4	18.3	1.2	PC	D	P
Carangidae	38.6	16.5	71.2	20.3	19.0	5.8	C	P	P
Chaenopsidae	15.0		15.0	4.4	34.1			D	P
Chaetodontidae	42.4	5.7	62.5	24.2	17.5	1.2	PC	D	P
Clupeidae	22.6	9.9	32.4	6.6	34.2	3.5	C	P	C
Diodontidae	6.7		6.7	5.6	11.9		DV,C	D	T
Gerreidae	25.3		25.3	19.5	12.9		C	D	P
Haemulidae	33.8		33.8	21.0	16.1		C	D	P
Holocentridae	85.5	8.2	100.8	23.5	36.4	0.8	C	D	B
Labridae	27.9		27.9	16.5	16.9		P	D	P
Lethrinidae	38.4		38.4	17.7	21.8		C	D	P
Lutjanidae	42.4	2.8	53.3	16.7	25.5	1.5	C	D	P
Monacanthidae	25.1	3.1	34.6	11.7	21.4	2.2	DV	D	T
Mullidae	47.0		47.0	9.5	49.3		PC	D	P
Nemipteridae	31.2	5.4	36.6	18.5	16.9	0.6	PC	D	P
Nomeidae	28.0		28.0	14.0	20.0		C	P	P
Ogcocephalidae	0.3		0.3	0.5	6.4		C	D	L
Ostraciidae	13.8	0.3	14.1	18.1	7.6	0.4	DV,C	D	T
Pomacanthidae	25.5	3.8	29.3	12.9	19.8	3.6	PC	D	P
Pomacentridae	34.1	1.6	54.9	23.6	14.4	0.7	PC	D	P
Pseudochromidae	27.2	2.3	29.9	15.5	17.5	0.2	PC	D	P
Serranidae	22.3	9.6	31.9	13.6	16.4	5.6	C	D	P
Siganidae	67.1		67.1	22.7	29.5		C	D	P
Sphyraenidae	24.1	5.7	29.8	11.6	20.8	2.2	C	P	P
Syngnathidae	3.8	1.6	5.4	1.1	33.6	9.6		D	G
Terapontidae	51.0		51.0	26.5	19.2		C	D	P
Tetraodontidae	23.2	3.9	34.6	11.2	20.6	4.6	DV	D	T

Very limited data on U-crit speeds in the youngest larval stages are available. Measurements on just hatched larvae (1.5–4.5 mm total length) of 10 species of coral reef fishes suggest that newly hatched larvae of at least some species can swim swam as fast as 4 cm s⁻¹ (mean 2 cm s⁻¹; up to 14 bl s⁻¹) (Fisher , 2005).

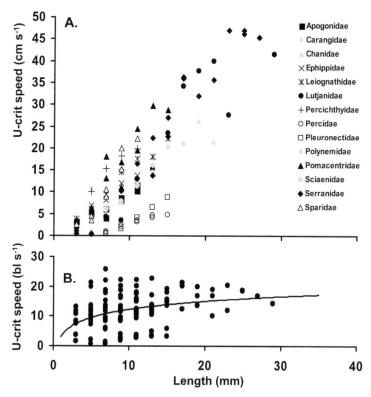

Fig. 11.3   The ontogeny of U-crit swimming speeds in cm s⁻¹ (A) and body lengths per second (bl s⁻¹) (B) with respect to larval length, for a variety fish families. Associated water temperatures (where reported), along with the data sources are for each family as follows: Apogonidae–29°C (Fisher *et al.*, 2000); Carangidae–29°C (Leis *et al.*, 2006a); Chanidae–29°C (Leis *et al.*, 2007a); Ephippidae–29°C (Leis *et al.*, 2007a); Leiognathidae 29°C (Leis *et al.*, in 2007a); Lutjanidae–29°C (Leis *et al.*, 2007a); Percichthyidae–19°C (Clark *et al.*, 2005); Percidae–13°C (Houde, 1969); Pleuronectidae–7°C (Ryland, 1963); Polynemidae–29°C (Leis *et al.*, 2007a); Pomacentridae–29°C (Fisher *et al.*, 2000); Sciaenidae–23°C (Clark *et al.*, 2005); Scombridae–24°C (Sepulveda and Dickson, 2000); Serranidae–29°C (Leis *et al.*,2007a); Sparidae–23°C (Clark *et al.*, 2005). Note that length estimates are a mix of standard and total length and this will introduce a small error.

Variation in the newly hatched U-crit speeds in these coral reef fish larvae could be explained reasonably well by their length at hatching (Fisher, 2005).

   Overall it appears that U-crit is a useful measurement of swimming ability in fishes that can be measured relatively easily using cheaply constructed and simple equipment and appears to be influenced to only a small extent by methodology. U-crit is a useful measure of *relative* speed, useful for comparisons among taxa or developmental stages. Furthermore, its close correlation with other measures of swimming ability, such as routine, *in situ* and sustained swimming speed, mean that it can be used to estimate these other more time intensive but ecologically relevant measures of speed. It is important to remember, however, that U-crit is not directly applicable to field situations, and that larval fish probably never actually swim at U-crit speeds in the field for any length of time. U-crit can best be described as measuring the maximum steady (as opposed to burst) swimming capacity of fishes. Values observed for coral reef fish larvae, clearly demonstrate the high level of among taxa variability in the maximum swimming capacity of larval coral reef fishes, especially those towards the end of the pelagic developmental stage. Although some of this variability can be reasonably explained by differences in their morphology (Fisher *et al.*, 2005; Fisher and Hogan, 2007), physiological differences among species are also likely to occur. Dramatic differences in the maximum swimming capacity of taxa reflect similar levels of taxonomic variability in other behaviourally relevant measures of swimming ability, and have considerable implications, determining the extent to which swimming behaviour may be important to their ecology.

Endurance swimming

Much of the initial work on swimming abilities in late stage larval coral reef fishes was based on an endurance swimming technique, carried out by forcing unfed larvae captured using light traps to swim in flumes at a fixed speed until exhaustion (Stobutzki, 1997a, 1998; Stobutzki and Bellwood, 1997). This type of swimming performance does not fit into any of the three categories described earlier—burst, prolonged or sustained (Videler, 1993), which all measure the maximum speeds attained during each activity. Rather, endurance swimming measures the capacity of larvae to maintain swimming at a constant speed. Although critically dependent on the speeds swum (Fisher and Bellwood, 2002), the endurance swimming technique adopted by Stobutzki and co-workers showed definitively for the first time that late stage larval fishes were capable of maintaining relatively fast swimming speeds (over 10 body lengths per second for some species) for long periods of time (days) and covering very large distances (100's of kilometers in some cases) (Fig. 11.4).

These results for settlement stage coral reef fish larvae were later confirmed for fish from a temperate reef system, which also showed remarkable swimming endurance (Dudley *et al.*, 2000). Swimming times and associated distances varied among the species examined, but ranged from as high as 200 km (Scorpidae) down to 50 km (Monocanthidae) (Dudley *et al.*, 2000). Results were relatively similar for both reef and non-reef larvae of similar developmental stage. These initial studies have transformed our impression of the swimming capabilities of these relatively small fishes, and raised awareness of the potential effect of larval behaviour

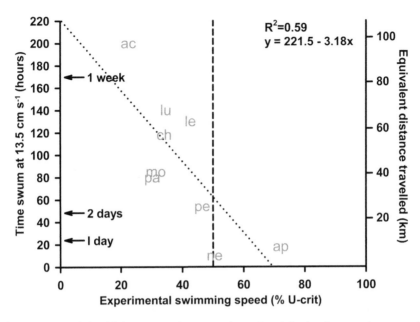

Fig. 11.4 The relationship between endurance and experimental swimming speed expressed as a percentage of U-crit (maximum speed) for late stage larvae of several families of coral reef fishes. Swimming times from (Stobutzki and Bellwood, 1997). U-crit data from (Fisher *et al.*, 2005). Endurance is expressed in hours (left axis) and the equivalent distance travelled in kilometres (right axis). The dotted line shows a fitted least squares linear regression. The dashed line indicates the point at which experimental speed equals 50% of U-crit speed. Symbols: ac–Acanthuridae, lu–Lutjanidae, le–Lethrinidae, ch–Chaetodontidae, mo–Monocanthidae, pa–Pomacanthidae, pe–Pomacentridae, ne–Nemipteridae and ap–Apogonidae.

on dispersal patterns, particularly in the latter stages of development (Wolanski *et al.*, 1997; Armsworth *et al.*, 2001). However, despite the importance of this endurance swimming data in increasing our understanding of larval capabilities, such data in their initial form are probably not directly appropriate for inclusion into dispersal models, because: 1) swimming times and distances vary substantially with swimming speed (Fisher and Bellwood, 2002; Fisher and Wilson, 2004), 2) swimming times and distances can be greatly increased in the presence of food (Fisher and Bellwood, 2001; Leis and Clark, 2005), and 3) studies on a variety of adult fishes have shown that spontaneous swimming, is, on average more expensive energetically than forced swimming (Boisclair and Tang, 1993; Lowe, 1996).

Because endurance swimming experiments take a long time to complete, invariably such studies (at least those that concentrate on more than one species) are conducted only at a limited number of speeds. Much of the existing measurements of endurance swimming performance of coral reef fishes (Stobutzki, 1997a, 1998; Stobutzki and Bellwood, 1997) were carried out at a speed of 13.5 cm s^{-1} (the mean current speed around Lizard Island on the Great Barrier Reef during the recruitment season (Frith *et al.*, 1986), where Stobutzki's work took place). If we plot the endurance swimming times of late stage coral reef fish larvae (Stobutzki and Bellwood, 1997) against their experimental swimming speed (13.5 cm s^{-1}) expressed as a percentage of the U-crit for each family (Fisher, 2005), it is clear that swimming endurance drops off rapidly as larvae are swum at higher and higher percentages of their critical speed. Appreciable swimming endurance (more than 24 hours) only occurs for larvae that are forced to swim at a speed of less than 50–60% of U-crit (Fig. 11.4). A similar finding was also observed for larval coral reef fishes throughout ontogeny (Fisher and Bellwood, 2002) and clearly demonstrates the crucial importance of speed on estimates of swimming endurance measures. Furthermore, when larvae are periodically fed during swimming endurance trials at a fixed speed of 50% of their U-crit, some larvae appear to be able to maintain swimming activity indefinitely, and actually grew at a similar rate to larvae that were not swum at all (Fisher and Bellwood, 2001).

Sustained swimming

An understanding of sustained swimming capability is essential for defining the upper limit on the potential effects of larval behaviour on dispersal patterns. In adult fish Ums (maximum swimming speed) is usually defined as the speed fish can maintain for at least 200 min (Videler, 1993). However, for larvae to influence their dispersal patterns, swimming needs to be maintained for longer periods. Therefore, for performance to be ecologically relevant, given the effects of speed and food on endurance, perhaps the most appropriate measure of sustained swimming performance is the maximum speed that can be maintained by larvae over a predetermined, ecologically relevant time scale. One study has determined the speeds several species of late stage larval coral reef fishes are able to maintain for 24 hours (Fisher and Wilson, 2004). This time period was selected because starvation is not expected to be an issue, yet the duration is long enough to show convincingly that the experimental speed is sustainable over long enough time periods to be relevant to dispersal. Interestingly, these sustainable speeds scale much more closely to the U-crit of larvae than they do to larval size, and 50% U-crit appears to provide a relatively precise means of estimating sustainable speeds across taxa (Fisher and Wilson, 2004). This is very similar to another measure, coined "50% fatigue velocity" (Meng, 1993; Jenkins and Welsford, 2002), which aims to identify the speed that 50% of individuals could maintain for a given period, usually 1–2 h. Two drawbacks of this maximum sustainable swimming speed method are the fact that they require a pre-determined and arbitrary definition of "sustainable" and that a relatively large number of larvae (60–80) are required at a similar stage of development over a considerable time frame (1 week or more) to determine this speed in any one species at any stage of development.

Studies on sustainable speed are important in defining the true "maximum" potential of these larvae to influence dispersal patterns in the open ocean using swimming behaviour. Results so far, particularly for the later larval stages of coral reef fishes, clearly show that many of these fishes exhibit considerable swimming endurance, and are able to swim very long distances, as least when swum at appropriately slow speeds. Although further work is needed to determine how fast long distance migration might be achieved by larvae, the excellent work by Stobutzki (and co-workers) has clearly demonstrated that many coral reef-associated taxa have the potential to achieve considerable control over their horizontal position in the ocean and may substantially influence dispersal and recruitment patterns. Furthermore, this also appears true for some species at temperate locations (Dudley *et al.*, 2000), and is not entirely a tropical phenomenon. Late stage larvae of a range of families clearly have the swimming capabilities necessary to take control of their settlement locations on relatively large spatial scales.

Burst Swimming

Burst swimming can be maintained over fractions of seconds to seconds, and is a high-performance spurt of speed based on anaerobic metabolism, that is used to escape from predators or adverse stimuli, or to capture food (Domenici and Blake, 1997; Plaut, 2001; Wakeling, 2001). This is the fastest speed of which an individual is capable, and is usually initiated in a C-type fast start response. Burst speeds are typically measured in the laboratory, and because the speeds involved are so great and take place over very short periods of time high-speed photography is required. Most commonly, burst speeds are measured as a startle response—a fast start behaviour that most larvae exhibit when startled by a perceived threat, and typically reflect the "escape" speed of larvae, that is, the speed at which larvae move away from a predator. A detailed review of escape responses in adult fishes is provided by another chapter in this volume (Domenici, Chapter 5). For fishes, this is one of the earliest behavioural patterns to appear during development (Eaton and DiDomenico, 1986; Sugisaki *et al.*, 2001). Compared to escape responses, there are very few studies of burst swimming performance with respect to food capture.

When examining escape responses, startle behaviour by the larva is typically provoked either by the presence of predators (Masuda, 2006) or by artificial stimuli, such as the approach of a probe (Fisher *et al.*, 2007), or pendulum (Fuiman and Cowan, 2003). Although the speeds are high, the distances travelled are short, because the time over which the speed is maintained is very limited. Because of their short time frame (typically < 20 seconds), burst speeds are inappropriate for direct considerations of dispersal, although they have been used as such (Bradbury *et al.*, 2003). Burst speeds are however, directly relevant to the ability of larvae to avoid predation and capture food, and burst swimming is therefore an important measure of swimming ability relevant to the ecology of larval fishes. Although burst speeds are important, overall escape potential is related to a wide range of factors in addition to speed, including the distance of the startle response, latency, reaction distance, turning rates and overall responsiveness (Fuiman *et al.*, 2006) as well as acceleration and turning ability (Walker *et al.*, 2005).

Comparisons of measurements of burst speeds can be confounded by methodology. In particular, burst speeds can often be reported either as the "integrated speed" of the burst, integrated over the entire duration of the burst sequence (Williams *et al.*, 1996; Fuiman *et al.*, 1999; Sugisaki *et al.*, 2001; Masuda *et al.*, 2002; Fisher *et al.*, 2007) or the maximum speed attained (McCormick and Molony, 1993; Gibson and Johnston, 1995; Williams *et al.*, 1996; Fisher *et al.*, 2007). Either is potentially a useful measure to report, but both have potential problems. Integrated speed may be confounded somewhat by the arbitrary choice of both the distance over which to measure the burst activity, and the end of the burst "event", and will always be slower than the maximum speed attained. On the other hand, the maximum attained speed that can be measured will depend somewhat on the frame frequency of the video technique being used, because the reported "maximum" will be an average over the time frame of the recording frequency. True "maximum" escape speeds can only be measured using very high frame rate video (100's frame per second). Very few studies report both measures of burst speed. Data from one study on newly extruded rockfish larvae (*Sebastes* spp) (Fisher *et al.*, 2007) suggests that integrated burst speed is 49 to 56% lower than the maximum speed (across 4 species). This general trend is confirmed in another study on several temperate marine fishes suggesting integrated burst speeds are 50–60% lower than maximum recorded velocities (Williams *et al.*, 1996).

Data are available on maximum velocity estimates of burst speed for a small size range (20–23 mm) of recently settled juveniles of the family Mullidae on the Great Barrier Reef (McCormick and Molony, 1993). Maximum speeds for this species were comparable to similar sized individuals of the family Pleuronectidae from temperate regions (McCormick and Molony, 1993), and faster than similar sized salmon (Hale, 1996) (Fig. 11.5). There are no coral reef fish studies reporting integrated burst speed, but

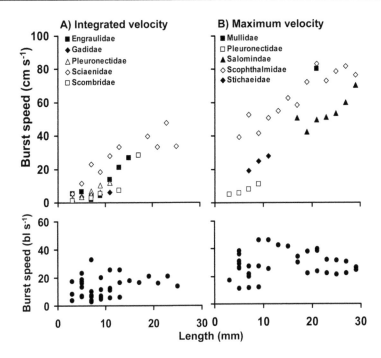

Fig. 11.5 The ontogeny of burst swimming speeds in cm s⁻¹ (top panels) and body lengths per second (bl s⁻¹, lower panels) with respect to larval length, for 5 fish families. Data are for (A) integrated burst speed (over the duration of the burst event, times vary) or (B) the maximum velocity attained during the burst event. Associated water temperatures (where reported), along with the data sources for each family for average velocity are as follows: Engraulidae– 17.5°C (Hunter, 1972); Gadidae–6°C (Sugisaki *et al.*, 2001); Pleuronectidae–7°C (Ryland, 1963), 7°C (Williams and Brown, 1992); Sciaenidae–26°C (Fuiman *et al.*, 1999); Scombridae– (Masuda *et al.*, 2002). Associated water temperatures and data sources for maximum velocity are: Mullidae–28°C (McCormick and Molony, 1993); Pleuronectidae–7°C (Williams and Brown, 1992); Salmonidae–12°C (Hale, 1996); Scophthalmidae–18°C (Gibson and Johnston, 1995); Stichaeidae–8°C (Williams *et al.*, 1996). Note that length estimates are a mix of standard and total length and this will introduce a small error.

data for one family (Sciaenidae) studied at warmer temperatures (26°C) (Fuiman *et al.*, 1999) showed the fastest development of burst speed with respect to length among available estimates (Fig. 11.5).

There are clear relationships between the startle speeds of larvae and their size across species (Fisher *et al.*, 2007). A relationship of increasing escape capabilities with increasing body size appears common for larval fishes (Batty *et al.*, 1993; Williams *et al.*, 1996; Fuiman *et al.*, 1999; Wesp and Gibb, 2003) (Fig. 11.5). Burst speed in absolute terms increases relatively steadily as larvae grow; although in relative terms (expressed in body lengths per second, bl s⁻¹) burst performance appears to reach near maximum at relatively small larval sizes, and then tends to decrease as larvae become larger (Fig.

11.5). Although many researchers argue that absolute speed will better predict the outcome of escaping predators, this is not necessarily true because the ease with which predators are able to capture prey also increases with prey size (Van-Damme and Van-Dooren, 1999). Because of the increased risk of predation at small sizes, the rapid development of burst speed in relative terms may be important to survival in fish larvae.

Data on attack swimming speeds of larvae is relevant in terms of the swimming performance necessary for feeding. Data for 5–8 mm Gadidae suggests speeds in the order of 1.8–3.04 bl s^{-1} (Hunt Von Herbing *et al.*, 2001) and are comparable to the integrated burst speeds recorded for this species (Bailey *et al.*, 1993). Speeds reported for juvenile Cottidae are considerably faster (6.5 bl s^{-1}, Cook, 1996), as are the speeds reported for Engraulid larvae (average 14 bl s^{-1}, Hunter, 1972) which appear only marginally slower than their burst escape speeds (average 19 bl s^{-1}, Webb and Corolla 1981). Depending on the group and developmental stage, attack speed will vary in importance in food capture success, as suction feeding is also an important element for many species.

Environmental Factors and Larval Swimming Ability

In this chapter we have examined the available information on several swimming ability measures of larval fishes across a broad range of taxa, with only passing regard for the potential importance of various environmental factors that may influence swimming ability. A wide range of factors can influence the swimming ability of fishes, including temperature, dissolved O_2 and CO_2, salinity, diet and the presence of toxicants. These issues are reviewed thoroughly for adult fishes in other chapters in this volume (Wilson *et al.*, Chapter 9; McKenzie and Claireaux Chapter 10) and will not be covered in detail here. Temperature, however, is particularly relevant to the present chapter, because it will be a confounding factor for many of the taxonomic comparisons made. Although we have taken care to ensure that the temperature at which each study was conducted are readily available to the reader, a formal comparison of the effect of temperature on the different types of swimming ability measures examined here are problematic. This is because of a paucity of data across a broad range of temperatures, and general taxonomic confounding between temperate and tropical locations. At present, the extent to which "tropical" larvae might swim better than "temperate" larvae is unclear. For most measures of swimming ability, tropical larvae tend to show the greatest abilities of the taxa examined (see particularly the data for U-crit, Fig. 11.3). However, there are also plenty of examples were tropical species are comparable to temperate species of the same size (e.g., Apogonidae, routine swimming, Fig. 11.2), or colder-water species which are comparable to the tropical larvae in their capabilities (e.g., sustained swimming, (Dudley *et al.*, 2000)).

The physics and physiology of swimming both predict that swimming by fish larvae will be influenced to some extent by water temperature. Water viscosity increases with decreasing temperature, making swimming more difficult and less energetically efficient (Fuiman and Batty, 1997; Hunt Von Herbing, 2002). Because issues related to water viscosity are only relevant at low Reynolds numbers, this effect is most pronounced for small larvae, or larvae that swim slowly (Fuiman, 1986), and are probably most likely to influence routine speeds, because these are considerably slower than other swimming speed measures (see later discussion in section: Collating swimming measurements—understanding the ecological importance of larval swimming). For larger juveniles, the cost of swimming appears to be independent of temperature (Beamish, 1990). Larvae swimming in colder water also operate over an inefficient range of muscle-fibre shortening velocities (Wieser and Kaufmann, 1998), and this will impact on swimming efficiency and limit swimming performance (Hunt Von Herbing, 2002). In larval herring and plaice, muscle contraction speed increases by about one-third when temperature increases from 5 to 12–15°C (Blaxter, 1992), for example, and in herring larvae this leads to a doubling of swimming speed (Batty *et al.*, 1991). Temperature also has a marked effect on the ultrastructure, number and phenotype of larval muscle fibres (Johnston and Hall, 2004). Aside from potential effects on muscle activity, larvae from higher temperatures show faster growth rates (Houde, 1989a), which will shorten developmental times and increase the daily rate of development of swimming speed. Although growth is faster in warmer water, there is no effect of temperature on growth efficiency, and it appears that faster growth rates must be supported by increased food consumption (Houde, 1989a).

Despite the potential for temperature to alter swimming ability among tropical and temperate locations, general conclusions are complicated by taxonomic differences among locations, and by the potential for temperature adaptation by local species (Johnston, 1985). If temperature does have a substantial effect on swimming performance and/or larval activity levels, this has important implications for the parameterisation of foraging based individual models, and suggests there may be large differences in the fundamental mechanistic processes behind larval growth and survival at different latitudes. Furthermore, temperature influences on growth and developmental rates mean that larvae from tropical regions may increase swimming capabilities at a much faster rate than temperate larvae, and may be expected to have a greater influence on their dispersal patterns earlier in ontogeny.

Collating Swimming Measurements—Understanding the Ecological Importance of Larval Swimming

A wealth of information exists on the swimming ability of larval fishes, and together this suggests they have the potential to influence their dispersal patterns on relatively large scales, as well as rapidly alter vertical positioning in the water column and relocate horizontally on small spatial scales. At the settlement stage, many coral reef fishes could have considerable control of their settlement locations and timing. Furthermore, we are beginning to attain an appreciation for how swimming speeds may change during ontogeny, vary in the ocean with environmental cues, and the speeds used in foraging and swimming on a regular basis.

All measures of swimming speed increase positively with larval size in absolute terms (Fig. 11.6), especially when data are available for taxa ontogenetically. Size is a much better indicator of developmental stage than age in interspecific comparisons (Fuiman *et al.*, 1998), so it is useful to examine developmental rates of swimming ability with respect to larvae length, rather than time since hatching. As would be expected, burst speeds show the greatest rate of increase with body length during larval ontogeny, with maximum velocity estimates of burst speed increasing approximately 1.5 times faster per mm of larval length than U-crit speeds (Fig. 11.6). Integrated speeds reported over the burst duration are only slightly faster than are seen for U-crit (Fig. 11.6). These estimates of length specific increases in burst speed are likely to be conservative because the burst swimming data used in this meta-analysis include only one representative coral reef taxon, which tends to show faster development of swimming speeds. Across the taxa for which data are available, maximum burst velocity increases at a rate of 2.6 cm s^{-1} per mm of larval length. U-crit speeds increase at a rate of approximately 1.8 cm s^{-1} per mm of larval length. Information on the ontogeny of *in situ* speed is too limited at present for an overall analysis, but values of 0.4 to 2.0 cm s^{-1} per mm increase in larval length have been reported for four warm temperate and tropical species in four families (Leis *et al.*, 2006a, b). Routine swimming speeds, on the other hand, are considerably slower than burst and U-crit speeds, increasing at a rate of only 0.6 cm s^{-1} per mm of larval length (Fig. 11.6).

With the range of methods for estimating swimming speed that are available in the literature, and the large differences in resultant estimates of speed, it is important to discuss how these measures relate to one another, and how they should be interpreted in an ecological context, and their use as parameters in modelling studies of larval growth, dispersal and survival. In general, the ecological relevance of swimming ability to larvae can be divided into three important aspects of larval behaviour: escaping, foraging and their behavioural influence on spatial location (i.e., dispersal).

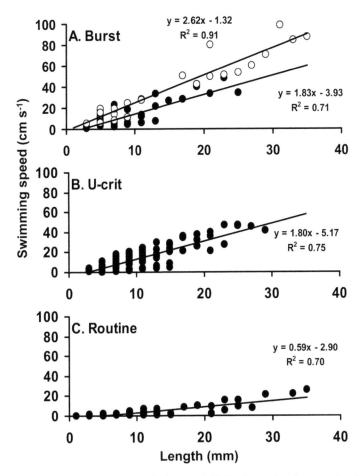

Fig. 11.6 The ontogeny of A) Burst, B) U-crit and C) Routine swimming speeds with larval length. For burst swimming speed data are shown for the integrated speed averaged over the duration of the burst event (closed circles) and the maximum velocity recorded (open circles). Data comprise the same sources as in Figs. 1–3. For comparisons axes have been plotted on the same scales for each type of swimming measure, and fitted linear regression (solid lines) results are shown. Note that length estimates are a mix of standard and total length and this will introduce a small error.

Escaping predators: Measurements of burst swimming performance are directly relevant to the ability of larvae to escape predators, including plankton nets. An understanding of how burst swimming performance varies among taxa, during ontogeny and under different environmental conditions such as light and turbidity, will be key to understanding the potential natural variability in the vulnerability of larval fishes to predation. Such information may be critical to developing useful individual based models of larval survival in the open ocean. Burst speeds attained by larvae may vary substantially among species, and increase with larval size. The differences in escape performance may be important drivers of larval survival. In contrast to absolute speeds, burst speeds in relative terms (body lengths per second) reach a maximum very early in larval life, perhaps offsetting the risk of increased predation associated with small size. Aside from burst swimming speeds, a multitude of other factors will be important in determining the susceptibility of larvae to predation, including visibility (Morgan and Christy, 1996) and behaviour (Zhou and Weis, 1998), the distance of the startle response, latency, reaction distance, turning rates and overall responsiveness (Fuiman *et al.*, 2006) as well as acceleration and turning ability (Walker *et al.*, 2005).

Foraging: Measurements of spontaneous swimming such as routine and *in situ* measures of swimming speed provide valuable data relevant to the feeding behaviour and search swimming speeds of larvae. In general, laboratory measurements of routine speeds of larval fishes are scarce, particularly for tropical species. The one study that has been carried out for coral reef fishes suggests these speeds change during ontogeny, among species, and diurnally. Although this information may provide a starting point for parameterizing larval foraging models, more data are urgently needed on this type of swimming ability in larval coral reef fishes. In particular, we need to explore some of the previously discussed factors that may influence routine swimming speed estimates. How do speeds vary in response to food concentration, turbidity and turbulence? Even small changes in speed can lead to relatively large differences in realised search volumes of larvae, having a considerable impact on larval food encounter rates.

Of particular interest is the idea of "optimal" foraging speeds of larvae, which is the speed that maximises food encounter rates whilst minimizing energetic expenditure. Spontaneous measures of swimming speed, such as routine and *in situ*, are probably the closest we can get to empirical estimates of optimal speed, although there is no way of empirically determining if these are in fact optimal. Assuming larvae adapt their behaviour to differing environmental situations; theoretically optimal swimming speeds will vary depending on prey density, turbidity (which might influence perceptive distance), turbulence and other similar factors. It is, therefore, not surprising that routine and *in situ* speeds can vary widely with respect to methodology. Theoretical simulations show that optimal swimming speed should decrease with increasing turbulence (Pitchford *et al.*, 2003). For a 5 mm larva with a prey density of 1.5×10^4 copepods m^{-3}, optimal swimming speed should be in the order of 12 bl s^{-1} (Pitchford *et al.*, 2003). This is considerably faster than commonly reported routine speeds, and exceeds U-crit in a range of species. It is also considerably faster than the optimal cruising speed for transport purposes calculated by Weihs (1973), although these calculations are based on data for adult salmonids, and larvae can swim at much faster speeds in bl s^{-1} relative to adults (Bellwood and Fisher, 2001). Considerable work needs to be done to explore the idea of optimal foraging speeds in larval fishes, and how these may be influenced by a variety of environmental parameters.

Although laboratory experiments can be ideal for determining functional relationships between swimming behaviour and various environmental parameters under controlled conditions, we should ideally obtain measures of undisturbed routine speeds in the open ocean, in the presence of natural stimuli, to see how these compare to *in situ* measures (see below). Spatial variation in swimming speeds and behaviour as observed *in situ* by divers suggests larval behaviour will change in response to a wide range of external environmental cues. Such observations in an undisturbed context may soon become available due to newly developed technology for *in situ* measurements of larval behaviour using underwater video equipment (Paris *et al.*, 2008).

Along with the search behaviour of larvae, attack speed will also be important to foraging success of larvae. Attack speeds are relatively less well studied when compared to escape speeds, but the limited data suggests these are comparable. In reducing the distance between larva and prey during an attack, the relative importance of sucking the prey towards the mouth and swimming forward is variable (Drost and Boogaart, 1986). Attack speeds will be most important for species and developmental stages for which suction feeding is not well developed.

Spatial location and dispersal: There are a wide range of swimming measurement techniques that provide useful information in terms of the spatial location behaviour of larval fishes. The direct relevance of each measure depends on the scale over which spatial location is being considered, and whether the researcher is more interested in considering the maximum or the minimum potential of the larvae to influence dispersal patterns. The range of potentially relevant swimming measures are summarized in Fig. 11.7, which provides a schematic diagram of how the different measures of swimming ability are related, at least in terms of the work that has been done on larval coral reef fishes.

The ability of larvae to rapidly alter their position in the water column, move between water masses and/or food patches on a relatively small scale and swim away from predators and/or other potential threats, is likely related to their maximum speeds that can be maintained for minutes, i.e., U-crit, rather than to U-burst, which is completely anaerobic and cannot be sustained beyond seconds.

However, in order for larvae to significantly influence their spatial location and dispersal patterns at larger scales using horizontal swimming behaviour, swimming activity needs to be maintained for considerable lengths of time (days). In an extended swimming chamber mean swimming speed of late

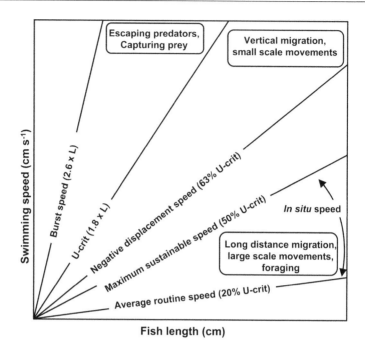

Fish length (cm)

Fig. 11.7 Schematic diagram of the relationship between different measurements of swimming ability in larval coral reef fishes. The fastest speeds are burst speeds used in escape and prey capture and scale roughly as 26 larval length (from Fig. 11.6A). U-crit represents the maximum swimming speeds of larvae maintained over short time periods that may be used for vertical migration and/or small-scale horizontal movements, and scale as around 18 larval length (from Fig. 11.6B). The negative displacement speed represents the speed at which fish larvae are no longer motivated to maintain position, although this value has only been determined for one species thus far (63% U-crit)(Hogan and Mora, 2005). *In situ* speed represents the range of speeds over which larvae swim in the ocean (Leis and Carson-Ewart, 1997, 1998; Leis *et al.*, 2006b) and are bounded by the maximum sustainable speeds of larvae (50-60% U-crit)(Fisher and Bellwood, 2002; Fisher and Wilson, 2004) and the routine swimming speeds maintained in the absence of stimuli on a daily basis (20% U-crit)(Fisher and Bellwood, 2003).

stage larvae of one coral reef fish species (a pomacentrid) kept pace with increasing mean current until about 20 cm s⁻¹ (ca 12 bl s⁻¹) and remained at this speed as currents increased, resulting in negative displacement at the faster current speeds (Hogan and Mora, 2005). In this species, asymptotic swimming speed was about 60% of the critical speed, so larvae chose to swim slower than they were able and did not attempt to keep up with the current flow when it became too strong. That larvae are not motivated to swim as fast as U-crit in an un-forced situation is to be expected, since U-crit incorporates some anaerobic muscle activity and is not sustainable over long periods.

However, U-crit is a useful measure because it is related to a number of key variables (Fig. 11.6).

Endurance studies on coral reef fish larvae show that maximum sustainable swimming speeds lie in the region of 50% of the U-crit speeds of late stage larvae of a range of species (Fisher and Bellwood, 2002; Fisher and Wilson, 2004), and this appears to be consistent during ontogeny for the few species for which data are available (Fisher and Bellwood, 2002). Interestingly, a value of 50% U-crit for the sustainable speeds (over 24 hours) of late stage larval coral reef fishes appears to hold for the Pomacentridae, Apogonidae, and Lethrinidae (Fisher and Wilson, 2004), even though these families exhibit different swimming modes as adults (the Pomacentridae are primarily pectoral fin swimmers, whereas the Apogonidae and Lethrinidae are pectoral-caudal fin swimmers). For these groups at least, the relationship between U-crit, and sustainable speed (perhaps a more dispersal relevant measure of swimming ability) appears to be consistent across taxa with different swimming modes. Thus a value of around 50% of their U-crit values (maximum capacity) provides an upper limit to the speeds likely to be used by larvae to influence dispersal patterns on relatively large scales. Faster speeds than these (approaching U-crit) could only be used for short lengths of time, and thus are only likely to be useful for relocating over small spatial scales. A value of 50–60% of U-crit speed may be appropriate in dispersal modelling in the context of examining the maximum potential impact of larval behaviour, and in situations were it is advantageous for larvae to cover the greatest possible distance in the shortest length of time.

On the other end of the scale, average routine swimming speeds appear to be around 20% of U-crit (Fisher and Bellwood, 2003) and provide an important estimate of the minimum undisturbed speeds at which larvae swim on a daily basis. This value is based on a single laboratory study including only a handful of species, and more work is needed to examine the generality of this relationship over a wider range of species and under varying laboratory conditions. Because these routine speed measurements were conducted in the absence of external stimuli, in an enclosed environment, and in the presence of quite high food concentrations, we suggest that 20% of U-crit probably represents a conservative estimate of the minimum speeds larvae would swim at on a routine basis. Such values may be useful for considering the minimum impacts larval behaviour may have on dispersal, but it is likely larvae will swim in the ocean at higher speeds. Of particular interest is the idea of optimal foraging speed, as this would be the theoretical speed larvae should maintain in nature. Although such speeds would be difficult to estimate empirically, theoretical simulations suggest they may be in the order of 12 bl s^{-1} for some species, and this is considerably faster than many routine speed estimates.

In situ speeds of larvae (as measured by observing divers) range somewhere between their maximum sustainable speeds and their routine speeds (Leis and Fisher, 2006), and provide important insight into how the swimming speeds of larvae vary in the open ocean in response to a range of environmental sensory cues (Leis and Carson-Ewart, 1998). There is a strong positive relationship between *in situ* speed and U-crit (Leis and Fisher, 2006), and U-crit may be used to predict *in situ* speed from laboratory measures of potential speed for stages or taxa that are not amenable to *in situ* methodology. In agreement with estimates of maximum sustainable speeds, the maximum *in situ* swimming speeds of larvae appear to be around 50% of their U-crit speeds (Leis and Fisher, 2006), suggesting that some larvae do swim *in situ* near, or at, their maximum sustainable speed. On average, however, *in situ* speeds are around 40% of U-crit, and they range substantially among species as well as among individuals within species (Leis and Fisher, 2006). A greater understanding of the factors influencing *in situ* speed will enable the development of more precise models of larval dispersal and the likely impacts of active behaviour under specific environmental and ecological scenarios. Given the importance of *in situ* swimming on the effective dispersal trajectories of larvae, a means of measuring swimming speed in the ocean without using divers would be very useful. To that end, it may be possible to use remote video (Holbrook and Schmitt, 1999; Paris *et al.*, 2008), acoustic telemetry (Bégout Anras and Lagardére, 2004) or specialized sonar (Jaffe *et al.*, 1995; Genin *et al.*, 2005; Onsrud *et al.*, 2005; Trevorrow, 2005) to determine what larvae are doing in the sea in the absence of divers.

CONCLUDING REMARKS

Here we have examined the range of different types of swimming measures relevant to the behaviour and ecology of larval fishes, concentrating to some extent on coral reef species. Although techniques vary substantially in their estimates of speed and their relevance to different aspects of larval behaviour, all methods provide valuable estimates of larval swimming ability, and confirm several important points about the behavioural capabilities of larval fishes. Firstly, it appears that the larvae of many taxa, particularly in coral reef environments are highly capable swimmers able to escape from predators from a very small size, search quite large volumes for food and rapidly alter their vertical distribution in the water column, and have a substantial influence on their horizontal location on relatively large spatial scales. Secondly, absolute swimming speeds, regardless of the type of measurement being made, develop more or less linearly with respect to length, but increase rapidly to a maximum value when considered in relative terms (body lengths per second). Maximal relative performance is thus observed quite early on in larval life, and may be a result of the necessity for fast development of swimming capabilities in order to enhance survival of vulnerable early stages. Lastly, considerable differences occur among taxa, and perhaps among tropical and temperate regions, in the swimming speeds exhibited by larvae (Leis, 2007). These differences imply considerable variance in larval behavioural capabilities, and the extent to which individual larvae may be capable of controlling their fate in the open ocean, in terms of their energetics and feeding ecology, their overall dispersal and connectivity patterns, and the degree to which recruitment is an active processes involving specific habitat choice and timing. Whilst larvae are clearly not the passive particles they were once considered, neither are all larval fish equivalent in their behaviour and ecology, despite occupying a similar pelagic habitat. Such differences need to be carefully considered when developing ecological and biophysical models involving larvae.

REFERENCES

Andrews, J.C. 1983. Water masses, nutrient levels and seasonal drift on the outer central Queensland shelf (Great Barrier Reef). *Australian Journal of Marine Freshwater Research* 34: 821–34.

Armsworth, P.R., M.K. James and L. Bode. 2001. When to press on or turn back: dispersal strategies for reef fish larvae. *American Naturalist* 157: 434–450.

Bailey, K.M., R.D. Brodeur, N. Merati and M.M. Yoklavich. 1993. Predation on walleye pollock (*Theragra chalcogramma*) eggs and yolk-sac larvae by pelagic crustacean invertebrates in the western Gulf og Alaska. *Fisheries Oceanography* 2: 30–39.

Batty, R.S. 1987. Effect of light intensity on activity and food-searching of larval herring, *Clupea harengus*: a laboratory study. *Marine Biology* 94: 323–327.

Batty, R.S., J.H.S. Blaxter and Q. Bone. 1991. The effect of temperature on the swimming of a teleost *(Clupea harengus)* and an ascidian larva *(Dendrodoa grossularia)*. *Comparative Biochemistry and Physiolology* A100: 297–300.

Batty, R.S., J.H.S. Blaxter and K. Fretwell. 1993. Effect of temperature on the escape responses of larval herring, *Clupea harengus*. *Marine Biology* 115: 523–528.

Beamish, F. 1990. Swimming metabolism and temperature in juvenile walleye, *Stizostedion vitreum vitreum*. *Environmental Biology of Fishes* 27: 309–314.

Bégout Anras, M.-L. and J.P. Lagardére. 2004. Measuring cultured fish swimming behaviour: first results on rainbow trout using acoustic telemetry in tanks. *Aquaculture* 240: 175–186.

Bellwood, D.R. and R. Fisher. 2001. Relative swimming speeds in reef fish larvae. *Marine Ecology Progress Series* 211: 299–303.

Bellwood, D.R., J.M. Leis and I.C. Stobutzki. 1998. Fishery and reef management. *Science* 279: 2021–2022.

Blake, R.W. 1983. *Fish Locomotion*. Cambridge University Press, Cambridge.

Blaxter, J.H.S. 1986. Development of sense organs and behaviour of Teleost larvae with special reference to feeding and predator avoidance. *9th Larval Fish Conference* 115: 98–114.

Blaxter, J.H.S. 1992. The effect of temperature on larval fishes. *Netherlands Journal of Zoology* 42: 336–357.

Boisclair, D. and M. Tang. 1993. Empirical analysis of the influence of swimming pattern on the net energetic cost of swimming in fishes. *Journal of Fish Biology* 42: 169–183.

Botsford, L., A. Hastings and S. Gaines. 2001. Dependence of sustainability on the configuration of marine reserves and larval dispersal distance. *Ecology Letters* 4: 144–150.

Bradbury, I.R., P.V.R. Snelgrove and P. Pepin. 2003. Passive and active behavioural contributions to patchiness and spatial pattern during the early life history of marine fishes. *Marine Ecology Progress Series* 257: 233–245.

Brett, J.R. 1964. The respiratory metabolism and swimming performance of young sockeye salmon. *Journal of the Fisheries Research Board of Canada* 21: 1183–1226.

Chick, J.H. and M.J. Van Den Avyle. 2000. Effects of feeding ration on larval swimming speed and responsiveness to predator attacks: implications for cohort survival. *Canadian Journal of Fisheries and Aquatic Sciences* 57: 106–115.

Clark, D.L., J.M. Leis, A.C. Hay and T. Trnski. 2005. Swimming ontogeny of larvae of four temperate marine fishes. *Marine Ecology Progress Series* 292: 287–300.

Cook, A. 1996. Ontogeny of feeding morphology and kinematics in juvenile fishes: a case study of the cottid fish *Clinocottus analis*. *Journal of Experimental Biology* 199: 1961–1971.

Cooke, S.J., K.P. Chandroo, T.A. Beddow, R.D. Moccia and R.S. McKinley. 2000. Swimming activity and energetic expenditure of captive rainbow trout *Oncorhynchus mykiss* (Walbaum) estimated by electromyogram telemetry. *Aquaculture Research* 31: 495.

Cowen, R.K. and L.R. Castro. 1994. Relation of coral reef fish larval distributions to island scale circulation around Barbados, West Indes. *Bulletin of Marine Science* 54: 228–244.

Cowen, R. K., K.M.M. Lwiza, S. Sponaguale, C.B. Paris and D.B. Olsen. 2000. Connectivity of marine populations: open or closed? *Science* 287: 857–859.

Cowen, R.K., C.B. Paris and A. Srinivasan. 2006. Scaling connectivity in marine populations. *Science* 311: 522–527.

Doi, M., H. Kohno, Y. Taki and A. Ohno. 1998. Development of swimming and feeding functions in larvae and juveniles of the red snapper, *Lutjanus argentimaculatus*. *Journal of Tokyo University of Fisheries* 85: 81–95.

Domenici, P. and R. Blake. 1997. The kinematics and performance of fish fast-start swimming. *Journal of Experimental Biology* 200: 1165–1178.

Dou, S., T. Sekai and K. Tsukamoto. 2000. Feeding behaviour of japanese flounder larvae under laboratory conditions. *Journal of Fish Biology* 56: 654–666.

Drost, M.R. and J.G.M. Boogaart. 1986. The energetics of feeding strikes in larval carp, *Cyprinus carpio*. *Journal of Fish Biology* 29: 371–379.

Drucker, E.G. 1996. The use of gait transition speed in comparative studies of fish locomotion. *American Zoologist* 36: 555–566.

Drucker, E.G. and J.S. Jensen. 1996. Pectoral fin locomotion in the striped surfperch I. kinematic effects of swimming speed and body size. *Journal of Experimental Biology* 199: 2235–2242.

Dudley, B., N. Tolimieri and J. Montgomery. 2000. Swimming ability of the larvae of some reef fishes from New Zealand waters. *Marine Freshwater Research* 51: 783–787.

Eaton, R.C. and R. DiDomenico. 1986. Role of the teleost escape response during development. *Transactions of the American Fisheries Society* 115: 128–142.

Farlinger, S. and F.W. Beamish. 1977. Effects of time and velocity increments on the critical swimming speed of largemouth bass (*Micropterus salmoides*). *Transactions of the American Fisheries Society* 106: 436–439.

Fisher, R. 2005. Swimming speeds of larval coral reef fishes: impacts on self-recruitment and dispersal. *Marine Ecology Progress Series* 285: 223–232.

Fisher, R. and D.R. Bellwood. 2001. Effects of feeding on the sustained swimming abilities of late-stage larval *Amphiprion melanopus*. *Coral Reefs* 20: 151–154.

Fisher, R. and D.R. Bellwood. 2002. The influence of swimming speed on sustained swimming performance of late-stage reef fish larvae. *Marine Biology* 140: 801–807.

Fisher, R. and D.R. Bellwood. 2003. Undisturbed swimming behaviour and nocturnal activity of coral reef fish larvae. *Marine Ecology Progress Series* 263: 177–188.

Fisher, R., D.R. Bellwood and S.D. Job. 2000. The development of swimming abilities in reef fish larvae. *Marine Ecology Progress Series* 202: 163–173.

Fisher, R. and J.D. Hogan. 2007. Morphological predictors of swimming speeds of fishes; a case study of pre-settlement juvenile coral reef fishes. *Journal of Experimental Biology* 210: 2436–2443.

Fisher, R., J.M. Leis, D.J. Clark and S.K. Wilson. 2005. Critical swimming speeds of late-stage coral reef fish larvae: variation within species, among species and between locations. *Marine Biology* 147: 1201–1212.

Fisher, R., S.M. Sogard and S.A. Berkeley. 2007. Trade-offs between size and energy reserves reflect alternative strategies for optimising larval survival potential. *Marine Ecology Progress Series* 344: 257–270.

Fisher, R. and S.K. Wilson. 2004. Maximum sustainable swimming speeds of nine species of late stage larval reef fishes. *Journal of Experimental Marine Biology and Ecology* 312: 171–186.

Frith, C.A., J.M. Leis and B. Goldman. 1986. Currents in the Lizard Island region of the great barrier Reef Lagoon and their relevance to potential movements of larvae. *Coral Reefs* 5: 81–92.

Fuiman, L.A. 1986. Burst swimming performance of larval *Zebra danios* and the effects of diel temperature fluctuations. *Transactions of the American Fisheries Society* 115: 143–148.

Fuiman, L.A. and R.S. Batty. 1997. What a drag it is getting cold: partitioning the physcial and physiological effects of temperature on fish swimming. *Journal of Experimental Biology* 200: 1745–1755.

Fuiman, L.A. and J.H.J. Cowan. 2003. Behavior and recruitment success in fish larvae: repeatability and covariation of survival skills. *Ecology* 84: 53–67.

Fuiman, L.A., K.R. Poling and D.M. Higgs. 1998. Quantifying developmental progress for comparative studies of larval fishes. *Copeia* 1998: 602–611.

Fuiman, L.A., K.A. Rose, J.H. Cowan and E.P. Smith. 2006. Survival skills required for predator evasion by fish larvae and their relation to laboratory measures of performance. *Animal Behaviour* 71: 1389–1399.

Fuiman, L.A., M.E. Smith and V.N. Malley. 1999. Ontogeny of routine swimming speed and startle responses in red drum, with a comparison of responses to acoustic and visual stimuli. *Journal of Fish Biology* 55: 215–226.

Fukuhara, O. 1985. Functional morphology and behavior of early life stages of red sea bream. *Bulletin of the Japanese Society of Scientific Fisheries* 51: 731–743.

Fukuhara, O. 1987. Larval development and behaviour in early life stages of black sea bream reared in the laboratory. *Nippon Suisan Gakkaishi* 53: 371–379.

Fulton, C.J. 2005. *Wave Energy and the Role of Swimming in Reef Fish Ecology.* James Cook Unversity, Townsville.

Fulton, C.J., D.R. Bellwood and P.C. Wainwright. 2005. Wave energy and swimming performance shape coral reef fish assemblages. *Proceedings of the Royal Society* B272: 827–832.

Gell, F. and C. Roberts. 2003. Benefits beyond boundaries: the fishery effects of marine reserves. *Trends in Ecology & Evolution* 18: 448–455.

Genin, A., J.S. Jaffe, R. Reef, C. Richter and P.J.S. Franks. 2005. Swimming against the flow: a mechanism of zooplankton aggregation. *Science* 308: 860–862.

Gibson, S. and I. Johnston. 1995. Scaling relationships, individual variation and the influence of temperature on maximum swimming speed in early settled stages of the turbot *Scophthalmus maximus*. *Marine Biology* 121: 401–408.

Green, B. and R. Fisher. 2004. Temperature influences swimming speed, growth and larval duration in coral reef fish larvae. *Journal of the Experimental Marine Biology and Ecology* 299: 115–132.

Hale, M. 1996. The development of fast start performance in fishes: Escape kinematics of the Chinook salmon (*Oncorhynchus tshawytscha*). *American Zoologist* 36: 695–709.

Halpern, B.S. and R.R. Warner. 2003. Matching marine reserve design to reserve objectives. *Proceedings of the Royal Society of London,* B270: 1871–1878.

Hamner, W.M. and I.R. Hauri. 1981. Effects of island mass: water flow and plankton pattern around a reef in the Great Barrier Reef Lagoon, Australia. *Limnology and Oceanography* 26: 1084–1102.

Hartwell, S.I. and R.G. Otto. 1978. Swimming performance of juvenile menhaden (*Brevoortia tyrannus*). *Transactions of American Fisheries Society* 107: 793–798.

Hartwell, S.I. and R.G. Otto. 1991. Critical swimming capacity of the Atlantic Silverside, *Menidia menidia* L. *Estuaries* 14: 218–221.

Hawkins, D. and T. Quinn. 1996. Critical swimming velocity and associated morphology of juvenile coastal cutthroat trout (*Onchorhynchus clarkii*), steelhead trout (*Oncorhynchus mykiss*), and their hybrids. *Canadian Journal of Fisheries and Aquatic Sciences* 53: 1487–1496.

Hecht, T., S. Battaglene and B. Talbot. 1996. Effect of larval density and food availability on the behaviour of pre-metamorphosis snapper, *Pagrus auratus* (Sparidae). *Marine and Freshwater Research* 47: 223–231.

Hillgruber, N., M. Kloppmann, E. Wahl and H.v. Westernhagen. 1997. Feeding of larval blue whiting and Atlantic mackerel: a comparison of foraging strategies. *Journal of Fish Biology* 51sa): 230–249.

Hindell, J.S., G.P. Jenkins, S.M. Moran and M.J. Keough. 2003. Swimming ability and behaviour of post-larvae of a temperate marine fish re-entrained in the pelagic environment. *Oecologia* 135: 158–166.

Hogan, J.D., R. Fisher and C. Nolan. 2007. Critical swimming abilities of late-stage coral reef fish larvae from the Caribbean: A methodological and intra-specific comparison. *Bulletin of Marine Science* 80: 219–232.

Hogan, J.D. and C. Mora. 2005. Experimental analysis of the contribution of swimming and drifting to the displacement of reef fish larvae. *Marine Biology* DOI10.1007/s00227-005-0006-5.

Holbrook, S.J. and R.J. Schmitt. 1999. *In situ* nocturnal observations of reef fishes using infrared video. In: *Proceedings of the 5th Indo-Pacific Fish Conference, Noumea*. B. Seret, and J.-Y. Sire, (Eds.). Societie Francaise Ichtyologie, Paris, Noumea. pp. 805–812.

Houde, E. 1969. Sustained swimming ability of larvae of walleye (*Stizostedion vitreum*) and yellow perch (*Perca flavescens*). *Journal of the Fisheries Research Board of Canada* 26: 1647–1659.

Houde, E. 1989a. Comparative growth, mortality and energetics of marine fish larvae: temperature and implied latitudinal effects. *Fisheries Bulletin, USA*, 87: 471–495.

Houde, E.D. 1989b. Subtleties and episodes in the early life history of fishes. *Journal of Fish Biology* 35(Supplement A): 29–38.

Hunt Von Herbing, I. 2002. Effects of temperature on larval fish swimming performance: the importance of physics to physiology. *Journal of Fish Biology* 61: 865–876.

Hunt Von Herbing, I., S.M. Gallager and W. Halteman. 2001. Metabolic costs of persuit and attack in early larval Atlantic cod. *Marine Ecology Progress Series* 216: 201–212.

Hunter, J.R. 1972. Swimming and feeding behaviour of larval anchovy *Engraulis mordax*. *Fisheries Bulletin* 70: 821–838.

Hunter, J.R. 1981. Feeding Ecology and predation of marine fish larvae. In: *Marine Fish larvae*, R. Laker (Ed.).University of Washington, Washington, pp. 33–79.

Hunter, J.R. and C.A. Kimbrell. 1980. Early life history of pacific mackerel, *Scomber japonicus*. *Fisheries Bulletin* 78: 89.

Jaffe, J.S., E. Ruess, D. McGehee and G. Chandran. 1995. FTV: a sonar for tracking macrozooplankton in three dimensions. *Deep Sea Research* 42: 1495–1512.

James, M.K., P.R. Armsworth, L.B. Mason and L. Bode. 2002. The structure of reef fish metapopulations: modelling larval dispersal and retention patterns. *Proceedings of the Royal Society of London* B269: 2079–2086.

Jenkins, G.P. and D.C. Welsford. 2002. The swimming abilities of recently settled post-larvae of Sillaginodes punctata. *Journal of Fish Biology* 60: 1043.

Johnston, I.1985. Temperature adaptation of enzyme function in fish muscle. *Society of the Experimental Biology* 49. pp. 95–122.

Johnston, I.A. and T.E. Hall. 2004. Mechanisms of muscle development and responses to temperature change in fish larvae. *American Fisheries Society Symposium* 40: 85–116.

Jones, D.R., J.W. Kiceniuk and O.S. Bamford. 1974. Evaluation of the swimming performance of several fish species from Mackenzie River. *Journal of the Fisheries Research Board of Canada* 31: 1641–1647.

Kasumyan, A., M. Ryg and K. Doving. 1998. Effect of amino acids on the swimming activity of newly hatched turbot larvae (*Sophthalmus maximus*). *Marine Biology* 131: 189–194.

Killen, S.S., J.A. Brown and A.K. Gamperl. 2007a. The effect of prey density on foraging mode selection in juvenile lumpfish: balancing food intake with the metabolic cost of foraging. *Journal of Animal Ecology* 76: 814–825.

Killen, S.S., I. Costa, J.A. Brown and A.K. Gamperl. 2007b. Little left in the tank: metabolic scaling in marine teleosts and its implications for aerobic scope. *Proceedings of the Royal Society* B274: 431–438.

Killen, S.S., A.K. Gamperl and J.A.Brown. 2007c. Ontogeny of predator-sensitive foraging and routine metabolism in larval shorthorn sculpin, *Myoxocephalus scorpius*. *Marine Biology* 152: 1249–1261.

Kolok, A.S. 1991. Photoperiod alters the critical swimming speed of Juvenile Largemouth Bass, *Micropterus salmoides*, Acclimated to cold water. *Copeia* 1991: 1085–1090.

Kolok, A.S. 1999. Interindividual variation in the prolonged locomotor performance of ectothermic vertebrates: a comparison of fish and herpetofaunal methodologies and a brief review of the recent fish literature. *Canadian Journal of Fisheries and Aquatic Sciences* 56: 700–710.

Leis, J.M. 2006. Are larvae of demersal fishes plankton or nekton? *Advances in Marine Biology* 51: 59–141.

Leis, J.M. 2007. Behaviour of fish larvae as an essential input for modelling larval dispersal: behaviour, biogeography, hydrodynamics ontogeny, physiology and phylogeny meet hydrography. *Marine Ecology Progress Series*. 347: 185–193.

Leis, J.M. and B.M. Carson-Ewart. 1997. *In situ* swimming speeds of the late pelagic larvae of some indo-pacific coral reef fishes. *Marine Ecology Progress Series* 159: 165–174.

Leis, J.M. and B.M. Carson-Ewart. 1998. Complex behaviour by coral reef fish larvae in open water and near-reef pelagic environments. *Environmental Biology of Fishes* 53: 259–266.

Leis, J.M. and B.M. Carson-Ewart. 2001. Behaviour of pelagic larvae of four coral reef fish species in the ocean and an atoll lagoon. *Coral Reefs* 19: 247–257.

Leis, J.M. and B.M. Carson-Ewart. 2003. Orientation of pelagic larvae of coral reef fishes in the ocean. *Marine Ecology Progress Series* 252: 239–253.

Leis, J.M. and D.L. Clark. 2005. Feeding greatly enhances swimming endurance of settlement-stage reef-fish larvae of damselfishes (Pomacentridae). *Ichthyological Research* 52: 185–188.

Leis, J.M. and R. Fisher. 2006. Swimming speed of settlement-stage reef-fish larvae measured in the laboratory and in the field: a comparison of critical speed and in situ speed. In: *Proceedings of 10th International Coral Reef Symposia*, Okinawa, pp. 438–445.

Leis, J.M., B.M. Carson-Ewart and D.H. Cato. 2002. Sound detection in situ by the larvae of a coral-reef damselfish (Pomacentridae). *Marine Ecology Progress Series* 232: 259–268.

Leis, J.M., A.C. Hay, D.A. Clark, I.-S. Chen and K.-T. Shao. 2006a. Behavioural ontogeny in larvae and early juveniles of the giant trevally, *Caranx ignobilis* (Pisces: Carangidae). *Fisheries Bulletin, USA.,* 104: 401–414.

Leis, J.M., A.C. Hay, M.M. Lockett, J.-P. Chen and L.-S. Fang. 2007a. Ontogeny of swimming speed in larvae of pelagic-spawning, tropical, marine fishes. *Marine Ecology Progress Series*. 349: 257–269.

Leis, J.M., A.C. Hay and T. Trnski. 2006b. *In situ* behavioural ontogeny in larvae of three temperate, marine fishes. *Marine Biology* 148: 655–669.

Leis, J.M. and M.M. Lockett. 2005. Localization of reef sounds by settlement-stage larvae of coral-reef fishes (Pomacentridae). *Bulletin of Marine Science* 76: 715–724.

Leis, J.M., R.F. Piola, A.C. Hay, C. Wen and K.-P. Kan. 2009. Ontogeny of behaviour relevant to dispersal and connectivity in larvae of two non-reef demersal, tropical fish species. *Marine and Freshwater Research* 60: 211–223.

Leis, J.M., H.P.A. Sweatman and S.E. Reader. 1996. What the pelagic stages of coral reef fishes are doing out in blue water: daytime field observations of larval behavioural capabilities. *Marine and Freshwater Research* 47: 401–411.

Leis, J.M., K.J. Wright and R.N. Johnson. 2007b. Behaviour that influences dispersal and connectivity in the small, young larvae of a reef fish. *Marine Biology* 153: 103–117

Letcher, B.H., J.A. Rice, L.B. Crowder and K.A. Rose. 1996. Variability in survival of larval fish: disentangling components with a generalized individual-based model. *Canadian Journal of Fisheries and Aquatic Sciences* 53: 787–801.

Lindsey, C.C. 1978. Form, function, and locomotory habits in fish. In: *Fish Physiology*, W. S. Hoar and D. J. Randall, (Eds.). Academic Press, New York. Vol. 2, pp. 1–100

Lowe, C.G. 1996. Kinematics and critical swimming speed of juvenile scalloped hammerhead sharks. *Journal of Experimental Biology* 199: 2605–2610.

MacKenzie, B.R. and T. Kiorboe. 1995. Encounter rates and swimming behaviour of pause-travel and cruise larval fish predators in calm and turbulent laboratory environments. *Limnology and Oceanography* 40: 1278–1289.

Man, A., R. Law and N.V.C. Polunin. 1995. Role of marine reserves in recruitment to reef fisheries: A metapopulation model. *Biological Conservation* 71: 197–204.

Marchand, F., P. Magnan and D. Boisclair. 2002. Water temperature, light intensity and zooplankton density and the feeding activity of juvenile brook charr (*Salvelinus fontinalis*). *Freshwater Biology* 47: 2153–2162.

Masuda, R. 2006. Ontogeny of anti-predator behavior in hatchery-reared jack mackerel *Trachurus japonicus* larvae and juveniles: patchiness formation, swimming capability, and interaction with jellyfish. *Fisheries Science* 72: 1225–1235.

Masuda, R., J.U.N. Shoji, M. Aoyama and M. Tanaka. 2002. Chub mackerel larvae fed fish larvae can swim faster than those fed rotifers and *Artemia nauplii*. *Fisheries Science* 68: 320–324.

McCormick, M.I. 1999. Delayed metamorphosis of a tropical reef fish (*Acanthurus triostegus*): a field experiment. *Marine Ecology Progress Series* 176: 25–38.

McCormick, M.I. and B.W. Molony. 1993. Quality of the reef fish *Upeneus tragula* (Mullidae) at settlement: is size a good indicator of condition? *Marine Ecology Progress Series* 98: 45–54.

McHenry, M.J. and G.V. Lauder. 2005. The mechanical scaling of coasting in zebrafish (*Danio rerio*). *JEB* 208: 2289–2301.

Meng, L. 1993. Sustainable swimming speeds of striped bass larvae. *Transactions of the American Fisheries Society* 122: 702–708.

Miller, T., L.B. Crowder, J.A. Rice and E.A. Marschall. 1988. Larval size and recruitment mechanisms in fishes: toward a conceptual framework. *Canadian Journal of Fisheries and Aquatic Sciences* 45: 1657–1670.

Morgan, S.G. and J.H. Christy. 1996. Survival of marine larvae under the countervailing selective pressures of photodamage and predation. *Limnology and Oceanography* 41: 498–504.

Muller, U.K., E J. Stamhus and J.J. Videler. 2000. Hydrodynamics of unsteady fish swimming and the effecs of body size: comparing the flow fields of fish larvae and adults. *Journal of Experimental Biology* 203: 193–206.

Munk, P. 1995. Foraging behaviour of larval cod (*Gadus morhua*) influenced by prey density and hunger. *Marine Biology* 122: 205–212.

Myrick, C.A. and J.J. Cech. 2000. Swimming performance of four California stream fishes: temperature effects. *Environmental Biology of Fishes* 58: 289–295.

Nelson, J.A., P.S. Gotwalt, S.P. Reidy and D.M. Webber. 2002. Beyond Ucrit: matching swimming performance tests to the physiological ecology of the animal, including a new fish "drag strip". *Comparative Biochemistry and Physiology* A133: 289–302.

Nilsson, G.E., S. Ostlund-Nilsson, R. Penfold and A.S. Grutter. 2007. From record performance to hypoxia tolerance: respiratory transition in damselfish larvae settling on a coral reef. *Proceedings of the Royal Society* B274: 79–85.

Norcross, B. and R. Shaw. 1984. Oceanic and estuarine transport of fish eggs and larvae: A review. *Transactions of the American Fisheries Society* 115: 153–165.

Onsrud, M.S.R., S. Kaartvedt and M.T. Breien. 2005. In situ swimming speed and swimming behaviour of fish feeding on the krill *Meganyctiphanes norvegica*. *Canadian Journal of Fisheries and Aquatic Sciences* 62: 1822–1832.

Paris, C.B. and R.K. Cowen. 2004. Direct evidence of a biophysical retention mechanism for coral reef fish larvae. *Limnology and Oceanography* 49: 1964–1979.

Paris, C.B., C. Guigand, J.O. Irisson, R. Fisher and E. D'Alessandro. 2008. Orientation With No Frame of References (OWNFOR) A novel system to observe and quantify orientation in reef fish larvae, Pages 52–62 in R. Grober-Dunsmore, and B.D. Keller, eds. Caribbean connectivity: Implications for marine protected area management. Proceedings of a Special Symposium, 9–11 November 2006, 59th Annual Meeting of the Gulf and Caribbean Fisheries Institute, Belize City, Belize. *Marine Sanctuaries Conservation Series* ONMS-08-07. Silver Spring, MD, U.S. Department of Commerce, National Oceanic and Atmospheric Administration, Office of National Marine Sanctuaries.

Peake, S. and R.S. McKinley. 1998. A re-evaluation of swimming performance in juvenile salmonids relative to downstream migration. *Canadian Journal of Fisheries and Aquatic Sciences* 55: 682–687.

Pitchford, J.W., A. James and J. Brindley. 2003. Optimal foraging in patchy turbulent environments. *Marine Ecology Progress Series* 256: 99–110.

Pitts, P.A. 1994. An investigation of near-bottom flow patterns along and across Hawk Chanel, Florida Keys. *Bulletin of Marine Science* 54: 610–620.

Plaut, I. 2000. Effects of fin size on swimming performance, swimming behaviour and routine activity of zebrafish *Danio rerio*. *Journal of Experimental Biology* 203: 813–820.

Plaut, I. 2001. Critical swimming speed: its ecological relevance. *Comparative Biochemistry and Physiology* A131: 41–50.

Post, J.R. and J.A. Lee. 1996. Metabolic ontogeny of teleost fishes. *Canadian Journal of Fisheries and Aquatic Sciences* 53(4): 910–923.

Puvanendran, V., L.L. Leader and J.A. Brown. 2002. Foraging behaviour of Atlantic cod (*Gadus morhua*) larvae in relation to prey concentration. *Canadian Journal of Zoology* 80: 689–699.

Reidy, S.P., S.R. Kerr and J.A. Nelson. 2000. Aerobic and anaerobic swimming performance of individual Atlantic cod. *Journal of Experimental Biology* 203: 347–357.

Roberts, C.M. 1997. Connectivity and management of Caribbean coral Reefs. *Science* 278: 1454–1457.

Ryland, J.S. 1963. The swimming speeds of plaice larvae. *Journal of Experimental Biology* 40: 285–299.

Sakakura, Y. and K. Tsukamoto. 1996. Onset and development of cannibalistic behaviour in early life stages of yellowtail. *Journal of Fish Biology* 48: 16–29.

Sale, P.F. and R.K. Cowen. 1998. Fishery and reef Management. *Science* 279: 2022.

Sepulveda, C. and K. Dickson. 2000. Maximum sustainable speeds and cost of swimming in juvenile Kawakawa Tuna (*Euthynnus affinis*) and chub mackerel (*Scomber japonicus*). *Journal of Environmental Biology* 203: 3089–3101.

Simpson, S.D., M. Meekan, J. Montgomery, R. McCauley and A. Jeffs. 2005b. Homeward sound. *Science* 308: 221.

Simpson, S.D., M.G. Meekan, R.D. McCauley and A. Jeffs. 2004. Attraction of settlement-stage coral reef fishes to reef noise. *Marine Ecology Progress Series* 276: 263–268.

Smith, M.E. and L.A. Fuiman. 2004. Behavioral performance of wild-caught and laboratory-reared red drum *Sciaenops ocellatus* (Linnaeus) larvae. *Journal of Experimental Marine Biology and Ecology* 302: 17–33.

Stobutzki, I.C. 1997a. Energetic cost of sustained swimming in the late pelagic stages of reef fishes. *Marine Ecology Progress Series* 152: 249–259.

Stobutzki, I.C. 1997b. *Swimming abilities and orientation behaviour of pre-settlement coral reef fishes*. Ph.D. Thesis James Cook University, Townsville.

Stobutzki, I.C. 1998. Interspecific variation in sustained swimming ability of late pelagic stage reef fish from two families (Pomacentrudae and Chaetodontidae). *Coral Reefs* 17: 111–119.

Stobutzki, I.C. and D.R. Bellwood. 1994. An analysis of the sustained swimming abilities of pre- and post-settlement coral reef fishes. *Journal of Experimental Biology and Ecology* 175: 275–286.

Stobutzki, I.C. and D.R. Bellwood. 1997. Sustained swimming abilities of the late pelagic stages of coral reef fishes. *Marine Ecology Progress Series* 149: 35–41.

Sugisaki, H., K. Bailey and R. Brodeur. 2001. Development of the escape response in larval walleye pollock (*Theragra chalcogramma*). *Marine Biology* 139: 19–24.

Theilacker, G. and K. Dorsey. 1980. Larval fish diversity, a summary of laboratory and field research. In: *UNESCO Intergovernmental Oceanography Committee Workshop Report 28*, pp. 105–142.

Tolimieri, N., A. Jeffs and J.C. Montgomery. 2000. Ambient sound as a cue for navigation by the pelagic larvae of reef fishes. *Marine Ecology Progress Series* 207: 219–224.

Trevorrow, M. 2005. The use of moored inverted echosounders for monitoring meso-zooplankton and fish near the ocean surface. *Canadian Journal of Fisheries and Aquatic Sciences* 62: 1004–1018.

Trnski, T. 2002. Behaviour in settlement-stage larvae of fishes with and estuarine juvenile phase: *in situ* observations in an warm-temperate estuary. *Marine Ecology Progress Series* 242: 205–214.

Utne-Palm, A.C. 2004. Effects of larval ontogeny, turbidity, and turbulence on attack rate and swimming activity of Atlantic herring larvae. *Journal of Experimental Marine Biology and Ecology* 310: 147–161.

Van-Damme, R. and T.J. Van-Dooren. 1999. Absolute versus per unit body length speed of prey as an estimator of vulnerability to predation. *Animal Behaviour* 57: 347–352.

Videler, J.J. 1993. *Fish Swimming*. Chapman and Hall, London.

von Westernhagen, H., and H. Rosenthal. 1979. Laboratory and in-situ studies on larval development and swimming performance of Pacific herring *Clupea harengus pallasi*. *Helgolander Wissenchaftliche Meeresuntersuchungen* 32: 539–549.

Wakeling, J.M. 2001. Biomechanics of fast-start swimming in fish. *Comparative Biochemistry and Physiology*. A131: 31–40.

Walker, J.A., C.K. Ghalambor, O.L. Griset, D. McKenney and D.N. Reznick. 2005. Do faster starts increase the probability of evading predators? *Functional Ecology* 19: 808–815.

Warner, R.R., S.E. Swearer and J.E. Caselle. 2000. Larval accumulation and retention: Implications for the design of marine reserves and essential fish habitat. *Bulletin of Marine Science* 66: 821–830.

Webb, P.W. 1984. Body form, locomotion and foraging in aquatic vertebrates. *American Zoologist* 24: 107–120.

Webb, P.W. 1994. The biology of fish swimming. In: *Mechanics and Physiology of Animal Swimming*. L. Maddock, Q. Bone and J.M.V. Rayner (Eds.). Cambridge University Press, Cambridge, pp. 45–62.

Webb, P. and R. Corolla. 1981. Burst swimming performance of northern anchovy, Engraulis mordax, larvae. *Fisheries Bulletin* 79: 143.

Weihs, D. 1973. Optimal fish cruising speed. *Nature* (London) 245: 48–50.

Wesp, H.M., and A.C. Gibb. 2003. Do endangered razorback suckers have poor larval escape performance relative to introduced rainbow trout? *Transactions of the American Fisheries Society* 132: 1166–1178.

Wieser, W. and R. Kaufmann. 1998. A note on interactions between temperature, viscosity, body size and swimming energetics in fish larvae. *Journal of Experimental Biology* 201: 1369–1372.

Williams, D.M., E. Wolanski and J. Andrews. 1984. Transport mechanisms and the potential movement of planktonic larvae in the central region of the Great Barrier Reef. *Coral Reefs* 3: 229–236.

Williams, I.V. and J. R. Brett. 1987. Critical swimming speed of Fraser and Thompson River pink salmon (*Oncorhynchus gorbuscha*). *Canadian Journal of Fisheries and Aquatic Sciences* 44: 348–356.

Williams, P.J. and J. A. Brown. 1992. Development changes in the escape response of larval winter flounder (*Pleuronectes americanus*) from hatch through metamorphosis. *Marine Ecology Progress Series* 88: 185–193.

Williams, P.J., J.A. Brown, V. Gotceita and P. Pepin. 1996. Developmental changes in escape response performance of five species of marine larval fish. *Canadian Journal of Fisheries and Aquatic Sciences* 53: 1246–1253.

Wolanski, E., P.J. Doherty and J. Carleton. 1997. Directional swimming of fish larvae determines connectivity of fish populations on the Great Barrier Reef. *Naturwissenschaften* 84: 262–268.

Wolanski, E. and G.L. Pickard. 1983. Upwelling by internal tides and kelvin waves at the continental shelf break on the Great Barrier Reef. *Australian Journal of Marine and Freshwater Research* 34: 65–80.

Zhou, T. and J.S. Weis. 1998. Swimming behavior and predator avoidance in three populations of *Fundulus heteroclitus* larvae after embryonic and/or larval exposure to methylmercury. *Aquatic Toxicology* 43: 131–148.

12

The Role of Swimming in Reef Fish Ecology

Christopher J. Fulton

INTRODUCTION

Swimming plays a crucial role in the ecology and evolution of fishes. Swimming is the primary means by which fish interact and move through their environment. As such, swimming performance determines how well a fish can overcome the daily challenges of acquiring food, avoiding predators, gaining access to different habitats and conducting reproductive activities (Webb, 1994; Batty and Domenici, 2000; Plaut, 2001; Blake, 2004). Since swimming can also be one of the largest costs in the daily energy budget of fishes (Feldmeth and Jenkins, 1973; Kitchell, 1983; Boisclair and Tang, 1993), even small variations in swimming activity can have profound implications for patterns of energy allocation among growth, maintenance and reproduction (Koch and Wieser, 1983; Boisclair and Sirois, 1993). In this way, swimming can influence the evolutionary fitness of fishes and be a primary target for natural selection (Huey and Stevenson, 1979; Hertz et al., 1988).

Fish are faced with numerous challenges for locomotion in shallow demersal systems such as coral and rocky reefs. These challenges can range from extracting food from complex benthic structures (Wainwright and Bellwood, 2002) and escaping high-speed attacks from hidden predators

Authors' address: School of Botany and Zoology, The Australian National University, Canberra, ACT 0200, Australia. E-mail: christopher.fulton@anu.edu.au

(Hixon and Beets, 1993), to negotiating wave-swept habitats characterised by extreme flow speeds and turbulence (Webb, 2004; Fulton and Bellwood, 2005). The magnitude of these challenges can vary substantially over space and time and interact with the swimming abilities of species to strongly influence their patterns of distribution and abundance (e.g., Bellwood and Wainwright, 2001; Fulton and Bellwood, 2004). Such environmental challenges may also represent an opportunity, with those species able to adapt their swimming abilities gaining access to new habitats and sources of food. While these direct links between swimming performance and fish ecology appear intuitive, until recently there has been a distinct lack of empirical evidence to support such direct biophysical interactions in demersal habitats such as reefs (Arnold, 1983; Wainwright and Reilly, 1994; Webb, 1994; Blake, 2004).

A major impediment to exploring links between swimming and fish ecology has been the doubt surrounding the ecological relevance of different performance metrics. While sustained speed (i.e., activity maintained for 200+ minutes; Blake, 2004) is generally thought to be the most relevant measure for questions concerning patterns of habitat-use in fish (Hammer, 1995; Plaut, 2001; Blake, 2004), there is less consensus on how best to measure this level of performance. One performance measure in particular, critical swimming speed (U_{crit}), has received considerable attention in this regard (Hammer, 1995; Plaut, 2001; Blake, 2004). First developed by Brett (1964), U_{crit} is obtained from fishes subjected to incremental changes in speed over time in an experimental flume. While Brett (1964) originally intended U_{crit} to provide a time-effective means of estimating maximum sustained speed, some studies have found U_{crit} to overestimate sustained speed as a result of anaerobically powered performance in the final stages of the critical speed test (Kolok, 1991; Hammer, 1995; Plaut, 2001), unnatural swimming behaviours (Peake, 2004), restrictive experimental flumes (Peake, 2004), and/or variations in the specific stepwise changes in speed and time used in each trial (Farlinger and Beamish, 1977; Kolok, 1999; Plaut, 2001). Despite this, many other studies have found U_{crit} to be a robust and reliable metric for comparing the relative swimming speed capabilities of fishes (Williams and Brett, 1987; Hartwell and Otto, 1991; Peake and McKinley, 1998; Hogan *et al.*, 2007). Moreover, U_{crit} has been found to relate directly to other aspects of swimming performance, with strong correlations to sustained (Johansen *et al.*, 2007a) and routine swimming speeds in a wide range of fish taxa (Fisher *et al.*, 2005; Fulton *et al.*, 2005; Fulton, 2007). As such, U_{crit} is now generally viewed as a robust measure of swimming speed performance, as measured over prolonged time periods (20 seconds–200 minutes; Hammer, 1995; Plaut, 2001; Fisher *et al.*, 2005; Fulton, 2007).

In adult reef fishes, we do find that U_{crit} overestimates sustained and routine field speeds by 40–100%, depending on the swimming mode employed (Fulton, 2007; Johansen *et al.*, 2007a). Nonetheless, once allometric effects are taken into account, U_{crit} also provides a strong predictor of routine field speeds measured in individuals swimming undisturbed on the reef (Fig. 12.1a; Fulton *et al.*, 2005; Fulton, 2007). Moreover, an almost isometric relationship between lab- and field-based measurements of swimming speed may be achieved by ensuring lab-based estimates are taken from individuals using the same gait as they would during their daily swimming activities. This can be accomplished by setting an alternative cut-off point in experimental speed tests, the gait transition speed (U_{trans}), which is the point where the fish shifts from either a pectoral-only to pectoral-caudal gait, or from a steady-caudal to burst-coast caudal gait (Fig. 12.1b; Drucker and Jensen, 1996; Walker and Westneat, 2002; Fulton *et al.*, 2005; Cannas *et al.*, 2006). While this concept may have limited utility for some groups such as larval fishes (Fisher and Leis, Chapter 12, this book), the method appears robust in adult reef fishes from the 10 families examined to date (Fulton *et al.*, 2005; Fulton, 2007). In discussing the ecological implications of swimming performance, I present both experimental (lab-based) and field-based measurements of speed performance, with U_{trans} being the metric of choice when available. While manoeuvrability is undoubtedly a crucial aspect of swimming performance for fishes interacting with demersal habitats (Gerstner, 1999; Webb, 2002, 2004; Webb *et al.*, Chapter 2, this book), my main focus here is on the implications of speed performance for reef fish ecology.

In this chapter I explore the role of swimming in shaping patterns of habitat-use in adult reef fishes. Firstly, I examine the diversity of swimming modes employed by reef fishes and how these may relate to the environmental challenges of living a reef-associated lifestyle. I then draw upon a decade of research to review our current understanding of labriform swimming and how this has provided new insights into the forces structuring coral and rocky reef fish assemblages around the globe. In particular, I explore the importance of fin morphology for patterns of performance, and how the acquisition of a unique fin shape may have been a key determinant in the success of labriform-swimming fishes on reefs. Finally, I discuss the importance of behaviour as a modulating factor in the link between fish swimming morphology, performance and patterns of resource use. Reef fishes are an ideal group to examine these concepts, given their wide diversity of body forms, modes of locomotion and evolutionary lineages, matched only by the diversity of available habitats and environmental conditions found in reef ecosystems. In several areas I am forced to make speculative assertions on the underlying mechanisms due to a lack of empirical evidence, particularly in the area of swimming energetics; many of these I expect to be adapted or rejected in light of future research.

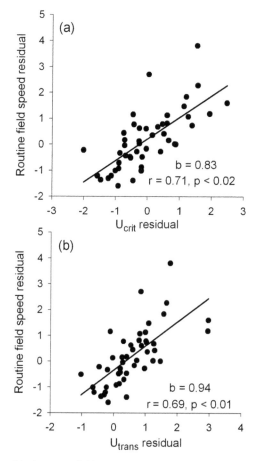

Fig. 12.1 Relationship between field- and lab-based measurements of swimming speed performance in 44 reef fish species. Each point represents the mean swimming performance of each species under forced swimming trials estimating (a) U_{crit} or (b) U_{trans} speed, against speeds measured from the same species swimming undisturbed on the reef (modified after Fulton *et al.*, 2005). All performance measures were corrected for differences in body size by calculating swimming speed residuals from the linear regression of speed against total length (TL), following Reist (1985) and Wainwright *et al.* (2002). Residualisation was chosen because alternatives such as body length ratios do not remove allometric effects on speed performance in these species (Fulton, 2007). High positive residuals represent high speeds for a given size.

Swimming Modes and the Reef-Associated Lifestyle

Reefs are occupied by an abundance of fishes that appear to take full advantage of the diversity of food resources, home sites and predator refuges provided by these complex ecosystems. While occupying reefs may seem an ideal lifestyle for fishes, both coral and rocky reefs can be demanding environments that pose substantial challenges for fish locomotion. Wave-induced hydrodynamic forces can often exceed the equivalent of cyclone-force winds on land (Denny, 1988), and even under relatively calm conditions (<1 metre significant wave height), flow speeds in shallow reef habitats (<4 metres depth) can average 40–60 cm s^{-1} and exceed 400 cm s^{-1} on a regular basis (Young, 1989; Symonds et al., 1995; Kench, 1998; Fulton and Bellwood, 2005; Gourlay and Colleter, 2005). Many reef fishes are simply unable to cope with the magnitude of these flows (Fulton, 2007); other species may be capable of such speeds, but excessive energy expenditure incurred in maintaining this level of performance may prove too costly for the long-term occupation of high-flow habitats (Clarke, 1992; Korsmeyer et al., 2002; Aseada et al., 2005; Fulton and Bellwood, 2005). Moreover, the unsteady and multidirectional nature of wave-induced flows can prove challenging for fishes to maintain stable swimming trajectories and complete fine-scale tasks such as extracting small food items from benthic microhabitats (Gerstner, 1999; Webb, 2002, 2004; Liao, 2007).

Why would reef fishes want to gain access to shallow wave-swept habitats? Some of the richest food resources can be found in the most wave-exposed reef habitats. For example, loadings of detritus, a nutritious dietary resource targeted by many reef fish taxa (Wilson et al., 2003), is often greatest in the wave-swept reef crest and flat habitat zones (Crossman et al., 2001; Purcell and Bellwood, 2001; Wilson et al., 2003). Similarly, the biomass and productivity of the epilithic algal matrix (EAM), a resource actively targeted by most herbivorous reef fishes, is highest in habitats of high water movement (Carpenter et al., 1991; Steneck and Dethier, 1994; Purcell and Bellwood, 2001). Moreover, planktivorous fishes may find their suspended food items delivered at greater rates in high-flow, wave-swept environments (Holzman and Genin, 2003; Clarke et al., 2005). Besides an apparent abundance of food resources, fishes may also find less interspecific competition, with the total number of fish species within wave-swept reef flat habitats often being the lowest of all habitat zones in a given locality (Bellwood and Wainwright, 2001; Fulton et al., 2001; Fulton and Bellwood, 2004). While gaining and maintaining access to high flow reef habitats can be a considerable biophysical challenge for fishes, the ecological advantages may also be substantial.

Recognizing both the apparent desirability and difficulty of occupying wave-swept habitats, we may ask which reef fishes can access such habitats and why? A good starting point in answering this question is to examine the different swimming modes utilized by reef fishes. Since the concept was first formalized by Breder (1926), swimming modes (otherwise known as gaits) have provided a primary framework for examining fish swimming behaviour and performance. Each mode involves the use of different fins

Table 12.1   Diversity of swimming modes employed during the daily swimming activities of reef fishes. Based on field observations of fin and body use following Fulton (2007). Swimming mode nomenclature and definitions follow Webb (1994), images adapted from Randall *et al.* (1997).

Swimming mode	Source of propulsion	Reef fish Family or Genera
Labriform	Pectoral fins	Acanthuridae, Labridae, Pomacentridae, *Pomacanthus*, Scaridae, Zanclidae
Subcarangiform	Body-caudal fin	Haemulidae, Kyphosidae, Lethrinidae, Lutjanidae, *Naso*, Serranidae, Siganidae, Mullidae, Scorpaenidae
Chaetodontiform	Pectoral-caudal fins	Apogonidae, Chaetodontidae, *Centropyge*, *Dischistodus*, Ephippidae, Nemipteridae, *Pygoplites*, *Stegastes*
Carangiform	Caudal fin	Carangidae, Caesionidae,
Balastiform	Dorsal-ventral fins (broad)	Balistidae, Monacanthidae, Syngnathidae
Tetraodontiform	Dorsal-ventral fins (narrow)	Tetraodontidae, Ostraciidae
Anguilliform	Body undulation	Muraenidae, Plotosidae

or body regions during swimming, such as the employment of either the body and caudal fin (BCF) or the median-paired fins (MPF) for propulsion (Webb, 1984, 1994; Blake, 2004). In reef fishes we find a wide range of swimming modes being used during routine activities, ranging from steady BCF undulation (e.g., subcarangiform), to pectoral-augmented BCF (e.g., chaetodontiform) and rigid-body MPF propulsion using just the pectoral (e.g., labriform) or dorso-ventral (e.g., balistiform) fins (Table 12.1; Webb,

1984, 1994; Blake, 2004; Fulton, 2007). Despite this diversity of behaviours, most reef fish taxa tend to use just one of three modes during routine swimming: labriform, subcarangiform, or chaetodontiform (Table 12.1; Fulton, 2007). Indeed, recent surveys of an entire reef fish assemblage (156 species, 7 families) revealed that more than 60% of reef-associated species use labriform propulsion during daily swimming activities (Fulton and Bellwood, 2005; Fulton, 2007). Furthermore, three of the most abundant reef fish families are composed primarily of labriform swimmers (the Labridae, Pomacentridae and Acanthuridae; Fulton, 2007). Consequently, reef-associated fish communities are characterized by labriform-swimming individuals, both in terms of numerical abundance and taxonomic diversity.

Each swimming mode has inherently different performance traits and energetic costs, largely as a consequence of using different body parts and/ or muscle groups to produce thrust (Webb, 1984, 1994; Weihs, 2002; Blake, 2004). Within the broad division of BCF (subcarangiform, chaetodontiform) and MPF (labriform) propulsion, we generally find greater slow-speed manoeuvrability in fishes using an MPF mode (Webb, 1984, 1994; Gerstner, 1999; Webb and Fairchild, 2001; Blake, 2004). Such high manoeuvrability is undoubtedly a useful trait for close interaction with complex reef habitats, and may help explain the prevalence of the labriform (MPF) mode in reef fish assemblages. However, as discussed above, swimming speed performance may also be crucial for gaining access to wave-swept reef habitats. Traditionally, BCF-swimming has been viewed as the superior mode for attaining high swimming speeds, since MPF taxa seem to have a body shape and musculature optimized towards slow-speed manoeuvres (Alexander, 1967; Webb, 1984; Blake, 2004). However, this paradigm appears to have developed from comparisons of fishes of disparate size and lifestyle – typically large pelagic or anadromous fishes (predominantly BCF swimmers) matched against small, bottom-associated taxa (a mix of MPF and BCF swimmers; Wardle, 1975; Lindsay, 1978; Walker and Westneat, 2002). When we compare the results of swimming performance trials in solely marine demersal taxa of equivalent size, we find that MPF-swimmers attain either similar or higher speeds than their BCF-swimming counterparts (Fig. 12.2). Moreover, these demersal MPF-swimmers maintain these high speeds while going about their daily swimming activities, with routine field speeds of 85% U_{crit} in labriform-swimming species, versus field speeds of below 50% U_{crit} in BCF-swimming taxa (Fig. 12.2; Fulton, 2007; Johansen *et al.*, 2007a). In addition to their taxonomic and numerical dominance of reef fish assemblages, labriform-swimming fishes also appear to be excellent swimming speed performers.

Fundamental differences in propulsion and swimming performance, such as those described above, can have profound ecological implications for reef fishes. Evidence of this can be found in the distribution of fishes according to swimming mode across a typical coral reef. Here we find

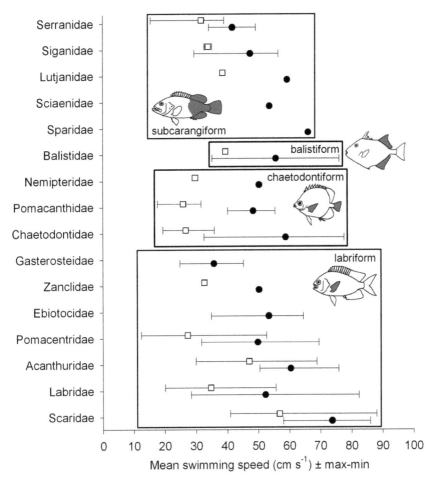

Fig. 12.2 Mean swimming speed of demersal fishes using different swimming modes. Each point is the mean U_{crit} (filled circles) or routine field speed (open squares) observed for species in each family, with the max-min ranges indicating mean fastest and slowest species recorded in each family. Note swimming speed performance of MPF taxa (labriform, balistiform) either matches or exceeds the speeds displayed by BCF taxa (subcarangiform, chaetodontiform). Data based on fishes of 5–25 cm total length, measured across a similar range of temperatures (11–28°C) in all 16 families. Sources: Scaridae, Labridae (Wainwright *et al.*, 2002; Fulton, 2007); Acanthuridae, Pomacentridae, Zanclidae, Chaetodontidae, Pomacanthidae, Nemipteridae, Lutjanidae, Siganidae, Serranidae (Fulton, 2007); Embiotocidae (Drucker and Jensen, 1996); Gasterosteidae (Schaarschmidt and Jurss, 2003; Tudorache *et al.*, 2007); Balistidae (Korsmeyer *et al.*, 2002; Fulton, unpublished data); Sparidae (Basaran *et al.*, 2007); Sciaenidae (Fitzgibbon *et al.*, 2007).

labriform-swimming fishes widely distributed across the full range of reef habitats, while BCF-swimmers are restricted mostly to sheltered, low-flow areas (Fig. 12.3; Fulton and Bellwood, 2005). A primary reason for this trend appears to be a direct interaction between wave-induced flows and the relative swimming speed performances of fishes. Indeed, the average swimming performance of fishes in a given habitat (regardless of species or mode) tends to be closely matched to the relative flow conditions in that habitat, such that habitats with higher flows are occupied by fishes with higher swimming performance (Fig. 12.4). Moreover, this interaction between flow velocity and swimming speed performance is found at multiple spatial scales. At the finer within-habitat scale, we find patterns of water-column use closely linked to variations in fish swimming mode and performance in a pattern that mimics the profile of flow velocities in the water column. Specifically, we find fishes occupying the upper layers of the water column employing a labriform mode and exhibiting higher levels of speed performance for their size (Fig. 12.5).

While this evidence suggests that swimming speed performance is a crucial determinant of habitat-use in reef fishes, such trends may not simply be due to differences in raw speed performance amongst modes. In fact, speed estimates taken in experimental trials indicate that fishes using both MPF and BCF swimming can attain the raw speeds needed to access wave-swept habitats (Fig. 12.2; Blake, 2004; Fulton, 2007). However, it seems that the critical performance aspect is their ability to maintain high routine speeds over daily timescales, with routine field speeds in labriform taxa regularly exceeding 40 cm s^{-1} versus less than 40 cm s^{-1} in BCF-swimming fishes (Fig. 12.2; Fulton, 2007). Incidentally, this apparent speed threshold around 40 cm s^{-1} appears to coincide with an environmental threshold in the levels of ambient flow found on reefs, this being the typical lower boundary of flow velocities recorded in shallow, wave-swept habitats (Fig. 12.5b; Young, 1989; Symonds et al., 1995; Kench, 1998; Fulton and Bellwood, 2005; Gourlay and Colleter, 2005). As such, fishes incapable of maintaining routine speeds above this level could be simply excluded from wave-swept locations. Collectively, this evidence suggests that swimming speed performance may set overriding limits on the habitat-use patterns of reef fishes. In particular, the ability to maintain high speeds over daily timescales appears to be a primary determinant of whether a reef fish can occupy habitats exposed to high levels of wave-induced water motion. In this, labriform fishes appear to be ideally matched to the challenges of a high-flow environment and have taken their place as the dominant occupants of wave-swept habitats and locations on reefs.

Notably, these links between swimming speed performance and habitat-use seem to apply regardless of any differences in fish trophic status. Indeed, fish from the same trophic group display markedly different swimming

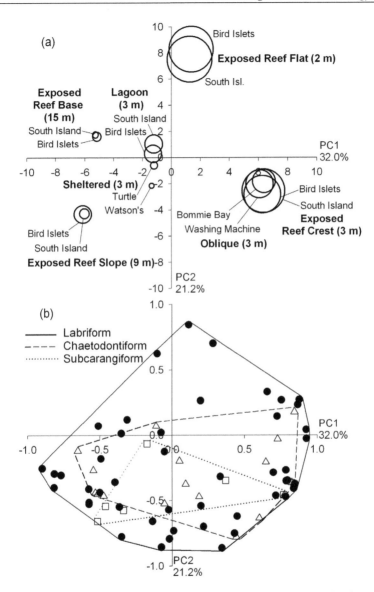

Fig. 12.3 Principal Component Analysis of the distribution and abundance of 71 fish species across exposed, oblique, lagoonal and sheltered reef locations (n = 4 transects per habitat/ site, two sites at each location) at Lizard Island, Great Barrier Reef. Four habitat types (base, slope, crest and flat) are included for the two wave exposed sites, while all other locations are reef crest habitats. (a) Habitat ordination plot with average wave-induced flow velocities indicated (proportional to bubble size after Fulton and Bellwood, 2005). (b) Species vector plot optimized by swimming mode for each species. Note that labriform-swimming taxa (solid line) occupy the entire range of habitats censused, whereas chaetodontiform and subcarangiform taxa are largely found in sheltered, low-flow habitats.

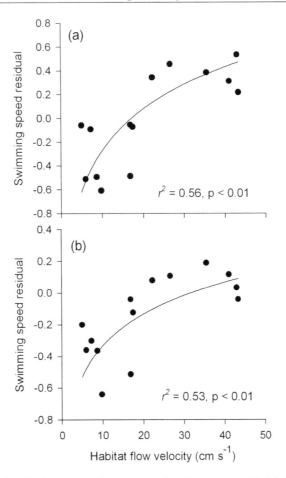

Fig. 12.4 Relationship between swimming speed performance and habitat-specific flow velocities for the same reef fish assemblage presented above (Fig. 12.3). Values are average speed residuals of all fish individuals in each habitat against the mean flow velocity recorded in that habitat (Fulton and Bellwood, 2005) for both (a) routine field speeds and (b) forced swimming speed (U_{crit}) measures of performance (Fulton, 2007). High positive residuals represent high speeds for a given size.

abilities that seem to match the flow conditions in their respective habitats. For instance, the most abundant planktivores on the relatively calm reef slopes, *Pomacentrus brachialis* and *Cirrhilabrus punctatus*, display average field speeds of 19.6 cm s^{-1} and 20.6 cm s^{-1} respectively, whereas *Chromis atripectoralis* and *Thalassoma amblycephalum* dominate wave-swept crest habitats with average speeds of 36.1 cm s^{-1} and 43.8 cm s^{-1} respectively (Fulton, 2007). Likewise, in the roving herbivores and detritivores we find *Ctenocheatus binotatus* occurring primarily on the reef slope and displaying a mean field speed of 29.3 cm s^{-1}, while *C. striatus* (43.1 cm s^{-1}) and *Acanthurus nigrofuscus*

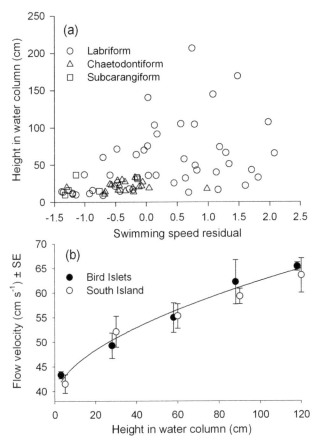

Fig. 12.5 Relationship between swimming speed performance and height in the water column in 71 fish species occupying the exposed reef crest at Lizard Island, Great Barrier Reef. (a) Mean swimming speed performance and height above the substratum for each species (Fulton *et al.*, 2001; Fulton, 2007). (b) Mean water flow velocity with height in the water column at the same two exposed sites as measured over 24 hours with a plaster dissolution method (n = 3 per site; Fulton and Bellwood, 2005). Swimming performance values are the routine field speeds of each species, residualised to minimize allometric effects (negative residuals indicate particularly poor swimming speeds for a given size).

(45.6 cm s^{-1}) occupy the crest in abundance, and *A. triostegus* (68.8 cm s^{-1}) dominate the most wave-swept reef flat habitat (Fulton, 2007). Spanning reef fish taxa from several perciform lineages (e.g., the Percoidei, Labroidei and Acanthuroidei; Nelson, 2006), it appears that the link between swimming mode, performance and patterns of habitat-use in reef fishes is a general phenomenon that transcends any differences in trophic status or taxonomic affiliation.

Similar relationships between swimming performance and habitat-use have been found in freshwater fish communities, where freshwater fishes of greater swimming performance tend to occupy habitats of higher flow (e.g., Sagnes *et al.*, 1997, 2000; Rosenfeld *et al.*, 2000; Kodric-Brown and Nicoletto, 2005). While this selection of either low or high-flow habitats has often been attributed to differential predation risk in freshwater systems (Power, 1984; Gilliam and Fraser, 1987; Lonzarich and Quinn, 1995), marked differences in fish growth and condition prompted several workers to examine the energetic implications of habitat choice. Certainly, the energetic cost of swimming increases exponentially with speed (Brett, 1964; Webb, 1975; Sepulveda *et al.*, 2003; Alexander, 2005), and swimming can consume substantial amounts of energy on a daily basis (Feldmeth and Jenkins, 1973; Kitchell, 1983; Boisclair and Tang, 1993), so that even small changes in the cost of locomotion could play a large role in shaping fish habitat selection. Accordingly, the application of bioenergetic cost-benefit analyses, which contrast the cost of swimming against energy intake from invertebrate drift, revealed that avoidance of high-flow riffle habitats by freshwater fishes could in fact be driven primarily by locomotor energetics (Hughes and Dill, 1990; Hill and Grossman, 1993; Nislow *et al.*, 2000; Rosenfeld and Boss, 2001). Moreover, similar bioenergetic analyses have indicated that many freshwater fishes may choose flow conditions that allow them to swim at an energetically optimal speed during upstream migrations and station-holding (Hinch and Rand, 1998; Haro *et al.*, 2004; Castro-Santos, 2005).

Could the energetic cost of locomotion be a determining factor in the habitat-use of reef fishes? While there is a paucity of information on the comparative energetics of reef fish swimming, relationships between cost-of-transport and speed in some representative MPF and BCF taxa provide some insight. Fish using a labriform (MPF) mode have been found to display minimum costs of transport across a broad range of swimming speeds (1.5–4.0 BL s^{-1}), while fish using a subcarangiform (BCF) mode display a narrow minima at relatively low speeds (0.8–1.5 BL s^{-1}) and steep increases in cost-of-transport with increasing speed (Korsmeyer *et al.*, 2002; Tolley and Torres, 2002; Lee *et al.*, 2003; Cannas *et al.*, 2006; Jones *et al.*, 2007). Based on rates of oxygen consumption, these cost-of-transport relationships provide a relative measure of energy consumption during continuous swimming for a set distance or time period. As such, these relationships indicate that labriform swimming provides the most efficient means of swimming continuously over a broad range of speeds, while fishes of similar size using a subcarangiform (BCF) mode would incur much greater energetic costs. Certainly, multi-species comparisons of speed performance have found that labriform taxa display a wide range of swimming speed performances, experience lower drag as a result of rigid-body locomotion, and show a propensity for higher field speeds than demersal BCF taxa of the same size (Fig. 12.2; Walker and Westneat, 2000; Wainwright *et al.*, 2002; Blake, 2004; Fulton, 2007). This ability to maintain high speeds over daily time periods with minimal energetic cost may be the crucial advantage that allows labriform fishes to successfully occupy high-flow habitats on reefs, while fishes using BCF modes of propulsion can not (Boisclair and Tang, 1993; Pettersson and Hedenström, 2000; Aseada *et al.*, 2005). Although more empirical data on reef fish locomotor energetics is needed, it appears that energetic efficiency may be a critical factor in the link between fish swimming performance and patterns of habitat-use on reefs. Bio-energetic analyses such as those conducted in freshwater systems may go a long way towards explaining the high prevalence of labriform-swimming fishes in wave-swept reef habitats. Indeed, cost-benefit analyses in planktivorous reef fishes has already provided some unique insights in patterns of habitat selection according to flow and food delivery (Clarke, 1992; Holzman and Genin, 2003; Clarke *et al.*, 2005). Acknowledging that much work remains to be done on the comparative energetics of reef fish swimming before we can take these assertions any further, I now move on to explore the apparent 'success' of labriform-swimming reef fishes in greater detail.

Labriform Swimming: the Ecological Implications of Form and Function

Fishes that employ labriform locomotion during their daily swimming activities encompass an astounding level of diversity. Over two-thirds of the non-cryptic fish species living on reefs adopt labriform locomotion during their routine swimming (Fulton and Bellwood, 2005; Fulton *et al.*, 2005; Fulton, 2007). Moreover, these species span several families (e.g., the Acanthuridae, Labridae, Pomacanthidae, Pomacentridae, Scaridae, Zanclidae; Table 12.1), an enormous range of body sizes (e.g. 3–229 cm total length; Wainwright *et al.*, 2002), trophic groups (e.g., corallivores, detritivores, herbivores, planktivores, ectoparasite cleaners; Bellwood *et al.*, 2006) and reproductive modes (e.g., haremic societies, pair-wise brooding, broadcast spawning; Randall *et al.*, 1997). Indeed, it appears that the only common attribute shared by these fishes is that they use solely their pectoral fins to produce thrust during routine swimming activities (Webb, 1994; Blake, 2004; Fulton, 2007). Using this unifying trait and a remarkable link between pectoral fin shape and swimming performance, we have been able to transcend these vast differences in lifestyle and uncover some of the general mechanisms shaping reef fish communities at a global scale.

Our understanding of labriform locomotion has increased dramatically over the past decade, largely due to the considerable attention devoted towards the consequences of pectoral fin shape for locomotor function

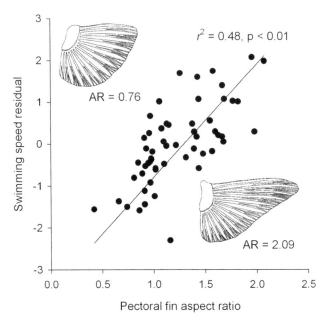

Fig. 12.6 Relationship between swimming speed performance and pectoral fin shape in 56 species of labriform-swimming fishes spanning three families (the Acanthuridae, Pomacentridae and Labridae). Pectoral fin aspect-ratio (AR) is a unitless number calculated as the length of the leading edge squared, divided by total fin area, (Wainwright *et al.*, 2002). Swimming performance values are based on U_{trans} speed for each species measured in a recirculating flow tank, residualised to minimize allometric effects (Fulton *et al.*, 2005). Inset pictures illustrate typical examples of a low and high AR fin.

and performance. Labriform-swimming fishes display a wide spectrum of pectoral fin shapes, ranging from deeply rounded to highly tapered fins (Fig. 12.6; Wainwright *et al.*, 2002; Fulton *et al.*, 2005). Kinematic analyses have revealed that these variations in fin shape are linked to two different forms of fin movement and thrust production: drag-based rowing using rounded fins, versus lift-based flapping using tapered fins (Blake, 1979, 1981; Drucker and Jensen, 1996; Walker and Westneat, 1997, 2002; Walker, 2004). As the names suggest, the former involves rowing the fin to exploit drag-based forces on a backward power stroke, while the latter involves flapping the fin in a figure-eight sweep to exploit lift-based thrust on both upward and downward strokes (Drucker and Lauder, 1999; Walker and Westneat, 2000, 2002). Since thrust is produced on only half of the fin strokes in drag-based locomotion (i.e., no thrust produced by forward recovery stroke), this form of labriform swimming is effective only over a limited range of slow speeds (Blake, 1981; Vogel 1994; Drucker and Lauder, 1999). Conversely, lift-based flapping produces thrust on all fin strokes and provides a highly efficient means of maintaining very high swimming speeds, albeit at the cost of less slow-speed power and acceleration (Vogel 1994; Walker and Westneat, 2000, 2002). As a result, we find that tapered-fin species tend to swim faster than rounded-fin species of the same size, both under experimental conditions and while swimming undisturbed on the reef (Walker and Westneat, 2000; Wainwright *et al.*, 2002). Representing two extremes at either end of a continuum, this trade-off in performance with fin shape produces a spectrum of swimming speeds among species, with even relatively small changes in fin shape having a marked impact on speed performance. Consequently, we find a consistent functional link between speed and pectoral fin shape in labriform fishes, regardless of their taxonomic affiliation (Fig. 12.6; Wainwright *et al.*, 2002; Fulton *et al.*, 2005). While highlighting the unifying importance of morphology for variations in swimming speed performance, this functional relationship also allows us to predict the relative performance of labriform fishes based solely on their pectoral fin shape.

Applying this functional relationship to patterns of ecology, we find that fin shape also provides a strong predictor of habitat-use across gradients of water motion. Whether it is height in the water-column or the occupation of reef habitats of different wave energy, we find a distinct trend in the distribution of labriform reef fishes according to their fin shape. Specifically, we find tapered-fin species occurring in high abundance in high-flow habitats, while rounded-fin species are restricted to sheltered, low-flow areas (Fig. 12.7; Bellwood and Wainwright, 2001; Fulton *et al.*, 2001). In a pattern that reflects the differential distribution of swimming modes described in the section above on "Swimming Modes and the Reef-Associated Lifestyle", this trend appears to be the result of differences in swimming performance according to pectoral fin type. Namely, tapered-fin species are capable of high levels of speed performance with high efficiency, whereas rounded-fin species are unable to maintain the high speeds needed to occupy high-flow habitats (Vogel, 1995; Walker and Westneat, 2000, 2002; Wainwright *et al.*, 2002;). Notably, this dichotomous distribution is found in reef fish assemblages throughout the world, spanning high diversity fish assemblages on the Great Barrier Reef to intermediate and low diversity systems in French Polynesia and the Caribbean (Fig. 12.7; Bellwood *et al.*, 2002). Despite distinct variations in species composition and a five-fold difference in species richness, labriform fishes in each of these locations are arranged in highly congruent patterns according to pectoral fin shape and the relative wave energy of habitats (Fig. 12.7). This phenomenon is driven primarily by the relative abundance of individuals, rather than a simple presence-absence of species (Bellwood *et al.*, 2002). This is because relatively few tapered-fin species occupy wave-swept habitats, but do so in extremely huge abundances. Indeed, population sizes of tapered-fin species are typically an order of magnitude higher than rounded-fin species in the same habitats. For example, the tapered-fin species *Thalassoma hardwicke* (aspect-ratio (AR)

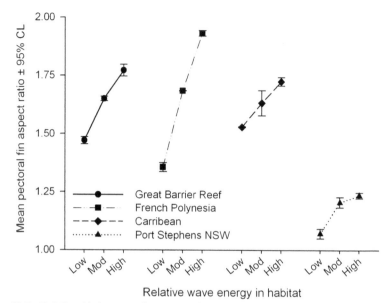

Relative wave energy in habitat

Fig. 12.7 Relationship between the pectoral fin shape of labriform-swimming fishes and the relative level of wave energy in their resident habitat, replicated across four different reef locations. Each wave energy level relates to a specific habitat zone: Low = reef back (1–5 m depth, leeward side), Mod = slope (8–14 m, windward), High = crest (1–4 m, windward) as described in Bellwood *et al.* (2002) and Fulton and Bellwood (2004). Note that the mean aspect-ratio of fishes in each habitat increases with increasing wave energy (i.e., fins are more tapered), although the temperate assemblage at Port Stephens tends to have much lower average fin AR values than the three other coral reef locations.

= 1.90) and *T. jansenii* (AR = 1.96) can attain average densities of 16 individuals per 100 m² on wave-exposed reef crests of the Great Barrier Reef, while rounded-fin species such as *Cheilinus chlorurus* (AR = 0.97) and *Halichoeres melanurus* (AR = 1.11) display densities of 0.8 per 100 m² in the same crest habitats (Bellwood and Wainwright, 2001; Fulton *et al.*, 2001). Even greater differences are found in open-ocean systems such as French Polynesia, where we find densities of up to 380 per 100 m² (*T. quinquevitattum*, AR = 1.97) versus 0.2 per 100 m² (*C. trilobatus*, AR = 0.90) in species occupying reef crests exposed to oceanic swells (Bellwood *et al.*, 2002). Taking population size as a measure of success, it would seem that labriform taxa with highly tapered pectoral fins are the most successful group of fishes found on reefs.

Reef fish assemblages in temperate latitudes display a notable variation on this general theme. At first glance, it appears that temperate fish assemblages exhibit a similar link between fin shape and wave energy to that found in coral reef locations (Fig. 12.7). However, increases in fin aspect-ratio with increasing wave energy appears to be less pronounced in the temperate reef fish assemblage occupying offshore reefs near Port Stephens, south-east Australia (Fig. 12.7). This is not due to a breakdown in the functional link between fin shape and speed performance in temperate species. Indeed, we find a similarly strong relationship between fin shape and speed performance to that seen in tropical reef fishes, despite the fact that the temperate assemblage contains several unique taxa (e.g., *Achoerodus, Ophthalmolepis, Notalabrus*; Fulton and Bellwood, 2004). Moreover, rounded-fin temperate species also display similar patterns of habitat-use to their tropical contemporaries, with low AR species largely restricted to low-flow, sheltered rocky habitats (Fulton *et al.*, 2001; Fulton and Bellwood, 2004; Denny, 2005). Where we do find marked differences between the tropical-temperate assemblages is in wave-exposed habitats: temperate reefs are almost devoid of fishes in wave-swept areas, while similar habitats on coral reefs are replete with tapered-fin individuals (Fulton and Bellwood, 2004; Fulton *et al.*, 2005). The reason for these differences appears to be that species with highly tapered pectoral fins are not present in temperate fish assemblages, with the maximum fin aspect-ratio in temperate species (AR = 1.43 in *Coris picta*; Fulton and Bellwood, 2004) being markedly lower then the maximum found in tropical species (AR = 2.1 in *Stethojulis bandanensis*; Wainwright *et al.*, 2002). Unlike tropical taxa, a highly tapered fin shape has not evolved within temperate reef fish assemblages. As a consequence, only temperate fishes which grow to a large size seem able to attain the high speeds needed to access wave-swept habitats (Fulton and Bellwood, 2004). Certainly, those few fishes found in exposed rocky reef habitats are much larger (26–30 cm total length) than individuals in more sheltered areas (14–18 cm total length; Fulton and Bellwood, 2004), which is in distinct contrast to the consistent body size of coral reef fishes distributed across wave energy environments (7–9 cm total length; Bellwood and Wainwright, 2001; Fulton and Bellwood, 2004). Despite this alternative mechanism, we see very few individuals in wave-swept habitats on rocky reefs, possibly because of the long timeframe and high degree of mortality incurred in reaching this large body size (Fulton and Bellwood, 2004; Denny, 2005). Ultimately, temperate systems provide an example of an assemblage devoid of tapered-fin fishes and the striking results—a complete lack of abundant fish populations in wave-swept habitats.

   This raises some pertinent questions about the evolution of labriform locomotion in reef fishes. Could the acquisition of a tapered fin and the lift-based mechanism of propulsion facilitated the expansion of reef fishes into wave-swept habitats, and if so, why has this not occurred in temperate reef fish assemblages? While the evolution of labriform swimming remains to be fully resolved (Drucker and Lauder, 2002; Thorsen and Westneat, 2005), ontogenetic evidence suggests that tapered pectoral fins are a relatively derived trait. All of the juvenile stages of labriform taxa examined to date have a generalized, rounded pectoral fin shape (Fig. 12.8; Fulton and

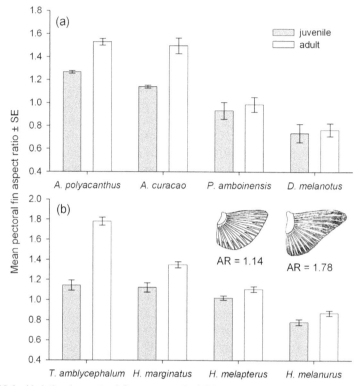

Fig. 12.8 Variation in pectoral fin aspect-ratio (AR) with ontogeny in reef fishes from the families (a) Pomacentridae and (b) Labridae. Fin AR means are based on a minimum of three individuals in each life-history stage (average n = 7.1). Juveniles were classed as individuals with a total length less than 50% of mean adult length (Randall *et al.*, 1997). Note the change in fin AR with ontogeny in species with high AR fins as adults (*Acanthochromis polyacanthus, Amblyglyphidodon curacao, Thalassoma amblycephalum* and *Halichoeres marginatus*), compared to little or no change in species with low fin AR as adults (*Pomacentrus amboinensis, Dischistodus melanotus, Hemigymnus melapterus* and *Halichoeres melanurus*). Inset pictures illustrate low and high AR fins from a juvenile and adult *Thalassoma amblycephalum*, respectively. Fin data in (a) collected as per Fulton and Bellwood (2002); data in (b) modified after Fulton and Bellwood (2002).

Bellwood, 2002; Wainwright *et al.*, 2002), with only those species having tapered fins as adults acquiring this fin shape through ontogenetic development (Fig. 12.8; Fulton and Bellwood, 2002; Wainwright *et al.*, 2002). Current phylogenetic evidence also indicates that tapered fins and the lift-based mode of locomotion are relatively derived traits (Drucker and Lauder, 2002), which appear to have arisen independently in at least two perciform lineages: the Acanthuroidei and Labroidei (Fulton *et al.*, 2005; Nelson, 2006). Moreover, there seems to be multiple origins of lift-based swimming within these lineages, with up to three in the Pomacentridae (*Abudefduf,*

Chromis-Dascyllus and *Neopomacentrus-Pomacentrus*; Quenouille *et al.*, 2004), six within the Labridae (*Cirrhilabrus, Chlorurus-Scarus, Halichoeres, Labroides, Thalassoma-Gomphosus* and *Stethojulis*; Wainwright *et al.*, 2002; Thorsen and Westneat, 2005; Westneat and Alfaro, 2005), and at least two in the Acanthuroidei (*Acanthurus-Ctenochaetus* and *Zanclus*; Clements *et al.*, 2003). Despite these multiple origins, all the lift-based lineages seem to have evolved within the tropics. Although temperate assemblages largely consist of relatively basal fish taxa (Westneat and Alfaro, 2005), it is unclear why lift-based lineages have not expanded into the seemingly empty wave-swept habitats found on temperate reefs. It is possible that the advantages associated with lift-based locomotion do not apply in cold temperate waters. Alternatively, the generally larger body sizes found in temperate fish taxa may make body size a more readily accessible trait for these fishes to attain the swimming speeds needed to access wave-swept habitats (Fulton and Bellwood, 2004; Denny, 2005). An examination of a sub-tropical fish assemblage, where tropical fish species occur in a relatively cool climate, may provide some more clarity to these issues. At least in tropical systems, the acquisition of a tapered fin appears to have been a key innovation in the evolution of reef fishes; an innovation that may ultimately have facilitated the ecological expansion of labriform fishes into wave-swept reef habitats.

Dominance of wave-swept habitats by relatively few unique fish species could have profound implications for the health and function of reef ecosystems. Wave-swept habitat zones such as the shallow crest and flat are often areas of high benthic diversity and live coral cover (Done, 1983; Steneck and Dethier, 1994; Hughes and Connell, 1999). Reef fishes play an important part in preserving this high diversity in the shallows via their role as predators of benthic organisms, helping to maintain a competitive balance among algae, corals and other benthic invertebrates (Hughes, 1994; Bellwood *et al.*, 2004; Mumby, 2006; Hughes *et al.*, 2007). In fact, the two dominant fish genera in wave-swept reef habitats, *Thalassoma* and *Scarus*, are tapered-fin fishes that feed upon a range of benthic invertebrates, dead coral and algae (Bellwood *et al.*, 2006; Hoey and Bellwood, 2008). Due to their extremely high abundances, these fishes have a substantial impact on the presence and abundance of benthic organisms in shallow reef habitats (Hoey and Bellwood, 2008). Given that access to such wave-swept areas is dependent on possession of a tapered fin shape and high speed performance, the loss of any fish species in these areas may not be easily replaced – certainly not by rounded-fin species coming from more sheltered areas. Consequently, if we were to lose our unique tapered-fin fishes from wave-swept reef habitats, we may see a breakdown in reef ecosystem function that results in a severe loss of biodiversity and habitat degradation (Bellwood *et al.*, 2004; Mumby, 2006; Hughes *et al.*, 2007).

Of course, the severity of this issue is greatly dependent on whether fin shape and swimming performance is fixed within each species according to their phylogenetic history. Can reef fishes express flexibility in their fin shape and performance? Locomotion has been found to be a highly plastic trait in many vertebrates, including frogs (Parichy and Kaplan, 1995), snakes (Jayne and Bennett, 1990; Aubret, 2005), lizards (Losos *et al.*, 1997, 2000) and some freshwater fishes (Nelson *et al.*, 2003; Peres-Neto and Magnan, 2004). Preliminary evidence in reef fishes also suggests the possibility of locomotor flexibility. In one widely distributed species, *Acanthochromis polyacanthus* (Spiny Damselfish), we find that juveniles display a similarly rounded fin morphology, regardless of their wave energy location (average fin aspect-ratio = 1.24 ± 0.03 SE). However, fully-grown adults inhabiting different wave energy environments display markedly different fin shapes (Fig. 12.9) in a pattern that matches their local environment. Specifically, adult fish of this species display more tapered fins (and potential for higher swimming speeds) in habitats of higher wave energy (Fig. 12.9). Presumably, these intraspecific variations in fin shape arise from variations in the fin ontogeny of individuals living in different habitats (Fulton and Bellwood, 2002). However, the question remains as to whether this may be due to phenotypic plasticity, or genetic differentiation among separate populations in each wave energy environment (i.e., local adaptation; Via, 2001; Bradshaw, 2006). These alternative scenarios of phenotypic plasticity versus local adaptation are highly contingent on dispersal mode and extent of gene flow across the spatial scale of different wave energy environments. Theory suggests that if there is poor dispersal and restricted gene flow, local adaptation may be the most likely explanation (Via 2001; Coyne and Orr, 2004). However, if spatial heterogeneity in the selective environments is sufficiently smaller than the dispersal capabilities of species, phenotypic plasticity can arise as an advantageous trait (West-Eberhard, 1989; Sultan and Spencer, 2002; Bradshaw, 2006). Indeed, *A. polyacanthus* displays relatively high levels of gene flow at the within-reef scale where these intraspecific variations in fin shape have been detected (Bay, 2007). While this suggests that phenotypic plasticity may be the most likely mechanism behind intraspecific variations in fin phenotype, this would need to be verified by a suite of common-garden experiments. If confirmed, such flexibility may provide further explanation for why labriform-swimming fishes occupy such a wide diversity of reef habitats in high abundance.

Behavioural Mechanisms: Fishes Swimming beyond their Limits

Matching morphology to patterns of ecology has provided some important insights into the forces structuring patterns of resource use in fishes (review in Wainwright and Reilly, 1994). However, the impact of behaviour must

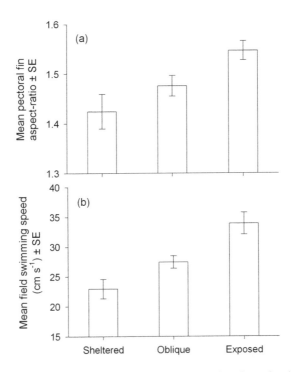

Fig. 12.9 Intraspecific variation in pectoral fin aspect-ratio and routine swimming field speeds in *Acanthochromis polyacanthus* individuals distributed among sheltered (16.9 ± 3.4 cm s⁻¹), oblique (37.8 ± 2.1 cm s⁻¹) and directly exposed (62.6 ± 6.1 cm s⁻¹) habitats around Lizard Island, Great Barrier Reef (mean water flow velocities ± SE in parentheses; Fulton and Bellwood, 2005). Mean fin aspect-ratios, based on four individuals per location, are significantly different among the three wave energy locations (One-way ANOVA $F_{df=2,9} = 8.592$, $p < 0.01$), as are mean swimming speeds (One-way ANOVA $F_{df=2,33} = 11.727$, $p < 0.01$), which were based on 12 individual field observations per location. Fin and speed data collected using the same methods and locations described in Fulton *et al.* (2005).

not be overlooked when examining the relationship between swimming and habitat-use in reef fishes. While morphology may set overriding limits on the range of resources a species can exploit, behaviour ultimately determines how this morphology is used to interact with their environment (Arnold, 1983; Wainwright and Reilly, 1994). As such, behavioural mechanisms can sometimes produce patterns of resource use that significantly depart from predictions based solely on morphology or physiology alone. For example, many labroid fishes have the pharyngeal jaw strength needed to crush hard-bodied prey, but behavioural selection of prey types often results in soft-bodied organisms being the dominant prey item consumed (Wainwright and Richard, 1995; Bellwood *et al.*, 2006). Similarly, small fishes may have relatively poor swimming capabilities when compared to their larger predators, but the use of anti-predator behaviours can substantially increase their likelihood of survival (Langerhans *et al.*, 2004; Lethiniemi, 2005; Masuda, 2006). Behaviour can also be used to reduce the environmental demands placed on fishes, helping them to overcome barriers that may have otherwise curtailed their patterns of ecology. Schooling is one such behaviour, where fish significantly increase their travel time/distance by staying close to conspecifics and exploiting downstream vortices to reduce costs of locomotion (Parker, 1973; Ross and Backman, 1992; Herskin and Steffensen, 1998; Liao *et al.*, 2003). Likewise, flow avoidance behaviour such as refuging can reduce the flows experienced by a fish to substantially decrease the energy expenditure needed to station-hold within fast-flowing habitats (Gerstner, 1998; Gerstner and Webb, 1998; Webb, 2006). Such behavioural mechanisms can provide a powerful means of shaping the boundaries of resource use, seemingly pushing fishes beyond the limits imposed by their morphology and physiology.

Behaviour is undoubtedly an important factor in shaping patterns of fish habitat-use across wave energy gradients. Despite a general trend that only faster swimming fishes occupy the most wave-swept reef habitats (Fig. 12.4, Fig. 12.5), we do find several species occupying wave-swept habitats which have a morphology and body size that suggests very poor swimming abilities. While these slow swimmers tend to occur in wave-swept habitats in low abundance, we might question how they can persist in such high flow habitats with a seemingly inadequate swimming ability? An answer may be found in their behavioural choice of water-column position, with all of the slow-swimming taxa in wave-swept habitats tending to stay very close to the substratum (Fig. 12.5a; Fulton *et al.*, 2001; Fulton and Bellwood, 2002; Johansen *et al.*, 2007b). By maintaining close proximity to the reef, these fishes would experience reduced flow speeds due to a phenomenon known as the boundary-layer effect. The boundary-layer effect is where friction between the water and the substratum reduces wave-induced water movements so that flows immediately adjacent to the reef are reduced by around 20–35% below ambient (Fig. 12.5b; Symonds *et al.*, 1995; Kench, 1998; Gourlay and Colleter, 2005; Reidenbach *et al.*, 2006). Such variations in water-column use and flow with swimming ability are not unique to reef fish assemblages, and have been found in a wide range of marine and freshwater systems (e.g., Chipps *et al.*, 1994; Gerstner and Webb, 1998; Gatwicke and Speight, 2005; Webb, 2006).

Flow avoidance, however, may not be the only reason for refuging behaviour in reef fishes. An alternative hypothesis is that refuging is primarily a predator-avoidance mechanism. Certainly, most refuging behaviour in reef fishes has been attributed to predation risk, with any reduction in flow stress considered to be adjunct and/or insignificant (Hixon and Beets, 1993; Beukers-Stewart and Jones, 1997; Lethiniemi, 2005). However, recent experimental work has found that not only do substratum refuges substantially reduce the environmental demands placed on the swimming abilities of fishes, such refuging may be the sole reason why slow-swimming fishes can occupy high-flow habitats (Fulton *et al.*, 2001; Fulton and Bellwood, 2002; Webb, 2006; Johansen *et al.*, 2007b, 2008). Even relatively simple structures such as sand ripples can significantly reduce the flow speeds experienced by fishes and increase their station-holding capabilities ten-fold (Gerstner, 1998; Gerstner and Webb, 1998). More complex benthic structures, such as the holes and coral heads commonly found in a coral reef matrix, have also been found to dramatically reduce flow speeds, with typically flow speed reductions of 60–80% below ambient (Fig. 12.10a; Johansen *et al.*, 2007b, 2008). Rather than maintaining a swimming speed of 60+ cm s^{-1} in the ambient flows of a wave-swept habitat (Fulton and Bellwood, 2005), fishes occupying such benthic refuges would only need to maintain a speed of 24 cm s^{-1} or less to hold their position. This substantial reduction, and the associated energy savings, may mean the difference between whether or not a slow-swimming species can persist in such high-flow locations. Indeed, this certainly seems the case in at least two reef fish species whose field swimming speed capabilities are significantly slower than the ambient flows found on the coral reef flats where they commonly reside. These two species, *Halichoeres margaritaceus* (average routine field speed of 48 cm s^{-1}) and *Pomacentrus chrysurus* (27 cm s^{-1}; Fulton, 2007), are incapable of holding station for longer than 1 hour when placed in an experimental flow of 60 cm s^{-1} without access to a refuge (Fig. 12.10b; Johansen *et al.*, 2007b). However, when presented with a hole refuge, these species were able to increase their station-holding time by more than 400% and maintain their swimming activity for 35+ hours (Fig. 12.10b; Johansen *et al.*, 2007b). Conducted in the absence of predators, these flow tank experiments suggest that such refuging behaviour is largely due to the need for flow-avoidance in order to hold their position in high flows (Johansen *et al.*, 2007b, 2008). Without such flow-refuging behaviour, it appears that these two species would simply be unable to occupy wave-swept habitats over the long term. Ultimately, behaviour has allowed these reef fish species to expand their ecological limits beyond what we may otherwise have predicted from their swimming speed capabilities.

CONCLUSIONS

Reefs are both challenging and rewarding environments for reef fishes. Wave-induced water motion creates gradients of environmental stress that directly interact with the swimming capabilities of species to shape their patterns of distribution and abundance on reefs. Swimming speed performance appears to be the critical trait that determines these patterns of habitat-use, with clear and consistent trends between habitat-specific levels of water motion and the

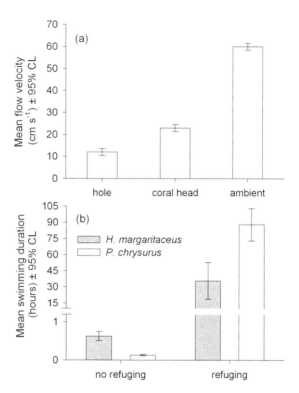

Fig. 12.10 Impact of refuging on (a) flow velocities experienced by fishes in two different types of benthic refuges relative to the ambient conditions found on a wave-swept reef flat, and (b) mean swimming duration in two species of labriform-swimming reef fishes swimming with and without a refuge in 60 cm s⁻¹ ambient flow. Mean flow values are based on n = 5 observations taken within each refuge and n=6 observations within the ambient flow (Fulton and Bellwood, 2005; Johansen *et al.*, 2008). Mean swimming duration based on swimming trials (n = 8 *Halichoeres margaritaceus*, n = 11 *Pomacentrus chrysurus*) conducted in a recirculating flow tank set at an ambient flow speed of 60 cm s⁻¹ and where fishes either did (refuging) or did not (no refuging) have access to a hole refuge (Johansen *et al.*, 2007b). Note that both species could only maintain swimming activity for less than an hour in an ambient flow of 60 cm s⁻¹ without access to a flow refuge.

swimming speed performance of resident fishes. Given the generality of this relationship in fishes from multiple perciform lineages and locations around the world, wave energy seems to impose a strong and overriding force on reef fishes, which applies regardless of any differences in lifestyle or taxonomy. As a consequence, reef fishes are both adapted, and restricted, to a particular suite of habitats according to their inherent swimming capabilities. Of course, this assertion would depend on the extent to which swimming morphology and performance is fixed within a species. While there is some evidence to suggest that reef fishes may show some flexibility in their swimming morphology and performance, the specific mechanism remains to be determined. Determining whether fishes can use the relatively rapid mechanism of phenotypic plasticity to change their swimming morphology and performance, or whether such changes have been acquired through longer-term genetic differentiation and local adaptation should be a primary focus for future research.

One group of reef fishes in particular have emerged as the dominant occupants of reefs. Using solely their pectoral fins to produce thrust, these so-called labriform-swimming fishes occupy the full range of reef habitats available to fishes, from the deepest reef slopes to the shallowest wave-swept areas of coral and rocky reefs alike. In particular, one sub-group of labriform-swimming fishes—those with highly tapered pectoral fins—seem most suited to the challenges of a wave-swept existence. Using the equivalent of underwater flight, these species use their tapered pectoral fins to exploit lift-based thrust and maintain very high speeds with apparently little energetic cost. In a pattern that is repeated in reef fish assemblages throughout the globe, these tapered-fin reef fishes completely dominate wave-swept habitats and attain the largest population sizes. Combining high speed performance with efficiency and flexibility, labriform swimming appears to provide a highly versatile mode of swimming that is ideally suited to the challenges of a reef-associated lifestyle. Undoubtedly, further research into this unique group, particularly in the area of locomotor energetics, should provide further insights into the mechanisms that have shaped the evolution and ecology of fishes on reefs.

ACKNOWLEDGMENTS

Many people have contributed to the vast body of research on the biology and ecology of swimming in reef fishes. In particular, I have benefited greatly from interactions with D. Bellwood, R. Blake, R. Fisher, J. Johansen, J. Steffensen, P. Wainwright, J. Walker, P. Webb, M. Westneat and their published research in this field. This work was supported by funding from the Australian Research Council, Lizard Island Reef Research Foundation and the Australian National University.

REFERENCES

Alexander, R. McN. 1967. *Functional Design in Fishes*. Hutchinson, London.
Alexander, R. McN. 2005. Models and the scaling of energy costs for locomotion. *Journal of Experimental Biology* 208: 1645–1652.
Arnold, S.J. 1983. Morphology, performance and fitness. *American Zoologist* 23: 347–361.
Aseada, T., T.K. Vu and J. Manatunge. 2005. Effects of flow velocity on feeding behaviour and microhabitat selection of the Stone Moroko *Pseudorasbora parva*: A trade-off

between feeding and swimming costs. *Transactions of the American Fisheries Society* 134: 537–547.

Aubret, F. 2005. Aquatic locomotion and behaviour in two disjunct populations of Western Australian tiger snakes, *Notechis ater occidentalis*. *Australian Journal of Zoology* 52: 357–368.

Basaran, F., H. Ozbilgin and Y.D. Ozbilgin. 2007. Comparison of swimming performance of farmed and wild gilthead sea bream, *Sparus aurata*. *Aquaculture Research* 38: 452–456.

Batty, R.S. and P. Domenici. 2000. Fish swimming behavior. In: *Biomechanics in animal behaviour*, P. Domenici and R. W. Blake. (Eds.). BIOS, Oxford, pp. 237–257.

Bay, L. 2007. *The population genetic structure of coral reef fishes on the Great Barrier Reef*. Ph.D. Thesis. James Cook University, Townsville.

Bellwood, D.R. and P.C. Wainwright. 2001. Locomotion in labrid fishes: implications for habitat use and cross-shelf biogeography on the Great Barrier Reef. *Coral Reefs* 20: 139–150.

Bellwood, D.R., P.C. Wainwright, C.J. Fulton and A.S. Hoey. 2002. Assembly rules and functional groups at global biogeographical scales. *Functional Ecology* 16: 557–562.

Bellwood, D.R., T.P. Hughes, C. Folke and M. Nystrom. 2004. Confronting the coral reef crisis. *Nature* (London) 429: 827–833.

Bellwood, D.R., P.C. Wainwright, C.J. Fulton and A.S. Hoey. 2006. Functional versatility supports coral reef biodiversity. *Proceedings of the Royal Society of London* B273: 101–107.

Beukers-Stewart, J. and J.P. Jones. 1997. Habitat complexity modifies the impact of piscivores on a coral reef fish population. *Oecologia* 114: 50–59.

Blake, R.W. 1979. The mechanics of labriform locomotion I. Labriform locomotion in the Angelfish (*Pterophyllum eimekei*): an analysis of the power stroke. *Journal of Experimental Biology* 82: 255–271.

Blake, R.W. 1981. Influence of pectoral fin shape on thrust and drag in labriform locomotion. *Journal of Zoology London* 194: 53–66.

Blake, R.W. 2004. Fish functional design and swimming performance. *Journal of Fish Biology* 65: 1193–1222.

Boisclair, D. and M. Tang. 1993. Empirical analysis of the swimming pattern on the net energetic cost of swimming in fishes. *Journal of Fish Biology* 42: 169–183.

Boisclair, D. and P. Sirois. 1993. Testing assumptions of fish bioenergetics models by direct estimation of growth, consumption, and activity rates. *Transactions of the American Fisheries Society* 122: 784–796.

Bradshaw, A.D. 2006. Unravelling phenotypic plasticity – why should we bother? *New Phytologist* 170: 644–648.

Breder, C.M. 1926. The locomotion of fishes. *Zoologica* 4: 159–297.

Brett, J.R. 1964. The respiratory metabolism and swimming performance of young sockeye salmon. *Journal of the Fisheries Research Board of Canada* 21: 1183–1226.

Cannas, M., J. Schaefer, P. Domenici and J.F. Steffensen. 2006. Gait transition and oxygen consumption in swimming striped surfperch *Embiotoca lateralis* Agassiz. *Journal of Fish Biology* 69: 1612–1625.

Carpenter, R.C., J.M. Hackney and W.H. Adey. 1991. Measurement of primary productivity and nitrogenase activity of coral reef algae in a chamber incorporating oscillatory flow. *Limnology and Oceanography* 36: 40–49.

Castro-Santos, T. 2005. Optimal swim speeds for traversing velocity barriers: an analysis of volitional high-speed swimming behaviour of migratory fishes. *Journal of Experimental Biology* 208: 421–432.

Chipps, S.R., W.B. Perry and S.A. Perry. 1994. Patterns of microhabitat use among four species of darters in three Appalachian streams. *American Midland Naturalist* 131: 175–180.

Clarke, R.D. 1992. Effects of microhabitat and metabolic-rate on food-intake, growth and fecundity of two competing coral-reef fishes. *Coral Reefs* 11: 199–205.

Clarke, R.D., E.J. Buskey and K.C. Marsden. 2005. Effects of water motion and prey behavior on zooplankton capture by two coral reef fishes. *Marine Biology* 146: 1145–1155.

Clements, K.D., R.D. Gray and J.H. Choat. 2003. Rapid evolutionary divergences in reef fishes of the family Acanthuridae (Perciformes: Teleostei). *Molecular Phylogenetics and Evolution* 26: 190–201.

Coyne, J.A. and H.A. Orr. 2004. *Speciation.* Sinauer Associates, Sunderland.

Crossman, D.J., J.H. Choat, K.D. Clements, T. Hardy and J. McConochie. 2001. Detritus as food for grazing fishes on coral reefs. *Limnology and Oceanography* 46: 1596–1605.

Denny, C. M. 2005. Distribution and abundance of labrids in northeastern New Zealand: the relationship between depth, exposure and pectoral fin aspect ratio. *Environmental Biology of Fishes* 72: 33–43.

Denny, M. W. 1988. *Biology and the Mechanics of the Wave-swept Environment.* Princeton University Press, Princeton.

Done, T.J. 1983. Coral zonation: its nature and significance. In: *Perspectives on Coral Reefs,* D.J. Barnes (Ed.). Brian Clouston, Sydney, pp. 107–147.

Drucker, E.G. and J.S. Jensen. 1996. Pectoral fin locomotion in the striped surfperch. I. Kinematic effects of swimming speed and body size. *Journal of Experimental Biology* 199: 2235–2242.

Drucker, E.G. and G.V. Lauder. 1999. Locomotor forces on a swimming fish: three-dimensional vortex wake dynamics quantified using digital particle image velocimetry. *Journal of Experimental Biology* 202: 2393–2412.

Drucker, E.G. and G.V. Lauder. 2002. Wake dynamics and locomotor function in fishes: interpreting evolutionary patterns in pectoral fin design. *Integrative and Comparative Biology* 42: 997–1008.

Farlinger, S. and F.W. Beamish. 1977. Effects of time and velocity increments on the critical swimming speed of largemouth bass (*Micropterus salmoides*). *Transactions of the American Fisheries Society* 106: 436–439.

Feldmeth, C.R. and T.M. Jenkins. 1973. An estimate of energy expenditure by rainbow trout (*Salmo gairdneri*) in a small mountain stream. *Journal of the Fisheries Research Board of Canada* 30: 1755–1759.

Fisher, R., J.M. Leis, D.J. Clark and S.K. Wilson. 2005. Critical swimming speeds of late-stage coral reef fish larvae: variation within species, among species and between locations. *Marine Biology* 147: 1201–1212.

Fitzgibbon Q.P., A. Strawbridge and R.S. Seymour. 2007. Metabolic scope, swimming performance and the effects of hypoxia in the mulloway, *Argyrosomus japonicus* (Pisces: Sciaenidae). *Aquaculture* 270: 358–368.

Fulton, C.J. 2007. Swimming speed performance in coral reef fishes: field validations reveal distinct functional groups. *Coral Reefs* 26: 217–228.

Fulton, C.J. and D.R. Bellwood. 2002. Ontogenetic habitat use in labrid fishes: an ecomorphological perspective. *Marine Ecology Progress Series* 236: 255–262.

Fulton, C.J. and D.R. Bellwood. 2004. Wave exposure, swimming performance, and the structure of tropical and temperate reef fish assemblages. *Marine Biology* 144: 429–437.

Fulton, C.J. and D.R. Bellwood. 2005. Wave-induced water motion and the functional implications for coral reef fish assemblages. *Limnology and Oceanography* 50: 255–264.

Fulton, C.J. and D.R. Bellwood and P. C. Wainwright. 2001. The relationship between swimming ability and habitat use in wrasses (Labridae). *Marine Biology* 139: 25–33.

Fulton, C.J., D.R. Bellwood and P. C. Wainwright. 2005. Wave energy and swimming performance shape coral reef fish assemblages. *Proceedings of the Royal Society of London* B272: 827–832.

Gatwicke, B. and M.R. Speight. 2005. Effects of habitat complexity on Caribbean marine fish assemblages. *Marine Ecology Progress Series* 292: 301–310.

Gerstner, C.L. 1998. Use of substratum ripples for flow refuging by Atlantic Cod, *Gadus morhua. Environmental Biology of Fishes* 51: 455–460.

Gerstner, C.L. 1999. Maneuverability of four species of coral-reef fish that differ in body and pectoral-fin morphology. *Canadian Journal of Zoology* 77: 1102–1110.

Gerstner, C.L. and P.W. Webb. 1998. The station-holding performance of the plaice *Pleuronectes platessa* on artificial substratum ripples. *Canadian Journal of Zoology* 76: 260–268.

Gilliam, J.F. and D.F. Fraser. 1987. Habitat selection under predation hazard: test of a model with foraging minnows. *Ecology* 68: 1856–1862.

Gourlay, M.R. and G. Colleter. 2005. Wave-generated flow on coral reefs - an analysis for two-dimensional horizontal reef-tops with steep faces. *Coastal Engineering* 52: 353–387.

Hammer, C. 1995. Fatigue and exercise tests in fish. *Comparative Biochemistry and Physiology* A112: 1–20.

Haro, A., T. Castro-Santos, J. Noreika and M. Odeh. 2004. Swimming performance of upstream migrant fishes in open-channel flow: a new approach to predicting passage through velocity barriers. *Canadian Journal of Fisheries and Aquatic Sciences* 61: 1590–1601.

Hartwell, S.I., and R.G. Otto. 1991. Critical swimming capacity of the Atlantic Silverside, *Menidia menidia* L. *Estuaries* 14: 218–221.

Herskin, J. and J.F. Steffensen. 1998. Energy savings in sea bass swimming in a school: measurements of tail beat frequency and oxygen consumption at different swimming speeds. *Journal of Fish Biology* 53: 366–376.

Hertz, P.E., R.B. Huey and Jr. T. Garland. 1988. Time budgets, thermoregulation, and maximal locomotor performance: are ectotherms olympians or boy scouts? *American Zoologist* 28: 927–938.

Hill, J. and G.D. Grossman. 1993. An energetic model of microhabitat use for rainbow trout and rosyside dace. *Ecology* 74: 685–698.

Hinch, S.G. and P.S. Rand. 1998. Swim speeds and energy use of upriver-migrating sockeye salmon (*Oncorhynchus nerka*)—role of local environment and fish characteristics. *Canadian Journal of Fisheries and Aquatic Sciences* 55: 1821–1831.

Hixon M.A. and J.P. Beets. 1993. Predation, prey refuges, and the structure of coral reef fish assemblages. *Ecological Monographs* 63: 77–101.

Hogan, J.D., R. Fisher and C. Nolan. 2007. Critical swimming abilities of late-stage coral reef fish larvae from the Caribbean: A methodological and intra-specific comparison. *Bulletin of Marine Science* 80: 219–232.

Hoey, A.S. and D.R. Bellwood. 2008. Cross-shelf variation in the role of parrotfishes on the Great Barrier Reef. *Coral Reefs* 27: 37–47.

Holzman, R. and A. Genin. 2003. Zooplanktivory by a nocturnal coral reef fish: Effects of light, flow, and prey density. *Limnology and Oceanography* 48: 1367–1375.

Huey, R.B. and R.D. Stevenson. 1979. Integrating thermal physiology and ecology of ectotherms: a discussion of approaches. *American Zoologist* 19: 357–366.

Hughes, T.P. 1994. Catastrophes, phase shifts, and large-scale degradation of a Caribbean coral reef. *Science* 265: 1547–1551.

Hughes, T.P. and J.H. Connell. 1999. Multiple stressors on coral reefs: a long-term perspective. *Limnology and Oceanography* 44: 932–940.

Hughes, N.F. and L.M. Dill. 1990. Position choice by drift-feeding salmonids: model and test for Artic grayling (*Thymallus arcticus*) in subarctic mountain streams, interior Alaska. *Canadian Journal of Fisheries and Aquatic Sciences* 47: 2039–2048.

Hughes, T.P., M. J. Rodrigues, D.R. Bellwood, D. Ceccarelli, O. Hoegh-Guldberg, L. McCook, N. Moltschanowsky, M.S. Pratchett, R.S. Steneck and B. Willis. 2007. Phase shifts, herbivory, and the resilience of coral reefs to climate change. *Current Biology* 17: 360–365.

Jayne, B.C. and A.F. Bennett. 1990. Selection on locomotor performance capacity in a natual population of garter snales. *Evolution* 44: 1204–1229.

Johansen, J.L., C.J. Fulton and D.R. Bellwood. 2007a. Estimating the sustained swimming ability of coral reefs fishes. *Marine and Freshwater Research* 58: 233–239.

Johansen, J.L., C.J. Fulton and D.R. Bellwood. 2007b. Avoiding the flow: refuges expand the swimming potential of coral reef fishes. *Coral Reefs* 26: 577–583.

Johansen, J.L., D.R. Bellwood and C.J. Fulton. 2008. Do coral reef fishes exploit flow refuges in high-flow habitat? *Marine Ecology Progress Series* 360: 219–226.

Jones, E.A., K.S. Lucey and D.J. Ellerby. 2007. Efficiency of labriform swimming in the bluegill sunfish (*Lepomis macrochirus*). *Journal of Experimental Biology* 210: 3422–3429.

Kench, P.S. 1998. Physical processes in an Indian Ocean atoll. *Coral Reefs* 17: 155–168.

Kitchell, J.F. 1983. Energetics. In: *Fish biomechanics*, P.W. Webb and D. Weihs. (Eds). Praeger, New York, pp. 312–338.

Koch, F. and Q. Wieser. 1983. Partitioning of energy in fish: can reduction of swimming activity compensate for the cost of production. *Journal of Experimental Biology* 107: 141–146.

Kodric-Brown, A. and P.F. Nicoletto. 2005. Courtship behavior, swimming performance, and microhabitat use of Trinidadian guppies. *Environmental Biology of Fishes* 73: 299–307.

Kolok, A.S. 1991. Photoperiod alters the critical swimming speed of Juvenile largemouth bass, *Micropterus salmoides*, Acclimated to cold water. *Copeia* 1991: 1085–1090.

Kolok, A.S. 1999. Inter-individual variation in the prolonged locomotor performance of ectothermic vertebrates: a comparison of fish and herpetofaunal methodologies and a brief review of the recent fish literature. *Canadian Journal of Fisheries and Aquatic Sciences* 56: 700–710.

Korsmeyer K E., J.F. Steffensen and J. Herskin. 2002. Energetics of median and paired fin swimming, body and caudal fin swimming, and gait transition in parrotfish (*Scarus schlegeli*) and triggerfish (*Rhinecanthus aculeatus*). *Journal of Experimental Biology* 205: 1253–1263.

Langerhans, R.B., C.A. Layman, A.M. Shokrollahi and T.J. DeWitt. 2004. Predator-driven phenotypic diversification in *Gambusia affinis*. *Evolution* 58: 2305–2318.

Lee, C.G., A.P. Farrell, A. Lotto, M.J. MacNutt, S.G. Hinch and M.C. Healey. 2003. The effect of temperature on swimming performance and oxygen consumption in adult sockeye (*Oncorhynchus nerka*) and coho (*O. kisutch*) salmon. *Journal of Experimental Biology* 206: 3239–3251.

Lethiniemi, M. 2005. Swim or hide: Predator cues cause species-specific reactions in young fish larvae. *Journal of Fish Biology* 66: 1285–1299.

Liao, J.C. 2007. A review of fish swimming mechanics and behaviour in altered flows. *Philosophical Transactions of the Royal Society of London* B362: 1973–1993.

Liao, J.C., D.N. Beal, G.V. Lauder and M.S. Triantafyllou. 2003. Fish exploiting vortices decrease muscle activity. *Science* 302: 1566–1569.

Lindsey, C.C. 1978. Form, function, and locomotory habitats in fish. In: *Fish Physiology*, W.S. Hoar and D.J. Randall. (Eds.). Academic Press, New York, Vol . 7: pp. 1–100.

Lonzarich, D.G. and T.P. Quinn. 1995. Experimental evidence for the effect of depth and structure on the distribution, growth, and survival of fishes. *Canadian Journal of Zoology* 73: 2223–2230.

Losos, J.B., K.H. Warhelt and T.W. Schoener. 1997. Adaptive differentiation following experimental island colonization in *Anolis* lizards. *Nature* (London) 387: 70–73.

Losos, J.B., D.A. Creer, D. Glossip, R. Goellner, A. Hampton, G. Roberts, N. Haskell, P. Taylor and J. Ettling. 2000. Evolutionary implications of phenotypic plasticity in the hindlimb of the lizard *Anolis sagrei*. *Evolution* 54: 301–305.

Masuda, R. 2006. Ontogeny of anti-predator behavior in hatchery-reared jack mackerel *Trachurus japonicus* larvae and juveniles: patchiness formation, swimming capability, and interaction with jellyfish. *Fisheries Science* 72: 1225–1235.

Mumby, P.J. 2006. The impact of exploiting grazers (Scaridae) on the dynamics of Caribbean coral reefs. *Ecological Applications* 16: 747–769.

Nelson, J.A., P.S. Gotwalt and J.W. Snodgrass. 2003. Swimming performance of blacknose dace (*Rhinichthys atratulus*) mirrors home-stream current velocity. *Canadian Journal of Fisheries and Aquatic Sciences* 60: 301–308.

Nelson, J.S. 2006. *Fishes of the World*. John Wiley & Sons, Hoboken. 4th Edition.

Nislow, K.H., C.L. Folt and D.L. Parrish. 2000. Spatially explicit bioenergetic analysis of habitat quality for age-0 Atlantic salmon. *Transactions of the American Fisheries Society* 129: 1067–1081.

Parichy, D.M. and R.H. Kaplan. 1995. Maternal investment and developmental plasticity: functional consequences for locomotor performance on hatchling frog larvae. *Functional Ecology* 9: 606–617.

Parker, F.R. Jr. 1973. Reduced metabolic rates in fishes as a result of induced schooling. *Transactions of the American Fisheries Society* 102: 125–130.

Peake, S. 2004. An evaluation of the use of critical swimming speed for determination of culvert water velocity criteria for smallmouth bass. *Transactions of the American Fisheries Society* 133: 1472–1479.

Peake, S., and R.S. McKinley. 1998. A re-evaluation of swimming performance in juvenile salmonids relative to downstream migration. *Canadian Journal of Fisheries and Aquatic Science* 55: 682–687.

Peres-Neto, P.R. and P. Magnan. 2004. The influence of swimming demand on phenotypic plasticity and morphological integration: a comparison of two polymorphic charr species. *Oecologia* 140: 36–45.

Pettersson, L.B. and A. Hedenström. 2000. Energetics, cost reduction and functional consequences of fish morphology. *Proceedings of the Royal Society of London* B267: 759–764.

Plaut, I. 2001. Critical swimming speed: its ecological relevance. *Comparative Biochemistry and Physiology* A131: 41–50.

Power, M.E. 1984. Depth distributions of armoured catfish: predator induced resource avoidance? *Ecology* 65: 523–528.

Purcell, S.W. and D.R. Bellwood. 2001. Spatial patterns of epilithic algal and detrital resources on a windward coral reef. *Coral Reefs* 20: 117–125.

Quenouille, B., E. Bermingham and S. Planes. 2004. Molecular systematics of the damselfishes (Teleostei: Pomacentridae): Bayesian phylogenetic analyses of mitochondrial and nuclear DNA sequences. *Molecular Phylogenetics and Evolution* 31: 66–88.

Randall, J.E., G.R. Allen and R.C. Steene. 1997. *Fishes of the Great Barrier Reef and Coral Sea.* Crawford House Publishing, Bathurst. 2nd Edition.

Reidenbach, M.A., S.G. Monismith, J.R. Koseff, G. Yahel, and A. Genin. 2006. Boundary layer turbulence and flow structure over a fringing coral reef. *Limnology and Oceanography* 51: 1956–1968.

Reist, J.D. 1985. An empirical evaluation of several univariate methods that adjust for size variation in morphometric data. *Canadian Journal of Zoology* 63: 1429–1439.

Rosenfeld, J.S. and S. Boss. 2001. Fitness consequences of habitat use for juvenile cutthroat trout: energetic costs and benefits in pools and riffles. *Canadian Journal of Fisheries and Aquatic Science* 58: 585–593.

Rosenfeld, J.S., M. Porter and E. Parkinson. 2000. Habitat factors affecting the abundance and distribution of juvenile cutthroat trout (*Oncorhynchus clarki*) and coho salmon (*Oncorhynchus kisutch*). *Canadian Journal of Fisheries and Aquatic Science* 57: 766–774.

Ross, R.M. and T.W.H. Backman. 1992. Group-size-mediated metabolic rate reduction in American Shad. *Transactions of the American Fisheries Society* 121: 385–390.

Sagnes, P., P. Gaudin and B. Statzner. 1997. Shifts in morphometrics and their relation to hydrodynamic potential and habitat use during grayling ontogenesis. *Journal of Fish Biology* 50: 846–858.

Sagnes, P., J.Y. Champagne and R. Morel. 2000. Shifts in drag and swimming potential during grayling ontogenesis: relations with habitat use. *Journal of Fish Biology* 57: 52–68.

Schaarschmidt, T. and K. Jurss. 2003. Locomotory capacity of Baltic Sea and freshwater populations of the threespine stickleback (*Gasterosteus aculeatus*). *Comparative Biochemistry and Physiology* A135: 411–424.

Sepulveda, C.A., K.A. Dickson and J.B. Graham. 2003. Swimming performance studies on the eastern Pacific bonito *Sarda chiliensis*, a close relative of the tunas (family Scombridae) I. Energetics. *Journal of Experimental Biology* 206: 2739–2748.

Steneck, R.S. and M.N. Dethier. 1994. A functional group approach to the structure of algal-dominated communities. *Oikos* 69: 476–498.

Sultan, S.E. and H.G. Spencer. 2002. Metapopulation structure favors plasticity over local adaptation. *American Naturalist* 160: 271–283.

Symonds, G., K.P. Black and I.R. Young 1995. Wave-driven flow over shallow reefs. *Journal of Geophysical Research* 100: 2639–2648.

Thorsen, D.H. and M.W. Westneat. 2005. Diversity of pectoral fin structure and function in fishes with labriform propulsion. *Journal of Morphology* 263: 133–150.

Tolley, S.G. and J.J. Torres. 2002. Energetics of swimming in juvenile common snook *Centropomus undecimalis*. *Environmental Biology of Fishes* 63: 427–433.

Tudorache, C., R. Blust and G. De Boeck. 2007. Swimming capacity and energetics of migrating and non-migrating morphs of three-spined stickleback *Gasterosteus aculeatus* L. and their ecological implications. *Journal of Fish Biology* 71: 1448–1456.

Via, S. 2001. Sympatric speciation in animals: the ugly duckling grows up. *Trends in Ecology and Evolution* 16: 381–290.

Vogel, S. 1994. *Life in Moving Fluids: the Physical Biology of Flow.* Princeton University Press, Princeton. 2nd Edition.

Wainwright, P.C. and D.R. Bellwood. 2002. Ecomorphology of feeding in coral reef fishes. In: *Coral reef fishes: dynamics and diversity in a complex ecosystem,* P. F. Sale (Ed.). Academic Press, San Diego, pp. 33–56.

Wainwright, P.C. and B.A. Richard. 1995. Predicting patterns of prey use from morphology of fishes. *Environmental Biology of Fishes* 44: 97–113.

Wainwright, P.C. and S.M. Reilly. 1994. *Ecological Morphology: Integrative Organismal Biology.* University of Chicago Press, Chicago.

Wainwright, P.C., D.R. Bellwood and M.W. Westneat. 2002. Ecomorphology of locomotion in labrid fishes. *Environmental Biology of Fishes* 65: 47–62.

Walker, J.A. 2004. Dynamics of pectoral fin rowing in a fish with an extreme rowing stroke: the threespine stickleback (*Gasterosteus aculeatus*). *Journal of Experimental Biology* 207: 1925–1939.

Walker, J.A. and Westneat, M.W. 1997. Labriform propulsion in fishes: kinematics of flapping aquatic flight in the Bird Wrasse *Gomphosus varius* (Labridae). *Journal of Experimental Biology* 200: 1549–1569.

Walker, J.A. and Westneat, M.W. 2000. Mechanical performance of aquatic rowing and flying. *Proceedings of the Royal Society of London* B267: 1875–1881.

Walker, J.A. and Westneat, M.W. 2002. Performance limits of labriform propulsion and correlates with fin shape and motion. *Journal of Experimental Biology* 205: 177–187.

Wardle, C.S. 1975. Limit of fish swimming speed. *Nature* (London) 255: 725–727.

Webb, P.W. 1975. Hydrodynamics and energetics of fish propulsion. *Bulletin of the Fisheries Research Board of Canada* 190: 109–119.

Webb, P.W. 1984. Body form, locomotion and foraging in aquatic vertebrates. *American Zoologist* 24: 107–120.

Webb, P.W. 1994. The biology of fish swimming. In: *Mechanics and Physiology of Animal Swimming,* L. Maddock, Q. Bone and J.M.V. Rayner. (Eds.). Cambridge University Press, Cambridge, pp. 45–62.

Webb, P.W. 2002. Control of posture, depth, and swimming trajectories of fishes. *Integrative and Comparative Biology* 42: 94–101.

Webb, P.W. 2004. Response latencies to postural disturbances in three species of teleostean fishes. *Journal of Experimental Biology* 207: 955–961.

Webb, P.W. 2006. Use of fine-scale current refuges by fishes in a temperate warm-water stream. *Canadian Journal of Zoology* 84: 1071–1078.

Webb, P.W. and A.G. Fairchild. 2001. Performance and maneuverability of three species of teleostean fishes. *Canadian Journal of Zoology* 79: 1866–1877.

Weihs, D. 2002. Stability versus maneuverability in aquatic locomotion. *Integrative and Comparative Biology* 42: 127–134.

West-Eberhard, M.J. 1989. Phenotypic plasticity and the origins of diversity. *Annual Review of Ecology and Systematics* 20: 249–278.

Westneat, M.W. and M.E. Alfaro. 2005. Phylogenetic relationships and evolutionary history of the reef fish family Labridae. *Molecular Phylogenetics and Evolution* 36: 370–390.

Williams, I.V. and J.R. Brett. 1987. Critical swimming speed of Fraser and Thompson River pink salmon (*Oncorhynchus gorbuscha*). *Canadian Journal of Fisheries and Aquatic Science* 44: 348–356.

Wilson, S.K., D.R. Bellwood, J.H. Choat and M.J. Furnas. 2003. Detritus in the epilithic algal matrix and its use by coral reef fishes. *Oceanography and Marine Biology Annual Review* 41: 279–309.

Young, I.R. 1989. Wave transformation over coral reefs. *Journal of Geophysical Research* 94: 9779–9789.

13

Swimming Behaviour and Energetics of Free-ranging Sharks: New Directions in Movement Analysis

David W. Sims

INTRODUCTION

There is now a general agreement that sharks, and pelagic sharks in particular, are facing widespread declines in population level due to fishing activity (Pauly *et al.*, 2003). Recent studies suggest dramatic reductions in relative abundance of up to 80% have occurred in as little as 15 years for some species (Baum *et al.*, 2003; Myers *et al.*, 2006), trends most likely linked to greater target catch and bycatch rates fuelled by increased demand for shark fins and meat for human consumption. Although some fisheries assessments indicate less pronounced declines for large pelagic (Sibert *et al.*, 2006) and coastal sharks (Burgess *et al.*, 2006), undoubtedly they are particularly susceptible to over-harvesting on account of slow growth rates, late age at maturity and relatively low fecundity. Many pelagic sharks are now red-listed by the World

Authors' address: Marine Biological Association of the United Kingdom, The Laboratory, Citadel Hill, Plymouth PL1 2PB, UK, and School of Biological Sciences, University of Plymouth, Drake Circus, Plymouth PL4 8AA, UK. E-mail: dws@mba.ac.uk

Conservation Union, with some now at a fraction of their historical biomass (Jackson *et al.*, 2001).

Ideally the sustainable management of shark populations requires detailed knowledge of how individuals within a species distribute themselves in relation to their environment, that is with respect to physical and biological resources. The spatial distribution of a shark species will be influenced by how swimming behaviour (linked to motivational and energy requirements) affects rates of key movements, such as search strategies and encounter rates with prey, the location of mates and timing of courtship, and occupation times in preferred habitats. Therefore, day-to-day changes in swimming movement and behaviour will influence broader scale patterns in distribution and population structure, but also signal a shark's response to environmental fluctuations. An ability firstly to understand and then to predict shark space-use patterns and responses to important variables should help identify the extent and dynamics of population distributions and essential habitats, ultimately providing spatial and temporal foci for management in relation to fishing activity and distribution.

A key problem in determining the appropriateness of management or conservation measures for highly mobile and wide-ranging species such as sharks however, is that their horizontal and vertical swimming movements, behaviour and distribution patterns remain largely unknown for the majority of species. Even the behavioural ecology of shark species' that are encountered relatively frequently (e.g., grey reef sharks) are still only partially described. Thus, without knowing where and what habitats sharks occupy, over what time scales, and what behaviour patterns drive the resulting observable changes, it is extremely difficult to determine a population's spatial dynamics and 'stock' boundaries, or to predict how they will respond to future environmental changes. For example, how sharks (and their prey) will respond to climate change that is predicted to increase sea surface temperatures by about 4°C over the next century is not clear, principally because swimming movements and behaviour leading to thermal habitat selection is insufficiently well documented over the appropriate spatio-temporal scales. Equally, how sharks are distributed in relation to fishing locations and effort is important to determine, not least because if remaining aggregations occur in areas where fishing activity is most intense, this will further exacerbate declines. Understanding these processes, therefore, relies on accurate studies of how swimming movements and behaviour in relation to environment influence population structuring and abundance changes of sharks.

Central to this problem of not knowing where sharks are is that they spend most, if not all, their time below the sea surface where they cannot be observed directly or followed to elucidate movements. Until recently only very coarse and simplistic data on fish movements and activities were available for identifying putative fish 'stocks' and possible migrations (Harden Jones, 1968; Guenette *et al.*, 2000). What we are learning from the application of new technologies is that fish such as sharks have complicated spatial and temporal behavioural dynamics, characterised by daily and seasonal migrations, regional differences in behaviour, distinct habitat preferences, and age and sexual segregation. The picture is much more behaviourally complex than previously thought (Harden Jones, 1968). Linked to the deficiency in knowing little about shark movements and behaviour in relation to environment is the difficulty in identifying why such behaviours occur when and where they do.

In behavioural ecology much has been learnt from optimality models seeking to reveal the decision-making processes of animals (Krebs and Davies, 1997), that is, why animals behave a certain way when faced with a set of (necessarily) changing conditions. In formulating these models it is usual to define the likely decision being made by the animal (e.g., time spent in a prey patch), then to define particular currencies (e.g., maximise net rate of gain) and constraints (e.g., travel or handling time, environmental factors) for testing hypotheses about how these different constraints influence the costs and benefits of decision-making. The problem with applying this framework to the behavioural study of sharks is that such decisions, currencies, and constraints may be relatively straightforward to formalise theoretically, but in practice they are very difficult to measure accurately (if at all) for long enough time periods over which natural changes in behaviour and energetics can be recorded. Therefore, empirical tests of optimality models for shark behaviour are not in widespread use (Sims, 2003). So how can we begin to understand what behavioural mechanisms underlie shark movements given these logistical difficulties?

The purpose of this chapter is to introduce a new approach for analysing, interpreting and thinking about the movement and behaviour patterns of free-ranging sharks. In so doing, this paper will (i) briefly describe what typical generalised movement patterns of free-ranging sharks have been recorded using electronic tags, and how this new technology has revolutionised shark behavioural ecology. The chapter will (ii) progress to identify how movement types may be closely linked to habitat types and why, and how foraging and behavioural trade-off models have been used recently to test habitat selection processes in sharks. (iii) A new approach of analysing shark movement data will be described that uses methods from statistical physics to evaluate behavioural response in relation to environment. The chapter closes by providing some future perspectives for better understanding the behaviour patterns of wild sharks.

Shark Swimming Behaviour

Recording Movements

Before the advent of acoustic, datalogging and satellite-linked transmitter tags the study of wild shark behaviour was limited to brief descriptions from divers or from observers aboard vessels. It was in the 1960s when electronic devices capable of emitting or transponding sound energy first came to be used to track individual sharks (Carey, 1992). Sound is the only practical means of transmitting a signal through seawater over distances greater than a few metres as radio waves do not propagate sufficiently in such a conducting medium, at least at the frequencies used in biotelemetry (Nelson, 1990; Priede, 1992). Radio transmitters can be used to track fish in shallow fresh water but through-the-water telemetry of marine fish must be made by acoustic transmission (Nelson, 1990). Ultrasonic frequencies are used (40–78 kHz) because since frequency is proportional to transducer diameter, this is an important factor in reducing the size of the transmitter (thus the 'package' to be carried by a fish). Lower frequency transmitters of greater size due to a larger transducer have been used to track large fish (e.g., Carey and Scharold, 1990), whereas higher frequency tags are reserved for the tracking of smaller species (Sims *et al.*, 2001). A wide array of sensors were made available for acoustic tags to transmit data on water temperature, muscle temperature, depth, cranial temperature, speed, tail beat, heart rate, and multiplexed tags were capable of transmitting sensor data for up to three channels (Carey and Scharold, 1990).

In practical terms, sound pulses emitted from a transmitter attached to a fish are received using a directional hydrophone and portable receiver such that a vessel can follow the moves made by the transmitter, and hence the fish, and receive any encoded signals from sensors. The main problem with this technique for tracking large, highly migratory species that traverse 10s of km per day is that tracking individual sharks continuously in the open sea far from land becomes prohibitively expensive, especially since only one at a time can be tracked by a ship. Because of this, even by the 1990s there were relatively few horizontal and vertical tracks of sharks available for analysis, and these were understandably of short duration (hours to a few days) (e.g., Carey *et al.*, 1982; Klimley, 1993).

The widespread availability of miniaturised data-logging computers in the 1980s revolutionised the study of wild fish behaviour. During the early 1990s dataloggers were developed that were small enough not to impede the natural swimming behaviour of the fish but with powerful batteries and memory sizes capable of recording and storing large amounts of high-quality data (termed archival data). They were also relatively cheap so large numbers could be deployed, and for commercially important species at least, tags could be returned to scientists through the fishery. Early tags were programmed to record pressure (depth) and later models also incorporated temperature and ambient light level sensors (Arnold and Dewar, 2001). Large-scale deployments on North Sea plaice (*Pleuronectes platessa*) (Metcalfe and Arnold, 1997) and on Atlantic bluefin tuna (*Thunnus thynnus*) (Block *et al.*, 2001) were among the first studies to successfully use this technology for tracking movements and behaviour of hundreds of individual fish. Studies followed that fitted depth/temperature/light-logging tags, also known as data storage tags (DSTs), to sharks and other large pelagic fish (e.g., Schaefer and Fuller, 2002) revealing new patterns of behaviour. However, one drawback was that data retrieval was unpredictable since it relied on tags being returned through developed fisheries where return rates can vary widely depending on fishing activity in a given region, and may even be quite low (~5–10%) (Metcalfe and Arnold, 1997).

New tracking methods using satellites were developed to provide a means of collecting data at precise times and independent of fisheries returns. A study in the early 1980s showed that direct satellite tracking of a shark was feasible (Priede, 1984). In a pioneering study in the Clyde Sea off Scotland, a Platform Transmitter Terminal (PTT) moulded into a large buoyant float was attached to a large, 7-m long basking shark (*Cetorhinus maximus*) via a 10-m long tether. Whenever the shark swam near the surface the tag was able to break the surface and transmit to Argos receivers borne on polar-orbiting satellites that estimated the tag's geographical location from these transmissions with reasonable accuracy (Taillade, 1992). The shark was tracked moving in an approximately circular course for 17 days near a biologically productive thermal front (Priede, 1984). Despite this early success, there were no further studies to satellite track sharks for about another 10 years. Tags were large therefore only the largest species such as whale shark, *Rhincodon typus* (Eckert and Stewart, 2001) and basking shark were capable of towing these tags through water. Despite a successful attempt to satellite track smaller-bodied predatory sharks (a blue shark, *Prionace glauca*) using a satellite transmitter mounted on a dorsal fin 'saddle' (Kingman 1996), by the late 1990s the problem still largely remained: how could open ocean fish species' movements and behaviour be tracked at reasonable spatial resolution over longer time periods?

A solution to this problem was developed in the form of a 'hydrid' electronic tag that combined sensor datalogging with satellite transmission (Block *et al.*, 1998; for review see Arnold and Dewar, 2001). This tag, termed a pop-off satellite archival transmitter (PSAT) was attached to a fish 'host' like an external parasite, during which time it recorded 61 hourly or daily water temperature measurements (Arnold and Dewar, 2001). After a pre-programmed time the tag would release from the fish, that is 'pop-off', float to the surface and begin transmitting to Argos receivers that geolocated the tag's position and received temperature data. Later generations of these tags were increasingly capable of recording and storing larger amounts of data and from more sensors; pressure (depth) and light level sensors were added in subsequent models. The narrow Argos receiver bandwidth however means transmitted message lengths from PSAT tags are necessarily short (360 ms) with relatively small amounts of data capable of being transmitted (32 bytes) per message. The transmission times of these tags after pop-off is limited due to rapid battery exhaustion, usually ranging from 0.5–1 month, so different methods have been employed by the manufacturers to increase rates of data recovery. Hence, for one tag type (see www.wildlifecomputers.com) comprehensive summary data in the form of histogram messages (swimming depth, water temperature, profiles of water temperature with depth, times of sunrise/sunset for use in geolocation, see below) derived from archival sensor records are transmitted remotely via satellites, whereas if a tag is physically recovered the entire archived dataset can be downloaded. For another tag type (see www.microwavetelemetry.com), times of sunrise and sunset together with hourly temperature and pressure readings can be recorded for over a year then transmitted after pop-off. In this tag, a special duty cycle timer extends transmission to one month for uploading of this archival dataset by activating the PTT only when Argos satellites are mostly likely to be in view of the tag. The development of PSAT tags, with the ability to geolocate a single position of a fish remotely and track its behaviour and habitat changes (depth and temperature) between tagging and 'pop-off, was nothing short of a revolution in large fish ecology (see below).

Although the early findings from PSATs were impressive, in terms of the details they gave about horizontal movement trajectories between tagging and pop-off, they were still quite rudimentary. PSATs at this stage were capable of providing essentially the same locational information as that obtained from conventional number tagging studies: that is, where the fish was located when tagged, and where it was after a known time period (although conventional tag return rates were low by comparison). Around the end of the 90s, PSATs and DSTs became available that could record not only pressure and temperature, but also ambient light level. This was important because longitude can be estimated by comparing the time of local midnight or midday with that at Greenwich, and latitude from estimates of day length (Wilson *et al.*, 1992). Therefore, electronic tags capable of recording light level could provide data amenable to calculations of geolocation anywhere on the Earth's surface, thereby allowing reconstruction of a fish's movement track from data retrieved from remote locations by Argos satellite (Block *et al.*, 2001).

Much recent research effort has been aimed at improving accuracy of the estimates of light-level-derived locations of sharks and other large fish during their free-ranging movements (e.g., Teo *et al.*, 2004; Nielsen *et al.*, 2006). However, it appears that there are distinct limits to the spatial accuracy of these estimates (60 to 180 km), such that a spatial error of only 10% of a shark's daily movement distance renders detection of specific behaviours within the track prone to error (Bradshaw *et al.*, 2007). Hence, light-level geolocation appears appropriate for tracking large scale movements, but is less able to resolve the specific pattern of smaller scale movements typical of most sharks during much of their annual cycle. In view of this, researchers have re-discovered Priede's idea of direct satellite tracking of sharks, and like F.G. Carey pioneered in the 1990s, have begun attaching satellite transmitting tags directly to the first dorsal fin of large sharks (Weng *et al.*, 2005). Because many large species of shark often come to the surface (see below), as do many large bony fishes such as swordfish (*Xiphias gladius*) and ocean sunfish (*Mola mola*), this presents a new opportunity for resolving more accurately the horizontal movements of sharks across a broader range of space-time scales.

Horizontal Movements

Acoustic tracking provided the first data on the movements and behaviour of free-ranging sharks in relation to environment. For example, the movements of transmitter-tagged scalloped hammerhead sharks (*Sphyrna lewini*) were monitored using datalogging acoustic receivers moored along the plateau of a seamount (Klimley *et al.*, 1988). The departures, arrivals and occupancy times of individual sharks were recorded and showed that individuals grouped at the seamount during the day, but departed before dusk moving separately into the pelagic environment at night and returning near dawn the next day. Acoustic tracking of individual *S. lewini* using a vessel resolved those nocturnal trajectories (Klimley, 1993). Individuals generally moved away from the seamount on a straight line or meandering course into deep water before heading back in the general direction of the seamount. Trackings however, were limited to 1–2 days so the persistence of this central place behaviour within an individual and among individuals across the annual cycle was not known. Studies using the same type of technology but on smaller shark species have shown interesting parallels with these general behaviours of larger species. Small spotted catsharks (*Scyliorhinus canicula*) fitted with acoustic transmitters and tracked continuously for up to 2 weeks show similar patterns of central place refuging (Sims *et al.*, 2001, 2006). Here, males and females remain generally inactive during the day, leaving to forage before dusk and returning to the preferred daytime locations near dawn. The pattern of daytime inactivity in a small activity space centred on rock refuges and nocturnal foraging behaviour in the wider environment was a pattern sustained over many months (Sims, 2003; Sims *et al.*, 2006).

The short duration of acoustic tracking of large pelagic sharks is perhaps even more of a problem when tracking individuals that were captured on lines prior to tagging. Klimley (1993) attached tags underwater to free-swimming sharks without the need for capture, but other studies requiring fish to be hooked on lines are naturally dealing with more distressed fish. Horizontal acoustic trackings of blue shark off the U.S. northwest coast showed that in late summer and autumn they generally moved south and southeast offshore following tagging and release (Carey and Scharold, 1990). It was suggested this consistent pattern of movement may have been related to seasonal migration, but it was also recognised that in part these movements may be a general reaction of pelagic fish to move into deep water offshore after being captured for tagging (Carey and Scharold, 1990). For longer trackings where the shark was tagged without the need for capture, horizontal movements are more likely to provide insights into natural foraging or ranging movements. For example, a white shark (*Carcharodon carcharias*) that was tagged feeding in the vicinity of a whale carcass off Long Island, U.S.A., continued feeding and swimming within a 3 km radius of the whale for 1.5 days (Carey *et al.*, 1982). The shark then moved southwest following quite closely the 25 m isobath for a further 2 days.

Before light level sensors were available on electronic tags, tagging and 'pop-off' locations provided information on dispersal distances over timescales much greater than was possible with acoustic telemetry. For example, bluefin tuna (*Thunnus thynnus*) fitted with PSATs off the eastern U.S. coast were found to travel over 3,000 km east across the North Atlantic Ocean in 2–3 months (Block *et al.*, 1998; Lutcavage *et al.*, 1999). In addition, the longer range movements of the white shark (*Carcharodon carcharias*) were generally unknown, but those aggregating around small islands in the coastal waters of California were generally thought to remain in shelf waters, moving up and down the coast according to season. However, PSAT tagging revealed some incredible movements of these apex predators; tagged white sharks moved into the open ocean away from the coast and one shark made the journey from California to Hawaii (some 3,800 km across the North Pacific Ocean) in less than 6 months, making dives down to 700 m and experiencing temperatures as low as 5°C (Boustany *et al.*, 2002).

Among the first studies to use light-level intensity changes to calculate large-scale horizontal movements of pelagic fish were those with bluefin tuna (Block *et al.*, 2001) and basking shark (*Cetorhinus maximus*) (Sims *et al.*, 2003). Reconstructed tracks of bluefin tuna tagged off North Carolina, U.S.A., showed movements between foraging grounds in the western Atlantic prior to transatlantic migrations into the Mediterranean Sea for spawning. Bluefins were also shown to dive to over 1,000 m depth during these excursions (Block *et al.*, 2001). The question of diving activity and what

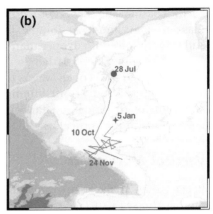

Fig. 13.1 Seasonal movements of basking sharks (*Cetorhinus maximus*) from archival tagging. (a) Reconstructed track of a 4.5-m long shark moving from the tagging location in the Western English Channel, around the west of Ireland and into Scottish waters over 77 days and covering an estimated minimum distance of 1,878 km. (b) Movement of a 7.0-m long shark from the Clyde Sea in Scotland, through the Irish Sea and into waters off southwest Britain, a journey tracked over 162 days and an estimated minimum distance of 3,421 km. Adapted from Sims *et al.* (2003).

large fish in temperate waters might do when, with changing season, food levels decline, is particularly relevant to the plankton-feeding basking shark. This species, the world's second largest fish, is most frequently seen in coastal waters feeding at the waters' surface during summer months (Sims and Quayle, 1998). In the northwest and northeast Atlantic, surface foraging occurs from around April to October usually with a peak in sightings from May until August. The seasonal increase in the surface sightings of *C. maximus* coincides with increased zooplankton abundance at this time (Sims *et al.*, 1997; Cotton *et al.*, 2005). Basking sharks tagged with PSATs were tracked moving between waters off south-west England to Scotland, and *vice versa*, sometimes over periods of only a few weeks (Sims *et al.*, 2003) (Fig. 13.1). Sharks travelled long distances (390 to 460 km) to locate temporally discrete productivity 'hotspots' along tidal fronts and on shelf-break fronts. It was also shown from basking shark trackings over seasonal scales (up to 7.5 months) that they were active during winter and do not hibernate, as was once supposed (Matthews 1962). Instead, they conduct extensive horizontal (up to 3,400 km) and vertical (> 200 m depth) movements to utilise productive continental-shelf and shelf-edge habitats during summer, autumn and winter (Sims *et al.*, 2003).

The long-range movements of the salmon shark (*Lamna ditropis*) also appear linked to seasonal changes in conditions. Using fin-mounted Argos satellite transmitters, salmon sharks were tracked from summer feeding locations in Alaskan coastal waters southward to overwintering areas encompassing a wide range of habitats from Hawaii to the North American

Pacific coast (Weng *et al.*, 2005). Some individuals, however, overwintered in Alaskan waters and in so doing experienced water temperatures between 2 and 8°C and dived no deeper than 400 m. Those sharks migrating south by contrast, occupied depths where sea water temperatures were up to 24°C and regularly dived into cooler waters to depths over 500 m (Weng *et al.*, 2005). This illustrates that in our attempts to understand the movements and behaviour of free-ranging sharks it is as important to consider swimming movements in the vertical as well as horizontal plane.

Vertical Movements

Pelagic sharks, like many other marine vertebrates, spend relatively little time at the sea's surface but instead may remain for long periods in the near-surface layers or at depth, and almost always make frequent dives through the water column, sometimes to very deep depths. What we are learning from the deployments of electronic tags on different species of shark is that although the different patterns of vertical movement can be as varied as they are complex, some general patterns appear similar among different species (Shepard *et al.*, 2006) (Fig. 13.2).

The blue shark was one of the first pelagic shark species to be tracked regularly in the open sea by acoustic telemetry yielding insights into the pattern and range of its vertical movements (for overview see Carey and Scharold, 1990). In general it was found that blue sharks tracked on the Northeast U.S. continental shelf and shelf edge exhibited depth oscillations and that this diving pattern was seasonal. During late summer, autumn and winter, sharks showed remarkably regular vertical oscillations, particularly during daylight hours, from the surface or near surface to about 400 m depth and back, whereas at night oscillations were usually confined to the top 100 m. Interestingly, sharks tracked in summer further inshore (water depth < 80 m) did not show these large, regular changes in swimming depth. In early and mid summer blue sharks remained in the upper 10 m, seldom diving deeper into colder water below the thermocline at 15 m (Carey and Scharold, 1990). In contrast, acoustic transmitter-tagged tiger sharks (*Galeocerdo cuvier*) tracked off the south coast of Oahu, Hawaii, across deep water to the Penguin Banks did not show night-day differences in depth oscillations or overall swimming depth (Holland *et al.*, 1999). In water >100 m depth they swam predominantly close to the seabed, whereas in deeper water (>300 m) they remained within the top 100 m in the warm, mixed layer above the thermocline (at 60–80 m depth) irrespective of time of day. Nevertheless, tiger sharks did undertake oscillations in swimming depth during their horizontal movements, albeit of relatively small amplitude (~50 m) compared with blue sharks swimming over deep water.

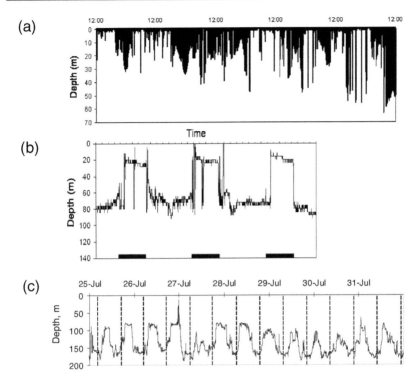

Fig. 13.2   Diving behaviours of basking shark (*Cetorhinus maximus*). (a) Vertical movements appear complex, such as this archival dive time-series over 5 days in shelf waters of the Western English Channel. (b) Some behaviours are readily recognisable and are linked to habitat type, such as this normal diel vertical migration (DVM) of a 6.5-m shark in strongly stratified waters of the Clyde Sea, Scotland. Reverse DVM exhibited by basking sharks is associated with tidal fronts, the biologically productive boundaries between stratified and mixed shelf waters. Dark bars denote nighttime. (c) Circa-tidal vertical movement of a basking shark in a deep gully in the English Channel. The date is shown on the x-axis. Times of low water are indicated by dashed vertical lines to demonstrate the consistency of the shark's movements with the tidal phase. Adapted from Sims *et al.* (2005) and Shepard *et al.* (2006).

These studies raise two principal questions: first, why do sharks show regular diving oscillations through the water column, or at least through a portion of it? And second, when they do, why are there diel and seasonal changes in these patterns? It was suggested for blue shark that regular dives through the water column were related to searching for prey. Vertical excursions are probably the most efficient way of sampling olfactory sources that spread largely horizontally in the ocean due to current shear between layers of differing density (Carey and Scharold, 1990). Frequent oscillations in swimming depth have also been recorded for other acoustically tracked species including the whale shark (*Rhincodon typus*) during foraging movements off Ningaloo Reef, Australia (Gunn *et al.*, 1999). Tracked sharks made regular dives through the water column from the surface to near the bottom at 60–80 m. It was suggested the sharks were searching the water column for food since the dives were not related closely to changes in hydrographic features.

The use of data storage tags (DSTs) and satellite-linked archival transmitters on sharks has provided more detailed temporal information on their depth utilisations over longer time periods. Viewed at this high resolution over days to months, longer term and sometimes quite different patterns in shark swimming behaviour are evident. For example, DST tracking of school shark (*Galeorhinus galeus*) showed vertical rhythms in swimming depth over the diel cycle. In continental shelf waters off southern Australia *G. galeus* spent the day at depth before ascending at night into shallower water often for several hours (West and Stevens, 2001). When *G. galeus* was in deeper water off the continental shelf this pattern was maintained, with descent at dawn to depths of up to 600 m before ascent to shallower waters at dusk. In maintaining this pattern it appears the school shark exhibits what is termed a 'normal' diel vertical migration (DVM) (dusk ascent, dawn descent) for much of the time when ranging and foraging off southern Australia. This pattern may represent prey tracking of vertically migrating populations of squid for example, that also undertake similar daily vertical movements as those observed in school sharks (West and Stevens, 2001).

By contrast, white sharks tagged and tracked off southern Australia appear to exhibit at least three different vertical movement patterns. A pattern of regular vertical oscillations from about 50 m depth to the surface irrespective of day or night was apparent during directed westerly movements of white sharks (Bruce *et al.*, 2006). These swimming movements were reminiscent of blue shark oscillations during straight-line movements. However, this pattern was not maintained when ranging within the Spencer Gulf inlet where bottom-oriented behaviour with few ascents to the surface was the dominant pattern. Furthermore, a third pattern comprised diel vertical movements with shallow depths between the surface and 25 m depth occupied principally during the day, and depths around 50 m selected at night (Bruce *et al.*, 2006). This diel vertical movement pattern occurred when the white shark was present at offshore islands. Therefore, it appears that large sharks display different vertical movements as a function of habitat and also in response to differing prey distribution, abundance and availability.

Habitat Selection

Central to an understanding of shark movements is identification of the role of environment and habitat characteristics in shaping movement and behaviour patterns. For such an understanding, it becomes necessary to investigate closely the links between different behaviour types and concomitant changes in the environment, and to try to separate habitat *selection* or preferences from habitat *correlations.* Detecting selection of sharks and other animals for particular habitats is difficult since movements generally convey little about why those habitats have been selected (Kramer *et al.,* 1997). Simple mapping of shark movements on to environmental fields, for example a horizontal trajectory overlaid on a satellite remote-sensing map of sea surface temperature, provides a view of where and in what type of habitat the individual was located, but actually tells us nothing in itself about the habitat selection processes underlying the movement; that is, why the habitat was selected from those habitats available. To understand why sharks go where they do, it becomes important to be able to move beyond simple shark-habitat correlations, and to monitor where the sharks are in comparison with where they are not. By comparing the habitat types where sharks are located to the types of other (presumably) equally available habitats where they are not located at a given time, it allows us to delve into the dynamics of habitat *selection.* As mentioned previously (see Introduction), an understanding of habitat choices moves us closer to being able to predict the movements and behaviour of sharks when faced with a particular set of environmental conditions.

There is a growing literature documenting the locations of sharks through time in relation to different habitat types encountered. Generally, the aim has been to describe the habitat correlations of the various species studied to try to gain a mechanistic insight into what may control movement patterns. Less attention however has been paid to comparing occupied habitat types with other habitats available to detect potential differences regarding biotic and abiotic factors, such as prey presence or water temperature. In addressing the former, horizontal trajectories have been plotted on maps of sea surface temperature (e.g., Priede, 1984; Skomal *et al.,* 2004; Weng *et al.,* 2005; Bonfil *et al.,* 2005), bathymetry (e.g., Holland *et al.,* 1999; Sims *et al.,* 2003), geomagnetic anomalies (Carey and Scharold, 1990) and primary (Sims *et al.,* 2003) and secondary productivity (Sims and Quayle, 1998; Sims *et al.,* 2006). Vertical movements of sharks have been mapped on to variations in vertical thermal structure (e.g., Carey and Scharold, 1990), in relation to seabed depth (e.g., Gunn *et al.,* 1999) and with respect to concentrations of prey species (Sims *et al.,* 2005) (Fig. 13.2b). Assessing movements in relation to prey is perhaps one of the most useful environmental fields to use, at least over the short-term, because the distribution of prey is a key factor influencing predator movements. It is expected that clear and persistent changes in swimming movements should be related to specific changes in prey abundance and availability.

Acoustic tracking studies of large pelagic fish have shown vertical movements linked closely to changes in prey distribution over the diel cycle. Depth movements of a large swordfish (*Xiphias gladius*) were related to an echogram made with a 50-kHz echosounder (Carey, 1992). A dense assemblage of prey organisms, probably squid and fish, produced a heavy band of echoes (termed a sound-scattering layer, SSL) near the surface at night, where the swordfish also occurred. About an hour before sunrise at around 04: 30h, the SSL gradually descended from surface waters to about 300 m. The swordfish appeared to track this change in SSL depth closely during the short tracking, probably preying on squid and fish both near the surface at night and at depth during the day (Carey, 1992). Similarly, a bigeye tuna (*Thunnus obsesus*) was tracked gradually moving with the downward migrating SSL from 100 to 350 m, and remaining within it until early morning after which the tuna moved upward into warm, shallow waters, perhaps for reasons related to behavioural thermoregulation rather than feeding opportunities (Holland *et al.*, 1990; see chapter 14 of this book).

Movements of sharks have been linked with long-term diel vertical migration of prey. The vertical movements of basking sharks recorded with archival tags in the Northeast Atlantic were found to be consistent with those associated with foraging on diel vertically migrating zooplankton prey (Sims *et al.*, 2005). In deep, thermally stratified waters sharks exhibited normal diel vertical migration (nDVM), comprising a dusk ascent into surface waters followed by a dawn descent to deeper depths. This corresponded closely with migrating sound-scattering layers made up of *Calanus* and euphausiids that moved into surface waters at dusk, returning to depths of 50–80 m at dawn where they remained during the day until dusk. Basking sharks occupying thermally stratified waters were recorded undertaking a nDVM pattern tracking zooplankton movements for up to a month before moving to new areas or changing their vertical movement pattern. Interestingly, the vertical pattern of movement was found to vary between different oceanographic habitat types. In contrast to basking sharks in stratified waters, individuals occupying shallow, inner-shelf areas near thermal fronts conducted reverse DVM, comprising a dusk descent to depths between 20–80 m before ascending at dawn into surface waters where they remained during the day (Sims *et al.*, 2005). This difference in shark swimming movements in fronts compared with thermally stratified waters was due to induction of reverse DVM in *Calanus* by the presence of high concentrations of chaetognaths (important predators of calanoid copepods) at the surface during the night, followed by downward migrations of chaetgnaths during the day (i.e., the normal DVM of the predators produced a behavioural switch to reverse DVM in the copepod prey). Other changes in vertical movements of basking sharks are known from the Northeast Atlantic. Dive patterns of one cycle per day (e.g., DVM) give way to depth oscillations of two cycles per day, in response to tidally-mediated migrations of zooplankton, when basking sharks moved into strongly tidal waters of the English Channel (Shepard *et al.*, 2006) (Fig. 13.2c).

Relating shark movements to environmental fields such as those in the aforementioned studies has revealed much about the diversity of behaviours present in the wild. It has provided insights into what environmental factors may influence observed patterns. However, the findings of such studies often represent simple habitat correlations that may or may not be the product of habitat selection. In such studies, random correlation may be an equally likely explanation of the observed patterns because other areas not occupied by the shark are not included in the analysis, only where the shark occurs. The question is, what is it about the location where the shark occurs that renders it different from the surrounding habitats that are not apparently selected? Explicit tests of habitat preferences of sharks are much less common in the literature, although some useful recent examples suggest what is possible to gain from such an approach.

The general approach used in preference-testing studies is to compare the amount of time spent or prey encountered in each habitat as a function of the movement track observed, compared with predicted values for that individual based on random walks through the same environment. The habitat preferences of tiger sharks in a shallow, seagrass ecosystem in Australia were examined by acoustic tracking and animal-borne cameras (Crittercams) as they moved through different habitats (Heithaus *et al.*, 2002). Tracks of actual sharks were used to estimate the proportion of time spent in two different habitats, shallow and deep. These were then compared with habitats visited during random walks. To produce a random walk, the distances moved between 5-minute position fixes in an actual tiger shark track, termed the move step lengths, were randomised to form a new track of the same length, but with a new direction of travel angle assigned to each move step taken from the distribution of commonly observed travel angles of tiger shark tracks. A particular strength of this approach is that actual shark tracks were each compared with 500 random tracks so significance levels of actual versus random could be calculated. The analysis showed that although there was individual variation in habitat use, tiger sharks preferred shallow habitat where prey was more abundant (Heithaus *et al.*, 2002).

Assessing the habitat selection of sharks in relation to dynamic prey 'landscapes' has been a more difficult goal to achieve. The principal reason for this is that temporally changing prey-density fields are not available for the vast majority of marine predators. One of the few cases where this is possible however, is for filter-feeding sharks. As mentioned previously, basking sharks feed selectively on large calanoid copepods in specific assemblages of zooplankton (Sims and Quayle, 1998). In the north Atlantic the Continuous Plankton Recorder (CPR) survey has for over 50 years undertaken broad-scale measurements of zooplankton abundance to species level (Richardson *et al.*, 2005). The large-scale spatial coverage of individual plankton samples (minimum spatial and temporal resolutions of 56 x 36 km and 14 days, respectively) means it was possible to relate the broad-scale foraging movements of satellite-tracked basking sharks to the spatio-temporal abundance of their copepod prey. Basking shark tracks were mapped onto time-referenced copepod abundance fields in a recent study, with the amount of zooplankton 'encountered' estimated for each movement track (Sims *et al.*, 2006). The total prey encountered along each track was compared with that encountered by 1000 random walks of model sharks, where the move steps taken by each random walker were drawn from the distribution of move steps observed for real sharks. The study showed that movements of adult and sub-adult basking sharks yielded consistently higher prey encounter rates than 90% of random-walk simulations. This suggests that the structure of movements undertaken by basking sharks were aimed at exploiting preferred habitats with the richest zooplankton densities available across their range.

Behavioural Trade-offs and Energetics

So far in this discussion we have generally considered only single factors at a time, such as prey or water temperature, to be influencing the behaviour pattern of sharks. Necessarily when these simple relationships break down, we suggest other factors to test. Assessing changes in shark swimming movements due to a single factor in piecemeal fashion is likely to be a simplification in our approach, one that makes it easier for us to identify apparent relationships more readily than if many factors are appraised. However, as we can all appreciate, sharks move through complex environments and their behaviours are in response to a myriad of changing influences. So, how can we investigate the effects of co-varying factors on shark behaviour? And if we can reliably, what do such studies show?

Although studies of the effects of multiple factors on habitat selection behaviour of birds and teleost fish under captive or semi-natural conditions have been investigated in carefully designed experiments, similar work on sharks is generally lacking. In part this is due to the logistical problems associated with keeping large sharks in sufficiently large aquaria such that experiments can be reliably undertaken. But in addition, it is due to the comprehensive knowledge needed of the natural habitats where sharks occur to begin such studies, how different factors may co-vary to influence

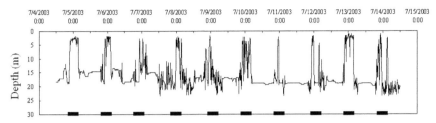

Fig. 13.3 The normal diel vertical migration (DVM) pattern of a male small spotted catshark (Scyliorhinus canicula) in a tidal sea lough in southwest Ireland. Movements into shallow water are usually made by ascending submarine slopes from deeper water. Dark bars denote nighttime. Adapted from Sims *et al.* (2006).

shark movement patterns, and also the energetic costs of movement and other functions, and of undertaking certain behaviours. Some progress has been made recently by linking tracking studies of wild sharks with environmental surveys and long-term monitoring, with laboratory experiments designed to test the field observations of behaviour, and combined with energetic modelling to help identify the costs of particular behaviour patterns. Nevertheless, this has so far only been attempted with small-bodied catsharks; we will come back to the problem of understanding the movements and behaviour of pelagic sharks in section 5.

The movements of the small-spotted catshark (*Scyliorhinus canicula*) have been closely studied in a tidal sea lough in the Republic of Ireland (for overview see Sims 2003). The lough is characterised by steep submarine slopes and it was found that male *S. canicula* spend the daytime resting in cool, deep water before moving up-slope at dusk into warm, shallow habitats for nocturnal foraging (Sims *et al.*, 2001, 2006). This approximates a normal diel vertical migration (nDVM) (Fig. 13.3); however, what factors determine this particular pattern of movement since similar nDVM of benthic prey is not evident in this environment? Long-term field deployments of water temperature-depth dataloggers showed that temperatures in the shallow habitat varied from 15.5 (at night) to 17.5°C (during daytime), whereas in their deep habitat the temperature remained between about 15–16°C (Sims *et al.*, 2006). Furthermore, deployments of baited traps showed high abundance of prey in warm, shallow areas (~2 m depth) during both day and night, but in deep water abundances were an order of magnitude less irrespective of light phase. This raised the possibility that this ectothermic catshark was exhibiting a trade-off between optimal foraging and thermal habitats. It appeared shallow, prey rich areas were entered only during periods when water temperatures were lowered during nighttime (when energy costs would be concomitantly lower), with sharks moving to deeper, cold water after foraging.

A laboratory manipulation of behaviour was designed to investigate the effects of water temperature and food availability on catshark habitat selection. Four male catsharks were trained to receive food in one compartment of a two-compartment choice chamber (water temperature, 14.4°C), the compartments being linked by a small tunnel through which the fish were free to move. After the training period, and for the next two weeks, the frequency of habitat (compartment) use strongly favoured that where food was made available (3 or more individuals were present there ~90% of the time). Throughout the following two weeks, the water temperature of the food-delivery compartment was maintained 0.9°C above the other compartment (i.e., elevated to 15.3°C), and habitat choice was recorded. By the end of the second week, the four catsharks were spending 90% of their time in the cooler side, only moving into the warmer habitat to obtain food, and which was consistent with wild behaviour of males. When the temperature gradient was removed for the next two weeks, habitat occupation switched back to the food-delivery compartment (Sims *et al.*, 2006). Thus, the thermal choice experiment using environmentally realistic temperatures supported the field observation that catsharks positively avoided warmer water even when it was associated with greater food availability.

To estimate the energetic costs of this behaviour extensive measurements of catshark metabolic rate (indirectly from oxygen consumption) were undertaken to determine the three principal components of a fish's energy budget: standard metabolism (metabolic rate at zero swimming speed), and feeding and active metabolic rates (Sims and Davies, 1994; Sims, 1996). From this data, potential energetic mechanisms underlying observed behaviour were approached by calculating the energy costs associated with occupying different-temperature habitat at varying levels of activity and feeding metabolism (termed specific dynamic action, SDA). Depth and temperature records from data-storage tagged catsharks comprised the empirical data used to build a model of energy expenditure which was then subjected to manipulations of thermal regime, that is, replacement of actual water temperature values with fixed (isothermal cold or warm temperatures) or biased-random values (within a range generated from warm or cold average 'seed' values) (Sims *et al.*, 2006). Modelling of energy budgets under these different realistic thermal choice scenarios determined by temperature and activity records from electronic tags attached to catsharks revealed that adopting a "hunt warm, rest cool" strategy could lower daily energy costs by just over 4%. Although in the short-term this saving appears small, this species is known to live for at least 15 years, during which time time these daily energy savings may contribute significantly to increasing lifetime reproductive success. Overall the findings were consistent with male *S. canicula* undertaking nDVM movements to increase energetic benefits through trade-offs between foraging opportunity and the effect of thermal resources on energy expenditure (Sims *et al.,* 2006). The results provide the first clear evidence that benthic sharks, at least, are capable of utilizing DVM as an energy conservation strategy that increases bioenergetic efficiency.

Macroscopic Patterns of Behaviour

It is relatively straightfoward to see how measurements of energy expenditure, manipulative laboratory experiments and field tracking of smaller-bodied sharks can be combined to test explicit hypotheses in behavioural ecology. Doing the same for large pelagic sharks however is logistically prohibitive. As we have seen (Horizontal Movements 2.2 and Vertical Movements 2.3), large sharks often undertake complicated horizontal and vertical movements but as has been mentioned already what is less well known is what they are doing during such movements and why. Generally speaking it has proved very difficult to identify what behaviours sharks are exhibiting, whether they are searching, feeding, commuting, resting, or migrating, and so on. Until electronic devices capable of providing a "daily diary" of shark activities and energy expenditure becomes a practical reality (for concept see Wilson *et al.,* 2008), we are reliant upon using inferential approaches to help tease out particular behaviours and, in some cases, the strategies potentially used by sharks to find resources (food, mates, refuge, etc). Here, I will describe some analysis techniques taken from the field of statistical physics and indicate where these have recently proved useful in helping to understand shark swimming movements and behaviour in the wild, in particular those associated with searching.

Searching Natural Environments

A central issue in behavioural ecology is understanding how organisms search for resources within heterogeneous natural environments (MacArthur and Pianka 1966; Stephens and Krebs 1986). Organisms are often assumed to move through an environment in a manner that optimises their chances of encountering resource targets, such as food, potential mates, or preferred refuging locations. For a forager searching for prey in a stable, unchanging environment, prior expectation of when and where to find items will inform a deterministic search pattern (Stephens and Krebs, 1986; Houston and McNamara, 1999). However, foragers in environments that couple complex prey distributions with stochastic dynamics will not be able to attain a universal knowledge of prey availability. This raises the question of how should a forager best search across complex landscapes to optimise the probability of encountering suitable prey densities? Because nearly all motile animals face this same problem it suggests the possibility that a general foraging rule for optimising search patterns has emerged in animals by natural selection. Pelagic sharks such as the blue shark (*Prionace glauca*) appear well designed for sustained cruising at low swimming speeds, presumably in part as a consequence of their need to search large areas and depths to locate sparse resources in sufficient quantities. Hence, pelagic sharks may be ideal candidates for testing ideas about optimal searching by applying Lévy statistics to movement patterns, and in so doing this may reveal insights about what governs aspects of their behaviour.

Optimal Lévy flights

Recent progress in optimal foraging theory has focused on probabilistic searches described by a category of random-walk models known as Lévy flights (Viswanathan *et al.*, 2000; Bartumeus *et al.*, 2005). Lévy flights are specialised random walks that comprise "walk clusters" of relatively short step lengths, or flight intervals (distances between turns), connected by longer movement 'jumps' between them, with this pattern repeated across all scales (Bartumeus *et al.*, 2005). In a Lévy flight the move step lengths are chosen from a probability distribution with a power-law tail, resulting in step lengths with no characteristic scale: $P(l_j) \sim l_j^{-\mu}$, with $1 < \mu \le 3$ where l_j is the flight length and μ is the exponent (or Lévy exponent) of the power law. Theoretical studies indicate Lévy flights represent an optimal solution to the biological search problem in complex landscapes where prey are sparsely and randomly distributed outside an organism's sensory detection range (Viswanathan *et al.*, 1999, 2000; Bartumeus *et al.*, 2005). Simulation studies indicate an optimal search has a Lévy exponent of $\mu \cong 2$ (Viswanathan *et al.*, 1996). The advantage to predators of selecting step lengths with a Lévy distribution compared with simple Brownian motion for example, is that Lévy flight increases the probability of encountering new patches compared with other types of searches (Viswanathan *et al.*, 2000; Bartumeus *et al.*, 2002). Recent studies (Benhamou, 2007; Edwards *et al.*, 2007; Sims *et al.*, 2007) contend Lévy flights have been wrongly ascribed to some species through use of incorrect methods, while others indicate Lévy-like behaviour with optimal power-law exponents (Bartumeus *et al.*, 2005), supporting the hypothesis that $\mu \cong 2$ may represent an evolutionary optimal value of the Lévy exponent (Bartumeus, 2007).

Here then, is a theoretical prediction about searching movements that can be tested with empirical data. The approach is to ask whether individual animals exhibit movement patterns that are consistent with Lévy behaviour, that is, whether the move step length frequency distribution is well described by a power law form with a heavy tail. If there is sufficient support for observed movements to be approximated by a Lévy distribution of move step lengths, where $1 < \mu \le 3$, then we can infer that the animal might be adopting a probabilistic, Lévy-like searching pattern. Furthermore, if the calculated exponent lies close to $\mu \cong 2$, it is possible the structure of movements undertaken by an animal may be optimal. Similarly, fluctuations or changes in the Lévy exponent of movements could signal ecologically important shifts in behaviour.

Testing Empirical Data

Marine vertebrates such as large sharks that feed on ephemeral resources like zooplankton and smaller pelagic fish typify the type of predator that might undertake such probabilistic searches described with Lévy distributions. This is principally because they have sensory detection ranges limited by the seawater medium and experience extreme variability in food supply over a broad range of spatio-temporal scales (Mackus and Boyd, 1979; Makris *et al.*, 2006). Although at near-distance scales sharks use sensory information of resource abundance and distribution, and at very broad scales some may have awareness of seasonal and geographical prey distributions, across the broad range of mesoscale boundaries (1 to 100s of km) pelagic sharks, in many instances such as during searching, are more like probabilistic or 'blind' hunters than deterministic foragers. Across such scales, the necessary spatial knowledge required for successful foraging will depend largely on the search strategy employed.

Lévy-like movement behaviour has apparently been detected among diverse organisms, including amoeba (Schuster and Levandowsky, 1996), zooplankton (Bartumeus *et al.*, 2003), insects (honeybees) (Reynolds *et al.*, 2007), social canids (jackals) (Atkinson *et al.*, 2002), arboreal primates (spider monkeys) (Ramos-Fernandez *et al.*, 2004) and even in human movements (Brockmann *et al.*, 2006; Gonzalez *et al.*, 2008). Recent studies indicated some methodological errors associated with early studies of organism movements (Edwards *et al.*, 2007; Sims *et al.*, 2007), perhaps resulting in false detections of Lévy behaviour; however, robust statistical methods have now been developed (Clauset *et al.*, 2007). It was during this recent period of methodological progress that marine vertebrate predator movements were tested rigorously for the first time (Sims *et al.*, 2008).

Most studies to date have analysed the horizontal trajectories of organisms to test for Lévy flights. However, when considering fully aquatic marine vertebrates such as sharks, this presents a problem since the horizontal tracks are subject to significant spatial errors. Inaccurate location determinations, either from direct Argos satellite tracking or from reconstructed tracks using light-level geolocation, are not necessarily important when considering large-scale movements such as migration. This is because the gross movement displacement is greater than the quantified error field. However, in testing for the presence of Lévy flights the move step lengths whether they be small or large are important to measure accurately as they form the frequency distribution, and it is from this that the Lévy exponent is determined. Simulation studies show that a Lévy exponent of a move step frequency distribution cannot be recovered from the original Lévy flight movement when locations are subjected to spatial errors of about 10% of the maximum daily movement distance (Bradshaw *et al.*, 2007). This means that for a shark moving say 50 km per day, the spatial location error during tracking can be no greater, and ideally much less, than 5 km, otherwise a Lévy flight that is present is unlikely to be reliably detected. This is clearly a problem if light-level geolocations of shark trajectories are used because error fields are large for this method (e.g. ~ 50–250 km). Although Argos satellite geolocations (class 1–3) of fin-mounted transmitters are much more accurate (Weng *et al.*, 2005), the gaps in transmissions caused by a shark not necessarily surfacing regularly enough are a problem since move step lengths cannot be determined accurately if locations in the trajectory are missing.

The limitations location errors or gaps put on using horizontal movement data from pelagic sharks has precluded, so far, their use in testing for Lévy flights. Recent progress however has been made in analysing vertical movements for Lévy behaviour (Sims *et al.*, 2008). Here, the change in consecutive depth measurements recorded by shark-attached electronic tags form a time-series of move steps suitable for analysis of macroscopic patterns across the long temporal scale (Fig. 13.4). A time series of consecutive vertical steps is reminiscent of a Lévy walk rather than a Lévy flight, since vertical move steps are determined across equal time intervals rather than between turns (turning points in a Lévy walk form a Lévy flight; see Shlesinger *et al.*, 1993). Using this approach it was demonstrated that move step frequency distributions of basking shark, bigeye tuna and Atlantic cod (*Gadus morhua*) tracked on their foraging grounds were consistent with Lévy-like behaviour (Sims *et al.*, 2008) (Fig. 13.4). A modification in the time-series was needed for air breathers such as leatherback turtles (*Dermochelys coriacea*) and Magellanic penguins (*Spheniscus magellanicus*), but they too showed move step patterns consistent with Lévy motion. Interestingly, the Lévy exponents for these five species were close to the theoretically optimal $\mu \cong 2$ exponent. Analysis of prey abundance time series and a predator-prey computer simulation was also undertaken and suggested that marine vertebrates in stochastic environments necessitating probabilistic searching may derive benefits from adapting movements described by Lévy processes (Sims *et al.*, 2008).

These results indicate that archival tag-derived data from pelagic species, including sharks, may be particularly amenable to movement analysis using the techniques described in various recent studies (Bartumeus *et al.*, 2003; Shepard *et al.*, 2006; Clauset *et al.*, 2007; Sims *et al.*, 2007; Sims *et al.*, 2008). Perhaps more importantly though, the recent investigation (Sims *et al.*, 2008) showed that in the absence of more direct information (experiments,

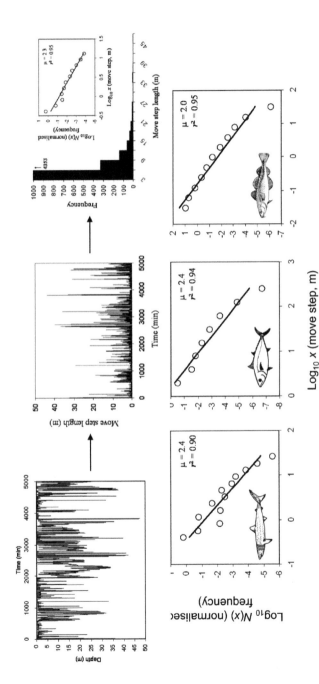

Fig. 13.4 Lévy-like scaling law among diverse marine vertebrates. (a) Movement time series recorded by electronic tags are analysed to determine the power law (Lévy) exponent to the heavy-tailed distribution. First panel: time series of swimming depths of a 4.5-m long basking shark (*Cetorhinus maximus*). Second panel: The vertical move (dive) steps ($n = 5,000$) for the same shark and time period showing an intermittent structure of longer steps. Third panel: The move step-length frequency distribution for the same data with, inset, the normalised log-log plot of move step frequency versus move step length giving an exponent (μ) within ideal Lévy limits ($\mu = 2.3$). (b) Normalised log-log plots of the move-step frequency distributions for (first panel) sub-adult and adult basking shark ($n = 503,447$ move steps), (second panel) bigeye tuna (*Thunnus obesus*) ($n = 222,282$ steps), (third panel) Atlantic cod (*Gadus morhua*) ($n = 94,314$ steps). Analysis shows these Lévy-like movements across species are close to the theoretical optimal for searching ($\mu_{opt} = 2$). Adapted from Sims *et al.* (2008).

observations) this approach provides a useful and insightful starting point concerning when and where sharks might be searching, the likely efficiency of such searches, and indeed, why a particular search pattern may be adopted under a particular set of environmental conditions. Determination of the potential movement 'rules' sharks (and other animals) have evolved to utilise their environment will allow a better predictive framework to develop. Hence, Lévy processes may be useful to consider for developing more realistic models of how sharks re-distribute themselves in response to environmental changes such as fishing and ocean warming (Sims *et al.*, 2008). Key to this conceptual advance will be the practical one of improvement in accuracy and regularity of locational data during tracking of shark movements so horizontal trajectories are open to rigorous spatial analyses.

Future Perspectives

The application of cutting edge tracking and datalogging technology, together with movement analysis and simulation modelling has recently pointed to potential mechanisms underlying the complex patterns of shark behaviour. There are several other examples that could have been discussed in this chapter that are equally important and interesting, and this illustrates that the field of shark movement ecology is entering a rapid phase of formal progress. So what lies ahead for us in the study of shark swimming movements in the wild? How can new technologies be applied to further enhance our understanding of free-ranging shark behaviour? Of central importance will be clear identification and testing of behavioural ecology hypotheses combining advanced movement analysis with simulations and modelling to better understand habitat preferences and hence re-distribution patterns. Patterns and dynamics of sexual segregation in sharks will be particularly important to elucidate, not least because of the potential for biased fishing effects on one sex over another and its likely deleterious population implications (Sims, 2005). Linked to this is the need for the next generation of instruments that will provide data of such good quality that much deeper insights into sharks behaving in their natural environments will be possible— we will then be capable of tackling those 'why' questions routinely. Some innovative advances will be global positioning system (GPS) tags for sharks for providing more accurate locations, the availability of datalogging sensors for signalling prey size and type ingested, and which will be needed to undertake natural experiments in shark foraging ecology. The advent of electronic "daily diaries" for sharks that reveal fine-scale data on movement, feeding events, energy expenditure and so on, over long time scales, is predicted to have a large impact in the field.

In this regard, over the last few years tests have been made of pop-up datalogging tags that instead of transmitting to satellites are capable of downloading data to mobile telephone networks. This has the advantage over satellite transmission, where message lengths are small, in that archival rather than summary data can be transmitted so long as the 'line' remains open. This year also sees the test of a new hybrid tag, this time a combination of an Argos satellite transmitter and a Fastloc GPS transmitter. This will improve the accuracy of locations of the fish by several orders of magnitude than can presently be obtained using Argos transmitters. However, fish carrying the tag must come to the surface to facilitate acquisition of the GPS satellite constellation for location fixing, which is very fast, about 50 ms (hence the 'Fastloc' name). Several recent studies show the potential of attaching a satellite transmitter to a shark's fin for example, to track it more accurately than is possible from light-level geolocation, so it seems likely that GPS tracking of large pelagic fish is just around the corner.

Tracking sharks in the wild is a new frontier in ecological science but one that has already resulted in significant advances in understanding behaviour. Development of ever smaller and more sophisticated tags to address clear science questions will ensure that this knowledge widens yet further to include more species, and will provide key data to improve the possibility for better shark management and conservation.

ACKNOWLEDGEMENTS

DWS was supported by a UK Natural Environment Research Council (NERC) MBA Research Fellowship and the NERC Oceans 2025 Strategic Research Programme Theme 6 (Science for Sustainable Marine Resources). The author thanks P. Domenici and three referees for helping to improve the chapter.

REFERENCES

Arnold, G. and H. Dewar. 2001. Electronic Tags in Marine Fisheries Research: A 30-Year Perspective. In: *Electronic Tagging and Tracking in Marine Fisheries*, J.R. Sibert and J.L. Nielsen (Eds.). Kluwer, Academic Publishers, Dordrecht, Netherlands, pp. 7–64.

Atkinson, R.P.D., C.J. Rhodes, D.W. Macdonald and R.M. Anderson. 2002. Scale-free dynamics in the movement pattern of jackals. *Oikos* 98: 134–140.

Baum, J.K., R.A. Myers, D.G. Kehler, B. Worm, S.J. Harley and P.A. Doherty. 2003. Collapse and conservation of shark populations in the northwest Atlantic. *Science* 299: 389–392.

Bartumeus, F. 2007. Lévy processes in animal movement: An evolutionary hypothesis. *Fractals* 15: 151–162.

Bartumeus, F., M.G.E. da Luz, G.M. Viswanathan and J. Catalan. 2005. Animal search strategies: a quantitative random-walk analysis. *Ecology* 86: 3078–3087.

Bartumeus, F., F. Peters, S. Pueyo, C. Marrasé and J. Catalan. 2003. Helical Lévy walks: adjusting search statistics to resource availability in microzooplankton. *Proceedings of the National Academy of Sciences of the United States of America.* 100: 12771–12775.

Benhamou, S. 2007. How many animals really do the Lévy walk? *Ecology* 88: 1962–1969.

Bertrand, A., E. Josse and J. Massé. 1999. *In situ* acoustic target-strength measurement of bigeye (*Thunnus obesus*) and yellowfin tuna (*Thunnus albacares*) by coupling split-beam echosounder observations and sonic tracking. *Journal of Marine Science* 56: 51–60.

Block, B.A., H. Dewar, C. Farwell and E.D. Prince. 1998. A new satellite technology for tracking the movements of Atlantic bluefin tuna. *Proceedings of the National Academy of Sciences of the United States of America.* 95: 9384–9389.

Block, B.A., H. Dewar, S.B. Blackwell, T.D. Williams, E.D. Prince, C.J. Farwell, A. Boustany, S.L.H. Teo, A. Seitz, A. Walli, D. Fudge. 2001. Migratory movements, depth preferences, and thermal biology of Atlantic bluefin tuna. *Science* 293: 1310–1314.

Block, B.A., S.L.H. Teo, A. Walli, A. Boustany, M.J.W. Stokesbury, C.J. Farwell, K.C. Weng, H. Dewar, T.D. Williams. 2005. Electronic tagging and population structure of Atlantic bluefin tuna. *Nature* (London) 434: 1121–1127.

Bonfil, R., M. Meÿer, M.C. Scholl, R. Johnson, S. O'Brien, H. Oosthuizen, S. Swanson, D. Kotze and M. Paterson. 2005. Transoceanic migration, spatial dynamics, and population linkages of white sharks. *Science* 310: 100–103.

Boustany, A., S. Davis, P. Pyle, S.D. Anderson, B. Le Boeuf and B.A. Block. 2002. Satellite tagging: Expanded niche for white sharks. *Nature* (London) 35–36.

Bradshaw, C.J.A., D.W. Sims and G.C. Hays. 2007. Measurement error causes scale-dependent threshold erosion of biological signals extracted from animal movement data. *Ecological Applications* 17: 628–38.

Brockmann, D, L. Hufnagel and T. Geisel. 2006. The scaling laws of human travel. *Nature* (London) 439: 462–465.

Bruce, B.D., J.D. Stevens and H. Malcolm. 2006. Movements and swimming behaviour of white sharks (*Carcharodon carcharias*) in Australian waters. *Marine Biology* 150: 161–172.

Burgess, G.H., L.R. Beerkircher, G.M. Cailliet, J.K. Carlson, E. Cortes, K.J. Goldman, R.D. Grubbs, J.A. Musick, M.K. Musyl and C.A. Simpfendorfer. 2005. Is the collapse of shark populations in the Northwest Atlantic Ocean and Gulf of Mexico real? *Fisheries* 30: 19–26.

Carey, F.G. 1992. Through the thermocline and back again: Heat regulation in big fish. *Oceanus* 35: 79–85.

Carey, F.G. and J.V. Scharold. 1990. Movements of blue sharks (*Prionace glauca*) in depth and course. *Marine Biology* 106: 329–342.

Carey, F.G., J.W. Kanwisher, O. Brazier, G. Gabrielson, J.G. Casey and Jr, H.L. Pratt. 1982. Temperature and activities of a white shark *Carcharodon carcharias. Copeia* 1982: 254–260.

Clauset, A., C.R. Shalizi and M.E.J. Newman. 2007. Power-law distributions in empirical data. E-Print, arXiv: 0706.1062v1.

Cotton, P.A., D.W. Sims, S. Fanshawe and M. Chadwick. 2005. The effects of climate variability on zooplankton and basking shark relative abundance off southwest Britain. *Fisheries Oceanography* 14: 151–155.

Eckert, S.A. and B.S. Stewart. 2001. Telemetry and satellite tracking of whale sharks, *Rhincodon typus,* in the Sea of Cortez, Mexico, and the north Pacific Ocean. *Environmental Biology of Fishes* 60: 299–308.

Eckert, S.A., L.L. Dolar, G.L. Kooyman, W. Perrin and R.A. Rahman. 2002. Movements of whale sharks (*Rhincodon typus*) in South-east Asian waters as determined by satellite telemetry. *Journal of the Zoological Society of London* 257: 111–115.

Edwards, A.E., R.A. Phillips, N.W. Watkins, M.P. Freeman, E.J. Murphy, V. Afanasyev, S.V. Buldyrev, M.G.E. da Luz, E.P. Raposo, H.E. Stanley, G.M. Viswanathan. 2007. Revisiting Levy flight search patterns of wandering albatrosses, bumblebees and deer. *Nature* 449: 1044–1048.

Gonzalez, M.C., C.A. Hidalgo and A.-L. Barabasi. 2008. Understanding individual human mobility patterns. *Nature* (London) 453: 779–782.

Graham, R.T., C.M. Roberts and J.C.R. Smart. 2006. Diving behaviour of whale sharks in relation to a predictable food pulse. *Journal of the Royal Society Interface* 3: 109–116.

Guénette, S., T.J. Pitcher and C.J. Walters. 2000. The potential of marine reserves for the management of northern cod in Newfoundland. *Bulletin of Marine Science* 66: 831–52.

Gunn, J.S., J.D. Stevens, T.L.O. Davis and B.M. Norman. 1999. Observations on the short-term movements and behaviour of whale sharks (*Rhincodon typus*) at Ningaloo Reef, Western Australia. *Marine Biology* 135: 553–559.

Harden Jones, F.R. 1968. *Fish Migration*. Edward Arnold, London.

Heithaus, M.R., L.M. Dill, G.L. Marshall and B. Buhleier. 2002. Habitat use and foraging behavior of tiger sharks (*Galeocerdo cuvier*) in seagrass ecosystem. *Marine Biology* 140: 237–248.

Holland, K.N., R.W. Brill, R.K.C. Chang, J.R. Sibert and D.A. Fournier. 1992. Physiological and behavioural thermoregulation in bigeye tuna (*Thunnus obesus*). *Nature* (London) 358: 410–412.

Holland, K.N., B.M. Wetherbee, C.G. Lowe and C.G. Meyer. 1999. Movements of tiger sharks (Galeocerdo cuvier) in coastal Hawaiian waters. *Marine Biology* 134: 665–673.

Jackson, J.B.C. *et al.* 2001. Historical overfishing and the recent collapse of coastal ecosystems. *Science* 293: 629–638.

Josse, E., P. Bach and L. Dagorn. 1998. Simultaneous observations of tuna movements and their prey by sonic tracking and acoustic surveys. *Hydrobiologia* 371/372: 61–69.

Kingman, A. 1996. Satellite tracking blue sharks. *Shark News* 7: 6.

Klimley, A.P. 1993. Highly directional swimming by scalloped hammerhead sharks, Sphyrna lewini, and subsurface irradiance, temperature, bathymetry, and geomagnetic field. *Marine Biology* 117: 1–22.

Klimley, A.P., S.B. Butler, D.R. Nelson and A.T. Stull. 1988. Diel movements of scalloped hammerhead sharks, Sphyrna lewini Griffith and Smith, to and from a seamount in the Gulf of California. *Journal of Fish Biology* 33: 751–761.

Klimley, A.P. and D.R. Nelson. 1984. Diel movement patterns of the scalloped hammerhead shark (*Sphyrna lewini*) in relation to El Bajo Espiritu Santo: a refuging central-position social system. *Behavioural Ecology and Sociobiology* 15: 45–54.

Kramer, D.L., R.W. Rangeley and L.J. Chapman. 1997. Habitat selection: patterns of spatial distribution from behavioural decisions. In: *Behavioural Ecology of Teleost Fishes* J.-G.J. Godin (Ed.). Oxford University Press, Oxford, pp. 37–80.

Krebs, J.R. and N.B. Davies (Eds.). 1997. *Behavioural Ecology: An Evolutionary Approach*. Blackwell Science, Oxford.

Lutcavage, M.E., R.W. Brill, G.B. Skomal, B.C. Chase and P.W. Howey. 1999. Results of pop-up satellite tagging of spawning class fish in the Gulf of Maine: do North Atlantic bluefin tuna spawn in the mid-Atlantic? *Candian Journal of Fisheries and Aquatic Science* 56: 173–177.

MacArthur, R.H. and E.R. Pianka. 1966. On optimal use of a patchy environment. *American Naturalist* 100: 603–609.

Mackus, D.L. and C.M. Boyd. 1979. Spectral analysis of zooplankton spatial heterogeneity. *Science* 204: 62–64.

Makris, N.C., P. Ratilal, D.T. Symonds, S. Jagannathan, S. Lee, R.W. Nero. 2006. Fish population and behaviour revealed by instantaneous continental shelf-scale imaging. *Science* 311: 660–663.

Matthews, L.H. 1962. The shark that hibernates. *New Scientist* 280: 756–759.

Metcalfe, J.D. and G.P. Arnold. 1997. Tracking fish with electronic tags. *Nature (London)* 387: 665–666.

Myers, R., J.K. Baum, T.D. Shepherd, S.P. Powers and C.H. Peterson. 2007. Cascading effects of the loss of apex predatory sharks from a coastal ocean. *Science* 315: 1846–1850.

Nelson, D.R. 1990. Telemetry studies of sharks: A review, with applications in resource management. In: *Elasmobranchs as Living Resources: Advances in the Biology, Ecology, Systematics, and Status of the Fisheries*, H.L. Pratt, S.H. Gruber and T. Taniuchi. (Eds.). NOAA Technical Report 90: 239–256.

Nelson, D.R., J.N. McKibben, Jr, W.R. Strong, C.G. Lowe, J.A. Sisneros, D.M. Schroeder and R.J. Lavenberg. 1997. An acoustic tracking of a megamouth shark, Megachasma pelagios: a crepuscular vertical migrator. *Environmental Biology of Fishes* 49: 389–399.

Nielsen, A., K.A. Bigelow, M.K. Musyl and J.R. Sibert. 2006. Improving light-based geolocation by including sea surface temperature *Fisheries Oceanography* 15: 314–325.

Pauly, D., V. Christensen, S. Guenette, T.J. Pitcher, U.R. Sumaila, C.J. Walters, R. Watson, D. Zeller. 2002. Towards sustainability in world fisheries. *Nature* 418: 689–695.

Priede, I.G. 1984. A basking shark (*Cetorhinus maximus*) tracked by satellite together with simultaneous remote-sensing. *Fisheries Research* 2: 201–216.

Priede, I.G. 1992. Wildlife telemetry: An introduction. In: *Wildlife Telemetry: Remote Monitoring and Tracking of Animals,* I.G. Priede and S.M. Swift (Eds.). Ellis Horwood, Chichester, pp. 3–25.

Ramos-Fernández, G., J.L. Mateos, O. Miramontes, G. Cocho, H. Larralde and B. Ayala-Orozco. 2004. Lévy walk patterns in foraging movements of spider monkeys (*Ateles geoffroyi*). *Behavioural Ecology and Sociobiology* 55: 223–230.

Reynolds, A.M., A.D. Smith, R. Menzel, U. Greggers, D.R. Reynolds and J.R. Riley. 2007. Displaced honey bees perform optimal scale-free search flights. *Ecology* 88: 1955–1961.

Richardson, A.J., A.W. Walne, A.W.G. John, T.D. Jonas, J.A. Lindley, D.W. Sims, D. Stevens and M.J. Witt. 2006. Using continuous plankton recorder data. *Progress in Oceanography* 68: 27–74.

Schaefer, K.M. and D.W. Fuller. 2002. Movements, behavior, and habitat selection of bigeye tuna (*Thunnus obesus*) in the eastern equatorial Pacific, ascertained through archival tags. *Fisheries Bulletin* 100: 765–788.

Schuster, F.L. and M. Levandowsky. 1996. Chemosensory responses of *Acanthamoeba castellani*: Visual analysis of random movement and responses to chemical signals. *Journal of Eukaryotic Microbiology* 43: 150–158.

Shepard, E.L.C., M.Z. Ahmed, E.J. Southall, M.J. Witt, J.D. Metcalfe and D.W. Sims. 2006. Diel and tidal rhythms in diving behaviour of pelagic sharks identified by signal processing of archival tagging data. *Marine Ecology Progress Series* 328: 205–213.

Shlesinger, M.F., G.M. Zaslavsky and J. Klafter. Strange kinetics. *Nature* (London) 363: 31–37 (1993).

Sibert, J., J. Hampton, P. Kleiber and M. Maunder. 2006. Biomass, size, and trophic status of top predators in the Pacific Ocean. *Science* 314: 1773–1776.

Sims, D.W. 1996. The effect of body size on the standard metabolic rate of lesser spotted dogfish, *Scyliorhinus caniculа. Journal of Fish Biology* 48: 542–544.

Sims, D.W. 1999. Threshold foraging behaviour of basking sharks on zooplankton: life on an energetic knife edge? *Proceedings of the Royal Society* B266: 1437–1443.

Sims, D.W. 2003. Tractable models for testing theories about natural strategies: Foraging behaviour and habitat selection of free-ranging sharks. *Journal of Fish Biology* 63 (Supplement A): 53–73.

Sims, D.W. 2005. Differences in habitat selection and reproductive strategies of male and female sharks. In: *Sexual Segregation in Vertebrates: Ecology of the Two Sexes* K. E. Ruckstuhl and P. Neuhaus (Eds.). Cambridge University Press, Cambridge, pp. 127–147.

Sims, D.W. and S.J. Davies. 1994. Does specific dynamic action (SDA) regulate return of appetite in the lesser spotted dogfish, *Scyliorhinus caniculа? Journal of Fish Biology* 45: 341–348.

Sims, D.W. and V.A. Quayle. 1998. Selective foraging behaviour of basking sharks on zooplankton in a small-scale front. *Nature* 393: 460–464.

Sims, D.W., A.M. Fox and D.A. Merrett. 1997. Basking shark occurrence off south-west England in relation to zooplankton abundance. *Journal of Fish Biology* 51: 436–440.

Sims, D.W., J.P. Nash and D. Morritt. 2001. Movements and activity of male and female dogfish in a tidal sea lough: alternative behavioural strategies and apparent sexual segregation. *Marine Biology* 139: 1165–1175.

Sims, D.W., E.J. Southall, A.J. Richardson, P.C. Reid and J.D. Metcalfe. 2003. Seasonal movements and behaviour of basking sharks from archival tagging: no evidence of winter hibernation. *Marine Ecology Progress Series* 248: 187–196.

Sims, D.W., E.J. Southall, G.A. Tarling and J.D. Metcalfe. 2005. Habitat-specific normal and reverse diel vertical migration in the plankton-feeding basking shark. *Journal of Animal Ecology* 74: 755–761.

Sims, D.W., M.J. Witt, A.J. Richardson, E.J. Southall and J.D. Metcalfe. 2006. Encounter success of free-ranging marine predator movements across a dynamic prey landscape. *Proceedings of the Royal Society* B273: 1195–1201.

Sims, D.W., V.J. Wearmouth, E.J. Southall, J. Hill, P. Moore, K. Rawlinson, N. Hutchinson, G.C. Budd, D. Righton, J.D. Metcalfe, J.P. Nash, D. Morritt. 2006. Hunt warm, rest cool: bioenergetic strategy underlying diel vertical migration in a benthic shark. *Journal of Animal Ecology* 75: 176–190.

Sims, D.W., D. Righton and J.W. Pitchford. 2007. Minimizing errors in identifying Lévy flight behaviour of organisms. *Journal of Animal Ecology* 76: 222–229

Sims, D.W., E.J. Southall, N.E. Humphries, G.C. Hays, C.J.A. Bradshaw, J.W. Pitchford, A. James, M.Z. Ahmed, A.S. Brierley, M.A. Hindell, D. Morritt, M.K. Musyl, D. Righton, E.L.C. Shepard, V.J. Wearmouth, R.P. Wilson, M.J. Witt, J.D. Metcalfe. 2008. Scaling laws of marine predator search behaviour. *Nature* 451: 1098–1102.

Skomal, G.B., G. Wood and N. Caloyianis. 2004. Archival tagging of a basking shark, Cetorhinus maximus, in the western North Atlantic. *Journal of the Marine Biological Association of the United Kingdom* 84: 795–799.

Stephens, D.W. and J.R. Krebs. 1986. *Foraging Theory*. Princeton University Press, Princeton.

Taillade, M. 1992. Animal tracking by satellite. In: *Wildlife Telemetry: Remote Monitoring and Tracking of Animals*, I.G. Priede and S.M. Swift (Eds.). Ellis Horwood, Chichester, pp. 149–160.

Teo, S.L.H., A. Boustany, S. Blackwell, A. Walli, K.C. Weng and B.A. Block. 2004. Validation of geolocation estimates based on light level and sea surface temperature from electronic tags. *Marine Ecology Progress Series* 283: 81–98.

Weng, K.C., P.C. Castilho, J.M. Morrissette, A.M. Ladeira-Fernandez, D.B. holts, R.J. Schallert, K.J. Goldman, B.A. Block. 2005. Satellite tagging and cardiac physiology reveal niche expansion in salmon sharks. *Science* 310: 104–106.

Satellite tagging and cardiac physiology reveal niche expansion in salmon sharks. *Science* 310: 104–106.

West, G.J. and J.D. Stevens. 2001. Archival tagging of school shark, Galeorhinus galeus, in Australia: initial results. *Environmental Biology of Fishes* 60: 283–298.

Wilson, R.P., E.L.C. Shepard and N. Liebsch. 2008. Prying into the intimate details of animal lives: use of a daily diary on animals. *Endangered Species Research* 4: 123–137.

Wilson, R.P., J.J. Ducamp, W.G. Rees, B.M. Culik and K. Niekamp. 1992. Estimation of location: Global coverage using light intensity. In: *Wildlife Telemetry: Remote Monitoring and Tracking of Animals* I.G. Priede and S.M. Swift (Eds.). Ellis Horwood, Chichester, pp. 131–134.

Wilson, S.G., B.S. Stewart, J.J. Polovina, M.G. Meekan, J.D. Stevens and B. Galuardi. 2007. Accuracy and precision of archival tag data: a multiple-tagging study conducted on a whale shark (*Rhincodon typus*) in the Indian Ocean. *Fisheries Oceanography* 16: 547–554.

Viswanathan, G.M., V. Afanasyev, S.V. Buldyrev, E.J. Murphy, P.A. Prince, and H.E. Stanley. 1996. Lévy flight search patterns of wandering albatrosses. *Nature* 381: 413–415.

Viswanathan, G.M., S.V. Buldyrev, S. Havlin, M.G.E. da Luz, E.P. Raposo, and H.E. Stanley. 1999. Optimizing the success of random searches. *Nature* 401: 911–914.

Viswanathan, G.M. V. Afanasyev, S.V. Buldyrev, S. Havlin, M.G.E. da Luz, E.P. Raposo and H.E. Stanley. 2000. Lévy flights in random searches. *Physica* A 282: 1–12.

14

The Eco-physiology of Swimming and Movement Patterns of Tunas, Billfishes, and Large Pelagic Sharks

Diego Bernal[1,*], Chugey Sepulveda[2],
Michael Musyl[3] and Richard Brill[4]

INTRODUCTION

In this chapter we will describe the species-specific swimming and movement patterns of tunas, billfishes, and large pelagic sharks derived from extensive data sets obtained using acoustic telemetry and electronic data-archiving tags. We then endeavor to interpret and explain these results based on our current understanding of species-specific physiological abilities (e.g.,

Authors' addresses: [1]Department of Biology, University of Massachusetts, Dartmouth, 285 Old Westport Road North Dartmouth, MA 02747-2300, USA. E-mail: dbernal@umassd.edu
[2]Pfleger Institute of Environmental Research, 315 N Clementine, Oceanside, CA 92054, USA. E-mail: chugey@pier.org
[3]Joint Institute for Marine and Atmospheric Research, University of Hawaii, 1125B Ala Moana Blvd. Honolulu, HI 96815, USA. E-mail: Michael.Musyl@noaa.gov
[4]Cooperative Marine Education and Research Program, Northeast Fisheries Science Center, National Marine Fisheries Service, mailing address: Virginia Institute of Marine Science, PO Box 1346, Gloucester Point, VA 23062, USA. E-mail: rbrill@vims.edu
*Corresponding author: E-mail: dbernal@umassd.edu

tolerances to hypoxia and rapid changes in ambient temperature) while integrating these ideas with other important ecological factors (e.g., prey movements and availability). Understanding the physiological ecology underlying the vertical and horizontal movement patterns of large pelagic fishes is, however, much more than a purely academic exercise. It has important implications for pelagic ecosystem models (e.g., Kitchell *et al.*, 2006), population assessments (e.g., Maunder and Punt, 2004; Bigelow and Maunder, 2007) and ultimately effective fishery management (e.g., Young *et al.*, 2006b).

Various authors have reached dramatically different conclusions as to the status of the stocks of apex-level pelagic predators (i.e., tunas, billfishes, and sharks) in the world's oceans. Some have argued that heavy commercial fishing pressure has: (1) reduced the abundances of these fishes to only 10% or less of pre-exploitation levels, (2) seriously impacted the trophic structure of the marine ecosystem, (3) brought the stocks to the brink of collapse, and (4) resulted in a situation that requires immediate (and perhaps draconian) reductions in catch and effort (Baum *et al.*, 2003; Myers and Worm, 2003; Ward and Myers, 2005a; Worm *et al.*, 2006). In contrast, other authors have stated that: (1) reductions in abundance (especially of the larger individuals) are as expected in exploited fish populations, (2) some large pelagic fish species are fully exploited whereas others are not, and (3) the situation is not dire, although effective international fishery management efforts are needed (Burgess *et al.*, 2005; Hampton *et al.*, 2005; Sibert *et al.*, 2006).

The drastic differences in the predicted state of the world's fisheries targeting large pelagic apex predators stem, in large measure, from the difficulties of estimating changes in spatio-temporal species abundance based on catch and catch-per-unit-effort data (Maunder and Punt, 2004). Estimating status of the stocks of apex pelagic predators requires correcting for changes in catch rates due to alterations in the areas commonly fished and to gear modifications (e.g., adjustment in the depths targeted by longline gear) (Bigelow *et al.*, 2002; Hinton and Maunder, 2003; Ward and Myers, 2005a; Maunder *et al.*, 2006). Exactly the best ways to accomplish this and accurately estimate abundance remains highly contentious (Goodyear, 2003; Hinton and Maunder, 2003; Ward and Myers, 2005b; Maunder *et al.*, 2006; Bigelow and Maunder 2007). We argue (as have others; Dagorn *et al.*, 2001; Kirby, 2001; Humston *et al.*, 2004) that corrections for species-specific changes in gear vulnerability must be based on a robust and sophisticated understanding of species-specific physiological abilities and tolerances, as well as cognitive and spatial learning abilities. This information is necessary to predict species-specific movement patterns and their relationship to oceanographic conditions, prey distribution, and fishery dynamics (Brill *et al.*, 2005; Sims *et al.*, 2006, 2008; Young *et al.*, 2006b; Bigelow and Maunder

2007). One of the primary intentions of this chapter is to help increase our understanding of the ecophysiological constraints and/or adaptations that govern species-specific movement patterns for pelagic fishes.

Although much has been written about the "depth preferences" of pelagic fishes (e.g., Block *et al.*, 2001; Weng and Block, 2004; Ward and Myers, 2005b; Luo *et al.*, 2006; Horodysky *et al.*, 2007), we will construct this chapter around the simple idea that tunas, billfishes, and large sharks do not (indeed they cannot) have specific or exact "depth preferences". To date, we know of no plausible sensory mechanism that would permit tunas, billfishes, and pelagic sharks to be cognizant of the specific depth they are occupying (i.e., fish have no mechanism to perceive if they are at 10m, 200m or 800m depth). Many bony fish do have stretch receptors in the walls of their gas-filled swim bladders that could potentially respond to changes in bladder volume accompanying depth change and recent work suggests that some marine organisms (e.g., crabs, *Carcinus* spp, dogfish, *Scyliorhinus canicula*) may even have hydrostatic pressure receptors (Fraser and Shelmerdine, 2002; Fraser *et al.*, 2008). These receptors, however, appear to be responsive only to acute changes in pressure (Blaxter and Tytler, 1978) or show rapid accommodation (Fraser and Shelmerdine, 2002; Fraser *et al.*, 2008), and as far as we can determine there is no evidence that these stretch receptors can or do provide bony fishes with a mechanism to determine absolute depth. In contrast to bony fishes, the lack of a swim bladder in sharks makes it impossible for them to use any changes in gas volume to detect variations (both slow and acute) in pressure (i.e., depth), although recent work on the vestibular hair cells in the labyrinth of dogfish suggests a potential mechanism to sense hydrostatic cues (Fraser and Shelmerdine, 2002). Thus, some species may have a sense of a change in pressure but lack the capacity to determine an absolute sense of exact pressure. We can think of no way that depth *per se* could be ecologically relevant to large pelagic fishes that inhabit an environment with continually changing oceanographic conditions [e.g., revolving water masses that affect water clarity, light penetration, dissolved oxygen (DO) and nutrient gradients, and the depth of the thermocline, all of which potentially alter prey distribution and availability]. Rather, we argue that there are four primary environmental parameters that influence the horizontal and vertical movements of large pelagic fishes. These are hydrostatic pressure, temperature, DO, and prey availability.[1] The first three we consider to be potentially "limiting" factors and the last "permissive-attractive". In this chapter, we will discuss hydrostatic pressure and prey availability only briefly and delve further into temperature and DO.

---

[1] Salinity gradients in the marine coastal environment (especially near the outflow of major rivers) and in high-latitude, polar waters may well be sufficient to affect the movements and distributions of pelagic fishes. In the vast majority of the open ocean environment, however, the small salinity gradients of obvious relevance to physical oceanography are too small to be directly physiologically relevant to, or detectable by, pelagic fishes.

Hydrostatic pressure (or more accurately the pressure change accompanying changes in depth) can affect swim bladder volumes in those species that possess this organ. Because tunas and billfishes have closed swim bladders (i.e., they are physoclists), the time required for significant gas secretion and resorbtion (Fänge, 1983) well exceeds the time periods over which rapid voluntary vertical movements occur (e.g., Carey, 1990; Musyl et al., 2003). Changes in hydrostatic pressure and swim bladder volume could be limiting for vertical movements but we strongly suspect that species which regularly undertake extensive changes in depth maintain maximum swim bladder volumes (and presumably near neutral buoyancy) only at the top of their vertical ranges (Stensholt et al., 2002). Otherwise they risk swim bladder rupture via gas expansion during rapid ascents (Jones, 1951, 1952). We likewise argue that the other direct effect of hydrostatic pressure, its influence on enzyme conformation and therefore biochemical function, can also be probably discounted as a factor limiting the vertical movements and depth distributions of large pelagic fishes. Significant pressure effects on the three dimensional conformations and functional properties of enzymes generally do not occur until pressures exceed 50 to 100 atmospheres (at approximately 500 to 1000 m depth, Hochachka and Somero, 1984). Therefore, it is unlikely that hydrostatic pressure per se will have any direct major influence on enzyme function over the depth ranges normally occupied by pelagic fishes for substantial periods of time (Arnold and Dewar, 2001; Gunn and Block, 2001).

The influences of the vertical movements of prey species on the vertical movements of large pelagic fishes (i.e., predators) are well documented (e.g., Carey, 1990; Josse et al., 1998). And in some instances, the vertical mobility of the prey species exceeds that of their predators, with the former showing greater physiological tolerances to lower water temperature and DO levels, thus clearly providing the ability to use the cold, hypoxic conditions occurring at depth as a refuge against predation (Childress, 1971). We argue, however, that the situation is more complex than it first appears in that the ability of large pelagic fishes to feed depends on both prey abundance and "catchablity", rather than on prey abundance alone. In this case we define "catchability" as the facility by which large predatory pelagic fishes can locate and capture prey. We assume that, for large pelagic predators, as in other fishes (e.g., Beauchamp et al., 1999; Liao and Chang, 2003), this is a complex function of the visual, olfactory, and acoustic environment, and species-specific sensory capabilities (Atema et al., 1980; Fritsches et al., 2003, 2005; Horodysky et al., 2007). Several of these factors are, in turn, related to the presence of cranial endothermy (see section Cranial endothermy) present in some tuna species, billfish, and lamnid sharks (Block and Carey, 1985; Block, 1986), because the ability to maintain elevated brain and eye temperatures will influence sensory capabilities and increase the overall ability to locate and catch prey (e.g., Fritsches et al., 2005). The overall ability to catch prey is, moreover, not only a function of the fishes' sensory repertoire but also the swimming capabilities involved in the search for prey and the rapid accelerations during the prey-capture event.

Open Ocean Movements

Studies on the movements of large pelagic fish are inherently difficult, costly, and labor intensive. But over the past 40 years there have been numerous successful efforts directed at understanding the movements of pelagic fishes despite the incredible logistical difficulties and technical complexities. This has been due, in part, to continuing technological advancements such as smaller and more affordable ultrasonic transmitters and data recording devices, and the linking of these systems with satellite technology (e.g., Block *et al.*, 1999; Arnold and Dewar, 2001; Dagorn *et al.*, 2001). An important factor fueling, or rather funding, these studies has also been the realization that many pelagic species represent global commodities that are exploited by several competing nations. Movement studies on pelagic fish have shown that a single pelagic species (stock) may be exploited by several nations over the life of the fish (e.g., Lutcavage *et al.*, 1999; Sibert *et al.*, 2006). This is especially true of highly derived species that have the capacity to undertake expansive horizontal annual migrations into high latitudes and yet return to the tropics to spawn [e.g., bluefin tuna (*Thunnus thynnus*), albacore tuna (*T. alalunga*), and swordfish (*Xiphias gladius*)].

Collectively, an extensive database now exists on the vertical movements of pelagic fishes which now includes movement information for both target and bycatch species of open ocean fisheries around the world. Many of the recent studies act to strengthen and further develop the hypotheses proposed by previous workers (Carey and Teal, 1969b; Carey and Robison, 1981), while others have disclosed movement patterns for large pelagic fishes that were, until recently, unknown (e.g., Nakano *et al.*, 2003; Weng and Block, 2004; Weng *et al.*, 2005, Schaeffer *et al.*, 2007).

Identifying species-specific differences in the movement patterns of pelagic fish can most easily be accomplished by examining the vertical and thermal distributions of the fish during daylight hours (Fig. 14.1). It is at this time that certain groups reveal their physiological specialization and tolerance to reduced temperatures and oxygen levels accompanying increases in depth. At night, however, there is substantial overlap in the vertical distribution of most pelagic fishes, in that they remain within the upper uniformed-temperature surface layer during the hours of darkness (e.g., Carey and Robison, 1981; Holland *et al.*, 1990; Schaefer and Fuller, 2002; Musyl *et al.*, 2003) (Fig. 14.1). This is most likely associated with the distribution of prey species comprising the deep-sound scattering layer (DSL), but also because many pelagic species rely heavily on visual cues for prey capture (e.g., Fritsches

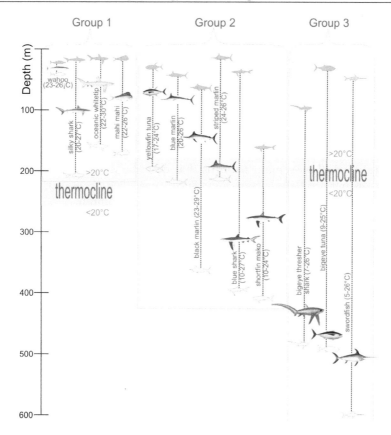

Fig. 14.1 Representative vertical movement patterns for pelagic fishes. Fish images represent the average depth (combined night and day) for each species. Gray-filled fish outlines represent the depth at which each species spent 95% of the time during the night. Open outlines represent the depth at which each species spent 95% of the time during the day. Values next to the common name show the temperature ranges encountered by each species. Orange shaded bar represents the thermocline, defined as depth range in which the water column is separated into the upper uniformed-temperature surface layer (i.e., water above 20°C) and the cooler deeper waters (i.e., below 20°C). Group 1: Fishes that spend the majority of their time in the upper uniformed-temperature surface layer. Group 2: Fishes that undertake short excursions below the thermocline. Group 3: Fishes that make frequent excursions below the thermocline. Fig. modified from Musyl *et al.* (2004). Wahoo[1] (*Acanthocybium solandri*), silky shark[2] (*Carcharhinus falciformis*), oceanic whitetip[2] (*Carcharhinus longimanus*), mahimahi[2] (*Coryphaena hippurus*), yellowfin tuna[3] (*Thunnus albacares*), blue marlin[4] (*Makaira nigricans*), black marlin[5] (*M. indica*), striped marlin[6] (*Tetrapturus audax*), blue shark[7] (*Prionace glauca*), shortfin mako shark[8] (*Isurus oxyrinchus*), bigeye thresher shark[9] (*Alopias superciliosus*), bigeye tuna[10] (*Thunnus obesus*), and swordfish[11] (*Xiphias gladius*).

[1]C. Sepulveda, S. Aalbers, D. Bernal, unpublished; [2]M. Musyl and R. Brill, unpublished; [3]Schaefer *et al.* (2007); [4] Block *et al.* (1992); [5]Gunn *et al.* (2003), Pepperell and Davies (1999); [6]Brill *et al.* (1993), Holts and Bedford (1990); [7]Carey and Scharold (1990); [8]Holts and Bedford (1993), Sepulveda *et al.* (2004); [9]Nakano *et al.* (2003), Weng and Block (2004); [10]Schaefer and Fuller (2002), Holland *et al.* (1992); [11]Carey and Robinson (1981), Carey (1990).

et al., 2003, 2005). Although we will discuss and highlight species-specific trends, we must also acknowledge that exceptions to these generalizations are common. For example, both the vertical and horizontal movements of pelagic fishes can also be influenced by other factors such as reproduction (Teo *et al.*, 2007a, b) and potential inter-specific interactions.

Although the vertical movement patterns of large pelagic fishes may be limited by discrete environmental parameters (e.g., thermocline depth, DO), we are just beginning to understand how oceanographic dynamics affect the eco-physiology of individual species. We have thus grouped the general vertical movement patterns that describe the behavior of pelagic fishes during a relatively short time scale (e.g., days to weeks) into three categories reflecting their utilization of the water column.

Group 1: Fishes that spend the majority of their time in the upper uniformed-temperature surface layer

The vertical movement patterns of this diverse group of pelagic fishes [e.g., skipjack tuna (*Katsuwonus pelamis*; Dizon *et al.* 1978), wahoo (*Acanthocybium solandri*), mahimahi (*Coryphaena hippurus*), sailfish (*Istiophorus platypterus*; Hoolihan, 2005), silky shark (*Carcharhinus falciformis*), oceanic white tip (*Carcharhinus longimanus*)] show that they remain both day and night in the warm, uniformed-temperature surface layer of the water column, where there is little or no thermal stratification (Fig. 14.1). Existing vertical movement data shows that, in general, these fishes only rarely descend deep enough to encounter the steep thermal gradients below the thermocline (Note: the depth of the thermocline is highly variable and strongly influenced by the ocean basin-specific oceanography ranging, for example, from as deep as 100–200 m off Hawaii to as shallow as 20 m off southern California). Although the absolute depth to which the each fish routinely penetrates is species-specific, we consider the unifying theme for this group to be that they rarely descend to water temperatures below 20°C and thus are most frequently distributed in the relatively warm (usually at or above 20°C), photic, and oxic layers of the ocean (Fig. 14.2).

Group 2: Fishes that undertake short excursions below the thermocline

Unlike Group 1, these fishes venture below the upper uniformed-temperature surface layer of the water columns and regularly undertake short-lived descents from the warm surface water to depths that are below the thermocline (Figs. 14.1 and 14.3). Although members of this group clearly descend to deeper depths than Group 1, the periodicity of their descents can vary from infrequent [e.g., up to a single event per day in striped marlin (*Tetrapturus audax*; Holts and Bedford, 1990; Brill *et al.*, 1993), blue marlin (*Makaira nigricans*; Block *et al.*, 1992), black

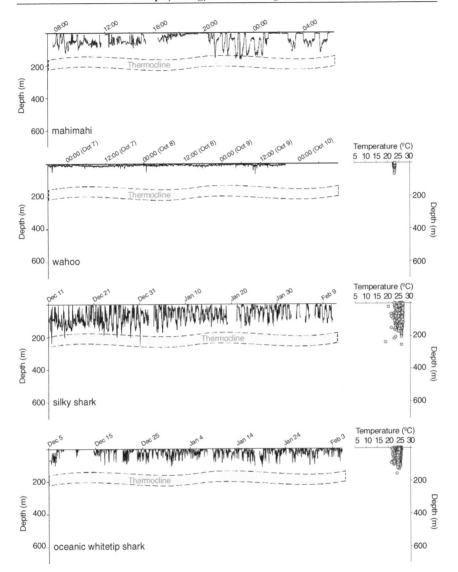

Fig. 14.2 Vertical movement patterns and temperature-depth profiles for Group 1 pelagic fishes [mahimahi[1] (*Coryphaena hippurus*), wahoo[2] (*Acanthocybium solandri*), silky shark[1] (*Carcharhinus falciformis*), and oceanic whitetip[1] (*Carcharhinus longimanus*)]. Date-marks indicate midnight of that day.

[1] M. Musyl and R. Brill, unpublished (data collected near the main Hawaiian Islands);
[2] C. Sepulveda, S. Aalbers, D. Bernal, unpublished (data collected off Baja California, Mexico).

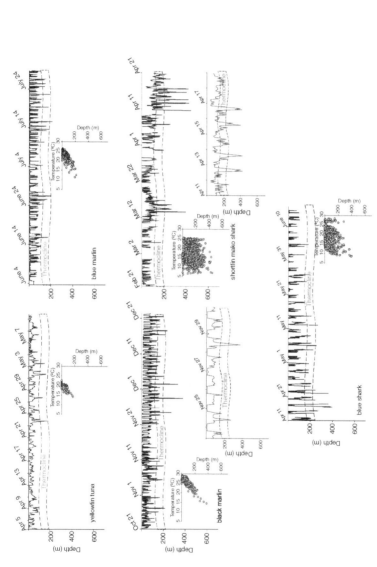

Fig. 14.3 Vertical movement patterns and temperature-depth profiles for Group 2 pelagic fishes [yellowfin tuna[1] (*Thunnus albacares*), blue marlin[1] (*Makaira nigricans*), black marlin[1] (*M. indica*), shortfin mako shark[1] (*Isurus oxyrinchus*), and blue shark[1] (*Prionace glauca*)]. Inserts in black marlin and shortfin mako show an expanded 7-day time period. Date-marks indicate midnight of that day. [1]M. Musyl and R. Brill, unpublished (data collected near the main Hawaiian Islands).

marlin (*M. indica;* Pepperell and Davis, 1999; Gunn *et al.,* 2003)] to more frequent [e.g., up to several excursions per day; yellowfin tuna (*Thunnus albacares;* Schaefer *et al.,* 2007), shortfin mako shark (*Isurus oxyrinchus;* Holts and Bedford, 1993; Sepulveda *et al.,* 2004) blue shark (*Prionace glauca;* Carey and Scharold, 1990)] (see Fig. 14.3). Due to the larger vertical range present in this group, these pelagic fishes are commonly exposed to a much wider thermal regime (i.e., they can briefly descend below 10°C) and certainly encounter both lower light and DO levels during their vertical excursions relative to the fishes of Group 1. In addition, some members of this group may even show diurnal vertical movements, in that they descend to deeper depths during the daylight hours and return to shallower depths during the nighttime (Fig. 14.3).

Group 3: Fishes that make frequent excursions below the thermocline

This group has the greatest vertical mobility and routinely penetrates colder water (Fig. 14.1). It is best exemplified by the bigeye tuna (*Thunnus obesus* Holland *et al.,* 1992; Schaefer and Fuller, 2002), swordfish (Carey and Robison, 1981; Carey, 1990; Takahashi *et al.,* 2003), and bigeye thresher shark (*Alopias superciliosus;* Nakano *et al.,* 2003; Weng and Block, 2004). Relative to Group 2, these fishes exhibit distinct diurnal behavior patterns, making regular deep descents during the day and remaining in the upper uniformed-temperature surface layer at night (Fig. 14.4). This group has the widest thermal range and routinely descends to ~500–800m, where water temperatures are below 5°C and DO levels approach 1 mgO_2 l^{-1} (Brill *et al.,* 2005). In addition, this group experiences continuous low light levels because they generally remain deep during daylight and occupy the upper water column at night. Other species that belong in this group are also capable of penetrating cold water during their extensive migrations to high latitudes [e.g., Atlantic bluefin tuna (Block *et al.,* 2001), porbeagle shark (*Lamna nasus;* Campana *et al.,* 2002), and salmon shark (*L. ditropis;* Weng *et al.,* 2005)].

We now will describe several morphological adaptations and physiological characteristics in tunas, billfishes, and pelagic sharks hypothesized to increase swimming performance, followed by a synopsis of the physiological specializations that allow the fishes in Groups 2 and 3 to routinely penetrate the colder water below the thermocline, which, in turn, effectively permits them to exploit a larger range of depths in search of prey.

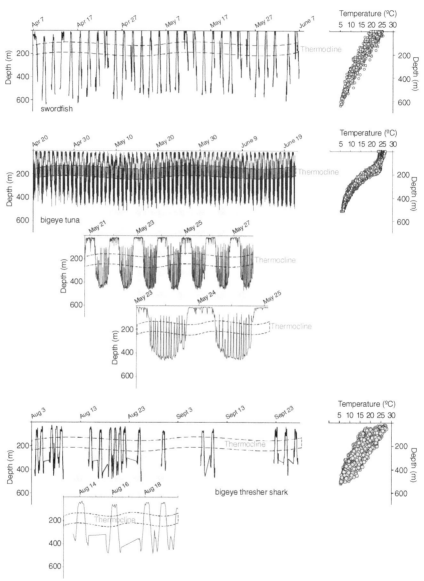

Fig. 14.4   Vertical movement patterns and temperature-depth profiles for Group 3 pelagic fishes [swordfish[1] (*Xiphias gladius*), bigeye tuna[1] (*Thunnus obesus*), and bigeye thresher shark[1] (*Alopias superciliosus*)]. Inserts in bigeye tuna and bigeye thresher shark show an expanded 7-day time period with an additional insert in bigeye tuna showing a 48hr time period. Date-marks indicate midnight of that day.

[1]M. Musyl and R. Brill, unpublished (data collected near the main Hawaiian Islands).

Physiology of Swimming

Body form, swimming muscle morphology, and swimming mode

Tunas, billfishes, and lamnid sharks rely on continuous forward motion to generate hydrodynamic lift to offset their negative buoyancy (Magnuson, 1973, 1978; Magnuson and Weininger, 1978; Alexander, 1993) and for ram ventilation (Roberts, 1978). These fishes have also evolved a suite of characters that act to increase body streamlining and decrease drag. Relative to other fishes that have the bulk of the locomotor muscles distributed more evenly along the length of the body [with the highest proportion of aerobic red muscle (RM) between 60 and 75% of fork length; Greer-Walker and Pull, 1975; Graham *et al.*, 1983], a small, but taxonomically diverse group of pelagic fishes [tunas, swordfish, lamnid sharks, and the common thresher shark (*Alopias vulpinus*)] possess a different locomotor muscle morphology in which the highest proportion of RM is located in a more anterior (from 40 to 55% of fork length) and in a medial body position (i.e., closer to the backbone; Fig. 14.5) (Graham *et al.*, 1983; Carey *et al.*, 1985; Bernal *et al.*, 2003; Sepulveda *et al.*, 2005). In all these fishes (with the exception of the common thresher shark) the anterior position of the RM allows the posterior portion of the body to taper (Fig. 14.5) into a narrow and laterally keeled caudal peduncle thereby increasing body streamlining. The caudal keels also act to reduce drag by decreasing lateral resistance during tail beats and potentially reducing macroturbulence in the flow of water across the rigid, high-aspect-ratio, nearly symmetrical caudal fin (Magnuson, 1973, 1978; Webb, 1998).

What are the selective advantages provided by this anterior-medial RM position? In tunas and lamnid sharks the unique anterior-medial RM position appears to provide a biomechanical performance advantage at sustained aerobic swimming speeds (Graham *et al.*, 1983; Westneat *et al.*, 1993; Block and Finnerty, 1994; Altringham and Shadwick, 2001; Katz, 2002; Donley *et al.*, 2004) by allowing strain in the medial RM to be uncoupled from local body bending. That is, the wave of midline curvature travels along the body in advance of the wave of muscle shortening. Therefore, when the bulk of the RM contracts in the anterior sections of the body, force is transmitted to the posterior sections without deforming the body at the site of contraction. The result is a more stiff-bodied swimming mode. In order to achieve this, a direct musculotendinous linkage between the anterior RM and the caudal fin is required, and this has been demonstrated to be present in tunas (Wainwright, 1983; Koval and Butuzov, 1986; Westneat *et al.*, 1993; Knower *et al.*, 1999; Shadwick *et al.*, 1999; Katz *et al.*, 2000), lamnid sharks (Bernal *et al.*, 2001b; Gemballa *et al.*, 2006), and recently also in swordfish and the common thresher shark (S. Gemballa, D. Bernal, C. Sepulveda, unpublished). Although a

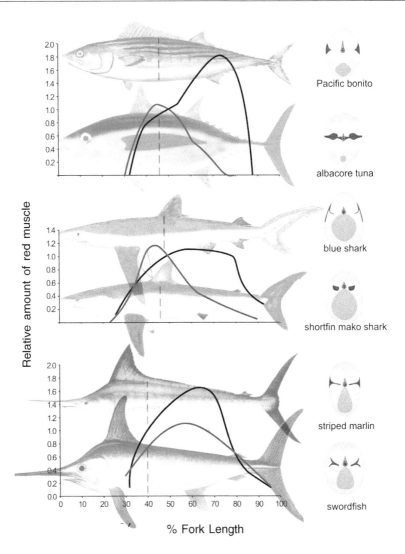

Fig. 14.5 Representative red muscle longitudinal distribution and transverse location in pelagic fishes. Black lines show the more posterior distribution of red muscle in Pacific bonito[1] (*Sarda chiliensis*), blue shark[2] (*Prionace glauca*), and striped marlin[3] (*Tetrapturus audax*). Red lines show the more anterior distribution of red muscle in albacore tuna[1] (*Thunnus alalunga*), shortfin mako shark[2] (*Isurus oxyrinchus*), and swordfish[4] (*Xiphias gladius*). The relative amounts of red muscle in the different positions along the body are expressed as a proportion of the red muscle cross-sectional area equal to 1 at 50% fork length (except for billfish where it is at 50% of lower jaw to fork length). Transverse sections images of each species show the representative location of red muscle at the position of the hatched lines across fish.

[1]Modified from Graham *et al.* (1983); [2]Modified from Bernal *et al.* (2003); [3]C. Sepulveda and D. Bernal, unpublished; [4]D. Bernal, C. Sepulveda, and S. Gemballa, unpublished.

stiff-bodied swimming mode is characteristic of these fast-cruising pelagic fishes, it may compromise their maneuverability and swimming performance (Blake *et al.*, 1995; Altringham and Shadwick, 2001; Blake, 2004).

In addition, the medially positioned RM in tunas and lamnids sharks is separated from the rest of the myotome by connective tissue. This system provides a loose connection between the RM and surrounding anaerobic, white muscle (WM) fibers (which are not recruited during slow speed swimming) and allows the shearing of these two muscle groups during sustained swimming (Shadwick *et al.*, 1999; Bernal *et al.*, 2001b; Donley *et al.*, 2004). Thus, in tunas and lamnid sharks the contraction of anterior-medial RM causes body deformation to more posterior segments, rather than locally, and confirms the hypothesis that these fishes produce thrust primarily by caudal oscillation ("thunniform" swiming mode) as opposed to body undulations (Fierstine and Walters, 1968; Lighthill, 1970; Magnuson, 1970; Westneat *et al.*, 1993; Dewar and Graham, 1994b; Knower, 1998; Webb, 1998; Shadwick *et al.*, 1999). There has been no work on the swimming kinematics of swordfish and other billfishes, as these species have never been maintained in captivity. The RM position and musculotendinous connections in the former (see Fig. 14.5), however, resemble the pattern present in thunniform swimmers, whereas other billfishes have RM that is both lateral (i.e., immediately subcutaneous) and more uniformly distributed along the length of the body (Carey, 1990; C. Sepulveda and D. Bernal, unpublished). The latter is a situation much like that of most fishes that utilize alternative modes of body-caudal fin propulsion (carangiform, subcarangiform, or anguilliform modes, Webb, 1998).

Vascular modifications arising from the medial RM position and regional RM endothermy

Most teleosts and sharks have a central circulation as the principal conduit for systemic transport of blood to and from the RM (and other tissues) (Harder, 1975). In bluefin and bigeye tunas, all lamnid sharks, and the common thresher shark, however, the central circulatory pathway is reduced and these species rely principally on a lateral circulation for RM perfusion. The lateral circulation gives rise to afferent and efferent RM-associated vessels that act to conserve heat produced by the RM during sustained swimming (Brill *et al.*, 1994; Graham and Dickson 2001). Consequently, this counter-current heat exchange system enables tunas, lamnids, and the common thresher shark to maintain RM temperatures that are significantly above that of the ambient water (i.e., regional endothermy) [although in tunas and lamnids the degree of RM elevation varies between the more tropical and temperature distributed species (the more cold-water distributed species maintain a higher relative RM temperature; Fig. 14.6) (Carey and Teal, 1966;

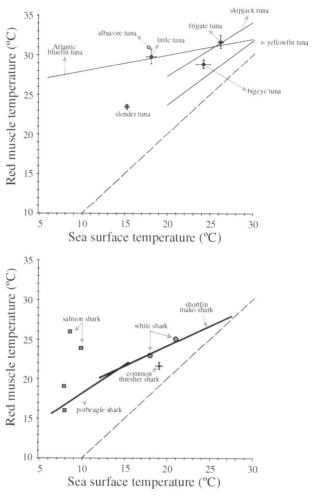

Fig. 14.6 Red muscle temperature (TRM) elevation as a function of sea surface temperature (SST) in selected tunas and sharks. Dashed line represents the line of equality (i.e., TRM=SST). Data are shown as mean ± SE for slender tuna[1] (*Allothunnus fallai*, n=30), frigate tuna[2] (*Auxis thazard*, n=7), little tunny[3] (*Euthynnus alleteratus*, n=16), bigeye tuna[3,4] (*Thunnus obesus*, n=17), and common thresher shark[5] (*Alopias vulpinus*, n=24). Individual data shown for albacore tuna[3] (*Thunnus alalunga*, n=1), white shark[6] (*Carcharodon carcharias*, n=2), and salmon shark[7] (*Lamna ditropis*, n=4). Linear regressions (i.e., TRM vs. SST) are shown for Atlantic bluefin tuna[8] (*Thunnus thynnus,* n=162; TRM = 25.84 + 0.206 SST), skipjack tuna[9] (*Katsuwonus pelamis*, n=53, TRM = 13.7 + 0.68 SST), yellowfin tuna[9] (*Thunnus albacares,* n=62, TRM = 7.47 + 0.81 SST), shortfin mako shark[6] (*Isurus oxyrinchus*, n=38, TRM = 13.8 + 0.51 SST), and porbeagle shark[10] (*Lamna nasus*, n=13, TRM = 10.8 + 0.72 SST).

[1]Sepulveda *et al.* (2008); [2]C. Sepulveda, unpublished, Schaefer (1985); [3]Linthicum and Carey (1972); [4]C. Sepulveda, unpublished; [5]Bernal and Sepulveda (2005); [6]Carey *et al.* (1985); [7]Bernal *et al.* (2001b), Rhodes and Smith (1983); [8]Graham and Dickson (2001); [9]Carey and Teal (1969b); [10]Carey and Teal (1969a).

Carey *et al.*, 1971; Bernal *et al.*, 2001b; Anderson and Goldman, 2003; Dickson and Graham, 2004)]. Neither tunas, lamnid sharks, nor the common thresher have specialized thermogenic tissues, and heat is metabolically generated by the continuous activity of the RM during swimming (Carey and Teal, 1969a, b; Linthicum and Carey, 1972; Block and Carey, 1985; Wolf *et al.*, 1988; Stevens *et al.*, 2000).

All tunas and lamnid sharks have the capacity for RM endothermy, and several members of these divergent groups have been shown to have the ability to exert physiological control over routes and rates of heat transfer presumably through control of vascular counter-current heat exchangers. Acoustic telemetry recordings of deep myotomal muscle temperature (i.e., most likely RM temperature) and water temperature in free-swimming tunas document the thermoregulatory capacity of the more derived species (i.e., Atlantic bluefin), while studies on the bigeye tuna show this species can adjust rates of heat loss and gain in response to changes in ambient temperature (Carey and Lawson, 1973; Holland *et al.*, 1992; Malte *et al.*, 2007). Laboratory studies on yellowfin tuna have also shown that, in response to changes in ambient water temperature, this species has the ability to modulate rates of heat production (i.e., aerobic swimming speed) or heat loss (by potentially altering the efficacy of vascular counter-current heat exchangers) (Dizon and Brill, 1979; Dewar *et al.*, 1994). Similarly, for lamnid sharks acoustic telemetry and laboratory swimming studies have shown that they have the capacity to change rates of heat loss and heat gain and thus maintain a relatively stable elevated RM operating temperature (Goldman, 1997; Bernal *et al.*, 2001a; Goldman *et al.*, 2004; Sepulveda *et al.*, 2004).

What is the biological and ecological significance of RM endothermy?

Rapid and repeated vertical movements (see Figs. 14.3 and 14.4) and movements across thermal fronts are a common trait of pelagic fishes. Because large changes in ambient water temperature are regularly encountered, it is not surprising that tunas and lamnid sharks have evolved the ability to maintain a more thermally stable operating environment for their aerobic locomotor muscles (Carey *et al.*, 1971; Carey, 1982; Holland *et al.*, 1992; Block *et al.*, 1998; Brill *et al.*, 1999b). Indeed, studies on the contractile mechanics of isolated muscle fibers in lamnid sharks have shown a dramatic detrimental effect on the RM if it cools slightly below its *in vivo* operating temperature. For example, the RM of the salmon shark, which inhabits water generally cooler than 10°C (but frequents water as cold as 2°C; Weng *et al.*, 2005), becomes ineffectual if exposed to a temperature below 20°C (Bernal *et al.*, 2005). Although not as extreme, similar muscle performance deterioration has been documented for shortfin mako shark RM fibers if cooled to below

15°C (Donley *et al.*, 2007), even though this species repeatedly descends below the thermocline to water temperatures cooler than 13°C (Holts and Bedford, 1993; Sepulveda *et al.*, 2004). By contrast, these thermal effects are not as prominent for other sharks not capable of RM endothermy (e.g., leopard shark, *Triakis semifasciata*) in which muscle function is still possible, albeit much slower (i.e., lower cycle frequencies) at cooler temperatures (below 15°C) (Donley *et al.*, 2007). Like lamnid sharks, tunas show a similar, but not as marked, effect of temperature on RM contractile mechanics (Johnston and Brill, 1984; Altringham and Block, 1997; Syme and Shadwick, 2002). Thus, for both lamnid sharks and tunas, RM endothermy provides an operating temperature that safeguards power output at high cycle frequencies when penetrating cooler waters. In addition, it has also been hypothesized that maintenance of elevated muscle temperatures allows tunas to swim faster and to recover faster from anaerobic activity (Stevens and Carey, 1981; Arthur *et al.*, 1992; Dickson, 1995, 1996; Brill, 1996) (see ahead Anaerobic metabolism).

There are, however, other pelagic fishes that share the vertical (i.e., temperature-depth) and horizontal (latitude) distribution of tunas and lamnid sharks that do not have the capacity for RM endothermy. For example, both bigeye thresher and blue sharks inhabit similar water temperatures (Figs. 14.3 and 14.4) and spend extensive periods below the thermocline (the former has been reported to have cranial endothermy, see section Cranial endothermy) (Carey and Scharold, 1990; Nakano *et al.*, 2003; Weng and Block, 2004). Vertical movement data for bigeye thresher, blue, and mako sharks do not reveal a large difference in the lower limit of water temperature that these sharks routinely penetrate, however, a striking difference becomes apparent in both the frequency and duration of their vertical oscillations (Figs. 14.3 and 14.4). It may be that the mako shark needs to stay within an optimal temperature range in order to maintain proper RM function and potentially benefit from an increased performance level while the other two species maintain adequate RM function at lower temperatures.

Acoustic telemetry data for a blue shark (Fig. 14.7) shows that a ~150 min incursion from the relatively warm surface waters (26°C) down to depth below the thermocline (9°C) results in a decrease in deep WM (i.e., body core) temperature from about 21 to 14°C. If the shark remained at this depth for an additional 300 min its WM (i.e., body core) temperature (extrapolated using k= -0.0051, T_e= 8.7°C; Carey and Scharold, 1990) would ultimately decrease until it essentially equaled water temperature. [Although no data were collected for RM temperature in this blue shark, the subcutaneous location of the RM would result in a more rapid temperature change relative to the deep WM.] By contrast, if a mako shark undertook the exact same vertical excursion, its physiological ability to alter rates of heat gain and heat loss (Bernal *et al.*, 2001a)

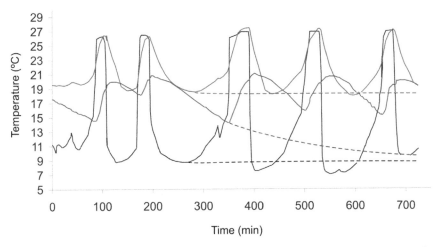

Fig. 14.7   Vertical movement profiles for a blue shark (*Prionace glauca*) tracked using ultrasonic telemetry (Carey and Scharold 1990). The black line shows temperature at depth and the blue line body temperature. The red line shows the predicted body temperature of a shortfin mako shark (*Isurus oxyrinchus*) undergoing similar changes in ambient temperature. Changes in mako shark body temperature were modeled using thermal rate coefficients and thermal equilibrium values from Bernal *et al.* (2001a) and were derived for every available data point from the blue shark vertical movement profile. The black hatched line shows a simulated descent, with the fish remaining at 8.73ºC for 300 minutes. The blue and red hatched lines represent the predicted body temperature for blue and mako sharks, respectively, resulting from the simulated descent. Notice the degree to which the mako shark body temperature remains elevated relative to that of the blue shark, with a very rapid increase during the time periods (i.e., between vertical sojourns) when the shark returns to warmer waters.

would provide it with an overall warmer RM operating temperature (Fig. 14.7). This outcome becomes even more pronounced when using long-term vertical movement data for a blue shark derived from satellite tags for which the deep WM (for blue sharks) and RM (for mako shark) temperatures are modeled using previously calculated thermal balance constants (Bernal *et al.*, 2001a) (Fig. 14.8A). Moreover, if a blue shark were to mimic the vertical movements of mako sharks (e.g., 30 min descents with 8 min periods of basking at the surface; Holts and Bedford, 1993; Sepulveda *et al.*, 2004), the body temperature of the former would decline progressively with each descent, while the mako shark would maintain a more stable and warmer RM operating temperature (Fig. 14.8B).

As first suggested by Neill *et al.* (1976), RM endothermy may, therefore, not provide an overall larger tolerance to colder surface water, but rather it may provide the ability to make longer or more frequent sojourns into cold water. The lower thermal limit for these fish may be a product of the ambient temperature at depth and the duration of the descent, which in turn reduce RM temperature and negatively affect muscle contractile mechanics. This potential thermal limit may play a role in bigeye tuna vertical movements, as

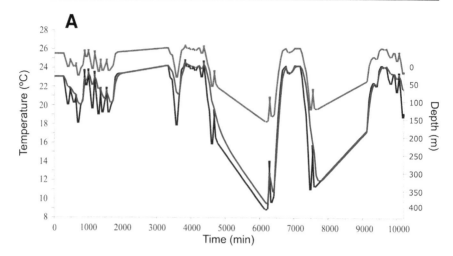

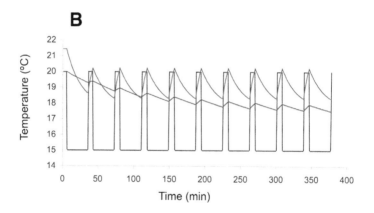

Fig. 14.8   **(A)** Vertical movement profiles for a blue shark (*Prionace glauca*) near the main Hawaiian Islands. The black line shows the temperature at depth data from a pop-up satellite archival tag carried by the shark. The blue line shows the derived body temperature for the blue shark and the red line shows the derived body temperature for a shortfin mako shark (*Isurus oxyrinchus*) following the same vertical movement pattern. Thermal rate coefficients (i.e., *k* values) used to model the changes in body temperature for both shark species were taken from Bernal *et al.* (2001a). Body temperatures were derived for every available water temperature data point. Notice the degree to which the mako shark body temperature remains elevated relative to that of the blue shark. **(B)** Body temperature changes in blue sharks (blue line) and shortfin mako sharks (red line) subjected to a series of modeled vertical movement patterns (i.e., rapid changes in ambient temperature, black line) following diving profiles observed for telemetry tracked mako shark by Holts and Bedford (1993). Thermal rate coefficients (i.e., *k* values) were taken from Bernal *et al.* (2001a).

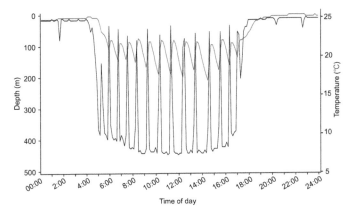

Fig. 14.9 Vertical movement patterns of a bigeye tuna (*Thunnus obesus*) near the main Hawaiian waters recorded by an archival tag implanted in the fish's dorsal musculature (Musyl *et al.*, 2003). Black line shows depth and ambient temperature and red line body temperature.

bigeye tuna descend from surface waters (26°C) to well below the thermocline (7°C) (Fig. 14.9) until the RM reaches (~15–17°C) and then return to the uniformed temperature surface layer to re-warm (Holland *et al.*, 1992; Musyl *et al.*, 2003). A similar thermal threshold is also expected in other vertically oscillating fishes with RM endothermy (e.g., mako shark, swordfish) but there are no data to show this. On the other hand, fishes (e.g., bluefin tuna, salmon shark, porbeagle shark) with RM endothermy that inhabit very cold, highly productive, subpolar waters (2–10°C) for prolonged periods (i.e., numerous months) do not experience rapid and large decreases in temperature during their vertical or horizontal movements, but rather experience cold water all the time. These species are able to maintain an almost constant RM temperature that is 20°C or more above ambient and thus a lower thermal threshold may not be reached and neither a degree of temperature decrease nor the duration at these cold temperatures may limit their muscle physiology as long there is an ample supply of metabolic heat (i.e., aerobic swimming) to maintain RM endothermy.

The cardio-respiratory system in tunas, billfish, and lamnid sharks

As described in section Open Ocean Movements, the ecology of actively swimming pelagic fishes demands continuous locomotion during long distance horizontal migrations (e.g., >50 miles day^{-1}; Lutcavage *et al.*, 2000; Schaefer and Fuller, 2002; Bonfil *et al.*, 2005) and throughout repeated vertical oscillations (>200 m; see Figs. 14.3 and 14.4). Because sustained swimming is powered by aerobic metabolism, the cardiorespiratory system needs to deliver ample supplies of oxygen and fuels to the working tissues, and at times even under hypoxic conditions.

In general, there are several differences between the cardiorespiratory systems of bony fish (i.e., billfish and tunas) and sharks (e.g., sharks hearts have a contractile conus arteriosus and lack sympathetic innervation, shark gills are medially anchored to a central septum; reviewed by Farrell and Jones, 1992; Bernal et al., 2001b). Tunas, billfishes, and lamnid sharks, nonetheless, share similar specializations that enhance oxygen (O_2) transfer across the gills, increase the quantity of O_2 transported by the blood, elevate the delivery rate of oxygenated blood to tissues, and facilitate the intracellular transport of O_2 to the mitochondria. All three rely on ram ventilation to force water into their mouth and past the gills. In order to enhance oxygen uptake at the gills these groups have a larger gill surface area and shorter blood-water diffusion distances than that of most other fishes. In additions, both tuna and billfish gills have the secondary lamellae fused to increase gill structural integrity and potentially improve water flow across the lamellae (Muir and Kendall, 1968; Muir and Hughes, 1969; Emery and Szczepanski, 1986; Oikawa and Kanda, 1997; Wegner et al., 2006).

When compared to other fishes, the hearts of tunas, billfishes, and lamnid sharks also have larger ventricles than other fishes, with thicker more muscular ventricular walls, and (in some species) more than twice the proportion of compact myocardium in the ventricle (features hypothesized to increase cardiac force, Brill and Bushnell 1991a; Farrell 1991; Bernal et al., 2001b; Brill and Bushnell, 2001) and have higher arterial blood pressures (Bushnell and Brill, 1992; Korsmeyer et al., 1997a, b; Lai et al., 1997). Because of the thick ventricular walls, the hearts of these three groups are necessarily perfused by an elaborate coronary (hypobranchial) blood supply providing myocardial cells with sufficient substrates to sustain elevated aerobic enzyme activities (reviewed by Dickson and Graham, 2004).

In most fishes, changes in cardiac output during times of increased aerobic demands (e.g., bouts of elevated sustained swimming speeds) are met by increases in both heart rate and stroke volume (Farrell, 1991; Lai et al., 1997). In tunas, and to a lesser degree in lamnid sharks, however, frequency modulation of cardiac output appears to be the major mechanism providing an elevated blood flow during times of increased aerobic swimming (Emery et al., 1985; Korsmeyer et al., 1997a, b; Lai et al., 1997).

Because sustained swimming activity requires a continuous supply of O_2 and metabolic fuels to power aerobic muscle metabolism, it would be expected that these fishes have a suite of characteristics that enhance the delivery and utilization of oxygen. The RM of tunas and lamnid sharks has a complex capillary-muscle fiber geometry that significantly increases the ratio of capillary surface area to muscle fiber volume (Mathieu-Costello *et al.*, 1992; Bernal *et al.*, 2003). RM capillary density (the number of capillaries surrounding a given fiber) can be up to four times higher in tunas than in other scombrids (e.g., mackerels, *Scomber* spp.) and up to twice higher in lamnids relative to other sharks (reviewed by Dickson, 1988; Mathieu-Costello *et al.*, 1996). It seems reasonable to conclude that this blood-fiber geometry increases oxygen delivery to the working tissues. Although the mitochondrial volume of the aerobic swimming muscle appears to be similar in most fishes (18–35%), tuna and lamnid sharks have greater mitochondrial cristae packing densities (reviewed by Block, 1991; Moyes *et al.*, 1992; Mathieu-Costello *et al.*, 1995; Bernal *et al.*, 2003). Compared to other fishes, tunas and lamnid shark RM also have an elevated concentration of myoglobin (Mb), which facilitates the diffusion of O_2 from the capillary blood to the mitochondria (Wittenberg, 1970 ; Wittenberg and Wittenberg, 1989; reviewed by Dickson, 1995, 1996). This may allow for a potentially greater O_2 flux to the mitochondria and is consistent with the need to maximize O_2 delivery to the RM fibers, which are warm and continually active. However, because tuna hearts are not warmer than ambient, reduced cardiac function in cool water has the potential to limit the delivery of O_2 and fuels to RM. Thus, during sojourns into cooler water, heart function may ultimately influence swimming capacity (Brill *et al.*, 1999a).

Taken together, these cardiovascular features appear to allow these pelagic fishes to supply the continuously active, aerobic locomotor muscles with sufficient O_2 and nutrients to sustain the rates of ATP production and muscle contraction required for continuous swimming.

Cranial endothermy

Cranial endothermy is the most widespread form of regional endothermy among fishes and has been documented to occur in three divergent groups of pelagic fish. All tuna, lamnid shark, and billfish genera possess the capacity to elevate cranial temperatures above ambient temperature, albeit using distinct heat production and retention mechanisms (Linthicum and Carey, 1972; Block and Carey, 1985; Block, 1986; Block and Finnerty, 1994; Dickson and Graham, 2004; Sepulveda *et al.*, 2007b). Other species also suspected of cranial endothermy include the butterfly mackerel (*Gasterochisma melampus*) and the bigeye thresher shark (Carey, 1982; Block and Finnerty, 1994; Weng and Block, 2004), although temperature data for these species are not available. The repeated convergence upon cranial endothermy suggests strong selection for this trait among pelagic fishes, as thermal effects on the central nervous system have been shown to conserve sensory and integrative functions while in the cooler and darker deep water (Linthicum and Carey, 1972; Block and Carey, 1985; Wolf *et al.*, 1988; Block, 1991; Alexander, 1998; Sepulveda *et al.*, 2007b).

Warming of the brain and eye region has been shown to enhance physiological processes such as synaptic transmission, postsynaptic integration, conduction, and, in the eye, temporal resolution (Konishi and Hickman, 1964; Friedlander *et al.*, 1976; Montgomery and Macdonald, 1990; Fritsches *et al.*, 2005; van den Burg *et al.*, 2005). Fritsches *et al.* (2005) recently showed that the swordfish eye is extremely temperature-sensitive; having a flicker fusion frequency thermal coefficient (Q_{10}) of approximately 5.1, and that warming the retina significantly improves temporal resolution. Warming the retina likely enhances the swordfish's ability to detect and capture fast moving prey at low temperatures (Block, 1986; Fritsches *et al.*, 2005). Other selective advantages associated with elevated cranial temperatures may include the buffering of the central nervous system from rapid temperature changes and minimization of temperature-induced decreases in cranial blood flow (van den Burg *et al.*, 2005).

Visceral endothermy

The capacity to elevate visceral temperatures above that of the surrounding sea water is present in all lamnid sharks, in at least five members of the most derived tuna genera (*Thunnus; T. obesus, T. alalunga, T. orientalis, T. maccoyii*, and *T. thynnus*) which inhabit the coldest waters (Gibbs and Collette, 1967; Carey and Lawson, 1973; Carey *et al.*, 1984; Fudge *et al.*, 1997; Goldman, 1997; Bernal *et al.*, 2001b; Anderson and Goldman, 2003; Sepulveda *et al.*, 2004), and is suspected in the common thresher shark (Fudge and Stevens, 1996). Although lamnid sharks and tunas differ in the structural specializations that allow for visceral heat-conservation both groups have been shown to utilize heat produced from digestion (protein catalysis) to maintain elevated gut temperatures. Telemetry records of free-swimming Atlantic bluefin tuna showed a sustained elevation of about 6°C above ambient, but when prey are ingested stomach temperature increases rapidly (within 12 to 20 h) to as much as 20°C above ambient, then steadily decreases back to the pre-feeding temperature over the next 20–30 h (Carey and Lawson, 1973; Carey *et al.*, 1984; Stevens and McLeese, 1984). The onset and offset of this temperature increase suggests that heat production results from increased aerobic metabolism during digestion and assimilation. The caecal mass is the warmest visceral organ and potentially the primary heat source (Carey *et al.*, 1984; Stevens and McLeese, 1984). Lamnid visceral temperatures range from 4 to 14°C above ambient, and unlike tunas, their stomach temperatures remain continuously elevated and appear to be independent of ambient temperature (Carey *et al.*, 1981; McCosker, 1987; Goldman, 1997; Anderson and Goldman, 2003; Goldman *et al.*, 2004; Sepulveda *et al.*, 2004). As in tunas, some of the lamnid visceral heat likely results from digestion and assimilation processes, with the spiral valve being the warmest organ and thus the likely main source of heat (Carey *et al.*, 1981, 1985; Bernal *et al.*, 2001b; Anderson and Goldman, 2003).

For both of these groups of pelagic fish, the presence of elevated visceral temperatures is a potential mechanism to enhance the rate of digestion and assimilation (Carey *et al.*, 1984; Stevens and McLeese, 1984) and may also be a significant contributor to the warming of the body core which may allow these fishes to penetrate and inhabit cool waters (reviewed by Dickson and Graham, 2004).

Metabolism

The capacity of tunas to sustain some of the world's largest commercial fisheries (landing approximately 2.3×10^9 kg annually) is in large part due to their high reproductive and growth rates. Tunas can support these energy demanding physiological processes, in spite of being apex level carnivores in an energy depauperate environment where foraging patches are widely scattered (Brill, 1996). If for no other reason than this, we argue that tunas deserve to be called "high energy demand fishes". Billfishes share many of the tunas' adaptations for life in the pelagic environment (Dickson, 1995). The same can also be said for lamind sharks, which show a remarkable degree of convergent evolution with tunas on features related to swimming performance (Bernal *et al.*, 2001b).

Aerobic metabolic rate

Metabolic studies on pelagic, obligate ram-ventilating species are inherently difficult as they require the manipulation of fragile open-ocean species and often the measurement of oxygen consumption in an un-natural environment (e.g., in a small tank or swimming tunnel respirometer; Gooding *et al.*, 1981; Graham and Laurs, 1982; Dewar and Graham, 1994a). There exists the possibility, therefore, that the values obtained by such studies may represent an overestimation of true metabolic rates because of the inability to measure metabolic rates of fishes that have been fully acclimated to the experimental apparatus (Steffensen, 1989). Blank *et al.* (2007a, b) acclimated tunas to the swim tunnel overnight and recorded significantly lower metabolic rates than in previous studies where the fish were given less time to recover (e.g., Dewar and Graham, 1994a). In contrast, employing a large 250,000 l mesocosm respirometer, Fitzgibbon *et al.* (2006) found the metabolic rates of free swimming southern bluefin tuna (*Thunnus maccoyii*) to be elevated compared to other teleosts and comparable to those of other tuna recorded by previous investigators (Gooding *et al.*, 1981; Graham and Laurs, 1982; Brill, 1987; Dewar and Graham, 1994a; Sepulveda and Dickson, 2000) employing a diverse array of techniques. Therefore, despite differences that may or may not be attributed to size effects, differences in temperature, or differences in experimental techniques, one central theme has emerged: tunas have elevated metabolic rates compared to most other teleosts (Brett 1972; Brill, 1996; Bernal *et al.*, 2001b; Korsmeyer and Dewar, 2001).

Standard metabolic rate (SMR, the metabolic rate of a fish at rest; Brill, 1987), routine metabolic rate (RMR; Fry, 1971) and maximum metabolic rate (MMR, the maximum measured metabolic; Korsmeyer and Dewar, 2001) are the metabolic estimates most commonly used to compare the swimming performance of tunas to other fishes. Tuna SMRs ($\sim$250–700 mgO$_2$kg^{-1}h^{-1}, corrected to 25°C assuming a Q_{10} = 1.65; Fitzgibbon *et al.*, 2006) are 2 to 6 times higher than those measured in other active, pelagic fishes [i.e., sockeye salmon (*Oncorhynchus nerka*); chub mackerel (*Scomber japonicus*); eastern Pacific bonito (*Sarda chiliensis*); yellowtail (*Seriola quinqeradata*); Brett and Glass, 1973; Yamamoto *et al.*, 1981; Brill, 1987; Dewar and Graham, 1994a; Sepulveda *et al.*, 2003]. The elevated SMR in tunas, documented in several independent studies on both swimming and paralyzed fish, is hypothesized to be a product of the maintenance costs associated with the cardiovascular and respiratory specializations that allow for high rates of energy turn-over (e.g., large gill surface areas, high mitochondrial densities) as well as the concomitant thermal effects of regional endothermy. However, size-matched comparisons of juvenile tunas below the minimum size necessary for regional endothermy (140-200 mm fork length; Dickson, 1994) and chub mackerel show that prior to having the capacity for temperature elevation, tunas still have SMRs that are 2–4 times higher than that of their close relatives (Sepulveda and Dickson, 2000). Although the thermal effects of regional endothermy have the potential to increase metabolic processes (e.g., RM contraction rates, digestion, synaptic transmission), the morphological and physiological mechanisms involved in heat production and retention do not necessarily reflect any additional metabolic cost, as the vascular heat exchangers merely harness aerobic heat that would otherwise be lost to the environment (species with thermogenic tissue excluded; reviewed by Block, 1986; Sepulveda *et al.*, 2007b).

Tuna MMRs are among the highest recorded for any fish species (reviewed by Brill, 1996; Korsmeyer and Dewar, 2001). Although most estimates of MMRs are from fish that are experiencing several simultaneous aerobic demands (i.e., swimming, recovery from anaerobic activity, digestion) these values, nonetheless, offer an estimate of the total aerobic capacity. Respirometry data and models predicting MMRs for skipjack and yellowfin tunas estimate them to be 2500 and 2700 mgO$_2$kg^{-1}h^{-1} (for $\sim$2 kg fish at 25°C), respectively (Brill, 1996), which are almost three times the MMRs of similarly sized salmonids (Brett and Glass, 1973; Gooding *et al.*, 1981; Dewar and Graham, 1994a; Korsmeyer *et al.*, 1996).

More biologically relevant than the absolute value of MMR is the difference between MMR and SMR. This defined as the aerobic scope (Priede, 1985) and represents the potential capacity to handle multiple simultaneous aerobic demands (e.g., continuous swimming, recovering from oxygen debt, somatic growth, digestion and assimilation) (Priede, 1985; Brill and Bushnell, 1991a; Brill, 1996; Korsmeyer *et al.*, 1996). The MMR of tunas is approximately 8–9 times the SMR (Bushnell and Jones 1994; Brill and Bushnell, 1991a, 2001), and this ratio is in the middle of the broad range of values reported for teleosts (4-15x) (Brett and Glass, 1973; Wood and Perry, 1985; Fitzgibbon *et al.* 2007; Jordon and Steffensen, 2007), rather than at its upper end. However, when corrected for differences in temperature, the SMR of tunas is 2–6 times higher than other teleosts (Brill and Bushnell 2001; Fitzgibbon *et al.* 2007). As a result, the actual increase in metabolic rate (i.e., the aerobic scope) of tunas (approximately 2500 mg $O_2kg^{-1}h^{-1}$) is approximately 2.5 to 18 times greater than the aerobic scope of other fishes (134–1050 mg $O_2kg^{-1}h^{-1}$), when data are corrected for differences in temperature and body mass (Fitzgibbon *et al.* 2007). The heightened aerobic scope in continuously swimming pelagic predators most probably reflects their physiological ecology, which revolves around the need for the capture of fast moving prey and its physiological consequences (e.g., rapid recovery from burst activity and rapid digestion) and high rates of somatic and gonadal growth (reviewed by Bushnell and Jones 1994; Brill, 1996; Korsmeyer *et al.*, 1996; Graham and Dickson, 2004; Blank *et al.*, 2007b). The inability thus far to conduct similar studies on other high energy demand pelagic fishes such as the swordfish and marlins allows us only to speculate. But given that they evolved under similar selective pressures, we hypothesize that billfishes probably also have high MMRs.

Work on lamnid sharks, another high energy demand species, has shown them to have MMRs that exceed those of other comparably sized elasmobranches (Graham *et al.*, 1990; Carlson *et al.*, 2004; Sepulveda *et al.*, 2007a). Swim tunnel studies on juvenile mako sharks have recorded an MMR of 540 mg $O_2kg^{-1}h^{-1}$ in a 6 kg fish (Sepulveda *et al.*, 2007a), which is comparable to that recorded for similar sized Pacific bluefin tuna (Blank *et al.*, 2007b), although lower than those of yellowfin and skipjack tunas (Brill, 1996). The recent work on the swimming energetics of mako sharks most likely underestimate their total aerobic capacity by not being able to truly measure their MMR, but rather measuring values that are limited by experimental conditions (e.g., maximum swim tunnel water velocities, adverse behavioral modification due to confined swimming). However, they still represent values that are much higher than those of most other teleosts and elasmobranches. The observed MMR for the mako shark is, as in the tunas, probably a result of their specialized cardiovascular and swimming muscle physiology (e.g., large gill surface area, relatively larger heart mass, increased muscle capillary density and myoglobin concentration, regional endothermy; see section: The cardio-respiratory system in tunas, billfish and lamnid sharks) supporting the hypothesized convergence in cardiovascular adaptations between tunas and lamnids (reviewed by Bernal et al., 2001b; Donley et al., 2004).

While additional studies on lamnid shark metabolic rates are needed, we predict that due to the striking convergence between lamnid sharks and tunas in features related to body morphology, swimming physiology, and continuous swimming behavior, a higher SMR, MMR, and overall aerobic metabolic scope is present in these groups relative to other cruise adapted swimming fishes. However, other tropical, pelagic species (e.g., mahimahi and marlin), share with tunas the properties of rapid somatic and gonadal growth, which may contribute to a high MMR (Brill, 1996). In contrast to teleosts, sharks generally have relatively low growth rates, mature relatively late in life, have long gestation times, and give birth to few pups (Gilbert, 1981; Compagno, 1990). In that context, lamnids are not different from other sharks (Stevens, 1983) and, unlike tropical, pelagic teleosts, the high MMR in lamnids is not reflective of a relatively fast rate of growth (Natanson et al., 2006).

Anaerobic metabolism

Coulson (1979) described anaerobic glycolysis as "The Smith and Wesson of heterotherms" referring to 19th century six-gun that advertised itself as the great equalizer. Coulson's contention was that heterotherms successfully compete with homeotherms, in spite of the latter's higher metabolic rates, as a result of heterotherms' ability to rapidly mobilize muscle energy stores anaerobically. If most fishes could be described as carrying an anaerobic "six-gun", tunas and billfishes should be considered much more heavily armed. As originally described by Hochacka et al. (1978), tuna white muscle achieves the highest white muscle lactate levels recorded in any vertebrate (> 100 μmol g^{-1}). Not surprisingly, tunas likewise exhibit the highest lactate dehydrogenase (LDH) activity of any vertebrate (Guppy and Hochachka, 1978; Guppy et al., 1979). Although less data are available on billfishes, blue marlin (and most likely the other billfishes) appear to share these adaptations (Dickson, 1995). Because lactate production is necessarily accompanied by a concomitant production of an equal molar volume of protons (Hochachka and Mommsen, 1983, 1995), tunas and marlin also have high muscle buffering capacities based primarily on histidine-containing buffers (Dickson, 1995), which are important enough that they are maintained even during periods of starvation (Abe et al., 1986). Likewise, the non-bicarbonate buffering capacity of tuna blood (-20.9 $\Delta[HCO_3^-]$ ΔpH^{-1}) is approximately twice that measured in other fishes (Brill et al., 1992).

Tunas (and presumably also billfishes) are also able to return blood and muscle lactate levels to pre-exercise more quickly than other fishes (Arthur *et al.*, 1992). This is a consequence of tunas' cardio-respiratory system being able to deliver oxygen and metabolic substrates at rates far above those needed at routine activity levels (Brill and Bushnell, 2001). As previously described (e.g., Dickson, 1995; Brill, 1996) these adaptations appear to be a direct result of the selective pressures in the pelagic environment where food resources are aggregated but widely scattered, and where no refuge exists where animals could hide and recover from a bout of strenuous activity.

Physiological Basis for Species-Specific Vertical and Horizontal Movement Patterns

We will now use the unique physiology of tunas, billfishes, and pelagic sharks to attempt to explain the species-specific movement patterns described in the previous sections. We have therefore organized the final section of this chapter around two central ideas:

1. As suggested by Childress and Nygaard (1974) the fishes and crustaceans of the DSL migrate downward during the day into the cold, oxygen minimum layer as a refuge against predation. That they are able to do so is a result of the evolution of a suite of morphological, physiological, and biochemical adaptations which allows them to tolerate both the hypoxic conditions and cold temperatures occurring at depth (Childress, 1971; Belman and Childress, 1976; Sanders and Childress, 1990).

2. As suggested by Josse *et al.* (1998), the vertical movements of large pelagic predators (i.e., tunas, billfishes, sharks, and squids) mirror the movements of their prey to the extent allowed by their individual physiological abilities and tolerances. Some species [e.g., bigeye tuna, swordfish, bigeye thresher sharks, and neon flying squid (*Ommastrephes bartrami*)] have evolved physiological abilities necessary to invade the DSL organisms' deep daytime predator refuge, whereas other pelagic species (e.g., skipjack tuna, mahimahi, wahoo, blue, black and striped marlins) have not.

Although many pelagic fishes have geographical ranges that often overlap in both time and space (Sund *et al.*, 1981) we propose that, because their depth distributions and vertical movement patterns are dramatically different, many of these species occupy ecosystems which are largely separate. Recent studies of the gut contents of tunas and swordfish support this conjecture. Bigeye tuna caught in the central Pacific were found generally to have selected mesopelagic prey from the DSL, while yellowfin tuna in the same area feed primarily on epipelagic species from the uniformed-temperature surface layer (Grubbs and Holland, 2003). Likewise, Potier *et al.* (2007) concluded that swordfish feed at deeper depth than yellowfin tuna in the Indian Ocean. While these studies are only examples, we argue that their conclusions are generally applicable both to other pelagic species (e.g., swordfish vs. marlins, shortfin mako and bigeye thresher shark vs. silky shark, see Figs. 14.1–14.4) and other areas.

For those tunas, billfishes, or large pelagic sharks whose extensive vertical mobility allows them to mirror the behavior of the organisms of DSL (i.e., to remain below the thermocline for more than brief periods), the cardio-respiratory system must be able to deliver dissolved oxygen and metabolic substrates at rates that match metabolic oxygen demands. This includes periods of increased energy demand, such as when chasing prey or escaping predators, during recovery from exhaustive exercise (Arthur *et al.*, 1992), or following food ingestion (Jobling, 1994). And the cardio-respiratory system must be able to do so even in deep, cold, hypoxic waters. In this section we therefore specifically (albeit briefly) concentrate on two areas. One is the way in which reductions of ambient temperature and oxygen affect cardio-respiratory function and thus potentially restrict the swimming ability and the vertical movements of tunas, billfishes, and large pelagic sharks. The second is a review of what is known about the abilities of the cardio-respiratory systems of specific species to tolerate these challenges.

Carey (1982) was one of the first to recognize an important difference between the vertical and horizontal thermal structuring within the oceans: "Temperature gradients of 15°C to 20°C are not uncommon within the depth ranges of pelagic fish. By moving a few hundred meters vertically, an animal may encounter a greater temperature change than it experiences seasonally or in moving thousands of miles horizontally." This difference implies that fish moving vertically must withstand large temperature changes occurring over minutes to tens of minutes, compared to the time scale of weeks or months usually required for fish moving horizontally (i.e., undertaking seasonal migrations) to undergo equivalent temperature changes. At the latter time scales, temperature acclimation (e.g., synthesis of isozymes, modification of cell membrane structure, etc.) is clearly possible, whereas under the former time scales it is not. Although there is a large body of literature on the ability of temperate fishes (primarily freshwater and inshore marine species) to acclimate to seasonal-scale temperature changes (e.g., Hochachka and Mommsen, 1995), there is no equivalent information on pelagic fishes. The remainder of this review will, therefore, deal primarily with the effects of short-term changes (i.e., those experienced over minutes or tens of minutes by fish moving vertically) and the ability of various species to tolerate these.

Acute temperature change and cardiac function

Fishes' hearts are located close to the ventral body wall and outside any areas maintained at elevated temperatures by thermoconserving species (Carey et al., 1971, 1984; Carey, 1982; Block and Carey, 1985; Bernal et al., 2001b). Hearts are also on the "downstream" side of the vascular heat exchangers which means venous blood returning to the heart is at ambient temperature (Graham and Dickson, 2001). The blood supply to the coronary (hypobranchial) circulation arises directly from the gills and is likewise at ambient temperature (Carey and Gibson, 1983; Carey et al., 1984; Bernal et al., 2001b). The result is that the cardiac muscle of all tunas, billfishes, and large pelagic sharks remains within 1°C of ambient temperature (Graham, 1983). Cardiac muscle is thus not afforded the temperature buffering effect of thermal inertia due to large body mass and the presence of vascular counter-current heat exchangers (Neill et al., 1974, 1976). Unlike the situation in swimming muscles—particularly the RM fiber portions of the mytomes in thermoconserving tuna and lamnid sharks—cardiac muscle temperature will immediately reflect changes in ambient temperature.

Direct experiential evidence has shown that an acute 10°C temperature change (from 25°C to 15°C) causes immediate and parallel decreases in heart rate and cardiac output (i.e., with a Q_{10} of approximately 2) in both yellowfin and skipjack tunas (Korsmeyer et al., 1997b; Blank et al., 2002). The reduction in heart rate accompanying acute temperature change in tunas is not neuronally mediated and cannot be counteracted by injection of vagolytic drugs such as atropine, (R. Brill, K. Cousins and T. Lowe, unpublished) which normally increase heart rate by 58 and 143% (respectively) in skipjack and yellowfin tunas (at 25°C) (Keen et al., 1995). Stroke volume in tunas is nearly fixed because their hearts normally operate under high venous returns which result in near-maximal ventricular filling volumes (i.e., the upper limb of the Starling curve) (Farrell et al., 1992; Korsmeyer et al., 1997a; Brill and Bushnell, 2001). Cardiac output is therefore primarily increased by changes in heart rate (Farrell, 1991; Brill and Bushnell, 2001). Taken together, these lines of evidence imply that yellowfin and skipjack tunas have little or no ability to increase cardiac output at low ambient temperatures. While these two tuna species might be able to meet oxygen delivery requirements at low swimming speeds, they would not be able to increase cardiac output to accommodate the doubling (or more) in oxygen demand accompanying increased swimming speeds, recovery from exhaustive exercise, and digestion and assimilation (i.e., specific dynamic action) (Bushnell and Jones, 1994; Jobling, 1994; Korsmeyer and Dewar, 2001; Blank et al., 2007b). Although there are no data on the effects of acute reduction in temperature on cardiac function of marlin and pelagic sharks that remain primarily in the uniformed-temperature surface layer, we hypothesize that the hearts of these species function similarly.

As described previously (see section: Open Ocean Movements), swordfish, several species of sharks, and bigeye tuna make extensive vertical movements, often to depths with water temperatures of 5°C or less; and both Atlantic and Pacific bluefin tuna and some lamnid sharks migrate to areas with similarly cold surface temperatures. While data on the effects of temperature on cardiac function are not available for swordfish or almost any pelagic shark species, there has been significant work on bluefin tuna cardiac function (Blank *et al.*, 2004) and a smaller amount on salmon shark hearts (Weng *et al.*, 2005). Pacific bluefin tuna hearts, like those of yellowfin and skipjack tunas, have a very limited ability to increase stroke volume during the reductions in heart rate accompanying acute reductions of ambient temperature (Blank *et al.*, 2004). As a result, cardiac output falls in almost direct proportion to heart rate. Where bluefin tuna hearts appear to differ is on the specific effect of temperature on heart rate. Blank *et al.* (2002, 2004), using *in situ* perfused heart preparations, found the Q_{10} for heart rate in Pacific bluefin tuna to be 2.1 between 25° C and 10° C, a value markedly smaller than the change in yellowfin tuna heart rate ($Q_{10} = 3.1$) over the same temperature range. [Note, however, that Korsmeyer *et al.* (1997b) found a Q_{10} of 2 for heart rate in swimming yellowfin tuna; unfortunately equivalent *in vivo* heart rate data are not available for bluefin tuna.] Pacific bluefin tuna hearts also continued contracting at temperatures as low as 2°C, whereas yellowfin tuna hearts ceased normal function at temperatures below 10°C (Blank *et al.*, 2002, 2004).

Impairment of heart function in eels (*Anguilla rostrata*) with decreasing temperature has been reported to be due to a slowing of active calcium extrusion from the myocyte resulting in a prolonged relaxation period (Bailey *et al.*, 1991), and it's likely that the same situation is occurring in yellowfin hearts. The ability of bluefin tuna to maintain cardiac function at low temperatures is, therefore, likely to be due to differences in calcium dynamics (e.g., excitation–contraction or EC coupling). Indeed, bluefin tuna ventricles do have faster calcium inactivation kinetics (Shiels *et al.*, 2004) than other teleosts due to high rates of calcium reuptake by the sarcoplasmic reticulum (SR) and elevated levels of SR Ca^{2+}-activated ATPase (Landeira-Fernandez *et al.*, 2004). Recent work on salmon sharks and blue sharks has shown the heart of the former to have an enhanced expression of EC coupling proteins which is likely responsible for high rates of SR Ca^{2+} release and reuptake. Salmon sharks routinely occupy cold north Pacific waters near Alaska, whereas blue sharks are more limited in their latitudinal distribution (Compagno, 1984).

In summary, enhanced cardiac Ca^{2+} kinetics appear key to maintaining adequate heart rates and cardiac outputs at low temperatures. Parallel studies on other pelagic species showing widely divergent abilities to tolerate changes in temperature with depth (e.g., swordfish vs. mahimahi; bigeye thresher shark vs. silky shark) should allow us to discern if fishes have evolved a convergent solution to the problem of maintaining adequate cardiac output at cold temperatures, or if they have evolved other mechanisms to do so. If the former is the case, we predict that the species with most limited vertical depth range (e.g., mahimahi, wahoo, blue marlin, silky shark) would be at one end of the continuum with respect to SR Ca^{2+} dynamics, with swordfish, bigeye and bluefin tuna, and bigeye thresher sharks being at the opposite end.

It should be noted, however, that at 5°C heart rates in bluefin tuna are 20 beats per minute, or one beat every three seconds (Blank *et al.*, 2004). As a result, there are necessarily significant periods of time (approximately 2 seconds duration) with both zero blood flow in the ventral aorta and through the gills, and probably zero (or near zero) arterial pressure. It is an open question, therefore, how the brain and other organs of bluefin tuna that are dependent on aerobic metabolism can withstand this low cardiac output, and possibly intermittent oxygen delivery. It is possible that, as in other diving animals (i.e., marine mammals), blood flow is shunted primarily to only a few specific organs (Hochachka, 1980), but bluefin tuna's ability to withstand extremely cold temperatures and possibly significant periods with low or no blood flow remains to be explained. Moreover, Pacific bluefin tuna in a swim tunnel begin to behave erratically at 8–10°C (Blank *et al.*, 2007a), implying that this may indeed be the lower temperature limit for this species, below which cardiac function is significantly compromised.

Hypoxia, blood oxygen affinity, and vertical mobility

When following the extensive vertical movements of organisms of the DSL, bigeye tuna frequently endure prolonged exposure to oxygen concentrations below 1.5 ml $O_2 l^{-1}$ at about 5°C (Hanamoto, 1987; Schaefer and Fuller, 2002; Musyl *et al.*, 2003). Nothing is known about the relative hypoxia tolerances of other large pelagic fishes with equivalent extensive vertical mobility (e.g., swordfish, bigeye thresher sharks), yet their abilities to follow the vertical movements of the DSL organisms and neon flying squid (e.g., Carey and Scharold, 1990; Nakamura, 1991, 1993; Dagorn *et al.*, 2001; Nakano *et al.*, 2003; Weng and Block, 2004; Young *et al.*, 2006a) implies that these species also have specific physiological adaptations that make them tolerant of exposure to low ambient oxygen. In contrast, oxygen levels of 3.5 ml $O_2 l^{-1}$ at about 18°C appear to limit the vertical movements of skipjack and yellowfin tunas (Bushnell and Brill, 1991; Cayre and Marsac, 1993; Brill, 1994; Korsmeyer *et al.*, 1996; Brill *et al.*, 1999b; Graham and Dickson, 2004) as well as marlins and sailfish (Prince and Goodyear, 2006). In laboratory studies, the onset of cardio-respiratory adjustments of bigeye tuna during acute hypoxia occurs at lower ambient O_2 levels than in those of skipjack or yellowfin tunas (Bushnell *et al.*, 1990) suggestive of the bigeye tuna's greater tolerance of low oxygen conditions.

The documented hypoxia tolerance of bigeye tuna, however, seems incongruous in light of tunas' well documented high energy demands and high performance cardio-respiratory systems (Dewar and Graham, 1994a; Bernal *et al.*, 2001b; Brill and Bushnell, 2001; Korsmeyer and Dewar, 2001; Blank *et al.*, 2007b). Hypoxia tolerant fishes are generally characterized by sluggish behaviors and low metabolic rates (e.g., Yang *et al.*, 1992; Almeida-Val *et al.*, 1995; Val *et al.*, 1998). Moreover, the association of low metabolic rate and hypoxia tolerance is also associated with blood oxygen binding characteristics. Marine and freshwater fishes tolerant of hypoxia typically have blood and/or hemoglobin with a higher oxygen affinity than less hypoxia tolerant species (Wood *et al.*, 1975; Weber *et al.*, 1976; Weber and Lykkeboe, 1978; Powers, 1980, 1985; Jensen *et al.*, 1993). However, blood with a high oxygen affinity also appears to compromise rates of oxygen delivery to the tissues, and thus can potentially reduce metabolic rates, because it does not as readily offload oxygen at the tissues as blood with lower oxygen affinity does. Fish with reduced MMRs forgo all accompanying selective advantages conferred by this adaptation (e.g., high somatic and gonadal growth rates, fast recovery from exhaustive exercise, elevated aerobic scope) (Brill, 1996). There are no data indicating that the metabolic rates of bigeye tuna are lower than those of yellowfin tuna.

Given their greater tolerance of hypoxic conditions, it would also be predicted that bigeye tuna blood would have a higher oxygen affinity than yellowfin tuna blood, and indeed this is the case (Brill and Bushnell, 1991a; Lowe *et al.*, 2000). Bigeye tuna have, however, evolved a unique ability which allows them to be a high-energy-demand species that is likewise tolerant of low ambient oxygen. As described previously, tunas are capable of maintaining RM fibers and some internal organ temperatures well above ambient temperature (see previous sections on endothermy) (reviewed by Brill *et al.*, 1994; Bernal *et al.*, 2001b; Graham and Dickson, 2001; Dickson and Graham, 2004). The presence of vascular counter-current heat exchangers also reduce rates of heat transfer to or from the environment following abrupt changes in ambient temperature and RM temperatures of even relatively small tunas lag significantly behind rapid alternations in water temperature encountered during rapid vertical movements (Neill *et al.*, 1976; Holland *et al.*, 1992; Dickson, 1994). Muscle temperatures can therefore differ from water temperature by >15°C (Gunn and Block, 2001). The blood of all thermoconserving fishes is, therefore, subjected to significant "closed-system" temperature changes (i.e., alterations in temperatures where the blood is not able to exchange gases or proton equivalents with another medium) as it passes from the gills (where it is at ambient temperature) through the vascular counter-current heat exchangers (where it is warmed to temperatures essentially matching those of the venous blood draining from RM fiber portions of the myotomes). During this situation, gas exchange between the blood within the arteries and veins of the vascular counter-current heat exchangers is limited. As a result blood O_2 content ($[O_2]$) remains essentially constant, whereas the partial pressures of O_2 and CO_2 (PO_2 and PCO_2, respectively) change in concert with temperature.

It is under these conditions that bigeye tuna blood displays its unique characteristics which enable this tuna species to be both tolerant of low ambient oxygen, and simultaneously able to maintain elevated metabolic rates characteristics of tunas. Under *in vitro* conditions matching those occurring during passage through the gills, bigeye tuna blood has an exceptionally high oxygen affinity (compared to that of yellowfin, skipjack, and Atlantic bluefin tuna) expected of hypoxia tolerant species (Brill and Bushnell, 1991b, 2006; Lowe *et al.*, 2000). But as the blood of the bigeye tuna is warmed during passage through the vascular counter-current heat exchangers, its oxygen affinity decreases dramatically to that equivalent to those of other tunas (Lowe *et al.*, 2000). This occurs because it experiences a large decrease in pH due to increase in pCO_2 which, in turn, is due to both the decrease in CO_2 solubility in the plasma and reciprocal titration of plasma proteins and plasma bicarbonate (Cech *et al.*, 1984; Perry *et al.*, 1985; Brill and Bushnell, 1991b; Brill *et al.*, 1992; Lowe *et al.*, 2000). The net result is that oxygen off-loading in the swimming muscles is not compromised. The high oxygen affinity seen in bigeye tuna blood during passage through the gills thus provides an effective system for extracting oxygen at high rates from the ventilatory water stream even during conditions of low ambient oxygen. In concert, the decrease in affinity accompanying warming in the vascular counter-current heat exchangers simultaneously ensures high rates of oxygen delivery and offloading in the muscle capillaries. This unusual characteristic, however, requires the maintenance of elevated muscle temperatures and may explain bigeye tuna's regular upward excursions into the warm surface layer (see Fig. 14.9). For bigeye tuna, maintenance of warm muscles appears to be obligatory for the functional characteristics of the blood to be expressed that support bigeye tuna's hypoxia tolerance.

The blood of all tunas, billfishes, and sharks showing extensive vertical mobility is also subject to rapid open-system temperature changes during its passage through the gills. Blood passing through their gills must, therefore, maintain its functional properties while being subjected to a wide range of temperatures, but where it is free to exchange gases and proton equivalents with the water passing over the gills. This is referred to as an "open-system" temperature change in contrast to the "closed-system" temperature changes occurring during the passage of blood through vascular counter-current heat exchangers. Likewise, Atlantic and Pacific bluefin tunas also subject themselves to long term ambient temperature changes of 15°C as a result of their extensive migrations from tropical to temperate waters (Lutcavage *et al.*, 2000; Block *et al.*, 2001, 2005; Gunn and Block, 2001; Stokesburry *et al.*, 2004). Blood from all tuna species studied to date show little or no shift in O_2 affinity when subjected to open-system temperature changes *in vitro* (Carey and Gibson, 1983; Cech *et al.*, 1984; Brill and Bushnell, 1991b, 2006). Rossi-Fanelli and Antonini (1960) were the first to propose that this temperature insensitivity "... would enable the animal to live in waters of very different temperatures without modification of the functional properties of its respiratory pigment", and we agree that this is most likely the reason for its commonality among all five tuna species.

Not surprisingly, findings of a decreased Hb thermal sensitivity have also been made for lamnid sharks. Andersen *et al.* (1973) found that 10°C warming of shortfin mako and porbeagle Hb solutions only minimally affected the O_2 affinity, while Larsen *et al.* (2004) reported that, in the presence of the Hb-affinity modulator ATP, a 16°C warming of purified porbeagle Hb actually increased O_2 affinity. Whole blood from mako sharks shows a similarly decreased closed-system temperature effect (D. Bernal, J. Cech, J. Graham, unpublished). This virtually non-existent thermal effect on blood oxygen affinity may insure that, for example, salmon shark RM receives fully saturated blood at its *in vivo* and stable operating temperature that can reach up to 20°C above ambient (Bernal *et al.*, 2001b, 2005; Goldman *et al.*, 2004).

CONCLUSIONS

Our current understanding of how temperature and oxygen conditions influence the swimming abilities, vertical movement patterns, and distribution of tunas, billfishes, and large pelagic sharks is incomplete. But by no means does it remain a complete mystery. We now know that species such as bigeye tuna and swordfish have evolved a suite of physiological and anatomical specializations necessary to effectively exploit fauna associated with the deep sound scattering layer, and are able to mirror their prey's extensive daily vertical migrations. In contrast, other large pelagic predators (e.g., mahimahi, silky sharks) lack such specializations and thus tend to remain in the upper uniformed-temperature surface layers of the ocean. Still other species (e.g., yellowfin tuna, striped marlin, mako and blue sharks) have vertical movement behaviors intermediate between the deep divers and the non-divers. The net result is that although groups of large pelagic predators often occupy the same geographic areas of the open ocean, and which may be exploited by the same fisheries, they really occupy largely separate ecosystems. This fact has significant implications for effective fishery management policies, resource conservation, and for the population assessments upon which they must ultimately be based. These species-specific differences also provide a wealth of opportunities (and significant challenges!) for developing a fundamental understanding of ecological physiology and for those of us fortunate enough to be afforded the opportunity to study these remarkable fishes.

ACKNOWLEDGEMENTS

The authors acknowledge support through NSF IOS-0617403 and IOS-0617384, Cooperative Agreements NA37RJ0199 and NA67RJ0154 between the National Oceanic and Atmospheric Administration (U.S. Department of Commerce) and the Pelagic Fisheries Research Program (University of Hawaii), and the Pfleger Foundation and Pfleger family. The authors or their agencies do not necessarily approve, recommend, or endorse any proprietary hardware or software mentioned in this publication. The views expressed herein are those of the authors and do not necessarily reflect the views of their agencies. We are indebted to S. Adams, J. Valdez and T. Tazo for logistical support.

REFERENCES

Abe, H., R.W. Brill and P.W. Hochacka. 1986. Metabolism of L-histidine, carnosine, and anserine in skipjack tuna. *Physiological Zoology* 59: 439–450.

Alexander, R.L. 1998. Blood supply to the eyes and brain of lamniform sharks (Lamniformes). *Journal of Zoology London* 245: 363–369.

Alexander, R.M. 1993. Buoyancy. In: *The Physiology of Fishes*, D.H. Evans (Ed.). CRC Press, Boca Raton. pp. 75-97 (In Press).

Almeida-Val, V.M.F., I.P. Farias, M.N.P. Silva, W.P. Duncan and A.L. Val. 1995. Biochemical adjustments to hypoxia by amazon cichlids. *Brazilian Journal of Medical and Biological Research* 28: 1257–1263.

Altringham, J.D. and B.A. Block. 1997. Why do tuna maintain elevated slow muscle temperatures? Power output of muscle isolated from endothermic and ectothermic fish. *Journal of Experimental Biology.* 200: 2617–2627.

Altringham, J.D. and R.E. Shadwick. 2001. Swimming and muscle function. In: *Fish Physiology*, B.A. Block and E.D. Stevens (Eds.). Academic Press: San Diego, Vol. 19, pp. 313–344.

Andersen, M.E., J.S. Olson, Q.H. Gibson and F.G. Carey. 1973. Studies on ligand binding to hemoglobins from teleosts and elasmobranchs. *Journal of Biological Chemistry* 248: 331–341.

Anderson, S.D. and K.J. Goldman. 2001. Temperature measurements from salmon sharks, Lamna ditropis, in Alaskan waters. *Copeia* 2003: 794–796.

Arnold, G. and Dewar, H. 2001. Electronic tags in marine fisheries research: A 30 year perspective. In: *Electronic Tagging and Tracking in Marine Fisheries*, J.R. Sibert and J.L. Nielsen (Eds.). Klwuer Academic Publishers. Dordrecht, pp. 7–64.

Arthur, P.G., T.G. West, R.W. Brill, P.M. Schulte and P.W. Hochachka. 1992. Recovery metabolism of skipjack tuna (*Katsuwonus pelamis*) white muscle rapid and parallel changes in lactate and phosphocreatine after exercise. *Canadian Journal of Zoology* 70: 1230–1239.

Atema, J., K. Holland and W. Ikehara. 1980. Olfactory responses of yellowfin tuna *Thunnus albacares* to prey odors chemical search image. *Journal of Chemistry Ecology* 6: 457–466.

Bailey, J., D. Sephto and W.R. Dreidzic. 1991. Impact of an acute temperature change on performance and metabolism of pickerel (*Esox niger*) and eel (*Anguilla rostrata*) hearts. *Physiological Zoology.* 64: 697–716.

Baum, J.K., R.A. Myers, D.G. Kehler, B. Worm, S.J. Harley and P.A. Doherty. 2003. Collapse and conservation of shark populations in the Northwest Atlantic. *Science* 299: 389–392.

Beauchamp, D.A., C.M. Baldwin, J.L. Vogel and C.P. Gubala. 1999. Estimating diel, depth-specific foraging opportunities with a visual encounter rate model for pelagic piscivores. *Canadian Journal of Fisheries and Aquatic Sciences* 56: 128–139.

Belman, B.W. and J.J. Childress. 1976. Circulatory adaptations to the oxygen minimum layer in the bathypelagic mysid *Gnathophausia ingens*. *Biological Bulletin.* 150: 15–37.

Bernal, D. and C.A. Sepulveda. 2005. Evidence for temperature elevation in the aerobic swimming musculature of the common thresher shark, *Alopias vulpinus*. *Copeia* 2005: 146–151.

Bernal, D., C. Sepulveda and J.B. Graham. 2001a. Water-tunnel studies of heat balance in swimming mako sharks. *Journal of Experimental Biology* 204: 4043–4054.

Bernal, D., K.A. Dickson, R.E. Shadwick and J.B. Graham. 2001b. Review: Analysis of the evolutionary convergence for high performance swimming in lamnid sharks and tunas. *Comparative Biochemistry and Physiology.* A129: 695–726.

Bernal, D., C. Sepulveda, O. Mathieu-Costello and J.B. Graham. 2003. Comparative studies of high performance swimming in sharks I. Red muscle morphometrics, vascularization and ultrastructure. *Journal of Experimental Biology* 206: 2831–2843.

Bernal, D., J.M. Donley, R.E. Shadwick and D.A. Syme. 2005. Mammal-like muscles power swimming in a cold-water shark. *Nature* (London) 437: 1349–1352.

Bigelow, K.A., J. Hampton and N. Miyabe. 2002. Application of a habitat-based model to estimate effective longline fishing effort and relative abundance of Pacific bigeye tuna (*Thunnus obesus*). *Fisheries Oceanography* 11: 143–155.

Bigelow, K.A. and M.N. Maunder. 2007. Does habitat or depth influence catch rates of pelagic species? *Canadian Journal of Fisheries and Aquatic Sciences* 64: 1581–1594.

Blake, R.W., L.M. Chatters and P. Domenici. 1995. Turning radius of yellowfin tuna (*Thunnus albacares*) in unsteady swimming manoeuvers *Journal of Fish Biology* 46: 536–538

Blake, R.W. 2004. Fish functional design and swimming performance. *Journal of Fish Biology* 65: 1193–1222.

Blank, J.M., J.M. Morrissette, P.S. Davie and B.A. Block. 2002. Effects of temperature, epinephrine and Ca^{2+} on the hearts of yellowfin tuna (*Thunnus albacares*). *Journal of Experimental Biology* 205: 1881–1888.

Blank, J.M., J.M. Morrissette, C.J. Farwell, M. Price, R.J. Schallert and B.A. Block. 2007a. Temperature effects on metabolic rate of juvenile Pacific bluefin tuna *Thunnus orientalis. Journal of Experimental Biology* 210: 4254–4261.

Blank, J.M., C.J. Farwell, J.M. Morrissette, R.J. Schallert and B.A. Block. 2007b. Influence of swimming speed on metabolic rates of juvenile Pacific bluefin tuna and yellowfin tuna. *Physiological and Biochemical Zoology* 80: 167–177.

Blank, J.M., J.M. Morrissette, A.M. Landeira-Fernandez, S.B. Blackwell, T.D. Williams and B.A. Block. 2004. *In situ* cardiac performance of Pacific bluefin tuna hearts in response to acute temperature change. *Journal of Experimental Biology* 207: 881–890.

Blaxter, J.H. and P. Tytler. 1978. Physiology and function of the swimbladder. *Advances in Comparative Physiology and Biochemistry* 7: 311–367.

Block, B., D. Booth and F.G. Carey. 1992. Direct measurement of swimming speeds and depth of blue marlin. *Journal of Experimental Biology* 166: 267–284.

Block, B.A. 1986. Structure of the brain and eye heater tissue in marlins, sailfish, and spearfishes. *Journal of Morphology* 190: 169–189.

Block, B.A. 1991. Endothermy in fish: Thermogenesis ecology and evolution. In: *Biochemistry and Molecular Biology of Fishes,* P. W. Hochachka and T. Mommsen (Eds.). Elsevier, Amsterdam, Vol. 1, pp. 269–311.

Block, B.A. and F.G. Carey. 1985. Warm brain and eye temperatures in sharks. *Journal of Comparative Physiology* B156: 229–236.

Block, B.A. and J.R. Finnerty. 1994. Endothermy in fishes: A phylogenetic analysis of constraints, predispositions, and selection pressures. *Environmental Biology of Fishes* 40: 283–302.

Block, B.A., H. Dewar, S.B. Blackwell and A. Boustany. 1999. Archival and pop-off satellite tags reveal new information about Atlantic bluefin tuna thermal biology and feeding behaviors. *Comparative Biochemistry and Physiology* 124.

Block, B.A., H. Dewar, T. Williams, E.D. Prince, C. Farwell and D. Fudge. 1998. Archival tagging of Atlantic bluefin tuna, *Thunnus thynnus. Marine Technology Society Journal* 32: 37–46.

Block, B.A., S.L.H. Teo, A. Walli, A. Boustany, M.J.W. Stokesbury, C.J. Farwell, K.C. Weng, H. Dewar and T.D. Williams. 2005. Electronic tagging and population structure of Atlantic bluefin tuna. *Nature* (London) 434: 1121–1127.

Block, B.A., H. Dewar, S.B. Blackwell, T.D. Williams, E.D. Prince, C.J. Farwell, A. Boustany, S.L.H. Teo, A. Seitz, A. Walli and D. Fudge. 2001. Migratory movements, depth preferences, and thermal biology of Atlantic bluefin tuna. *Science* 293: 1310–1314.

Bonfil, R., M. Meyer, M.C. Scholl, R. Johnson, S. O'Brien, H. Oosthuizen, S. Swanson, D. Kotze and M. Paterson. 2005. Transoceanic migration, spatial dynamics, and population linkages of white sharks. *Science* 310: 100–103.

Brett, J.R. 1972. The metabolic demand for oxygen in fish particularly salmonids and a comparison with other vertebrates. *Respiration Physiology* 14: 151–170.

Brett, J.R. and N.R. Glass. 1973. Metabolic rates and critical swimming speeds of sockeye salmon (*Oncorhynchus nerka*) in relation to size and temperature. *Journal of the Fisheries Research Board of Canada* 30: 379–387.

Brill, R.W. 1987. On the standard metabolic rates of tropical tunas including the effect of body size and acute temperature change. *Fishery Bulletin* 85: 25–36.

Brill, R.W. 1994. A review of temperature and oxygen tolerance studies of tunas, pertinent to fisheries oceanography, movement models and stock assessments. *Fisheries Oceanography* 3: 204–216.

Brill, R.W. 1996. Selective advantages conferred by the high performance physiology of tunas, billfishes, and dolphin fish. *Comparative Biochemistry and Physiology* 113: 3–15.

Brill, R.W. and P.G. Bushnell. 1991a. Metabolic and cardiac scope of high energy demand teleosts the tunas. *Canadian Journal of Zoology* 69: 2002–2009.

Brill, R.W. and P.G. Bushnell. 1991b. Effects of open and closed-system temperature changes on blood oxygen dissociation curves of skipjack tuna (*Katsuwonus pelamis*) and yellowfin tuna (*Thunnus albacares*). *Canadian Journal of Zoology* 69: 1814–1821.

Brill, R.W. and P.G. Bushnell. 2001. The cardiovascular system of tunas. In: *Fish Physiology, Tuna—Physiology, Ecology and Evolution*, B.A. Block and E.D. Stevens (Eds.). Academic Press, San Diego, Vol. 19, pp. 79–120.

Brill, R.W. and P.G. Bushnell. 2006. Effects of open- and closed-system temperature changes on blood O_2-binding characteristics of Atlantic bluefin tuna (*Thunnus thynnus*). *Fish Physiology and Biochemistry* 32: 283–294.

Brill, R.W., P.G. Bushnell, D.R. Jones and M. Shimizu. 1992. Effects of acute temperature change *in-vivo* and *in-vitro* on the acid-base status of blood from yellowfin tuna (*Thunnus albacares*). *Canadian Journal of Zoology* 70: 654–662.

Brill, R.W., D.B. Holts, R.K.C. Chang, S. Sullivan, H. Dewar and F.G. Carey. 1993. Vertical and horizontal movements of striped marlin (*Tetrapturus audax*) near the Hawaiian Islands, determined by ultrasonic telemetry, with simultaneous measurement of oceanic currents. *Marine Biology* 117: 567–574.

Brill, R.W., T.E. Lowe and K.L. Cousins. 1999a. How water temperature really limits the vertical movements of tunas and billfishes—it's the heart stupid. *Proceedings. International Congress on the Biology of Fish. Fish Cardiovascular Function: Control Mechanisms and Environmental Influences* pp. 57–62.

Brill, R.W., B.A. Block, C.H. Boggs, K.A. Bigelow, E.V. Freund and D.J. Marcinek. 1999b. Horizontal movements and depth distribution of large adult yellowfin tuna (*Thunnus albacares*) near the Hawaiian Islands, recorded using ultrasonic telemetry: Implications for the physiological ecology of pelagic fishes. *Marine Biology* 133: 395–408.

Brill, R.W., H. Dewar and J.B. Graham. 1994. Basic concepts relevant to heat transfer in fishes, and their use in measuring the physiological thermoregulatory abilities of tunas. *Environmental Biology of Fishes* 40: 109–124.

Brill, R.W., K.A. Bigelow, M.K. Musyl, K.A. Fritsches and E.J. Warrant. 2005. Bigeye tuna behavior and physiology... their relevance to stock assessments and fishery biology. *Collective Volume of Scientific Papers ICCAT* 57: 142–161.

Bushnell, P.G. and R.W. Brill. 1991. Responses of swimming skipjack (*Katsuwonus pelamis*) and yellowfin (*Thunnus albacares*) tunas to acute hypoxia and a model of their cardiorespiratory function. *Physiological Zoology* 64: 787–811.

Bushnell, P.G. and R.W. Brill. 1992. Oxygen transport and cardiovascular responses in skipjack tuna (*Katsuwonus pelamis*) and yellowfin tuna (*Thunnus albacares*) exposed to acute hypoxia. *Journal of Comparative Physiology* B162: 131–143.

Bushnell, P.G. and D.R. Jones. 1994. Cardiovascular and respiratory physiology of tuna: Adaptations for support of exceptionally high metabolic rates. *Environmental Biology of Fishes* 40: 303–318.

Bushnell, P.G., R.W. Brill and R.E. Bourke. 1990. Cardiorespiratory responses of skipjack tuna (*Katsuwonus pelamis*), yellowfin tuna (*Thunnus albacares*), and bigeye tuna (*Thunnus obesus*) to acute reductions of ambient oxygen. *Canadian Journal of Zoology* 68: 1857–1865.

Burgess, G.H., L.R. Beerkircher, G.M. Cailliet, J.K. Carlson, E. Cortés, K.J. Goldman, R.D. Grubbs, J.A. Musick, M.K. Musyl and C.A. Simpfendorfer. 2005. Is the collapse of shark populations in the Northwest Atlantic Ocean and Gulf of Mexico real? *Fisheries* 30: 19–26.

Campana, S.E., W. Joyce, L. Marks, L.J. Natanson, N.E. Kohler, C.F. Jensen, J.J. Mello, Jr. H.L. Pratt and S. Myklevoll. 2002. Population dynamics of the porbeagle in the northwest Atlantic Ocean. *North American Journal of Fisheries Management* 22: 106–121.

Carey, F.G. 1982. Warm fish. In: *A Companion to Animal Physiology, International Conference on Comparative Physiology, Sandbjerg, Denmark*. C.R. Taylor, K. Johansen and L. Bolis (Eds.). Cambridge University Press, Cambridge, UK, Vol. 5th, pp. 216–234.

Carey, F.G. 1990. Further acoustic telemetry observations of swordfish. In: *Planning the Future of Billfishes*, R.H. Stroud (Ed.). National Coalition for Marine Preservation, Savannah, Vol. 2, pp. 103–131.

Carey, F.G. and J.M. Teal. 1966. Heat conservation in tuna fish muscle. *Proceedings of the National Academy of Sciences of the United States of America* 56: 1464–1469.

Carey, F.G. and J.M. Teal. 1969a. Mako and porbeagle warm-bodied sharks. *Comparative Biochemistry and Physiology* 28: 199–204.

Carey, F.G. and J.M. Teal. 1969b. Regulation of body temperature by the bluefin tuna. *Comparative Biochemistry and Physiology* 28: 205–213.

Carey, F.G. and K.D. Lawson. 1973. Temperature regulation in free swimming bluefin tuna. *Comparative Biochemistry and Physiology* 44: 375–392.

Carey, F.G. and B.H. Robison. 1981. Daily patterns in the activities of swordfish (*Xiphias gladius*) observed by acoustic telemetry. *Fishery Bulletin* 79: 277–292.

Carey, F.G. and Q.H. Gibson. 1983. Heat and oxygen exchange in the rete mirabile of the bluefin tuna (*Thunnus thynnus*). *Comparative Biochemistry and Physiology* 74: 333–342.

Carey, F.G. and J.V. Scharold. 1990. Movements of blue sharks *Prionace glauca* in depth and course. *Marine Biology* 106: 329–342.

Carey, F.G., J.M. Teal, J.W. Kanwisher, K.D. Lawson and J.S. Beckett. 1971. Warm bodied fish. *American Zoologist* 11: 135–143.

Carey, F.G., J.M. Teal and J.W. Kanwisher. 1981. The visceral temperatures of mackerel sharks Lamnidae. *Physiological Zoology* 54: 334–344.

Carey, F.G., J.W. Kanwisher and E.D. Stevens. 1984. Bluefin tuna (*Thunnus thynnus*) warm their viscera during digestion. *Journal of Experimental Biology* 109: 1–20.

Carey, F.G., J.G. Casey, H.L. Pratt, D. Urquhart and J.E. McCosker. 1985. Temperature, heat production and heat exchange in lamnid sharks. *Memoirs of Society of the California Academy of Sciences* 9: 92–108.

Carlson, J.K., K.J. Goldman and C.G. Lowe. 2004. Metabolism, energetic demand, and endothermy. In: *Biology of Sharks and Their Relatives*, J.C. Carrier J.A. Musick and M.R. Heithaus (Eds.). CRC Press, Boca Raton, pp. 203–224.

Cayre, P. and F. Marsac. 1993. Modeling the yellowfin tuna (*Thunnus albacares*) vertical distribution using sonic tagging results and local environmental parameters. *Aquatic Living Resources* 6: 1–14.

Cech, J.J. Jr., R.M. Laurs and J.B. Graham. 1984. Temperature induced changes in blood gas equilibria in the albacore (*Thunnus alalunga*) a warm bodied tuna. *Journal of Experimental Biology* 109: 21–34.

Childress, J.J. 1971. Respiratory adaptations to the oxygen minimum layer in the bathypelagic Mysid *Gnatophausia ingens*. *Biological Bulletin* 141: 109–121.

Childress, J.J. and M.H. Nygaard. 1974. The chemical composition of mid-water fishes as a function of depth of occurrence off Southern California. *Deep Sea Research Ocean Abstracts* 20: 1093–1109.

Compagno, L.J.V. 1984. An annotated and illustrated catalog of shark species known to date 1. Hexanchiformes to Lamniformes. *Food and Agriculture Organisation* 1–249.

Compagno, L.J.V. 1990. Alternative life-history styles of cartilaginous fishes in time and space. *Environmental Biology of Fishes* 28: 33–76.

Coulson, R.A. 1979. Anaerobic glycolysis: the Smith and Wesson of the heterotherms. *Perspectives in Biology and Medicine E* 22: 465–479.

Dagorn, L., A. Bertrand, P. Bach, M. Petit and E. Josse. 2001. Improving our understanding of tropical tuna movements from small to large scales. In: *Electronic Tagging and Tracking in Marine Fisheries*, J.R. Sibert and J.L. Nielsen (Eds.). Kluwer Academic Publishers, Dordrecht, pp. 385–407.

Dewar, H. and J.B. Graham. 1994a. Studies of tropical tuna swimming performance in a large water tunnel: I. Energetics. *Journal of Experimental Biology* 192: 13–31.

Dewar, H. and J.B. Graham. 1994b. Studies of tropical tuna swimming performance in a large water tunnel: III. Kinematics. *Journal of Experimental Biology* 192: 45–59.

Dewar, H., J.B. Graham and R.W. Brill. 1994. Studies of tropical tuna swimming performance in a large water tunnel: II. Thermoregulation. *Journal of Experimental Biology* 192: 33–44.

Dickson, K.A. 1988. Why are some fishes endothermic? Interspecific comparisons of aerobic and anaerobic metabolic capacities in endothermic and ectothermic scombrids. *Scripps Institution of Oceanography*, Ph.D. Thesis, University of California, San Diego.

Dickson, K.A. 1994. Tunas as small as 207mm fork length can elevate muscle temperatures significantly above ambient water temperature. *Journal of Experimental Biology* 190: 79–93.

Dickson, K.A. 1995. Unique adaptations of the metabolic biochemistry of tunas and billfishes for life in the pelagic environment. *Environmental Biology of Fishes* 42: 65–97.

Dickson, K.A. 1996. Locomotor muscle of high-performance fishes: What do comparisons of tunas with ectothermic sister taxa reveal? *Comparative Biochemistry and Physiology* 113: 39–49.

Dickson, K.A. and J.B. Graham. 2004. Evolution and consequences of endothermy in fishes. *Physiological Biochemistry and Zoology* 77: 998–1018.

Dizon, A.E. and R.W. Brill. 1979. Thermoregulation in yellowfin tuna (*Thunnus albacares*). *Physiological Zoology* 52: 581–593.

Dizon, A.E., R.W. Brill and H.S.H. Yuen. 1978. Correlations between environment, physiology and activity and the effects on thermoregulation in skipjack tuna. In: *The Physiological Ecology of Tunas*, A.E. Dizon and G.D. Sharp (Eds.). Academic Press, New York, pp. 233–260.

Donley, J.M., R.E. Shadwick, C.A. Sepulveda and D.A. Syme. 2007. Thermal dependence of contractile properties of the aerobic locomotor muscle in the leopard shark and shortfin mako shark. *Journal of Experimental Biology* 210: 1194–1203.

Donley, J.M., C.A. Sepulveda, P. Konstantinidis, S. Gemballa and R.E. Shadwick. 2004. Convergent evolution in mechanical design of lamnid sharks and tunas. *Nature* (London) 429: 61–65.

Emery, S.H. and A. Szczepanski. 1986. Gill dimensions in pelagic elasmobranch fishes. *Biological Bulletin* 171: 441–449.

Emery, S.H., C. Mangano and V. Randazzo. 1985. Ventricle morphology in pelagic elasmobranch fishes. *Comparative Biochemistry and Physiology* A82: 635–643.

Fänge, R. 1983. Gas exchange in fish swim bladder. *Reviews of Physiology Biochemistry and Pharmacology* 97: 111–158.

Farrell, A.P. 1991. From hagfish to tuna a perspective on cardiac function in fish. *Physiological Zoology* 64: 1137–1164.

Farrell, A.P. and D.R. Jones. 1992. The heart. In: *Fish Physiology*, W.S. Hoar D.J. Randall and A.P. Farrell (Eds.). Academic Press, San Diego, Vol. 12A: pp. 1–88.

Farrell, A.P., P.S. Davie, C.E. Franklin, J.A. Johansen and R.W. Brill. 1992. Cardiac physiology in tunas I. *in-vitro* perfused heart preparations from yellowfin and skipjack tunas. *Canadian Journal of Zoology* 70: 1200–1210.

Fierstine, H.L. and V. Walters. 1968. Studies in the locomotion and anatomy of scombrid fishes. *Memoirs of Society of the California Academy of Sciences* 6: 1–31.

Fitzgibbon, Q.P., R.V. Baudinette, R.J. Musgrove and R.S. Seymour. 2006. Routine metabolic rate of southern bluefin tuna (*Thunnus maccoyii*). *Comparative Biochemistry and Physiology* doi: 10.1016/j.cbpa.2006.08.046

Fitzgibbon, Q.P., A. Strawbridg and R.S. Seymor. 2007. Metabolic scope, swimming performance and the effects of hypoxia in the mulloway, *Argyrosomus japonicus* (Pices: Sciaenidae). *Aquaculture* 270: 359–368.

Fraser, P.J., S.F. Cruickshank, R.L. Shelmerdine and L.E. Smith. 2008. *Hydrostatic pressure receptors and depth usage in crustacea and fish. Journal of the Institute of Navigation* 55: 159–165.

Fraser P.J. and R.L. Shelmerdine. 2002. Dogfish hair cells sense hydrostatic pressure. *Nature* 415: 495–496.

Friedlander, M.J., N. Kotchabhakdi and C.L. Prosser. 1976. Effects of cold and heat on behavior and cerebellar function in goldfish. *Journal of Comparative Physiology* 112: 19–45.

Fritsches, K.A., N.J. Marshall and E.J. Warrant. 2003. Retinal specializations in the blue marlin: Eyes designed for sensitivity to low light levels. *Marine and Freshwater Research* 54: 333–341.

Fritsches, K.A., R.W. Brill and E.J. Warrant. 2005. Warm eyes provide superior vision in swordfishes. *Current Biology* 15: 55–58.

Fry, F.E. 1971. The effect of environmental factors on the physiology of fish. In: *Fish Physiology*, W.S. Hoar, D.J. Randall (Eds.). Academic Press, New York, Vol. 6, pp. 1–98.

Fudge, D.S. and E.D. Stevens. 1996. The visceral *retia mirabilia* of tuna and sharks: an annotated translation and discussion of the Eschricht and Müller 1835 paper and related papers. *Guelph Ichthyology Reviews* 4: 1–328.

Fudge, D.S., E.D. Stevens and J.S. Ballantyne. 1997. Enzyme adaptation along a heterothermic tissue: The visceral retia mirabilia of the bluefin tuna. *American Journal of Physiology* 272: R1834–R1840.

Gemballa, S., P. Konstantinidis, J.M. Donley, C. Sepulveda and R.E. Shadwick. 2006. Evolution of high-performance swimming in sharks: Transformations of the musculotendinous system from subcarangiform to thunniform swimmers. *Journal of Morphology* 267: 477–493.

Gibbs, R.H. and B.B. Collette. 1967. Comparative anatomy and systematics of the tunas, genus *Thunnus. Fishery Bulletin* 66: 65–130.

Gilbert, P.W. 1981. Patterns of shark reproduction. *Oceanus* 24: 30–39.

Goldman, K.J. 1997. Regulation of body temperature in the white shark, *Carcharodon carcharias. Journal of Comparative Physiology* B167: 423–429.

Goldman, K.J., S.D. Anderson, R.J. Latour and J.A. Musick. 2004. Homeothermy in adult salmon sharks, *Lamna ditropis. Environmental Biology and Fishes* 71: 403–411.

Gooding, R.M., W.H. Neill and A.E. Dizon. 1981. Respiration rates and low-oxygen tolerance limits in skipjack tuna, *Katsuwonus pelamis. Fishery Bulletin* 79: 31–48.

Goodyear, C.P. 2003. Tests of the robustness of habitat-standardized abundance indices using simulated blue marlin catch-effort data. *Marine and Freshwater Research*. 54: 369–381.

Graham, J.B. 1983. Heat transfer. In: *Fish Biomechanics*, P.W. Webb and D. Weihs (Eds.)., Praeger, New York, pp. 248–278.

Graham, J.B. and R.M. Laurs. 1982. Metabolic rate of the albacore tuna *Thunnus alalunga*. *Marine Biology* 72: 1–6.

Graham, J.B. and K.A. Dickson. 2001. Anatomical and physiological specializations for endothermy. In: *Fish Physiology*, B.A. Block and E.D. Stevens (Eds.). Academic Press, San Diego, Vol. 19, pp. 121–165.

Graham, J.B. and K.A. Dickson. 2004. Tuna comparative physiology. *Journal of Experimental Biology* 207: 4015–4024.

Graham, J.B., F.J. Koehrn and K.A. Dickson. 1983. Distribution and relative proportions of red muscle in scombrid fishes consequences of body size and relationships to locomotion and endothermy. *Canadian Journal of Zoology* 61: 2087–2096.

Graham, J.B., H. Dewar, N.C. Lai, W.R. Lowell and S.M. Arce. 1990. Aspects of shark swimming performance determined using a large water tunnel. *Journal of Experimental Biology* 151: 175–192.

Greer-Walker, M. and G. Pull. 1975. A survey of red and white muscle in marine fish. *Journal of Fish Biology* 7: 295–300.

Grubbs, D. and K. Holland. 2003. Yellowfin and bigeye tuna in Hawai'i: dietary overlap, prey diversity and the trophic cost of associating with natural and man-made structures. In *Proceedings of the 54th Annual Interantional Tuna Conference*. Lake Arrowhead, CA: SWFSC, NMFS, La Jolla, CA.

Gunn, J. and B. Block. 2001. Advances in acoustic, archival, and satellite tagging of tunas. In: *Fish Physiology*, B.A. Block and E.D. Stevens (Eds.). Academic Press, San Diego, Vol. 19, pp. 167–224.

Gunn, J.S., T.A. Patterson and J.G. Pepperell. 2003. Short-term movement and behaviour of black marlin (*Makaira indica*) in the Coral Sea as determined through a pop-up satellite archival tagging experiment. *Marine and Freshwater Research* 54: 515–525.

Guppy, M. and P.W. Hochachka. 1978. Controlling the highest lactate dehydrogenase activity known in nature. *American Journal of Physiology* 234: 1978.

Guppy, M., C. Hulber and P.W. Hochachka. 1979. Metabolic sources of heat and power in tuna muscles. II. Enzymes and metabolite profiles. *Journal of Experimental Biology* 82: 303–320.

Hampton, J., J.R. Sibert, P. Kleiber, M.N. Maunder and S.J. Harley. 2005. Fisheries decline of Pacific tuna populations exaggerated? *Nature* (London) 434: E1–E2.

Hanamoto, E. 1987. Effect of oceanographic environment on bigeye tuna distribution. *Bulletin of Japanese Society of Fisheries and Oceanography* 51: 203–216.

Harder, W. 1975. *Anatomy of Fishes*. Schweizerbart, Stuttgart.

Hinton, M.G. and M.N. Maunder. 2003. Methods for standardizing CPUE and how to select among them. *Collective Volume of Scientific Papers, ICCAT* SCRS/2003/034.

Hochachka, P.W. 1980. *Living without Oxygen. Closed and Open Systems in Hypoxia Tolerance*. Harvard University Press, Cambridge, MA.

Hochachka, P.W. and T.P. Mommsen. 1983. Protons and anaerobiosis. *Science* 219: 1391–1397.

Hochachka, P.W. and G.N. Somero. 1984. Biochemical Adaptation. Princeton University Press, Princeton.

Hochachka, P.W. and T.P. Mommsen. 1995. Environmental and ecological biochemistry. In: *Biochemistry and Molecular Biology of Fishes*, P.W. Hochachka and T.P. Mommsen (Eds.). Elsevier Science Publishers, Amsterdam, Netherlands, Vol. 5, pp. 455.

Hochachka, P.W., W.C. Hulbert and M. Guppy. 1978. The tuna power plant and furnace. In: *The Physiological Ecology of Tunas*. G.D. Sharp and A.E. Dizon (Eds.). Academic Press, New York, pp. 485.

Holland, K.N., R.W. Brill and R.K.C. Chang. 1990. Horizontal and vertical movements of yellowfin and bigeye tuna associated with fish aggregating devices. *Fishery Bulletin* 88: 493–508.

Holland, K.N., R.W. Brill, R.K.C. Chang, J.R. Sibert and D.A. Fournier. 1992. Physiological and behavioral thermoregulation in bigeye tuna (*Thunnus obesus*). *Nature* (London) 358: 410–412.

Holts, D. and D. Bedford. 1990. Activity patterns of stripped marlin in the southern California Bight. In: *Planning the Future of Billfishes*, R.H. Stroud (Ed.). Savannah, Georgia: National Coalition for Marine Conservation, pp. 81–93.

Holts, D.B. and D.W. Bedford. 1993. Horizontal and vertical movements of the shortfin mako, *Isurus oxyrinchus*, in the southern California Bight. *Australian Journal of Marine and Freshwater Research* 44: 45–60.

Hoolihan, J.P. 2005. Horizontal and vertical movements of sailfish (*Istiophorus platypterus*) in the Arabian Gulf, determined by ultrasonic and pop-up satellite tagging. *Marine Biology* 146: 1015–1029.

Horodysky, A.Z., D.W. Kerstetter, R.J. Latour and J.E. Graves 2007. Habitat utilization and vertical movements of white marlin (*Tetrapturus albidus*) released from commercial and recreational fishing gears in the western North Atlantic Ocean: inferences from short duration pop-up archival satellite tags. *Fisheries Oceanography* 16: 240–256.

Humston, R., D.B. Olson and J.S. Ault. 2004. Behavioral assumptions in models of fish movement and their influence on population dynamics. *Transactions of American Fisheries Society* 133: 1304–1328.

Jensen, F.B., M. Nikinmaa and R.E. Weber. 1993. Environmental perturbations of oxygen transport in teleost fishes: causes, consequences and compensations. In: *Fish Ecophysiology*, J.C. Rankin and F.B. Jensen (Eds.). Chapman Hall, London, pp. 161–179.

Jobling, M. 1994. *Fish Bioenergetics*. Chapman and Hall, London.

Johnston, I.A. and R. Brill. 1984. Thermal dependence of contractile properties of single skinned muscle fibers from Antarctic and various warm water marine fishes including skipjack tuna (*Katsuwonus pelamis*) and kawakawa (*Euthynnus affinis*). *Journal of Comparative Physiology* B155: 63–70.

Jones, F.R.H. 1951. The swimbladder and the vertical movement of teleostean fishes: I. Physical factors. *Journal of Experimental Biology* 28: 553–566.

Jones, F.R.H. 1952. The swimbladder and the vertical movements of teleostean fishes: II The restriction to rapid and slow movements. *Journal of Experimental Biology* 29: 94–109.

Jordon, A.D. and J.F. Steffensen. 2007. Effects of ration size and hypoxia on specific dynamic action in the cod. *Physiological Biochemistry and Zoology* 80: 178–185.

Josse, E., P. Bach and L. Dagorn. 1998. Simultaneous observations of tuna movements and their prey by sonic tracking and acoustic surveys. *Hydrobiologia* 372: 61–69.

Katz, S.L. 2002. Design of heterothermic muscle in fish. *Journal of Experimental Biology* 205: 2251–2266.

Katz, S.L., D.A. Syme and R.E. Shadwick. 2000. High-speed swimming. Enhanced power in yellowfin tuna. *Nature* (London) 410: 770–771.

Keen, J.E., S. Aota, R.W. Brill, A.P. Farrell and D.J. Randall. 1995. Cholinergic and adrenergic regulation of heart rate and ventral aortic pressure in two species of tropical tunas, *Katsuwonus pelamis* and *Thunnus albacares*. *Canadian Journal of Zoology* 73: 1681–1688.

Kirby, D.S. 2001. On the integrated study of tuna behaviour and spatial dynamics: Tagging and modelling as complementary tools In: *Electronic Tagging and Tracking in Marine Fisheries*, S.J.R. and J. Nielsen (Eds.). Kluwer Academic Publishers, Dordrecht, pp. 407–421.

Kitchell, J.F., S.J.D. Martell, C.J. Walters, O.P. Jensen, I.C. Kaplan, J. Watters, T.E. Essington and C.H. Boggs. 2006. Billfishes in an ecosystem context. *Bulletin of Marine Science* 79: 669–682.

Knower, T. 1998. Biomechanics of thunniform swimming. In: *Scripps Institution of Oceanography*, Ph.D. Thesis, University of California, San Diego.

Knower, T., R.E. Shadwick, S.L. Katz, J.B. Graham and C.S. Wardle. 1999. Red muscle activation patterns in yellowfin (*Thunnus albacares*) and skipjack (*Katsuwonus pelamis*) tunas during steady swimming. *Journal of Experimental Biology* 202: 2127–2138.

Konishi, J. and C.P. Hickman. 1964. Temperature acclimation in the central nervous system of trout (*Salmo gardneri*) *Comparative Biochemistry and Physiology* 13: 433–442.

Korsmeyer, K.E. and H. Dewar. 2001. Tuna metabolism and energetics. In: *Fish Physiology*, B.A. Block and E.D. Stevens(Eds.). Academic Press, San Diego, Vol. 19, pp. 35–78.

Korsmeyer, K.E., H. Dewar, N.C. Lai and J.B. Graham. 1996. Tuna aerobic swimming performance: Physiological and environmental limits based on oxygen supply and demand. *Comparative Biochemistry and Physiology* 113: 45–56.

Korsmeyer, K.E., N.C. Lai, R.E. Shadwick and J.B. Graham. 1997a. Heart rate and stroke volume contributions to cardiac output in swimming yellowfin tuna: Response to exercise and temperature. *Journal of Experimental Biology* 200: 1975–1986.

Korsmeyer, K.E., N.C. Lai, R.E. Shadwick and J.B. Graham. 1997b. Oxygen transport and cardiovascular responses to exercise in the yellowfin tuna *Thunnus albacares. Journal of Experimental Biology* 20: 1987–1997.

Koval, A.P. and S.V. Butuzov. 1986. Features of the structure of the internal tendon system in some Scombridae species. *Vestnik Zoologii* 6: 59–65.

Lai, N.C., K.E. Korsmeyer, S. Katz, D.B. Holts, L.M. Laughlin and J.B. Graham. 1997. Hemodynamics and blood properties of the shortfin mako shark (*Isurus oxyrinchus*). *Copeia* 1997, 424–428.

Landeira-Fernandez, A.M., J.M. Morrissette, J.M. Blank and B.A. Block. 2004. Temperature dependence of the Ca^{2}z -ATPase (SERCA2) in the ventricles of tuna and mackerel. *American Journal of Physiology* 286: R398–R404.

Larsen, C., H. Malte and R.E. Weber. 2004. ATP-induced reverse temperature effect in isohemoglobins from the endothermic porbeagle shark, *Lamna nasus. Journal of Biological Chemistry* 278: 30741–30747.

Liao, I.C. and E.Y. Chang. 2003. Role of sensory mechanisms in predatory feeding behavior of juvenile red drum *Sciaenops ocellatus. Fisheries Sciences* 69: 317–322.

Lighthill, M.J. 1970. Aquatic animal propulsion of high hydromechanical efficiency. *Journal of Fluid Mechanics* 44: 265–301.

Linthicum, D.S. and F.G. Carey. 1972. Regulation of brain and eye temperatures by the bluefin tuna. *Comparative Biochemistry and Physiology* 43: 425–433.

Lowe, T.E., R.W. Brill and K.L. Cousins. 2000. Blood oxygen-binding characteristics of bigeye tuna (*Thunnus obesus*), a high-energy-demand teleost that is tolerant of low ambient oxygen. *Marine Biology* 136: 1087–1098.

Luo, J., E.D. Prince, C.P. Goodyear, B.E. Luckhurst and J.E. Serafy. 2006. Vertical habitat utilization by large pelagic animals: a quantitative framework and numerical method for use with pop-up satellite tag data. *Fisheries Oceanography* 15: 208–229.

Lutcavage, M.E., R.W. Brill, G.B. Skomal, B.C. Chase and P.W. Howey. 1999. Results of pop-up satellite tagging of spawning size class fish in the Gulf of Maine: Do North Atlantic bluefin tuna spawn in the mid-Atlantic? *Canadian Journal of Fisheries and Aquatic Sciences* 56: 173–177.

Lutcavage, M.E., R.W. Brill, G.B. Skomal, B.C. Chase, J.L. Goldstein and J. Tutein. 2000. Tracking adult North Atlantic bluefin tuna (*Thunnus thynnus*) in the northwestern Atlantic using ultrasonic telemetry. *Marine Biology* 137: 347–358.

Magnuson, J.J. 1970. Hydrostatic equilibrium of *Euthynnus affinis* a pelagic teleost without a gas bladder. *Copeia* 1970: 56–85.

Magnuson, J.J. 1973. Comparative study of adaptations for continuous swimming and hydrostatic equilibrium of scombroid and xiphoid fishes. *Fishery Bulletin* 71: 337–356.

Magnuson, J.J. 1978. Locomotion by scombrid fishes hydromechanics morphology and behavior. In: *Fish Physiology*, W.S. Hoar and D.J. Randall (Eds.). Academic Press, New York, NY, Vol. 7: Locomotion, pp. 239–313.

Magnuson, J.J. and D. Weininger. 1978. Estimation of minimum sustained speed and associated body drag of scombrids. In: *The Physiological Ecology of Tunas*. G.D. Sharp and A.E. Dizon (Eds.). Academic Press, New York, NY, pp. 293–311.

Malte, H., C. Larsen, M. Musyl and R. Brill. 2007. Differential heating and cooling rates in bigeye tuna (*Thunnus obesus* Lowe): a model of non-steady state heat exchange. *Journal of Experimental Biology* 210: 2618–2626.

Mathieu-Costello, O., R.W. Brill and P.W. Hochachka. 1995. Design for a high speed path for oxygen: Tuna red muscle ultrastructure and vascularization. In: *Biochemistry and Molecular Biology of Fishes, Metabolic biochemistry*, P.W. Hochachka and T.P. Mommsen(Eds.). Elsevier Science Publishers, Amsterdam, Netherlands, pp. 1–13.

Mathieu-Costello, O., R.W. Brill and P.W. Hochachka. 1996. Structural basis for oxygen delivery: Muscle capillaries and manifolds in tuna red muscle. *Comparative Biochemistry and Physiology* 113: 25–31.

Mathieu-Costello, O., P.J. Agey, R.B. Logemann, R.W. Brill and P.W. Hochachka. 1992. Capillary-fiber geometrical relationships in tuna red muscle. *Canadian Journal of Zoology* 70: 1218–1229.

Maunder, M.N. and A.E. Punt. 2004. Standardizing catch and effort data: a review of recent approaches. *Fisheries Research* 70: 141–159.

Maunder, M.N., M.G. Hinton, K.A. Bigelow and A.D. Langley. 2006. Developing indices of abundance using habitat data in a statistical framework. *Bulletin of Marine Sciences* 79: 545–559.

McCosker, J.E. 1987. The white shark, *Carcharodon carcharias*, has a warm stomach. *Copeia* 1987: 195–197.

Montgomery, J.C. and J.A. Macdonald. 1990. Effects of temperature on nervous system: implications for behavioral performance. *American Journal of Physiology* 259: 191–196.

Moyes, C.D., O.A. Mathieu-Costello, R.W. Brill and P.W. Hochachka. 1992. Mitochondrial metabolism of cardiac and skeletal muscles from a fast (*Katsuwonus pelamis*) and a slow (*Cyprinus carpio*) fish. *Canadian Journal of Zoology* 70: 1246–1253.

Muir, B.S. and J.I. Kendall. 1968. Structural modifications in the gills of tunas and some other oceanic fishes. *Copeia* 1968: 388–398.

Muir, B.S. and G.M. Hughes. 1969. Gill dimensions for three species of tunny. *Journal of Experimental Biology* 51: 271–285.

Musyl, M.K., L.M. McNaughon, Y. Swimmer and R.W. Brill. 2004. Convergent evolution of vertical movement behavior in swordfish, bigeye tuna, and bigeye thresher sharks, Vertical niche partitioning in the pelagic environment as shown by electronic tagging studies. *Pelagic Fisheries Research Program, University of Hawaii at Manoa, Newsletter* 9: 1–4.

Musyl, M.K., R.W. Brill, C.H. Boggs, D.S. Curran, T.K. Kazama and M.P. Seki. 2003. Vertical movements of bigeye tuna (*Thunnus obesus*) associated with islands, buoys, and seamounts near the main Hawaiian Islands from archival tagging data. *Fisheries Oceanography* 12: 152–169.

Myers, R.A. and B. Worm. 2003. Rapid worldwide depletion of predatory fish communities. *Nature* (London) 423: 280–283.

Nakamura, Y. 1991. Tracking of the mature female of flying squid *Ommastrephes bartrami* by an ultrasonic transmitter. *Bulletin of Hokkaido National Fisheries Research Institute* 55: 205–208.

Nakamura, Y. 1993. Vertical and horizontal movements of mature females of *Ommastrephes bartramii* observed by ultrasonic telemetry. In: *Recent Advances in Fishery Biology*, T. Okutani, R.K. O'Dor and T. Kubodera (Eds.). Tokai University Press, Tokyo, pp. 331–336.

Nakano, H., H. Matsunaga, H. Okamoto and M. Okazaki. 2003. Acoustic tracking of bigeye thresher shark, *Alopias superciliosus,* in the eastern Pacific Ocean. *Marine Ecology Progress Series* 265: 255–261.

Natanson, L.J., N.E. Kohler, D. Ardizzone, G.M. Cailliet, S. Wintner and H.F. Mollet. 2006. Validated age and growth estimates for the shortfin mako, Isurus oxyrinchus, in the North Atlantic Ocean. *Environmental Biology of Fishes* 77: 367–383.

Neill, W.H., R.K.C. Chang and A.E. Dizon. 1976. Magnitude and ecological implications of thermal inertia in skipjack tuna *Katsuwonus pelamis. Environmental Biology of Fishes* 1: 61–80.

Neill, W.H., E. Donstevens, F.G. Carey and K.D. Lawson. 1974. Thermal inertia vs thermoregulation in warm turtles and tunas. *Science* 184: 1008–1010.

Oikawa, S. and T. Kanda. 1997. Some features of the gills of a megamouth shark and a shortfin mako, with reference to metabolic activity. In: *Biology of the Megamouth Shark*, K. Yano, J.F. Morrissey, Y. Yabumoto and K. Nakaya (Eds.).Tokai University Press, Tokyo, pp. 93–104.

Pepperell, J.G. and T.L.O. Davis. 1999. Post-release behaviour of black marlin, *Makaira indica*, caught off the Great Barrier Reef with sportfishing gear. *Marine Biology* 135: 369–380.

Perry, S.F., C. Daxboeck, B. Emmett, P.W. Hochachka and R.W. Brill. 1985. Effects of temperature change on acid-base regulation in skipjack tuna (*Katsuwonus pelamis*) blood. *Comparative Biochemistry and Physiology* 81: 49–54.

Potier, M., F. Marsac, Y. Cherel, V. Lucas, R. Sabatié, O. Muary and F. Ménard. 2007. Forage fauna in the diet of three large pelagic fishes (lancetfish, swordfish and yellowfin tuna) in the western equatorial Indian Ocean. *Fisheries Research* 83: 60–72.

Powers, D.A. 1980. Molecular ecology of teleost fish hemoglobins strategies for adapting to changing environments. *American Zoologist* 20: 139–162.

Powers, D.A. 1985. Molecular and cellular adaptations of fish hemoglobin-oxygen affinity to environmental changes. In: *Respiratory Pigments in Animals: Relation Structure-Function. First International Congress of Comparative Physiology and Biochemistry; Liege, Belgium*, J. Lamy, J.P. Truchot and R. Gilles (Eds.). Springer-Verlag, New York, NY, pp. 97–124.

Priede, I.G. 1985. Metabolic scope in fishes. In: *Fish Energetics New Perspectives*, T.P. Calow(Ed.). John Hopkins University Press, Baltimore.

Prince, E.D. and C.P. Goodyear. 2006. Hypoxia-based habitat compression of tropical pelagic fishes. *Fisheries and Oceanography* 15: 451–464.

Rhodes, D. and R. Smith. 1983. Body temperature of the salmon shark, *Lamna ditropis*. *Journal of the Marine Biological Association of the United Kingdom*. U.K. 63: 243–244.

Roberts, J.L. 1978. Ram gill ventilation in fish. In: *The Physiological Ecology of Tunas*. G.D. Sharp and A.E. Dizon (Eds.). Academic Press, New York, NY, pp. 83–87.

Rossi-Fanelli, A. and E. Antonini. 1960. Oxygen equilibrium of hemoglobin from *Thunnus thynnus*. *Nature* (London) 186: 895–896.

Sanders, N.K. and J.J. Childress. 1990. Adaptations to the deep-sea oxygen minimum layer oxygen binding by the hemocyanin of the bathypelagic mysid *Gnathophausia Ingens* Dohrn. *Biological Bulletin* 178: 286–294.

Schaefer, K., D. Fuller and B. Block. 2007. Movements, behavior, and habitat utilization of yellowfin tuna (*Thunnus albacares*) in the northeastern Pacific Ocean, ascertained through archival tag data. *Marine Biology* 152: 503–525.

Schaefer, K.M. 1985. Body temperatures in troll-caught frigate tuna, *Auxis thazard*. *Copeia* 1985, 231–233.

Schaefer, K.M. and D.W. Fuller. 2002. Movements, behavior, and habitat selection of bigeye tuna (*Thunnus obesus*) in the eastern equatorial Pacific, ascertained through archival tags. *Fishery Bulletin* 100: 765–788.

Sepulveda, C.A. and K.A. Dickson. 2000. Maximum sustainable speeds and cost of swimming in juvenile kawakawa tuna (*Euthynnus affinis*) and chub mackerel (*Scomber japonicus*). *Journal of Experimental Biology* 203: 3089–3101.

Sepulveda, C.A., J.B. Graham and D. Bernal. 2007a. Aerobic metabolic rates of swimming juvenile mako sharks, *Isurus oxyrinchus*. *Marine Biology* 152: 1087–1094.

Sepulveda, C.A., N.C. Wegner, D. Bernal and J.B. Graham. 2005. The red muscle morphology of the thresher sharks (family Alopiidae). *Journal of Experimental Biology* 208: 4255–4261.

Sepulveda, C.A., K.A. Dickson and J.B. Graham. 2003. Swimming performance studies on the eastern Pacific bonito *Sarda chiliensis*, a close relative of the tunas (family Scombridae) I. Energetics *Journal of Experimental Biology* 206: 2739–2748.

Sepulveda, C.A., K.A. Dickson, D. Bernal and J.B. Graham. 2008. Elevated red myotomal muscle temperatures in the most basal tuna species, *Allothunnus fallai*. *Journal of Fish Biology* 73: 241–249.

Sepulveda, C.A., K.A. Dickson, L.R. Frank and J.B. Graham. 2007b. Cranial endothermy and a putative brain heater in the most basal tuna species, *Allothunnus fallai*. *Journal of Fish Biology* 70: 1720–1733.

Sepulveda, C.A., S. Kohin, C. Chan, R. Vetter and J.B. Graham. 2004. Movement patterns, depth preferences, and stomach temperatures of free-swimming juvenile mako sharks, *Isurus oxyrinchus*, in the Southern California Bight. *Marine Biology* 145: 191–199.

Shadwick, R.E., S.L. Katz, K.E. Korsmeyer, T. Knower and J.W. Covell. 1999. Muscle dynamics in skipjack tuna: Timing of red muscle shortening in relation to activation and body curvature during steady swimming. *Journal of Experimental Biology* 202: 2139–2150.

Shiels, H.A., J.M. Blank, A.P. Farrell and B.A.Block. 2004. Electrophysiological properties of the L-type Ca²⁺ current in cardiomyocytes from bluefin tuna and Pacific mackerel. *American Journal of Physiology* 286: R659–668.

Sibert, J., J. Hampton, P. Kleiber and M. Maunder. 2006. Biomass, size, and trophic status of top predators in the Pacific Ocean. *Science* 314: 1773–1776.

Sims, D.W., M.J. Witt, A.J. Richardson, E.J. Southall and J.D. Metcalfe. 2006. Encounter success of free-ranging marine predator movements across a dynamic prey landscape. *Proceedings of the Royal Society* B273: 1195–1201.

Sims, D.W., E.J. Southall, N.E. Humphries, G.C. Hays, C.J.A. Bradshaw, J.W. Pitchford, A. James, M.Z. Ahmed, A.S. Brierley, M.A. Hindell, D. Morritt,M.K. Musyl, D. Righton, E.L.C. Shepard, V.J. Wearmouth, R.P. Wilson, M.J. Witt and J.D. Metcalfe. 2008. Scaling laws of marine predator search behaviour. *Nature* (London) 451: 1098–1103.

Steffensen, J.F. 1989. Some errors in respirometry of aquatic breathers: how to avoid and correct them. *Fish Physiology and Biochemistry* 6: 49–59.

Stensholt, B.K., A. Aglen, S. Mehl and E. Stensholt. 2002. Vertical density distributions of fish: A balance between environmental and physiological limitation. *ICES Journal of Marine Science* 59: 679–710.

Stevens, E.D. and F.G. Carey. 1981. Why of the warmth of warm bodied fish. *American Journal of Physiology* 240: R151–R155.

Stevens, E.D. and J.M. McLeese. 1984. Why bluefin tuna *Thunnus thynnus* have warm tummies temperature effect on trypsin and chymotrypsin. *American Journal of Physiology* 246: R487–R494.

Stevens, E.D., J.W. Kanwisher and F.G. Carey. 2000. Muscle temperature in free-swimming giant Atlantic bluefin tuna (*Thunnus, thynnus* L.). *Journal of Thermal Biology* 25: 419–423.

Stevens, J.D. 1983. Observations on reproduction in the shortfin mako *Isurus oxyrinchus*. *Copeia* 1983, 126–130.

Stokesbury, M.J.W., S.L.H. Teo, A. Seitz, R.K. O'Dor and B.A. Block. 2004. Movement of Atlantic bluefin tuna (*Thunnus thynnus*) as determined by satellite tagging experiments initiated off New England. *Canadian Journal of Fisheries and Aquatic Sciences* 61: 1976–1987.

Sund, P.N., M. Blackburn and F. Williams. 1981. Tunas and their environment in the Pacific Ocean: a review. *Oceanography and Marine Biology Annual Reviews* 19: 443–512.

Syme, D.A. and R.E. Shadwick. 2002. Effects of longitudinal body position and swimming speed on mechanical power of deep red muscle from skipjack tuna (*Katsuwonus pelamis*). *Journal of Experimental Biology* 205: 189–200.

Takahashi, M., H. Okamura, K. Yokawa and M. Okazaki. 2003. Swimming behaviour and migration of a swordfish recorded by an archival tag. *Marine and Freshwater Research* 54: 527–534.

Teo, S.L.H., A.M. Boustany and B.A. Block. 2007a. Oceanographic preferences of Atlantic bluefin tuna, *Thunnus thynnus*, on their Gulf of Mexico breeding grounds. *Marine Biology* 152: 1105–1119.

Teo, S.L.H., A. Boustany, H. Dewar, M.J.W. Stokesbury, K.C. Weng, S. Beemer, A.C. Seitz, C.J. Farwell, E.D. Prince and B.A. Block. 2007b. Annual migrations, diving behavior, and thermal biology of Atlantic bluefin tuna, *Thunnus thynnus*, on their Gulf of Mexico breeding grounds. *Marine Biology* 151: 1–18.

Val, A.L., M.N.P. Silva and V.M.F. Almeida-Val. 1998. Hypoxia adaptation in fish of the Amazon: A never-ending task. *South African Journal of Zoology* 33: 107–114.

van den Burg, E.H., Peeters, R.R., Verhoye, M., Meek, J., Flik, G. and Van der Linden, A. 2005. Brain responses to ambient temperature fluctuations in fish: Reduction of blood volume and initiation of a whole-body stress response. *Journal of Neurophysiology* 93: 2849–2855.

Wainwright, S.A. 1983. To bend a fish. In: *Fish Biomechanics*, P.W. Webb and D. Weihs (Eds.). Praeger, New York, pp. 68–91.

Ward, P. and R.A. Myers. 2005a. Shifts in open-ocean fish communities coinciding with the commencement of commercial fishing. *Ecology* 86: 835–847.

Ward, P. and R.A. Myers. 2005b. Inferring the depth distribution of catchability for pelagic fishes and correcting for variations in the depth of longline fishing gear. *Canadian Journal of Fisheries and Aquatic Sciences* 62: 1130–1142.

Webb, P.W. 1998. Swimming. In: *The Physiology of Fishes*, D.H. Evans (Ed.). CRC Press, New York, pp. 3–24.

Weber, R.E. and G. Lykkeboe. 1978. Respiratory adaptations in carp blood influences of hypoxia red cell organic phosphates divalent cations and carbon dioxide on hemoglobin oxygen affinity. *Journal of Comparative Physiology* B128: 127–138.

Weber, R.E., G. Lykkeboe and K. Johansen. 1976. Physiological properties of eel hemoglobin hypoxic acclimation phosphate effects and multiplicity. *Journal of Experimental Biology* 64: 75–88.

Wegner, N.C., C.A. Sepulveda and J.B. Graham. 2006. Gill specializations in high-performance pelagic teleosts, with reference to striped marlin (*Tetrapturus audax*) and wahoo (*Acanthocybium solandri*). *Bulletin of Marine Sciences* 79: 747–759.

Weng, K.C. and B.A. Block. 2004. Diel vertical migration of the bigeye thresher shark (*Alopias superciliosus*), a species possessing orbital retia mirabilia. *Fishery Bulletin* 102: 221–229.

Weng, K.C., P.C. Castilho, J.M. Morrissette, A.M. Landeira-Fernandez, D.B. Holts, R.J. Schallert, K.J. Goldman and B.A. Block. 2005. Satellite tagging and cardiac physiology reveal niche expansion in salmon sharks. *Science* 310: 104–106.

Westneat, M.W., W. Hoese, C.A. Pell and S.A. Wainwright. 1993. The horizontal septum: Mechanisms of force transfer in locomotion of scombrid fishes (Scombridae, Perciformes). *Journal of Morphology* 217: 183–204.

Wittenberg, B.A. and J.B. Wittenberg. 1989. Transport of Oxygen in Muscle. In: *Annual Review of Physiology*, J.F. Hoffman(Ed.). Annual Reviews Inc, Palo Alto, CA, Vol. 51, pp. 857–878.

Wittenberg, J.B. 1970. Myoglobin facilitated oxygen diffusion: role of myoglobin in oxygen entry into muscle. *Physiological Zoology* 50: 559–636.

Wolf, N.G., P.R. Swift and F.G. Carey. 1988. Swimming muscle helps warm the brain of lamnid sharks. *Journal of Comparative Physiology* B157: 709–716.

Wood, S.C., K. Johansen and R.E. Weber. 1975. ATP modulated adaptations of fish blood to environmental oxygen tension. *Federation Proceedings — Federation of American Society for Experimental Biology* 34: 452.

Wood, C.M. and S.F. Perry. 1985. Respiratory, circulatory, and metabolic adjustment to exercise in fish. In: *Circulation, Respiration, and Metabolism—Current Comparative Approaches*. R. Gilles (Eds.). Springer-Verlag, Berlin, pp. 2–22.

Worm, B., E.B. Barbier, N. Beaumont, J.E. Duffy, C. Folke, B.S. Halpern, J.B.C. Jackson, H.K. Lotze, F. Micheli, S.R. Palumbi, E. Sala, K.A. Selkoe, J.J. Stachowicz and R. Watson. 2006. Impacts of biodiversity loss on ocean ecosystem services. *Science* 314: 787–790.

Yamamoto, K., Y. Itazawa and H. Kobayashi. 1981. Relationship between gas content and hematocrit value in yellowtail blood. *Journal of the Faculty of Agriculture, Kyushu University* 26: 31–37.

Yang, T.H., N.C. Lai, J.B. Graham and G.N. Somero. 1992. Respiratory, blood, and heart enzymatic adaptations of *Sebastolobus alascanus* (Scorpaenidae; Teleostei) to the oxygen minimum zone: A comparative study. *Biological Bulletin* 183: 490–499.

Young, J., M. Lansdell, S. Riddoch and A. Revill. 2006a. Feeding ecology of broadbill swordfish, *Xiphias gladius*, off eastern Australia in relation to physical and environmental variables. *Bulletin of Marine Sciences* 79: 793–809.

Young, J.L., Z.B. Bornik, M.L. Marcotte, K.N. Charlie, G.N. Wagner, S.G. Hinch and S.J. Cooke. 2006b. Integrating physiology and life history to improve fisheries management and conservation. *Fish and Fisheries* 7: 262–283.

15

Swimming Capacity of Marine Fishes and its Role in Capture by Fishing Gears

Pingguo He

INTRODUCTION

The swimming ability of fish may be one of the most important factors affecting their capture by or escape from fishing gears. It has a bearing on capture processes by both active (such as trawls and purse seines) and passive (such as gillnets, longlines and traps) fishing gears. Swimming ability of fish varies from species to species and size of the fish and can be influenced by environmental factors such as water temperature. As a result, swimming ability of fish may affect the size and species selectivity of the gear and affect the seasonal and spatial differences in availability and catchability. This chapter will review the swimming ability of some commercial marine fish species and discuss the swimming behavior of these species in relation to their capture by, and, escape from, otter trawls and bottom-set gillnets.

Swimming Capability of Marine Fishes

The Atlantic mackerel (*Scomber scombrus*) may be by far the most extensively studied commercial marine fish species regarding its swimming characteristics

Author's address: University of New Hampshire, Institute for the Study of Earth, Oceans and Space, Durham, NH 03824, USA. E-mail: Pingguo.He@unh.edu

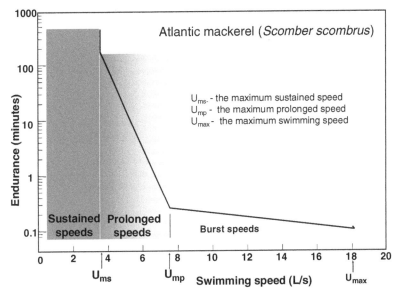

Fig. 15.1 Swimming speed spectrum in the Atlantic mackerel (*Scomber scombrus*) indicating the sustained swimming, endurance and burst swimming. Shades of color indicate the use of red (dark color) and white muscle (light color) (modified from He, 1993a).

and performances. Here mackerel is taken as an example to define different swimming speeds and related parameters. Swimming speed is usually expressed either in absolute terms (e.g., m/s) or in relative terms (body lengths per second, or L/s). Recorded swimming speeds of mackerel range from 0.4 L/s to 18 L/s for fish of about 0.3 m long. Fig. 15.1 illustrates the range of swimming speeds, swimming endurance, and the use of swimming muscles at different speeds (He, 1993a). The definition is given in Table 15.1 with values for mackerel as an example.

Fish swimming at Low Speeds

Atlantic mackerel has no swimbladder and is heavier than sea water; therefore it has to swim continuously in order to generate lift to maintain hydrostatic equilibrium (He and Wardle, 1986). The minimum swimming speed recorded was 0.4 L/s (He and Wardle, 1986). However, the fish had to swim with its head tilted up as much as 27°C to generate additional lift to keep itself afloat at these low speeds (He and Wardle, 1986). Without tilting its body, mackerel has to swim at a minimum speed of 1 L/s in order to maintain hydrostatic equilibrium. The preferred swimming speed (U_p) of mackerel is between 0.9 and 1.2 L/s when they cruise a 10-m diameter annular tank at 11.7°C (He and Wardle, 1988). Mackerel swims at speeds below U_p when space is reduced in a laboratory

Table 15.1 Definition of swimming speeds of fish with the Atlantic mackerel (*Scomber scombrus*) as an example. See He (1993a) for source of value for the species.

Types of speed	Definition	Mackerel (L/s)
Minimum speed	The minimum speed observed. In neutrally buoyant fish, it can be zero	0.4
Preferred speed (in captivity)	Range of swimming speed observed when freely cruising in a large tank	0.9–1.2
Sustained speeds	The range of speeds at which fish can swim for an indefinite period of time such as during migrations	0.4–3.5
Maximum sustained swimming speed	The maximum speed fish can swim without leading to exhaustion (in practice, ≥ 200 min).	3.5
Prolonged swimming speeds	The range of speeds at which endurance reduces with increasing speed	3.5–7.5
Maximum prolonged swimming speed	The maximum speed at which the red muscle has contribution to swimming power, with swimming endurance of 15 s or more	7.5
Burst swimming speeds	Swimming speed with endurance less than 15 s and utilize white muscle as sole power source. Red muscle may still active but may not contribute to swimming power due to overlapping contraction	7.5–18
Maximum swimming speed	The observed or predicted maximum swimming speed	18

tank or when the light level is very low (He and Wardle, 1986; Glass *et al.*, 1986). Tilted swimming was also found in negatively buoyant herring (*Clupea harengus*) swimming in the sea during the night when light level was reduced (Huse and Ona, 1996). Negatively buoyant winter flounder (*Pseudopleuronectes americanus*) and other flounder species adopt swim-and-rest mode (He, 2003; Winger *et al.*, 2004). Flounders can sit on the seabed, but when they swim, the lowest swimming speed is about 0.4 L/s at around 0°C measured at sea using a video camera (He, 2003). Swimming at low speeds and associated lift generation may use up more energy as measured in Pacific Bonito (*Sarda Chiliensis chiliensis*) (Sepulveda *et al.*, 2003). Many neutrally buoyant species such as Atlantic cod (*Gadus morhua*) can stop their swimming motion and their minimum swimming speed is effectively equal to zero.

Atlantic mackerel can swim at sustained swimming speeds between 1 and 3.5 L/s indefinitely (He and Wardle, 1988). At this speed range, mackerel were observed to cruise voluntarily in large tanks, feed at sea, during migration, or swimming with towed fishing gears (Wardle, 1986).

The Maximum Sustained Swimming Speed

When mackerel swim at speeds above 3.5 L/s, their limited endurance will lead to exhaustion. This 3.5 L/s is known as the maximum sustained swimming speed (U_{ms}). U_{ms} marks the upper limit of sustained swimming and the lower

Table 15.2   Maximum swimming speed (Ums) and endurance at prolonged speeds of some marine fishes. T–temperature, FT–flume tank, AT–annular tank, RES–respirometer. Sources: 1–He (1991), 2–Beamish (1966), 3–He and Wardle (1988); 4–Castro-Santos (2005), 5–Breen *et al.* (2004), 6–Xu, (1989), 7–Beamish (1984).

Species	Length (m)	Ums (m/s)	E - U relation	Method	T (°C)	Ref
Atlantic cod, *Gadus morhua*	0.40	0.42		FT	0.8	1
	0.49	0.45			0.8	
Atlantic cod, *Gadus morhua*	0.36	0.75	Log E = −0.99·U+ 3.99	FT	5	2
	0.36	0.90	Log E= −1.13·U + 4.96		8	
Atlantic herring, *Clupea harengu*	0.25	1.02	Log E= −1.43·U + 8.37	AT	13.5	3
Atlantic mackerel, *Scomber scombrus*	0.31	1.10	Log E= −0.96·U + 5.45	AT	11.7	3
American shad, *Alosa sapidissima*	0.42		Ln E= −1.00·U + 10.70	FT		4
Haddock, *Melanogrammus aeglefinus*	0.17	0.44		AT	9.9	5
	0.24	0.53				
	0.31	0.58				
	0.34	0.57				
	0.41	0.60				
Jack mackerel, *Trachurus japonicus*	0.14-0.21	0.90	Log E= −7.2·LogU + 9.3	FT	19	6
Japanese mackerel, *scomber japonicus*	0.10	0.99	Log E= −0.62·U + 4.38	FT	19	7
Redfish, *Sebestes marinus*	0.17	0.52	Log E= −0.25·U + 1.71	FT	5	2
	0.16		Log E= −0.23·U + 1.70	FT	8	
	0.16		Log E= −0.42·U + 2.94	FT	11	
Saithe, *Pollachius virens*	0.25	0.88	Log E= −1.63·U + 5.60	AT	14.4	3
	0.34	1.00	Log E= −1.52·U + 5.91			
	0.42	1.06	Log E= −1.36·U + 6.16			
	0.50	1.10	Log E= −1.17·U + 5.95			
Striped bass, *Morone saxatilis*	0.42-0.57		Ln E= −0.39·U + 5.99	FT		4

limit of prolonged swimming. The value of U_{ms} varies with species, body size, water temperature and other biological and environmental factors (Table 15.2).

The maximum sustained swimming speed has been measured in several commercial marine species. U_{ms} (in m/s) increases with fish body length (L, in m) in all marine species, but U_{ms} for saithe (*Pollachius virens*), a demersal species in the Gadoid family, is consistently lower than that for pelagic species (mackerel, herring and jack mackerel) even though they are of similar lengths (Fig. 15.2). Large pelagic species such as bluefin tuna (*Thunnus thynnus*) can grow to more than 3 m in length. Bluefin tunas in farm cages with a mean length of 2.44 m swam at an average speed of 0.77 L/s or 1.88 m/s when observed by a video camera (Wardle *et al.*, 1989). If the line in Fig. 15.2 for pelagic species is extrapolated to 2.44 m, we can project U_{ms} for the 2.44 m tuna to be about 2.10 m/s (Fig. 15.2: Inset) or 0.86 L/s.

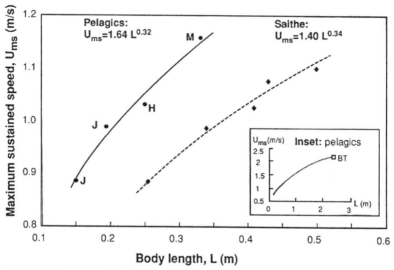

Fig. 15.2 Maximum sustained swimming speeds of some marine fish species in relation to their body length (He, 1993). Black circle (●): J–jack mackerel (*Trachurus japonicus*), H–herring (*Clupea harengus*). Square (□): BT–blufin tuna (*Thunnus thynnus*). Diamond (♦): saithe (*Pollachius virens*). Used with the kind permission of the International Council for the Exploration of the Sea.

In fish swimming literature, the term "critical swimming speed" (U_{crit}) is often used when measuring swimming speed in the laboratory (Brett, 1964; Brett and Glass, 1973; Plaut, 2001). U_{crit} is associated with a time value, usually 15, 20, 30 min or 60 min. In U_{crit} measurement, the swimming speed (or water velocity) is increased stepwise at predetermined fixed value (ΔU) at a fixed time interval (T). The final U_{crit} is calculated by the following formula derived from Brett (1964):

$$U_{crit} = U_f - \Delta U \left(1 - \frac{t}{T} \right)$$

Where t is the time duration the fish swam at the final speed (U_f) before exhaustion. The critical swimming speed has been used to evaluate prolonged swimming capabilities for different species and under different conditions in laboratory and for applications in fishway constructions (Brett and Glass, 1973; Peake, 2004). Measurements of critical swimming speed with 15, 20 and 30 min time intervals are shown in Table 15.3 for some commercial marine species.

Endurance at Prolonged Swimming Speeds

When swimming at speeds greater than U_{ms}, the faster the fish swim, the lesser their endurance. The maximum swimming speed involving power

Table 15.3 The critical swimming speed of some marine fishes. T–temperature. FT–flume tank, RES–respirometer, WC–water channel. Sources: 1–Joaquin *et al.* (2004), 2–Reidy *et al.* (2000), 3–Dickson *et al.* (2002), 4–Sepulveda and Dickson (2000), 5–Freadman (1979), 6–Graham *et al.* (1990), 7–Koumoundouros *et al.* (2002), 8–Claireaux *et al.* (2006).

Species	Length	Speed		Method	T	Ref.
	(m)	*m/s*	*L/s*		*(ºC)*	
15-min critical swimming speed						
Winter flounder,	0.36	0.23	0.65	FT	4	1
Pseudopleuronectes americanus	0.39	0.28	0.73	FT	10	
30-min critical swimming speed						
Cod, *Gadus morhua*	0.52	0.58	1.1	RES	5	2
Chub mackerel, *Scomber japonicus*	0.18	0.66	3.7	RES	18	3
	0.25	0.80	3.2			
Chub mackerel, *Scomber japonicus*	0.17	0.85	5.0	RES	24	4
	0.23	1.00	4.3			
Bluefish, *Pomatomus saltatrix*	0.22	1.02	4.64		15	5
Kawakawa, *Euthunnus affinis*	0.12	0.45	3.8	RES	24	4
	0.18	0.66	3.7			
	0.23	0.97	4.2			
Lemon shark, *Negaprion brevirostris*	0.7	0.77	1.1	FT	17–26	6
Leopard shark, *Triakis semifasciata*	0.37	0.53	1.43		19–23	6
	0.53	0.93	1.76		19–25	
Sea bass, *Dicentrarchus labrax*	0.025	0.15	6.0	WC	15	7
	0.025	0.20	8.0		20	
	0.025	0.22	8.8		25	
	0.025	0.21	8.4		28	
Sea bass, *Dicentrarchus labrax*	0.24	0.67	2.8	WC	7	8
	0.24	0.77	3.2		11	
	0.24	0.84	3.5		14	
	0.24	0.88	3.7		18	
	0.24	0.90	3.8		22	
	0.24	1.08	4.5		26	
	0.24	1.12	4.7		30	
Striped bass, *Morone saxatilis*	0.27	0.86	3.19	RES	15	5
Yellowfin tuna, *Thunnus albacares*	0.20	1.11	5.6	RES	24	4

from the red muscle may be predicted in the same way as Wardle (1975) did for the swimming involving the white muscle. The fastest red muscle contraction time is about twice that of the white muscle in mackerel (He *et al.*, 1990). The above authors predicted that at swimming speeds above 7.5 L/s, the red muscle would become ineffective as an energy source for swimming due to simultaneous contraction of the muscle on both sides. The forces from the opposing side would cancel one another, resulting in zero net contribution to swimming. The predicted value of 7.5 L/s is thus referred to as the maximum prolonged swimming speed (U_{mp}) and the range of speeds

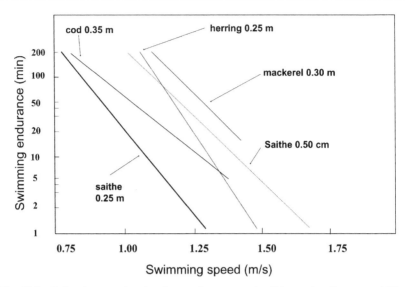

Fig. 15.3   Swimming speed and endurance in some marine fish species. Source: cod (*Gadus morhua*), He (1991), herring (*Clupea harengus*), mackerel (*Scomber scombrus*), and saithe (*Pollachius virens*), He and Wardle (1988).

between U_{ms} and U_{mp} is called the prolonged swimming speed at which the swimming speed is inversely related to its endurance.

Endurance at prolonged swimming speeds has been measured using various methods such as the "fish wheel" (Bainbridge, 1960), swimming flumes (e.g., Beamish, 1966; Xu, 1989; He, 1991), stationary annular tanks (He and Wardle, 1988; Breen *et al.*, 2004), and from at-sea observations near fishing gears (Wardle, 1983). He and Wardle (1988) measured swimming speed and endurance of mackerel, herring and saithe in a 10-m diameter annular tank. The fish were induced to swim through still water by moving a projected visual pattern, imitating fish swimming in the mouth area of a towed trawl. Measured endurance of some commercial marine species is plotted against swimming speed in Fig. 15.3. More detailed data on endurance swimming are given in Table 15.2. In all species swimming at prolonged speeds, endurance drops as speed is increased.

Larger fish can swim longer at the same absolute speed (in m/s) or they can swim faster at the same endurance (Fig. 15.3). For example, a 0.50 m long saithe can swim for 30 min at 1.25 m/s, while a 0.25 m saithe can only swim for 2 min at the same speed of 1.25 m/s. For the same endurance of 30 min, a 0.25 m saithe can only swim at 0.95 m/s compared with 1.25 m/s for a 0.50 m saithe. Pelagic species have longer swimming endurance than demersal species. At 14°C, a herring of 0.25 m can swim for over 15 min at 1.25 m/s, while a saithe of the same length can only sustain for 2 min at the same swimming speed.

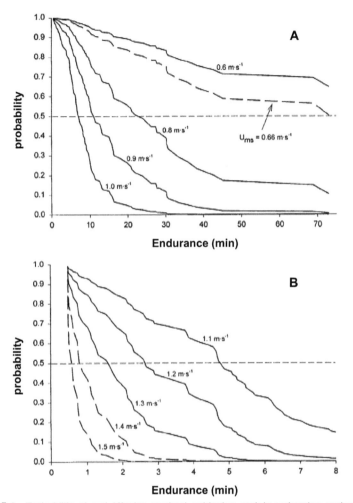

Fig. 15.4 Probability of cod (*Gadus morhua*) achieving certain swimming endurances at various swimming speeds. (A) 0.6 to 1.0 m/s, (B) 1.1 to 1.5 m/s. (From Winger *et al.*, 2000) © 2000 NRC Canada or its licensors. Reproduced with permission.

Winger *et al.* (1999) employed a hazard model to predict the probability of American plaice (*Hippoglossoides platessoides*) swimming at a certain speed using swimming endurance measurement obtained from a swimming flume. They later applied the same model to predict endurance probability of Atlantic cod swimming at different speeds (Winger *et al.*, 2000). Larger fish has higher probability of achieving certain endurance level than smaller fish. When swimming at the same speed, the probability of fish achieving certain endurance is much greater at lower swimming speeds. Figure 15.4 illustrates endurance probability of cod swimming at different speeds.

Several authors have modeled endurance (E) and swimming speed (U) relationships using with a log-linear relationship (Beamish, 1966; He and Wardle, 1988), e.g. Log E = $a \cdot$ U + b, where a and b are coefficients. Breen *et al.* (2004) examined the relationship for haddock (*Melanogrammus aeglefinus*) with an inverse linear model: 1/E = $a \cdot$ U + b. Regardless of the models used, decreased endurance has been shown to be associated with higher swimming speeds during prolonged swimming activities.

Maximum Swimming Speed of Marine Fishes

Swimming speeds beyond U_{mp} are called the burst speeds and are powered by fast, glycolytic white muscle. The fastest maximum swimming speed (U_{max}) of Atlantic mackerel ever recorded is 18 L/s, which is very close to the U_{max} of 19 L/s predicted from the white muscle contraction time (26 ms) measured at 12°C (Wardle and He, 1988).

It is extremely difficult to make accurate estimates of maximum swimming speed of fish and at the same time to prove that the speed recorded is the maximum for the species and the size under the prevailing conditions. Walters and Fierstine (1964) reported that they had measured maximum swimming speed of tuna in the wild by using a specially designed reel with a magnetized marked line running over a magnetic pick-up device. A swimming speed of 20.7 m/s or 21 L/s was recorded for a school of yellowfin tuna (*Thunnus albacares*) which had a mean length of 0.98 m; this is one of the fastest speeds ever recorded in fish. The recordings, however, showed a lot of variation as they were made on a small boat in the open sea with the fish dashing away from it. Even within the relative unit of L/s, the recorded fastest speed of yellowfin tuna is still one of the highest. Measurements of swimming speed (Bainbridge, 1958; Blaxter and Dickson, 1959; Hunter and Zweifel, 1971) and muscle physiology experiments (Wardle, 1975, 1977) have shown that a smaller fish should be able to achieve a higher relative speed (in L/s) than a larger fish. But only a few fish species have been shown to swim over 20 L/s (Walters and Fierstine, 1964; Wardle, 1975; Videler, 1993; Domenici and Blake, 1997). This suggests that most reported "maximum" swimming speeds may not be the maximum swimming speeds fish can actually achieve as suggested by Beamish (1978).

Maximum swimming speed (U_{max}, in m/s) has been shown to increase with body length (L) and to vary among fish species. Fig. 15.5 shows measured U_{max} (in m/s) of some marine fishes plotted against L (in m). Two solid lines represent 10 and 20 L/s. Notice that most fish have U_{max} between 10 and 20 L/s. The actual U_{max} of saithe and cod are probably higher than that recorded, as compared with that predicted from the contraction time of their white muscles (below).

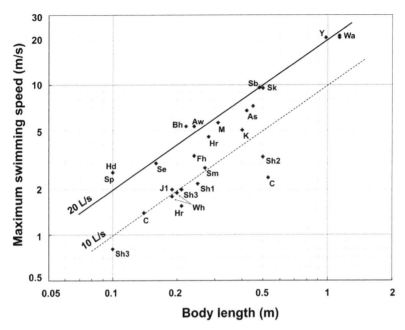

Fig. 15.5 Maximum swimming speed of some marine species in relation to their body length. Letter symbols, sources, and temperatures (when available) are: As–American shad, *Alosa sapidissima,* Castro-santos (2005), Aw–Alewife, *Alosa pseudoharengus,* Castro-santos (2005), Bh–blueback herring, *Alosa aestivalis,* Castro-santos (2005), C–cod, *Gadus morhua,* 9.5–12°C, Blaxter and Dickson (1959); Flathead, *platycephalus bassensis,* 20°C, Yanase *et al.* (2007), Hd–haddock, *Melanogrammus aeglefinus,* 12°C, Wardle (1975); Hr–herring, *Clupea harengus,* 10–15°C, Misund (1989); Jl–jack mackerel, *Trachurus japonicus,* 23°C, Xu (1989); J2–jack mackerel, *Trachurus symmetricus,* Hunter and Zweifel (1971); K–kawakawa, *Euthunnus affinis,* 25°C, cited in Beamish (1978); M–Atlantic mackerel, *Scomber scombrus,* 12°C, Wardle and He (1988); Sb–striped bass, *Morone saxatilis,* Castro-santos (2005), Se–seabass, *Dicentrarchus labrax,* 20 °C, Nelson and Claireaux (2005), Sh1–saithe, *Pollachius virens,* Videler (1993), Sh2–saithe,10.8°C, He (1986); Sh3–saithe, 14–16°C, Blaxter and Dickson (1959); Sk–skipjack tuna, *Katsuwonus pelamis,* cited in Magnuson (1978); Sp–sprat, *Sprattus sprattus,* 12°C, Wardle (1975); Wa–wahoo, *Acanthocybium solandrei,* >15°C, Fierstine and Walters (1968), Wh–whiting, *Gadus merlangus,* 9–13°C, Blaxter and Dickson (1959); Y–yellowfin tuna, *Thunnus albacares,* Fierstine and Walters (1968). Lines for 20 L/s (solid) and 10 L/s (dashed) are drawn to indicate swimming speed in L/s.

Maximum swimming speed can be predicted from contraction times of the white muscle (Wardle, 1975) and stride length (SL). Contraction time does not change among fish species, but varies with body length (Wardle, 1977), temperature (Brill and Dizon, 1979; Wardle, 1980) and location of muscle along the body (Wardle, 1985; Altringham and Ellerby, 1999). Stride length is species-specific with larger values for species with an efficient caudal fin and a low-drag streamlined body. Stride length values have been shown to vary between 0.60 and 1.04 L in marine fishes (Videler and Wardle, 1991).

Scombroid fishes such as Atlantic mackerel and tuna have a well-streamlined body and a large lunate-shaped caudal fin with a high aspect ratio (Magnuson, 1978). For example, Atlantic mackerel (SL=1.0 L, Wardle and He, 1988) will be able to swim 67% faster than a cod of the same length (SL=0.60 L, Videler and Wardle, 1991) if they beat their tails at the maximum frequency which is limited by muscle mechanics (Wardle, 1975).

Effect of Temperature on the Swimming Speed of Fish

Temperature has an interesting effect on the sustained and the prolonged swimming speeds of fish. Fishes, except some large tunas and large sharks, are poikilothermic and the temperature of their entire bodies is usually close to that of the water (Dean, 1976). Tunas, billfish and some large sharks have been shown to have higher temperatures in their internal red muscle, eyes and brain than that of the surrounding water (Carey and Teal, 1969, Stevens and Neill, 1978; Block and Carey, 1985; Block et al., 2001). In poikilothermic fish, the effect of water temperature on the prolonged swimming speed of fish has been well demonstrated (Brett et al., 1958; Beamish, 1966; He, 1991; Dickson et al., 2002). Within the range of thermal tolerance the prolonged swimming capacity usually increases with the temperature, but will eventually decrease when approaching their upper range of thermal tolerance.

There are no systematic, direct measurements of the maximum swimming speed at different water temperatures. But measurements of white muscle contraction time indicate a higher maximum swimming speed at higher temperatures in almost all marine fish species. Figure 15.6 shows the maximum swimming speed of cod at temperatures between 2 and 12°C calculated from the measured muscle contraction time, indicating a doubling of their theoretical maximum swimming speed with a temperature increase of 10°C, i.e. a Q_{10} of 2 (He, 1993a). These predicted swimming speeds may be compared with those of five Antarctic species of 0.21 to 0. 26 m in length with maximum swimming speeds between 0.91 and 1.39 m/s at −1°C (Franklin et al., 2003).

A reduction in temperature reduces prolonged swimming speed and the endurance of Atlantic cod (Beamish, 1966; He, 1991) (Fig. 15.7). Typical environmental temperature for the Atlantic cod in Newfoundland Grand Banks is between 0 and 5°C. It has been shown that a reduction in temperature from 5 to 0°C can cause a reduction in swimming speed of 0.35 m Atlantic cod from 1.05 m/s to 0.65 m/s for a swimming endurance of 30 min (He, 1991). If a cod has to swim at the speed of 1 m/s, e.g., in keeping with a towed trawl, the cod would be exhausted in 2 min at 0°C instead of 50 min at 5°C.

Experiments with American plaice of 0.35 m long indicate that the probability of achieving particular endurance level is increased at higher

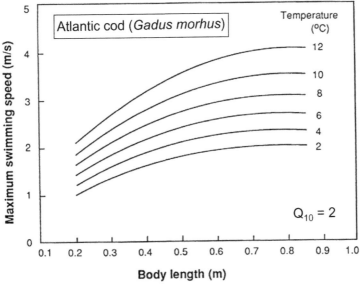

Fig. 15.6 Maximum swimming speed of Atlantic cod (*Gadus morhua*) in relation to body length and water temperature as predicted from muscle contraction time (from He, 1993a). Used with the kind permission of the International Council for the Exploration of the Sea.

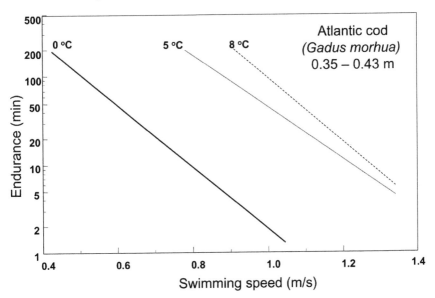

Fig. 15.7 Swimming endurance and swimming speed of Atlantic cod (*Gadus morhua*) 0.35–0.43 m long in relation to water temperature (He, 1991).

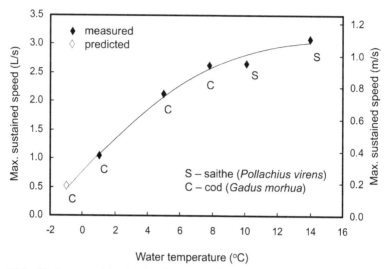

Fig. 15.8 Maximum sustained swimming speed of Gadidae 0.34–0.36 m long in relation to water temperature (He, 1993a). C–Atlantic cod (*Gadus morhua*), S–saithe (*Pollachius virens*). Filled diamond (◆)–measured, unfilled diamond (◇)–predicted. Used with the kind permission of the International Council for the Exploration of the Sea.

water temperatures (Winger *et al.*, 1999). Swimming at 0.3 m/s, a 0.35 m plaice will have 50% probability of swimming for 100 min at 9°C, compared with less than 20% probability when swimming at 0°C.

U_{ms} decreases with a drop in water temperature, especially at low temperatures near their low end of thermal tolerance (He, 1991). U_{ms} of Atlantic cod (0.35–0.42 m) at 0.8°C is only 0.42 m/s compared with 0.90 m/s at 8°C, a reduction of 54%. Figure 15.8 plots temperature (T) against the maximum sustained swimming speed of the saithe and cod. Notice that the increase in temperature from 5°C to 14.4°C results in an increase in U_{ms} from just over 2 L/s to nearly 3 L/s in the two species. The decrease of U_{ms} in Atlantic cod is comparable to a decrease in routine activity (by 75%) and frequency of opercula movement (by 50%) for the same species when temperature was dropped from 8.3 to 0.6°C (Brown *et al.*, 1989). Reduction of swimming and other related activity in some marine fish species as a result of a drop in water temperature was discussed by He (2003).

Fish Swimming in the Natural Environment

There are several methods for measuring or estimating swimming speed of fish in the sea. Walters and Fierstine (1964) used a fishing reel fitted with a magnetic recorder and measured swimming speed of yellowfin tuna and wahoo in the wild. Radio tags were used to track returning salmon in river systems (Hawkins and Smith, 1986). In a marine environment, fish fitted with an ultrasonic tag were tracked by a hydrophone mounted on a vessel in the sea (Hawkins and Urquhart, 1983). Automatic tracking systems using radio and acoustic positioning were used to reveal local movement of fish in three dimensions (Voegoli *et al.*, 2001). Echo sounders and scanning sonars were used to observe fish behavior in the open sea and near fishing gears. Arrhensius *et al.* (2000) compared swimming speed measured by a split-beam sonar and underwater video camera, and found there were no differences between measurements. Pedersen (2001) measured swimming speed of saithe using a split-beam sonar in the North Sea. He found that measurements were more accurate when the fish were swimming at higher speeds. Inoue and Arimoto (1989) tracked fish movements on several Japanese set net fishing grounds. A seabed-mounted scanning sonar and a video camera were used to observe movement of Atlantic cod near a Newfoundland cod trap (He, 1993b). Onsrud *et al.* (2005) used scanning sonar to measure swimming speed of Atlantic herring during feeding on copepods in a Norwegian fjord. Average swimming speed was between 0.02 to 0.61 m/s for herring 0.20 to 0.30 m in length. The mean speed was about 0.16 m/s in the day and 0.13 to 0.20 m/s during the night. Water temperature was not given in the paper. The swimming speed of migrating herring schools in Norwegian Sea measured by a scanning sonar was between 0.45 and 2.44 m/s (1.6–8.7 L/s) with an average of 1.2 m/s or 4.3 L/s for fish of mean length of 0.28 m at water temperature of 7°C near the surface to 15°C at 50 m water depth (Hafsteinsson and Misund, 1995). At the time period, the entire school was moving southwards at speed of 0.19 to 0.52 m/s. The mean swimming speed (4.3 L/s) is comparable to the maximum sustained swimming speed measured in the laboratory at 13.5°C by He and Wardle (1988).

Average swimming speed of capelin migrating in Barents Sea was 0.82 m/s or 6.3 L/s at 3°C for fish of 0.13 m as measured by Hafsteinsson and Misund (1995) using a scanning sonar during acoustic surveys. This is comparable to the maximum sustained speed of jack mackerel (*Trachurus japonicus*) of similar length measured in the laboratory (Xu, 1989). Swimming speed of herring of 4.61 m/s or 17 L/s was recorded during purse seine operations by a sector-scanning sonar sampled at 10-s intervals (Misund, 1989).

Large schools of striped mullet migrating along a shoreline were measured by a stopwatch. The swimming speed of 1.25 m/s or 9 L/s was recorded for fish of average length of 0.14 m (Peterson, 1975). This measured speed is greater than the sustained swimming speeds measured under laboratory conditions. It is questionable that how long these speeds can be sustained and the author (Peterson, 1975) suspected that there may be predators chasing them at the time of measurement. Peterson (1975) also found that fish in large schools (number >50) swam faster than those in smaller schools, suggesting that schooling may have contributed to locomotory performance as suggested by Weihs (1973).

Selective tidal transport, the use of a helpful tidal current, has been used by several marine species during their migration (Harden-Jones *et al.*, 1979; Arnold and Cook, 1984; Arnold *et al.*, 1994). Flatfish species such as sole (*Solea solea*) and plaice (*Pleuronectes platessa*) as well as roundfish such as Atlantic cod have been shown to utilize tidal current during their migrations in the North Sea (Arnold *et al.*, 1994) and in Newfoundland (Rose *et al.*, 1995). Less energy would be spent if these species were to rise up in the water column to travel with the tide and to stay close to bottom when the tidal direction changed. Swimming speed of Atlantic cod over the ground was more than 0.9 m/s when the fish was in midwater, and dropped to 0.32 m/s when they were near the seabed when they were tracked acoustically in the North Sea (Arnold *et al.*, 1994).

Yellowfin tuna 1.48 to 1.67 m long, tracked by an ultrasonic acoustic telemetry system over a period of several days revealed swimming speeds between 0.2 to 4.2 m/s over the ground at 26.3 to 27.9°C (Brill *et al.*, 1999), with 90% of the track period spent swimming less than 2 m/s. The mean swimming speed was between 0.72 and 1.54 m/s. In a similar study, juvenile bluefin tunas (0.74 to 1.06 m long) tracked off Virginia were recorded to swim at speeds up to 5 m/s with mean swimming speed between 1.28 and 1.68 L/s mostly in waters above 20°C (Brill *et al.*, 2002). Adult bluefin tuna kept in large farm cages in Mediterranean swam at speeds between 0.6 to 1.2 L/s at 20°C water temperature (Wardle *et al.*, 1989).

Conventional tags have been used to estimate long-term, long-distance swimming speed by measuring distance between the tagging and recovery positions. The estimated speed is always conservative as it is unlikely that the fish would have swum in straight lines over the entire period. The longest recorded migration may be that of a bluefin tuna traveling from Bahamas to Argentina covering a distance of 12250 km in 1335 days, representing a direct line rate of movement of 0.11 m/s (Mather *et al.*, 1995). Another Bluefin tuna traveled from Japan to Baja California, Mexico covering a distance of 9700 km and a time span of 323 days while the fish grew from 0.36 m to 0.68 m in length (Clemens and Flittner, 1969). Tags can also been used for monitoring short term movement of fish. Cadrin and Westwood (2004) tagged and recovered numerous yellowtail flounders (*Limanda ferruginea*) in the northeastern USA to understand stock structure and movement pattern. A direct line distance of 258 km was recorded for a yellowtail flounder during a 12-d period, an overall swimming speed of 0.25 m/s.

Archival tags record environmental and/or physiological parameters such as water temperature, depth, tail beat frequency and heart rates, in their internal memory and allow direct estimates of swimming activity or for reconstruction of swimming routes by interpreting celestial and oceanographic patterns upon retrieval of tags (Arnold *et al.*, 1994; Lutcavage *et al.*, 2000; Kawabe *et al.*, 2004). More recently, pop-up tags have been developed where tags automatic detach from the fish and return to the water surface after a prescribed time and transmit recorded data via a satellite (Wilson *et al.*, 2005).

Fish Swimming and Capture by Fishing Gears

Many types of fishing gears are employed in the world's commercial marine fisheries. Generally they can be divided into active gears and passive gears. Active gears include trawls, seines, and dredges. Passive gears include gillnets, loneliness, traps and pots. Here we discuss the effect of fish swimming on their capture by or escape from two types of fishing gears: otter trawl and set gillnets.

Fish Swimming and Capture by Trawls

A trawl is an active fishing gear that captures fish by herding and sieving through water. Trawls can be operated on bottom or in midwater, and can be towed by one vessel or two vessels. While some findings of fish behavior and underlying principles may be applicable to other types of trawls, the following discussion mainly refers to bottom otter trawls towed by one vessel (Fig. 15.9).

An otter trawl system is composed of warps, otter boards or doors, bridles, and the trawl net (Fig. 15.9). The net is usually composed of wings, square, body, and codend. Floats on the headline open the trawl vertically, and the groundgear keep the net on the bottom and overcome obstacles through a series of discs and bobbins. Typical otter trawlers for demersal species in the northeastern USA are 10 to 30 m in length, operating at depths between 40 to 300 m at towing speed between 1.2 to 1.7 m/s.

Fish capture by an otter trawl involves visual, acoustic and mechanical stimuli from the trawl and corresponding response of fish at different stages of capture (Wardle, 1986). Fish may hear the sound of the vessel and trawl components knocking on the seabed miles away, but precise reactions may start when the fish begin to see the fast approaching trawl door and trailing sand clouds. The fish may then react to various gear components from the bridles, the groundgear, and may eventually be retained by the codend. Fish may escape by skipping over the bridles, swimming over the headline or dashing through open meshes at the various points in the trawl net. Carefully selecting appropriate towing speeds of the trawl and utilizing species- and size-specific swimming capability may result in increased catch of target species and sizes, and reduce bycatch of undersized fish and

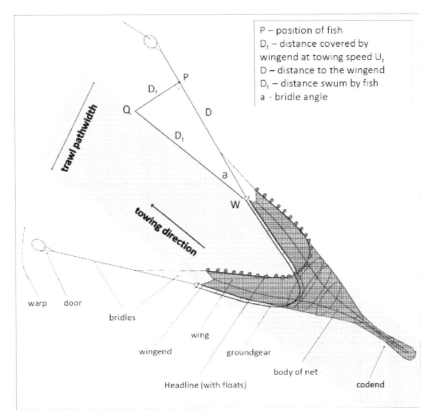

Fig. 15.9 Schematic illustration of an otter trawl and fish herding by trawl bridles. The swimming speed (U_f) required to cover D_f for a fish to be herded into the trawl path (shaded) is determined by whether the fish can swim from P to Q before the wingend moves from W to Q, covering the segment D_t at a speed U_t.

unwanted species. The following discusses fish swimming behavior and escape at the bridle area and in the codend of an otter trawl.

Fish Swimming and Herding by Bridles

Underwater observations of fish reaction to towed oblique cables indicate that fish, especially flounders, swim away from the approaching cable at the direct perpendicular to the cable (Main and Sangster, 1981). Roundfish, like Atlantic cod and haddock, swim with more varied angles, but they can also be herded into the trawl path if they are capable of swimming to avoid the advancing cables (Fig. 15.9). Herding of fish by the bridles connecting the wingend and the trawl door increases the width of the gear beyond the wing spread. Reaction of fish and probability of herding and escape have been modeled by various researchers attempting to understand fishing power of the trawl and these herding devices (Foster *et al.*, 1981; Wardle, 1983, 1986, 1989; Engås and Godø, 1989; Ramm and Xiao, 1995; Winger *et al.*, 2004). Engås and Godø (1989) compared length frequency of cod and haddock caught by survey trawls using different bridle lengths from 20 to 120 m. Trawls with longer bridle lengths caught more and larger fish for both species. Twice as many Atlantic cod greater than 0.50 m long were captured by the net with 120 m bridles compared with a net with 40 m bridles, suggesting that smaller cod may have escaped over the bridles due to their relatively poor swimming capability (Engås and Godø, 1989). Ramm and Xiao (1995) developed a model of fish herding by analyzing catch composition and length frequency of fish caught by trawls with different bridle lengths. They examined a number of Australian marine fish species and estimated the "effective pathwidth" of trawls for these species and gear configurations.

Winge *et al.* (1999) examined herding of American plaice of various lengths between 0.15 and 0.40 m by trawl bridles on assumption that most fish escaping advancing bridle cables by swimming away from it at the 90° angle (Fig. 15.9). For a fish at the position P at a distance of D from the wingend, the swimming speed required (U_f) to swim into the trawl path before being taken over by the bridle is:

$$U_f = U_t \cdot \sin \alpha$$

where U_t is the towing speed of the trawl and " is bridle angle (Fig. 15.9). The distance the fish has to swim (D_f) is:

$$D_f = D \cdot \tan \alpha$$

Using measured endurance of American plaice between 0.15 and 0.40 m swimming at 0.3 m/s at different temperatures, Winger *et al.* (1999) plotted a series of probability curves (Fig. 15.10) to predict whether plaice of various lengths can be herded into the trawl path at different temperatures when they are at various distances from the wingend. Towing speed of 1.5 m/s and bridle angle of 19.2° which are typical values of a Canadian survey trawl Campelen 1800 (McCallum and Walsh, 1995) were used in the simulation.

Figure 15.10 indicates that as the fish is farther away from the wingend, smaller fish are less and less likely to be herded into the trawl path while the effect to larger fish is less dramatic. This would result in relatively more large fish in the catch if longer bridles are used. Indeed, comparison of long (120 m) and short (40 m) bridles with a Norwegian survey trawl resulted in relatively more large fish in the long bridle rig as reported by Engås and Godø (1989). As most fish have longer endurance at higher water temperatures, more fish may be herded by the bridles at higher water temperatures; this would change trawl catch efficiency (He, 1991).

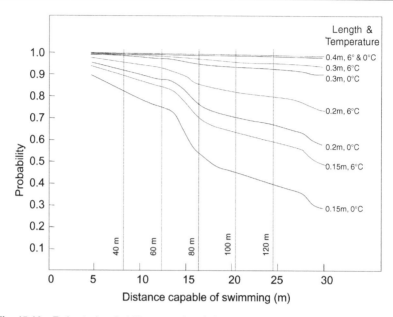

Fig. 15.10   Estimated probability curves in relation to the distances capable of swimming by American plaice (*Hippoglossoides platessoides*) of different fish lengths, swimming at 0.30 m/s at water temperature of 0 (solid line) and 6°C (dotted lines). Vertical dashed lines represent bridle length of an otter trawl with a sweep angle of 11.5° and a forward towing speed of 1.50 m/s. These lines indicate the distances that plaice would be required to swim in order to reach the trawl path after initially encountering the bridle at distances of 40, 60, 80, 100, and 120 m from the wingend. Corresponding probabilities of successfully reaching the trawl path are read from the Y-axis (modified from Winger *et al.*, 1999).

Under-standing the changes in capture efficiency in relation to environmental conditions such as water temperature is very important in resource survey trawls as this could result in different biomass indices.

Fish Swimming and Mesh selectivity of Trawl Codend

Size selectivity of fishing gear is defined as the proportion of fish of certain sizes retained by the gear. In otter trawls, size selectivity occurs mainly in three areas:

(1) Between the otter board and the wingend where fish may or may not be herded by bridles as discussed above;
(2) At the front part of the trawl where fish would swim with the trawl and may result in escape over the headline or under the groundgear; and
(3) In the codend where fish may escape by swimming through meshes.

All these are related to, in addition to other things, the swimming behaviour and swimming capacity of fish. Codend selectivity is the proportion of fish at each length retained in the codend as catch in relation to total number of fish of the same length group arrived at the codend prior to any escapement. Codend mesh size, mesh shape and riggings can affect the size selectivity of the codend (Reeves *et al.*, 1992).

However, the effect of towing speed and swimming ability of fish in relation to size selectivity in codend is less understood. He (1993a) developed a model to explore how the swimming ability of fish affects the codend mesh selection.

Whether a fish can escape from the codend is determined by the swimming speed required to escape and the achievable swimming speed of the fish. As escape through the mesh occurs in a short time period, the maximum swimming speed is considered to be important for escape.

Swimming speeds required to escape from codends of different mesh sizes (130, 150, 200 mm) towed at 1.75 m/s (Fig. 15.11) and from 130 mm mesh towed at different speeds of 1.25, 1.75 and 2.25 m/s (Fig. 15.12) were calculated by He (1993a). Plotted in the same graphs are the maximum swimming speed of the same species at three different water temperatures from Fig. 15.6. Those combinations of speed, length and mesh size above the maximum speed line at a particular temperature indicate the area where escape is impossible, and vice versa. Larger mesh sizes and slower towing speeds allow a wider size range of fish to escape from the codend. Higher temperature increases swimming speed of fish and makes escape through the codend meshes more likely. A 0.4 m Atlantic cod will be able to escape from a mesh of 200 mm mesh size at 2°C or higher when the gear is towed at 1.75 m/s (Fig. 15.11). But the same fish would only be able to escape the mesh of 150 mm at 12°C or higher temperature. At 12°C, a 0.42 m fish will be able to escape a 130 mm mesh towed at 1.25 m/s, but only fish below 0.35 m can escape the same sized mesh towed at 2.25 m/s (Fig. 15.12). When water temperature is decreased to 2°C, almost no Atlantic cod would be able to actively escape from codend meshes towed at 1.75 m/s or faster.

Fish Swimming and Capture by Stationary Gears

Stationary gears such as gillnets, longlines, pots and traps which rely heavily on the movement of fish can be impacted by the swimming speed of fish. Especially for gillnets set on the bottom or midwater, the faster a fish moves, the larger will be the fishing range, and the more likely the fish will get caught in the net when they are within the potential fishing area. Fish swimming at higher speeds may also result in greater probability of swimming into gillnets due to a reduction in visual response time when encountering the net (Engås and Løkkeberg, 1994).

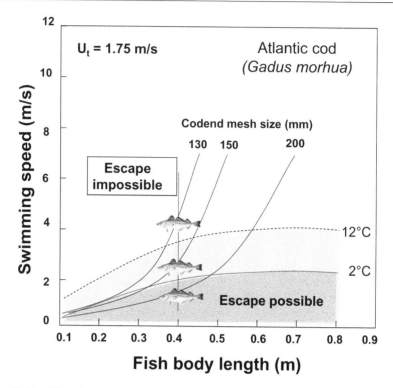

Fig. 15.11 Effect of mesh size (130, 150 and 200 mm) and water temperature (2 and 12°C) on the escape of Atlantic cod (*Gadus morhua*) from a trawl codend towed at 1.75 m/s. The point where fish length incepts mesh size curve determines if the fish can escape through the meshes. If the point falls in the shaded areas below the 2°C or 12°C line, escape is possible under these temperature conditions; otherwise escape is impossible. (modified from He, 1993a).

For bottom-set gillnets, the maximum fishing range is determined by the maximum distance the fish is capable of swimming during the time period that the gear is set (soaking time). The potential fishing area of the net is a zone formed by the maximum fishing range (Fig. 15.13). A portion of fish in the potential fishing area may be captured by gillnets. Increase in fishing range and potential fishing area for larger fish and its influence on size selectivity of gillnets was discussed by Rudstam *et al.* (1984). Here we discuss how water temperature may affect the fishing range and the potential fishing area of gillnets for winter flounder based on a model developed by He (2003).

A typical gillnet for groundfish and flounders in the eastern Canada and north eastern USA is 91 m long with mesh sizes ranging from 152 to 204 mm. Several gillnets are often tied together to form a fleet. Here we consider a fleet of gillnets containing 10 nets with a total length of 910 m and the net will fish for a period of 12 h (soaking time). As revealed in He (2003), the

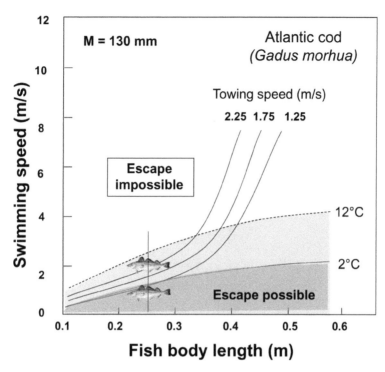

Fig. 15.12 Effect of towing speed and water temperature on the escape of Atlantic cod (*Gadus morhua*) from a trawl codend of 130 mm mesh size towed at different speed of 1.25, 1.75 and 2.25 m/s. The point where fish length incepts towing speed curve determines if the fish can escape through the meshes. If the point falls in the shaded areas below the 2°C or 12°C line, escape is possible under these temperature conditions; otherwise escape is impossible. (modified from He, 1993a).

rate of mover for winter flounder of 0.4 m long is 0.06 m/s at −1.2°C and 0.20 m/s at 4.4°C. The maximum fishing range would thus be 0.42 nautical miles (nm) at −1.2°C and 1.36 nm at 4.4°C (Fig. 15.13). Potential fishing area would thus be 0.96 nm² at −1.2°C and 7.11 nm² at 4.4°C, a more than 7-fold increase for a 5.6°C increase in water temperature. Potential fishing areas for the nets soaked for between 1 and 24 h at temperature of −1.2, 0, 2 and 4.4°C also plotted in Fig. 15.13 indicate that increases in the area are related to the square of soaking time. Longer soaking times usually produce greater catches in gillnets before the gear is saturated (Hovgård and Lassen, 2000). This may be a direct result of greater potential fishing area. It is expected that warmer waters may result in greater gillnet catch due to a greater potential fishing area and less net avoidance from faster swimming fish. However, capture of fish by a net is also related to many other factors such as temperature preference of the fish, prey availability and presence of predators, in addition to the potential fishing range related to

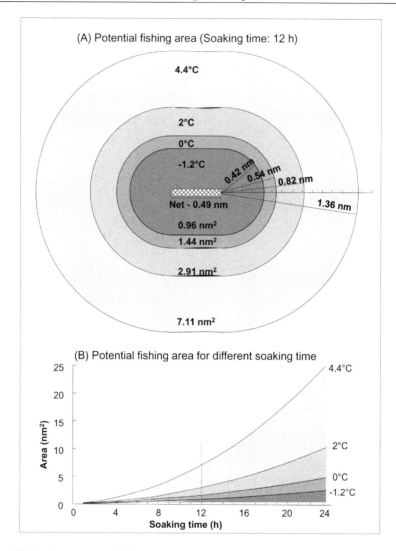

Fig. 15.13 The maximum fishing range and potential maximum fishing area of a fleet of 10 gillnets (910 m total length), set for 12 h at temperatures of −1.2, 0, 2, and 4.4°C, for winter flounder (*Pseudopleuronectes americanus*), 0.4m long (A). Potential fishing area for the same fleet of nets set for a period between 1 and 24 h (B). nm: nautical miles (1 nm = 1853m). (modified from He, 2003).

swimming speed. It is important though to consider temperature related fishing area changes when using fixed gears as stock assessment tools during fisheries surveys, or when considering distances between adjacent fixed gears in commercial operations.

CONCLUDING REMARKS

For commercial marine fish species, swimming capacity is not only important to maintain their life history related to foraging, migration, escape from predators and capture of prey, it is also very important in their capture by or escape from fishing gears. Swimming speed can be divided into sustained swimming speeds, prolonged swimming speeds and burst swimming speeds, each playing a role in their life history and interactions with fishing gears. Negatively buoyant fish without a swimbladder have to swim constantly to keep afloat, or adopt swim-and rest mode in bottom-dwelling flounder species. Endurance of swimming is inversely related to swimming speed during prolonged swimming. The maximum swimming speed is limited by muscle mechanics, morphology and hydrodynamics. The typical maximum swimming speed is between 10 and 20 L/s. Swimming capacity is positively correlated to water temperature in many marine fish species. Fish can swim faster and longer at higher water temperatures. Swimming capacity can affect capture in both active and stationary fishing gears. Size related swimming ability affect size selection of fishing gears and has an implication in commercial fisheries and stock assessment surveys.

REFERENCES

Altringham, J.D. and D.J. Ellerby. 1999. Fish swimming: patterns in muscle function. *Journal of Experimental Biology* 202: 3397–3403.

Arnold, G.P. and P.H. Cook. 1984. Fish migration by selective tidal stream transport: first result with computer simulation model for the European continental shelf. In: *Mechanisms of Migration in Fishes*, J.D. McCleave, G.P Arnold, J.J.Dodson, and W.H. Neill (Eds.). Plenum, New York, pp. 227–261.

Arnold, G.P., M. Greer Walker, L.S. Emerson and B.H. Holford. 1994. Movement of cod (*Gadus morhua* L.) in relation to the tidal stream in the southern North Sea. *ICES Journal of Marine Sciences* 51: 207–232.

Arrhensius, F., B.J.A.M. Benneheij, L.G. Rudstam and D. Boisclair. 2000. Can stationary bottom split-beam hydroacoustics be used to measure fish swimming speed in situ? *Fisheries Research* 45: 31–41.

Bainbridge, R. 1958. The speed of swimming as related to size and to the frequency and amplitude of the tail beat. *Journal of Experimental Biology* 35: 109–133.

Bainbridge, R. 1960. Speed and stamina in three fish. *Journal of Experimental Biology* 37: 129–153.

Beamish, F.W.H. 1966. Swimming endurance of some northwest Atlantic fishes. *Journal of the Fisheries Research Board of Canada* 23: 341–347.

Beamish, F.W.H. 1978. Swimming capacity. In: *Fish Physiology*, W.S. Hoar and D.J. Randall (Eds.). Academic Press, New York, Vol. 7, Locomotiom, pp. 101–187.

Beamish, F.W.H. 1984. Swimming performance of three southwest Pacific fishes. *Marine Biology* 79: 311–313.

Blaxter, J.H.S. and W. Dickson. 1959. Observations of the swimming speeds of fish. *Journal du Conseil Permanent International d'Exploration de la Mer* 24: 472–479.

Block, B.A. and F.G. Carey. 1985. Warm brain and eye temperature in sharks. *Journal of Comparative Physiology [B]* 156: 229–236.

Block, B.A., H. Dewar, S.B. Blackwell, T.D. Williams, E.D. Pricnce, C.J. Farewell. *et al.* 2001. Migratory movements, depth preferences, and thermal biology of Atlantic Bluefin tuna. *Science* 293: 1310–1314.

Breen, M., J. Dyson, F.G. O'Neill, E. Jones and M. Haigh. 2004. Swimming endurance of haddock (*Melanogrammus aeglefinus* L.) at prolonged and sustained swimming speeds, and its role in their capture by towed fishing gears. *ICES Journal of Marine Sciences* 61: 1071–1079.

Brett, J.R. 1964. The respiratory metabolism and swimming. performance of young sockeye salmon. *Journal of Fisheries and Research Board of Canada* 21: 1183–1226.

Brett, J.R. and N.R. Glass. 1973. Metabolic Rates and Critical Swimming Speeds of Sockeye Salmon (*Oncorhynchus nerka*) in Relation to Size and Temperature. *Journal of the Fisheries Research Board of Canada* 30: 379–387.

Brett, J.R., M. Holland and D.F. Alderdice. 1958. The effect of temperature on the cruising speed of young sockeye and coho salmon. *Journal of the Fisheries Research Board of Canada* 15: 587–605.

Brill, R.W. and A.E. Dizon. 1979. Effect of temperature on isotonic twitch of white muscle and predicted maximum swimming speeds of skipjack tuna, *Katsumonus pelamis.* *Environmental Biology of Fishes* 4: 199–205.

Brill, R., B. Block, C. Boggs, K. Bigelow, E. Freund and D. Marcinek. 1999. Horizontal movements and depth distribution of large adult yellowfin tuna (*Thunnus albacares*) near the Hawaiian Islands, recorded using ultrasonic telemetry: implications for the physiological ecology of pelagic fishes. *Marine Biology* 133: 395–408.

Brill, R., M. Lutcavage, G. Metzger, P. Bushnell, M. Arendt, J. Lucy, C. Watson and D. Foley. 2002. Horizontal and vertical movements of juvenile bluefin tuna (*Thunnus thynnus*), in relation to oceanographic conditions of the western North Atlantic, determined with ultrasonic telemetry. *Fisheries Bulletin* (US) 100: 155–167.

Brown, J.A., P. Pepin, D.A. Methven and D.C. Somerton. 1989. The feeding, growth and behaviour of juvenile cod, *Gadus morhua* L., in cold environment. *Journal of Fish Biology* 35: 373–380.

Cadrin, S.X. and A.D. Westwood. 2004. The use of electronic tags to study fish movement: a case study with yellowtail flounder off New England. *ICES CM 2004/K* 81: 34 p.

Carey, F.G. and J.M. Teal. 1969. Regulation of body temperature by the bluefin tuna. *Comparative Biochemistry and Physiology* 28: 205–214.

Castro-Santos, T. 2005. Optimal swim speeds for traversing velocity barrier: an analysis of volitional high-speed swimming behavior of migratory fishes. *Journal of Experimental Biology* 208: 421–432.

Claireaux G., C. Couturier and A-L. Groison. 2006. Effect of temperature on maximum sustainable speed and cost of swimming in juvenile European sea bass (*Dicentrarchus labrax*). *Journal of Experimental Biology* 209: 3420–3428.

Clemens, H. B. and G.A. Flittner. 1969. Bluefin tuna migrate across the Pacific Ocean. *California Fish Game* 55: 132–135.

Dean, J. M. 1976. Temperature of tissues in freshwater fishes. *Transactions of American Fisheries Society* 105: 709–711.

Dickson, K.A. 1994. Tunas as small as 207 mm fork length can elevate muscle temperatures significantly above ambient water temperature. *Journal of Experimental Biology* 190: 79–93.

Dickson, K.A., J.M Donley, C. Sepulveda and L. Bhoopat. 2002. Effects of temperature on sustained swimming performance and swimming kinematics of the chub mackerel *Scomber japonicus*. *Journal of Experimental Biology* 205: 969–980.

Domenici, P. and R.W. Blake. 1997.The kinematics and performance of fish fast-start swimming. *Journal of Experimental Biology* 200: 1165–1178.

Engås, A. and O.R. Godø. 1989. The effect of different sweep lengths on the length composition of bottom sampling trawl catches. *J. Cons. Int. Explor. Mer.* 45: 263–268.

Engås, A. and S. Løkkeberg. 1994. Abundance estimation using bottom gillnet and longline-The role of fish behaviour. In: *Marine Fish Behaviour*, A. Ferno and S. Olsen (Eds.). Fishing News Books, Oxford, pp. 134–165.

Fierstine, H.L. and V. Walters. 1968. Studies in locomotion and anatomy of Scombroid fishes. *Memoirs of Southern California Academy of Sciences* 6: 1–31.

Foster, J.J., C.M. Campbell and G.C.W. Sabin. 1981. The fish catching processes relevant to trawls. *Canadian Special Publication in Fisheries and Aquatic Sciences* 58: 229–246.

Franklin, C.E., R.S. Wilson and W. Davison. 2003. Locomotion at −1.0 C: burst swimming performance of five species of Antarctic fish. *Journal of Thermal Biology* 28: 59–65.

Freadman, M.A. 1979. Swimming energetics of striped bass (*Morone saxatilis*) and bluefish (Pomatamus saltatrix): gill ventilation and swimming metabolism. *Journal of Experimental Biology* 83: 217–230.

Glass, C.W., C.S. Wardle and W. Mojsiewicz. 1986. A light intensity threshold for schooling in the Atlantic mackerel, *Scomber scumbrus*. *Journal of Fish Biology* 29 (Suppl. A): 71–81.

Graham, J.B., H. Dewar, N.C. Lai, W.R. Lowell and S.M. Arce. 1990. Aspects of shark swimming performance determined using a large water ûume. *Journal of Fish Experimental Biology* 151: 175–192.

Hafsteinsson, M.T. and O.A. Misund. 1995. Recording the migration behavior of fish schools by multi-beam sonar during conventional acoustic surveys. *ICES Journal of Marine Sciences* 52: 915–924.

Harden-Jones, F.R., G.P. Arnold, M. Greer-Walker and P. Scholes. 1979. Selective tidal transport and the migration of plaice (*Pleuronectes platessa* L.) in the southern North Sea. *J. Cons. Int. Exp. Mer.* 38: 331–337.

Hawkins A.D. and G.W. Smith. 1986. Radio-tracking observations on Atlantic salmon ascending the Aberdeenshire Dee. *Scottish Fisheries Research Report* 36: 24 pp.

Hawkins, A.D. and G.G. Urquhart. 1983. Tracking fish in the sea. In: *Experimental Biology at Sea*, A.D. Macdonald and I.G. Priede (Eds.). Academic Press, London, pp.103–166.

He, P. 1986. *Swimming Performance of Three Species of Marine Fish and Some Aspects of Swimming in Fishing Gears*. Ph.D. Thesis. University of Aberdeen, Aberdeen, UK.

He, P. 1991. Swimming endurance of the Atlantic cod, *Gadus morhua* L. at low temperatures. *Fisheries Research* 12: 65–73.

He, P. 1993a. Swimming speeds of marine fish in relation to fishing gears. *ICES Marine Science Symposium* 196: 183–189.

He, P. 1993b. The behavior of cod around a cod trap as observed by an underwater camera and a scanning sonar. *ICES Marine Science Symposium* 196: 21–25.

He, P. 2003. Swimming behavior of winter flounder (*Pleuronectes americanus*) on natural fishing grounds as observed by an underwater video camera. *Fisheries Research* 60: 507–514.

He, P. and C.S. Wardle. 1986. Tilting behaviour of the Atlantic mackerel, *Scomber scombrus*, at low swimming speeds. *Journal of Fish Biology* 29 (Supplement A): 223–232.

He, P. and C.S. Wardle. 1988. Endurance at intermediate swimming speeds of Atlantic mackerel, *Scomber scombrus* L., herring, *Clupea harengus* L., and saithe, *Pollachius virens* L. *Journal of Fish Biology* 33: 255–266.

He, P., C.S. Wardle and T. Arimoto. 1990. Electrophysiology of the red muscle of mackerel, *Scomber scombrus* L. and its relation to swimming at low speeds. In: *The Second Asian Fisheries Forum*, R. Hirano and I. Hanyu (Eds.). Asian Fisheries Society, Manila, Philippines, pp. 469–472

Hemmings, C.C. 1969. Observation on the behavior of fish during capture by Danish seine net, and their relation to herding by trawl bridles. *FAO Fisheries Report* 62: 645–655.

Hovgård, H. and H. Lassen. 2000. *Manual on Estimation of Selectivity for Gillnet and Longline Gears in Abundance Surveys*. FAO Fish. Tech. Rep. 397. FAO, Rome.

Huse, I. and E. Ona. 1996. Tilt angle distribution and swimming speed of overwintering Norwegian spring spawning herring. *ICES Journal of Marine Sciences* 53: 863–873.

Hunter, J.R. and J.R. Zweifel. 1971. Swimming speed, tail beat frequency, tail beat amplitude and size in jack mackerel, *Trachurus symmetricus*, and other fish. *Fisheries Bulletin* (USA) 69: 253–266.

Inoue, Y. and T. Arimoto, T. 1989. Scanning sonar survey on the capturing process of trapnet. In: *Proceedings of World Symposium Fisheries Gear and Fish. Vessel Design*. Marine Institute, St. John's, Newfoundland, pp. 417–421.

Joaquin, N., G.N. Wagner and A.K. Gamperl. 2004. Cardiac function and critical swimming speed of the winter flounder (*Pleuronectes americanus*) at two temperatures. *Comparative Biochemistry and Physiology (A)*, 138: 277–285.

Kawabe, R., Y. Naito, K. Sato, K. Miyashita and N. Yamashita. 2004. Direct measurement of the swimming speed, tail beat, and body angle of Japanese flounder (*Paralichthys olivaceus*). *ICES Journal of Marine Sciences* 61: 1080–1087.

Koumoundouros, G., D.G. Sfakianakis, P. Divanach and M. Kentouri. 2002. Effect of temperature on swimming performance of sea bass juveniles. *Journal of Fish Biology* 60: 923–932.

Lindsey, C.C. 1978. Form, function and locomotory habits in fish. In: *Fish Physiology*, W.S. Hoar and D.J. Randall (Eds.). Academic Press, New York, Vol. 7: Locomotion, pp. 1–100.

Lutcavage, M.E., R.W. Brill, G.B. Skomal, B.C. Chase, J.L. Goldstein and J. Tutein. 2000. Movements and behavior of adult North Atlantic bluefin tuna (*Thunnus thynnus*) in the northwest Atlantic deter-mined using ultrasonic telemetry. *Marine Biology* 137: 347–358.

Magnuson, J.J. 1978. Locomotion by Scombrid fishes: hydromechanics, morphology and behaviour. In: *Fish Physiology*, W.S. Hoar and D.J. Randall (Eds.). Academic Press, New York, Vol. 7: Locomotion, pp. 239–313.

Main, J. and G.I. Sangster. 1981. A study of the fish capture process in a bottom trawl by direct observations from a towed underwater vehicle. *Scottish Fisheries Research Report* 23: 1–23.

Mather, F.M., III., Jr. J.M. Mason and A.C. Jones. 1995. Histrorical document: Life history and fisheries of Atlantic Bluefin tuna. *NOAA Tech. Memo, NMFS-SEFSC-370*. 172 p.

McCallum, B.R. and S.J. Walsh. 1995. Survey trawl standardization used in groundfish surveys. *ICES CM 1995/B* 25. 13 p.

Misund, O.A. 1989. Swimming behavior of herring (*Clupea harengus* L) and mackerel (*Scomber scombrus* L) in purse seine capture situations. In: *Proc. World Symp. Fish. Gear and Fish. Vessel Design*. Marine Institute, St. John's, Newfoundland, pp. 541–546.

Nelson, J.A. and G. Claireaux. 2005. Sprint swimming performance of juvenile European sea bass. *Transactions of American Fisheries Society* 134: 1274–1284.

Onsrud, M.S.R., S. Kaartvedt and M.T. Breien. 2005 In situ swimming speed and swimming behavior of fish feeding on the krill *Meganyctiphanes norvegica*. *Canadian Journal of Fisheries and Aquatic Sciences* 62: 1822–1832.

Peake, S. 2004. An evaluation of the use of critical swimming speed for determination of culvert water velocity criteria for smallmouth bass. *Transactions of American Fisheries Society* 133: 1472–1479.

Pedersen, J. 2001. Hydroacoustic measurement of swimming speed of North Sea saithe in the field. *Journal of Fish Biology* 58: 1073–1085.

Peterson, C.H. 1975. Cruising speed during migration of the striped mullet (*Mugil cephalus* L.): an evolutionary response to predation? *Evolution* 30: 393–396.

Plaut, I. 2001. Critical swimming speed: its ecological relevance. *Comparative Biochemistry and Physiology* A131: 41–50.

Ramm, D.C. and Y. Xiao. 1995. Herding in groundfish and effective pathwidth of trawls. *Fisheries Research* 24: 243–259.

Reeves, S.A., D.W. Armstrong, R.J. Fryer and K.A. Coull. 1992. The effects of mesh size, codend extension length and codend diameter on the selectivity of Scottish trawls and seines. *ICES Journal of Marine Sciences* 49: 279–288.

Reidy, S.P., S.R. Kerr and J.A. Nelson. 2000. Aerobic and anaerobic swimming performance of individual Atlantic cod. *Journal of Experimental Biology* 203: 347–357.

Rose G.A., B. deYoung and E.B. Colbourne. 1995. Cod(*Gadus morhua* L.) migration speeds and transport relative to currents on the north-east Newfoundland Shelf. *ICES Journal of Marine Sciences* 52: 903–913

Rudstam, L.G., J.J. Magnuson and W.M. Tonn. 1984. Size selectivity of passive fishing gear: A correction for encounter probability applied to gill nets. *Canadian Journal of Fisheries and Aquatic Sciences* 41: 1252–1255.

Sepulveda, C.A., K.A. Dickson and J.B. Graham. 2003. Swimming performance studies on the eastern Pacific bonito *Sarda chiliensis*, a close relative of the tunas (family Scombroidae). I. Energetics. *Journal of Experimental Biology* 206: 2739–2748.

Sepulveda, C. and K.A. Dickson. 2000. Maximum sustainable speeds and cost of swimming in juvenile kawakawa tuna (*Euthynnus affinis*) and chub mackerel (*Scomber japonicus*). *Journal of Experimental Biology* 203: 3089–3301.

Stevens, E.D. and W.H. Neill. 1978. Body temperature relations of tunas, especially skipjack. In: *Fish Physiology*, W.S. Hoar and D.J. Randall (Eds.). Academic Press, New York, Vol. 7: Locomotion, pp. 316–359.

Suuronen, P. 1989. Echo sounding observations of the behavior of Baltic herring in front of and inside midwater trawls. In: *Proc. World Symp. Fish. Gear and Fish. Vessel Design*. Marine Institute, St. John's, Newfoundland, pp. 422–424.

Videler, J.J. 1993. *Fish Swimming*. Chapman and Hall, London.

Videler, J.J. and C.S. Wardle. 1991. Fish swimming stride by stride: speed limits and endurance. *Reviews in Fish Biology and Fisheries* 1: 23–40.

Voegoli, F.A., M.J. Smale, D.M. Webber, Y. Andrade and R.K. O'Dor. 2001. Ultrasonic telemetry, tracking and automated monitoring technology for sharks. *Environmental Biology of Fishes* 60: 267–281.

Walters, V. and H.L. Fierstine. 1964. Measurements of swimming speeds of yellowfin tuna and wahoo. *Nature* (London) 202: 208–209.

Wardle, C.S. 1975. Limit of fish swimming speed. *Nature* (London) 255: 757.

Wardle, C.S. 1977. Effects of size on the swimming speeds of fish. In: *Scale Effects in Animal Locomotion*, T.J. Pedley (Ed.). Academic Press, London, pp. 299–313.

Wardle, C.S. 1980. Effect of temperature on the maximum swimming speed of fishes. In: *Environmental Physiology of Fish*, M.A. Ali (Ed.). Plenum, New York, pp. 519–531.

Wardle, C.S. 1983. Fish reaction to towed fishing gears. In: *Experimental Biology at Sea*. A.G. Macdonald and I.G. Priede (eds.), Academic Press, London, pp. 167–196.

Wardle, C.S. 1985. Swimming activity in marine fish. In: *Physiological Adaptation in Marine Animals. Symposium of Society of Experimental Biology* 39: 521–540.

Wardle, C.S. 1986. Fish behavior and fishing gear. In: *The Behavior of Teleost Fishes*, T.J. Pitcher (Ed.). Croom Helm, London, pp. 463–495.

Wardle, C.S. 1989. Understanding fish behaviour can lead to more selective fishing gears. In: *Proceedings of World Symposium on Fish Gear and Fish Vessel Design*. Marine Institute, St. John's, Newfoundland, pp. 12–18.

Wardle, C.S. and P. He. 1988. Burst swimming speeds of mackerel, *Scomber scombrus* L. *Journal of Fish Biology* 32: 471–478.

Wardle, C.S., J.J. Videler, T. Arimoto, J.M. Franco and P. He. 1989. The muscle twitch and the maximum swimming speed of giant bluefin tuna, *Thunnus thynnus* L. *Journal of Fish Biology* 35: 129–137.

Weihs, D. 1973. Hydromechanics of fish schooling. *Nature* (London) 245: 48–50.

Wilson, S.G., M.E. Lutcavage, R.W. Brill, M.P. Genovese, A.B. Cooper and A.W. Everly. 2005. Movements of Bluefin tuna (*Thunnus thynnus*) in the northwestern Atlantic Ocean recorded by pop-up satellite archival tags. *Marine Biology* 146: 409–423.

Winger, P., P. He and S. Walsh. 1999. Swimming endurance of American plaice (*Hippoglossoides platessoides*) and its role in fish capture. *ICES Journal of Marine Sciences* 56: 252–265.

Winger, P., P. He and S. Walsh. 2000. Factors affecting the swimming endurance and catchability of Atlantic cod (*Gadus morhua*). *Canadian Journal of Fisheries and Aquatic Sciences* 57: 1200–1207.

Winger, P.D., S. Walsh, P. He and J. Brown. 2004. Simulating trawl herding in flatfish: the role of fish length in behavior and swimming characteristics. *ICES Journal of Marine Sciences* 61: 1179–1185.

Xu, G. 1989. Study on the Fish Swimming Movement and its Application in Fishing By Trawls. Ph.D. Thesis. Tokyo University of Fisheries, Tokyo, Japan. (In Japanese with English summary).

Yanase, K., S. Eayrs and T. Arimoto. 2007. Influence of water temperature and fish length on the maximum swimming speed of sand flathead, *Platycephalus bassensis*: implications for trawl selectivity. *Fisheries Research* 84: 180–188.

Index

Forced swimming tests, 67
Fragmentation, 63, 64, 82
Framing rates, 143
Free-ranging sharks, 409
Free-riding behavior, 104
Freezing response, 124–126
French Polynesia, 389
Frequency distribution, 426
Freshwater angelfish, 214
Freshwater mussels, 63
Frictional plates, 42, 52
Frigate tun, 450
Froude numbers, 3
Functional groups, 27
Functional morphology, 158
Fundulus grandis, 314
Fusiform, 112, 175, 176

G

Gadidae, 356, 496
Gadids, 96
Gadiform, 339
Gadoid, 487
Gadus merlangus, 493
Gadus morhua, 53, 99, 112, 149, 237, 279, 280, 309, 428, 486–490, 493, 495, 496
Gait, 19, 23, 24, 31, 92–94
    transition, 93, 94, 300
    transition speed, 346, 347, 376
Galaxias meculatus, 22
Galeocerdo cuvier, 416
Galeorhinus galeus, 418
Gambusia, 218–220, 227, 228, 235, 236, 256
Gambusia affinis, 161, 206, 211, 222–224, 232
Gambusia amistadensis, 220, 236, 238
Gambusia atrora, 220, 236, 238
Gambusia eurystoma, 220, 236, 238
Gambusia gaigei, 220, 236, 238
Gambusia holbrooki, 283, 285, 287
Gambusia hubbsi, 207, 222–224
Gambusia hurtadoi, 220, 236, 238
Gambusia luma, 220, 236, 238
Gambusia manni, 220, 236, 238
Gambusia melapleura, 220, 236, 238
Gambusia nicaraguensis, 236
Gambusia oligosticta, 220, 236, 238
Gambusia rhizophorae, 220, 236, 238
Gametes, 77
Gape, 189
    limited predation, 158
    limited predators, 159
Gasterochisma melampus, 457
Gasterosteidae, 251, 381
Gasterosteus aculeatus, 213, 223, 237

Genetic quality, 253, 255, 259
Genyonemus lineatus, 237
Geolocation, 412
Geomagnetic anomalies, 419
Geometric morphometric, 212, 218
Gill, 456, 460, 469
Gill area, 158, 159
Gillnets, 499, 503, 504
Ginglymostoma cirratum, 144
Glide, 95, 96, 110
Gliding, 135
Global positioning system (GPS) tags, 430
Global warming, 303, 309
Glochidia, 63
Glycogen, 67, 298
Glyptothorax pectinopterus, 50, 51
Gobies, 30, 51, 178, 251, 255
Gobiesocidae, 50
Gobiidae, 50, 251
Gobio gobio, 21
Gobiosoma bosc, 279
Gobius niger, 154
Golden shiners, 99
Goldfish, 130, 135, 275
Gonads, 77
Gonopodium, 232, 256
GPS satellite constellation, 430
Gravity waves, 4
Grayling, 24, 26
Great Barrier Reef, 30, 31, 389
Green swordtail, 256
Grey mullet, 125
Ground effect, 41, 55, 56
Ground speed, 69, 302
Growth, 79
Growth rates, 357, 407
Gudgeon, 21, 24
Guidance, 74–76, 79, 82
Guide, 81
Guiding, 74
Gunnel, 125
Guppies, 141, 227, 228, 230, 234, 251, 253, 255, 257, 260, 261
Gypsum balls, 13, 17
Gyres, 21
Gyrinocheilid, 43
Gyrinocheilidae, 48, 49
Gyrinocheilus, 48
Gyrinocheilus aymonieri, 43, 44

H

Habitat, 173, 419
    changes, 412
    choices, 419

Color Plate Section

Chapter 1

Direction of wave

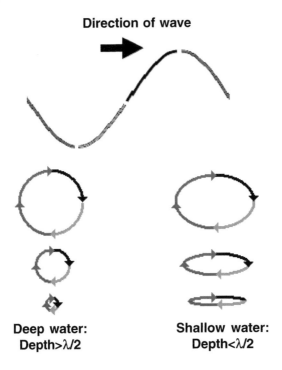

Deep water:
Depth>λ/2

Shallow water:
Depth<λ/2

Fig. 1.1 Diagrammatic representations of orbital trajectories, or orbits (circles and ellipses), followed by water particles as a non-breaking surface waves travels from left to right in deep and in shallow water. The vertical scale is exaggerated for clarity. (Based on Denny, 1988).

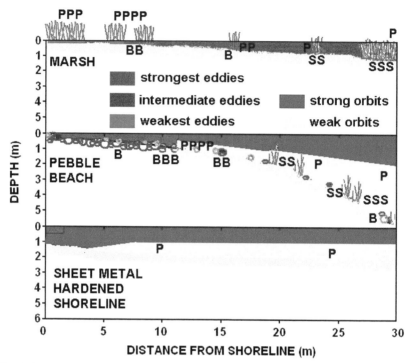

Fig. 1.8 A schematic representation of fish assemblages along three different wave-exposed shorelines. Three guilds of fishes are represented (Fig. 1.7); B = benthic fishes, P = pelagic fishes, and S = slow water fishes. One letter shows the fishes in a guild was a rare occurance, and four letters show that fishes were abundant. Prinipal turbulence features are indicated, as determined by waved gauges and PIV. Red tints representing areas where turbulence is largely derived from breaking waves, and blue tints indicate the strength of wave-driven orbits. Emergent macrophytes in marshes attenuate waves and reduce turbulence and shoreline assemblages include substantial numbers of fishes from the three guilds. Gradually sloping shorelines also dissipate wave energies, and in the breaking waves erode small substrate particles leaving primarily rocky materials. These provide habitat for benthic fishes. Pelagic fishes are common near shore where waves can suspend food items, but are found in the lower portion of the water column where orbit sizes are reduced by both depth and bottom effects (Fig. 1.1). Slow-water fishes are found in deeper water where wave effects become negligible and among macrophytes where orbits are attenuated. Sheet-pile reflects waves which interact with incoming waves. Wave effects result in bottom scour with little habitat attractive to fishes. Pelagic fishes were occasionally seen (P.W. Webb, A.J. Cotel and L. Meadows, unpublished observations).

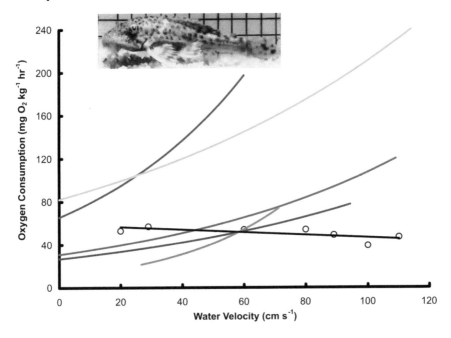

Chapter 2

Fig. 2.7 Relationship between mass specific oxygen consumption rate and water velocity for rainbow trout *Oncorhynchus mykiss* (green line), river sockeye salmon *O. nerka* (red line), coho salmon *O. kisutch* Walbaum (purple line), sockeye salmon *O. nerka* (orange line) and Atlantic cod *Gadus morhua* (blue line). Points for *Pterygoplichthys* spp. (inset) are given by open circles which were used to generate a regression (black line). Data from Blake *et al.*, (2007).

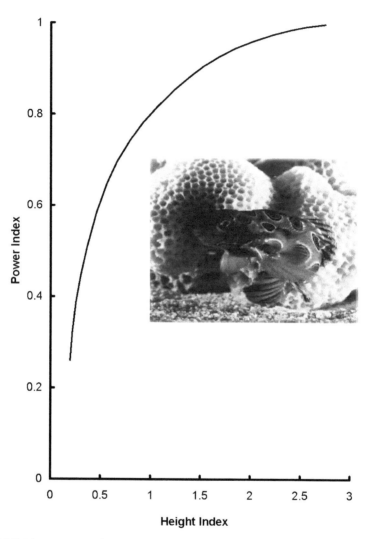

Fig. 2.8 Total power versus height for the mandarin fish *Synchiropus picturatus* (inset). The power required to hover out of ground effect divided by that in ground effect (power index) is plotted against the height index (height above the substrate divided by the span of a pectoral fin). Based on Blake (1979).

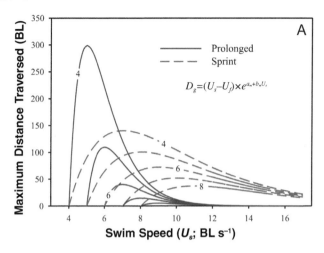

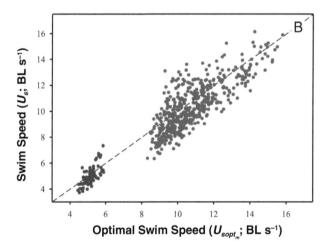

Chapter 3

Fig. 3.1 Relationship between swim speed, fatigue time, and flow velocity on distance of ascent through velocity barriers. Data are from American shad (*Alosa sapidissima*; Castro-Santos, 2005) and represent prolonged (blue) and sprint modes (red). (A): Theoretical maximum distance of ascent , from Equation 2 against flow velocities of 4–8 BL·s⁻¹ (U_f ; contours) and at swim speeds of 4–17 BL·s⁻¹ (see text for regression coefficients). Note the strong nonlinearity and clear distance-maximizing optima; speeds greater or less than these optima result in reduced distance of ascent. (B): Extent to which American shad approximated distance-maximizing groundspeeds ($U_{s\,opt_m}$, dashed line) for prolonged (blue) and sprint modes (red). Points represent actual swim speeds of 584 American shad swimming against flow velocities of 3.4–4.9 BL·s⁻¹ (prolonged mode) and 5.3–12.7 BL·s⁻¹ (sprint mode). Note that although most fish swam within 1 BL·s⁻¹ of $U_{s\,opt_m}$ in each mode, mean swim speeds were slightly lower than $U_{s\,opt_m}$ and considerable variability occurred. Consequences of this are shown in Fig. 3.2.

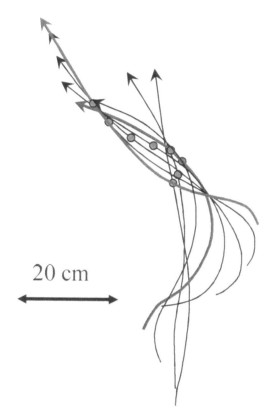

Chapter 5

Fig. 5.1 Escape response in dogfish (*Squalus acanthias*). Midline and centre of mass (red circles) of the fish at 40·ms intervals from the onset of the response. Head is indicated by the arrow. Red and blue lines indicate the end of stage 1 and stage 2, respectively. (Based on Domenici *et al.*, 2004 adapted with permission of the Journal of Experimental Biology).

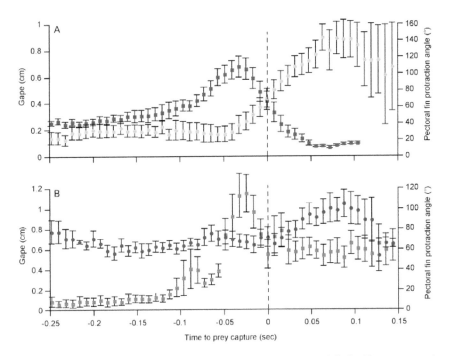

Chapter 6

Fig. 6.5 Kinematic plots of jaw movement (squares) and pectoral fin braking movement (circles) in two species of wrasse with very different feeding strategies: (A) *Scarus quoyi*, an herbivore and (B) *Oxycheilinus digrammus*, a piscivore. The moment of prey capture is indicated by the dashed vertical line. For *S. quoyi*, pectoral fin braking is initiated before it makes contact with the food item and the jaws close, while for *O. digrammus*, pectoral fin braking is initiated after it makes contact with the prey item and the jaws close. *S. quoyi* data are from Rice and Westneat (2005) and *O. digrammus* data are from Rice *et al.* (2008).

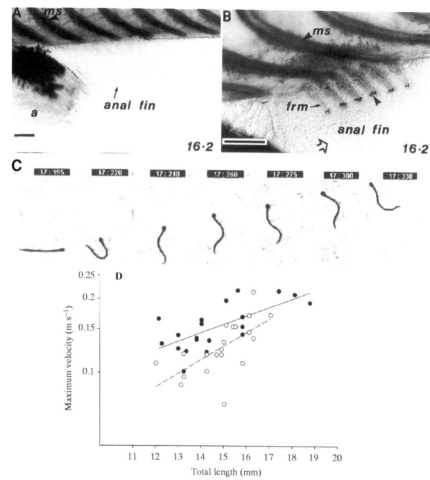

Chapter 9

Fig. 9.1 Influence of rearing temperature on dorsal and anal fin ray muscle development and fast-start performance in Atlantic herring (*Clupea harengus*). Embryos were incubated at 5°C (open circles) or 12°C (filled circles) until first feeding and then reared at ambient temperature. (A, B) Whole-mount larvae of 16.2 mm total length stained for acetylcholinesterase activity following the 5°C (A) and 12°C (B) rearing regimes. Abbreviations: *a*, anus; *frm*, fin ray muscles; *ms*, myosepta. (C) Escape response of Clyde herring larva (*Clupea harengus*) 18 mm total length, filmed at 200 frames s⁻¹. (D) Maximum velocity attained during escape responses in Clyde herring larvae reared at 5°C (open circles, dashed line) and 12°C (filled circles, solid line) until first feeding, when they were transferred to ambient seawater temperatures. Each point represents an escape response from one larva. First-order linear regressions were fitted to the data. 5°C, replace with: $U_{max}=-3.12TL^{1.92}$, $r^2=0.42$, $P=0.002$; 12°C, $U_{max}=2.20TL^{1.21}$, $r^2=0.48$, $P=0.001$), where U_{max} is maximum swimming velocity (m s⁻¹) and TL is total length (mm). From Johnston *et al.* (2001) with permission from Marine Ecology Progress Series.

Chapter 13

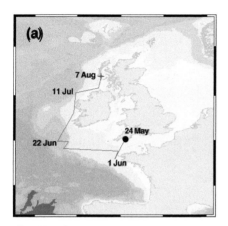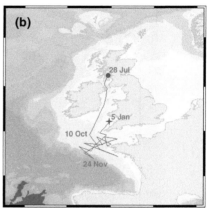

Fig. 13.1 Seasonal movements of basking sharks (*Cetorhinus maximus*) from archival tagging. (a) Reconstructed track of a 4.5-m long shark moving from the tagging location in the Western English Channel, around the west of Ireland and into Scottish waters over 77 days and covering an estimated minimum distance of 1,878 km. (b) Movement of a 7.0-m long shark from the Clyde Sea in Scotland, through the Irish Sea and into waters off southwest Britain, a journey tracked over 162 days and an estimated minimum distance of 3,421 km. Adapted from Sims *et al.* (2003).

Chapter14

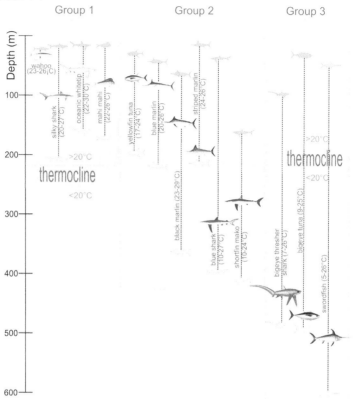

Fig. 14.1 Representative vertical movement patterns for pelagic fishes. Fish images represent the average depth (combined night and day) for each species. Gray-filled fish outlines represent the depth at which each species spent 95% of the time during the night. Open outlines represent the depth at which each species spent 95% of the time during the day. Values next to the common name show the temperature ranges encountered by each species. Orange shaded bar represents the thermocline, defined as depth range in which the water column is separated into the upper uniformed-temperature surface layer (i.e., water above 20°C) and the cooler deeper waters (i.e., below 20°C). Group 1: Fishes that spend the majority of their time in the upper uniformed-temperature surface layer. Group 2: Fishes that undertake short excursions below the thermocline. Group 3: Fishes that make frequent excursions below the thermocline. Fig. modified from Musyl *et al.* (2004). Wahoo[1] (*Acanthocybium solandri*), silky shark[2] (*Carcharhinus falciformis*), oceanic whitetip[2] (*Carcharhinus longimanus*), mahimahi[2] (*Coryphaena hippurus*), yellowfin tuna[3] (*Thunnus albacares*), blue marlin[4] (*Makaira nigricans*), black marlin[5] (*M. indica*), striped marlin[6] (*Tetrapturus audax*), blue shark[7] (*Prionace glauca*), shortfin mako shark[8] (*Isurus oxyrinchus*), bigeye thresher shark[9] (*Alopias superciliosus*), bigeye tuna[10] (*Thunnus obesus*), and swordfish[11] (*Xiphias gladius*).

[1]C. Sepulveda, S. Aalbers, D. Bernal, unpublished; [2]M. Musyl and R. Brill, unpublished; [3]Schaefer *et al.* (2007); [4] Block *et al.* (1992); [5]Gunn *et al.* (2003), Pepperell and Davies (1999); [6]Brill *et al.* (1993), Holts and Bedford (1990); [7]Carey and Scharold (1990); [8]Holts and Bedford (1993), Sepulveda *et al.* (2004); [9]Nakano *et al.* (2003), Weng and Block (2004); [10]Schaefer and Fuller (2002), Holland *et al.* (1992); [11]Carey and Robinson (1981), Carey (1990).

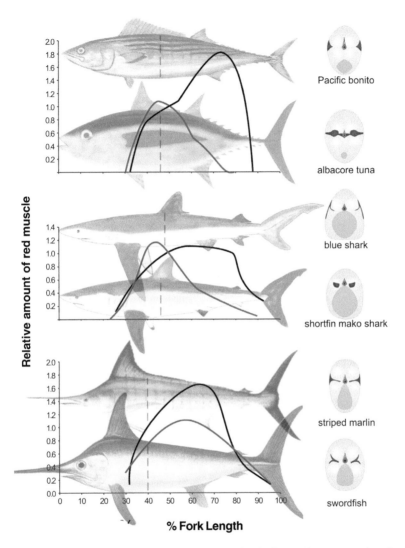

Fig. 14.5 Representative red muscle longitudinal distribution and transverse location in pelagic fishes. Black lines show the more posterior distribution of red muscle in Pacific bonito[1] (*Sarda chiliensis*), blue shark[2] (*Prionace glauca*), and striped marlin[3] (*Tetrapturus audax*). Red lines show the more anterior distribution of red muscle in albacore tuna[1] (*Thunnus alalunga*), shortfin mako shark[2] (*Isurus oxyrinchus*), and swordfish[4] (*Xiphias gladius*). The relative amounts of red muscle in the different positions along the body are expressed as a proportion of the red muscle cross-sectional area equal to 1 at 50% fork length (except for billfish where it is at 50% of lower jaw to fork length). Transverse sections images of each species show the representative location of red muscle at the position of the hatched lines across fish.

[1]Modified from Graham *et al.* (1983); [2]Modified from Bernal *et al.* (2003); [3]C. Sepulveda and D. Bernal, unpublished; [4]D. Bernal, C. Sepulveda, and S. Gemballa, unpublished.

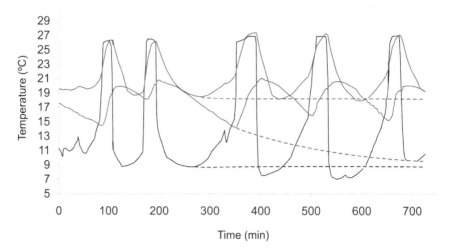

Fig. 14.7 Vertical movement profiles for a blue shark (*Prionace glauca*) tracked using ultrasonic telemetry (Carey and Scharold 1990). The black line shows temperature at depth and the blue line body temperature. The red line shows the predicted body temperature of a shortfin mako shark (*Isurus oxyrinchus*) undergoing similar changes in ambient temperature. Changes in mako shark body temperature were modeled using thermal rate coefficients and thermal equilibrium values from Bernal *et al.* (2001a) and were derived for every available data point from the blue shark vertical movement profile. The black hatched line shows a simulated descent, with the fish remaining at 8.73ºC for 300 minutes. The blue and red hatched lines represent the predicted body temperature for blue and mako sharks, respectively, resulting from the simulated descent. Notice the degree to which the mako shark body temperature remains elevated relative to that of the blue shark, with a very rapid increase during the time periods (i.e., between vertical sojourns) when the shark returns to warmer waters.

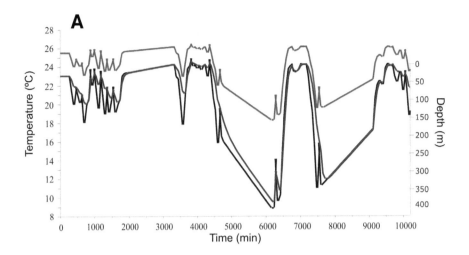

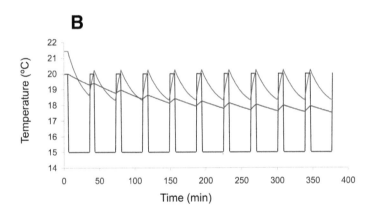

Fig. 14.8 **(A)** Vertical movement profiles for a blue shark (*Prionace glauca*) near the main Hawaiian Islands. The black line shows the temperature at depth data from a pop-up satellite archival tag carried by the shark. The blue line shows the derived body temperature for the blue shark and the red line shows the derived body temperature for a shortfin mako shark (*Isurus oxyrinchus*) following the same vertical movement pattern. Thermal rate coefficients (i.e., *k* values) used to model the changes in body temperature for both shark species were taken from Bernal *et al.* (2001a). Body temperatures were derived for every available water temperature data point. Notice the degree to which the mako shark body temperature remains elevated relative to that of the blue shark. **(B)** Body temperature changes in blue sharks (blue line) and shortfin mako sharks (red line) subjected to a series of modeled vertical movement patterns (i.e., rapid changes in ambient temperature, black line) following diving profiles observed for telemetry tracked mako shark by Holts and Bedford (1993). Thermal rate coefficients (i.e., *k* values) were taken from Bernal *et al.* (2001a).

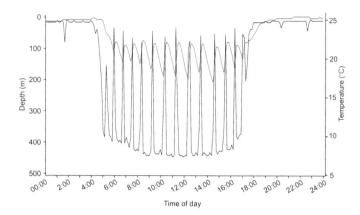

Fig. 14.9 Vertical movement patterns of a bigeye tuna (*Thunnus obesus*) near the main Hawaiian waters recorded by an archival tag implanted in the fish's dorsal musculature (Musyl *et al.*, 2003). Black line shows depth and ambient temperature and red line body temperature.

T - #0092 - 111024 - C564 - 229/152/26 - PB - 9780367452414 - Gloss Lamination